Long-Term Climate Monitoring by the Global Climate Observing System

International Meeting of Experts, Asheville, North Carolina, USA

Edited by

Thomas R. Karl

U.S. Department of Commerce, National Oceanic and Atmospheric Administration, National Climatic Data Center, Asheville, NC, U.S.A.

Reprinted from Climatic Change
Volume 31, Nos. 2–4, 1995

WMO   UNESCO   UNEP   ICSU

Springer Science+Business Media, B.V.

A C.I.P. Catalogue record for this book is available from the Library of Congress.

ISBN 978-94-010-4143-0   ISBN 978-94-011-0323-7 (eBook)
DOI 10.1007/978-94-011-0323-7

Printed on acid-free paper

Contents

THOMAS KARL / Foreword   v

Editorials

THOMAS SPENCE and JOHN TOWNSHEND / The Global Climate Observing
System (GCOS). An Editorial   1

THOMAS KARL, FRANCIS BRETHERTON, WILLIAM EASTERLING, CHRIS
MILLER, and KEVIN TRENBERTH / Long-Term Climate Monitoring by
the Global Climate Observing System (GCOS). An Editorial   5

CHRISTOPHER MILLER and FRANCIS BRETHERTON / Long-Term Climate
Monitoring by the Global Climate Observing System. Report of Breakout
Group A – Climate Forcings and Feedbacks. An Editorial   19

KEVIN E. TRENBERTH / Long-Term Climate Monitoring by the Global Climate
Observing System. Report of Breakout Group B – Climate Responses and
Feedbacks. An Editorial   33

WILLIAM E. EASTERLING / Long-Term Climate Monitoring by the Global
Climate Observing System (GCOS). Report of Breakout Group C – Climate
Impacts. An Editorial   51

General Aspects of Climate Observing

THOMAS R. KARL, VERNON E. DERR, DAVID R. EASTERLING, CHRIS K.
FOLLAND, DAVID J. HOFMANN, SYDNEY LEVITUS, NEVILLE
NICHOLLS, DAVID E. PARKER, and GREGORY W. WITHEE / Critical
Issues for Long-Term Climate Monitoring   55

J. D. MAHLMAN / Toward a Scientific Centered Climate Monitoring System   93

NEVILLE NICHOLLS / Long-Term Climate Monitoring and Extreme Events   101

J. HANSEN, W. ROSSOW, B. CARLSON, A. LACIS, L. TRAVIS, A. DEL
GENIO, I. FUNG, B. CAIRNS, M. MISHCHENKO, and M. SATO / Low-
Cost Long-Term Monitoring of Global Climate Forcings and Feedbacks   117

U. CUBASCH, J. WASZKEWITZ, G. HEGERL, and J. PERLWITZ / Regional
Climate Changes as Simulated in Time-Slice Experiments   143

Specific Aspects of Climate Observing

WILLIAM B. ROSSOW and BRIAN CAIRNS / Monitoring Changes of Clouds   175

WILLIAM P. ELLIOTT / On Detecting Long-Term Changes in Atmospheric
Moisture   219

CONTENTS

JOHN E. WALSH / Long-Term Observations for Monitoring of the Cryosphere 239

RAMAKRISHNA R. NEMANI and STEVEN W. RUNNING / Satellite Monitoring of Global Land Cover Changes and Their Impact on Climate 265

CHESTER F. ROPELEWSKI / Long-Term Observations of Land Surface Characteristics 285

KEVIN E. TRENBERTH / Atmospheric Circulation Climate Changes 297

JOHN R. CHRISTY / Temperature above the Surface Layer 325

NEVILLE R. SMITH, GEORGE T. NEEDLER, and THE OCEAN OBSERVING SYSTEM DEVELOPMENT PANEL / An Ocean Observing System for Climate. The Conceptual Design 345

SYDNEY LEVITUS and JOHN ANTONOV / Observational Evidence of Interannual to Decadal-Scale Variability of the Subsurface Temperature-Salinity Structure of the World Ocean 365

VIVIEN GORNITZ / Monitoring Sea Level Changes 385

P. D. JONES / Land Surface Temperatures – Is the Network Good Enough? 415

D. E. PARKER, C. K. FOLLAND, and M. JACKSON / Marine Surface Temperature: Observed Variations and Data Requirements 429

PAVEL YA. GROISMAN and DAVID R. LEGATES / Documenting and Detecting Long-Term Precipitation Trends: Where We Are and What Should Be Done 471

Climate Impacts and Climate Monitoring

W. E. EASTERLING and R. W. KATES / Indexes of Leading Climate Indicators for Impact Assessment 493

FOREWORD

One of the most challenging and critical issues facing climatologists during the next Century is related to the operation of a long-term climate monitoring system capable of delivering continuous and reliable data and information. As discussed in the editorial by Drs. Spence and Townshend, the international community is now developing the framework for a Global Climate Observing System (GCOS). The World Meteorological Organization, the Intergovernmental Oceanographic Commission of UNESCO, the United Nations Environment Programme, and the International Council of Scientific Unions are all sponsoring organizations of GCOS. Such a system will serve a variety of purposes, including climate change detection and response monitoring, national economic development; research toward improved understanding modeling and prediction, and a means to attribute natural and human induced factors to climate variations and change. During January 9–11, 1995, nearly 100 scientists gathered in Asheville, North Carolina to help develop the requirements for a long-term climate monitoring system. In the following collection of editorials and articles, a scientific framework is developed for the monitoring requirements of such a system.

NOAA, U.S. Department of Commerce/NOAA,   THOMAS KARL
National Climatic Data Center, 151 Patton Avenue,
Room 120, Asheville, NC 28801-5001, U.S.A.

THE GLOBAL CLIMATE OBSERVING SYSTEM (GCOS)

An Editorial

THOMAS SPENCE

Global Climate Observing System, Joint Planning Office, c/o World Meteorological Organization,
Case Postale 2300, CH-1211 Genève 2, Switzerland

and

JOHN TOWNSHEND

Department of Geography, University of Maryland, College Park, Maryland 20742, U.S.A.

Introduction

CLIMATE CONCERNS

It is now well recognized that the climate of the Earth results from a complex of interactions among the atmosphere, ocean, land surface, ecosystem, and human activity. The past few decades have seen a growing public awareness about the global environment, and in particular, the state of climate and its future evolution. With the continuing growth of population and the concomitant increases in agriculture, industry, and consumption of natural resources, the human influences on Earth are profound and far-reaching.

The climate varies over a vast range of time scales – seasonal to many thousands of years. Currently, attention has focused principally on two periods: (1) the seasonal-to-interannual time scales: and (2) the decadal-to-centennial time scales. The former are important, since recent progress has shown that with suitable initial observations, several phenomena are predictable, and consequently are of social and economic value. Recent research efforts to understand the El Niño phenomenon, a major tropical climate signal, have led to credible predictions for a few seasons in the future. This capability provides a tremendous benefit to those societal sectors critically dependent upon future weather events.

The latter are important since recent observations of the increase in anthropogenic greenhouse gases show they may be adequate to produce a measurable global climate change on these scales. Recently, attention has focused on the possibility of climate and global change which may result from increased concentration of greenhouse gases in the atmosphere due to human activity. Scientists recognize that increases of these gases may modify the climate on an scale unprecedented in human experience. In many cases, these changes could produce irreversible effects, some of which could be adverse and be detrimental to human health, human welfare, and sustainable development.

Climatic Change **31**: 131–134, 1995.

[1]

In the current debate about climate change, it is evident that adequate information is not available to governments to enable them to answer the critical scientific, economic and policy questions. Monitoring and observation are required to document the climate, to assess the impacts of change, and to adequately evaluate mitigation options.

While many observational programmes are currently underway and contributing information on climate issues, systematic global observations of key variables are urgently needed and should be made available to the nations to enable them to:
- detect and quantify climate change at the earliest possible time;
- document natural climate variability and extreme climate events;
- model, understand and predict climate variability and change;
- assess the potential impact on ecosystems and socio-economics;
- develop strategies to diminish potentially harmful effects;
- provide services and applications to climate-sensitive sectors;
- support sustainable development.

A GLOBAL CLIMATE OBSERVING SYSTEM

The Global Climate Observing System (GCOS) has recently been established to ensure that needed observations and information on climate-related issues are obtained and made available to the nations of the world. The GCOS is intended to be a long-term, user-driven operational system designed specifically to meet the comprehensive scientific requirements for monitoring the climate, and to provide the observational basis for detecting climate change, for predicting climate variations and change, and for observing the impacts of climate change. It will address the total climate system including physical, chemical and biological properties and atmospheric, oceanic, hydrologic, cryospheric and terrestrial processes. For this reason, the success of GCOS is critical –it will provide the observational underpinning for national and international climate programmes.

GCOS was established to ensure that needed observations and information on climate related issues are available to the nations of the world. The GCOS will not directly make observations or generate data products, but will rather encourage, coordinate and otherwise facilitate observations and data products which must be made by national or international organizations in support of their own requirements as well as common goals. The GCOS will, however, provide a operational framework for integrating observational components of the participating countries into a comprehensive system. In this way, a coherent observational capability may be established and be based on existing national and international operational and research programmes. When required observations or data management structures are unavailable, efforts to initiate them will be undertaken through GCOS using appropriate means.

Objectives of the Global Climate Observing System

To provide the data required to meet the needs for:
- Climate system monitoring, climate change detection and response monitoring especially in terrestrial ecosystems and mean sea-level;
- Application to national economic development;
- Research toward improved understanding, modelling and prediction of the climate system.

(The GCOS Memorandum of Understanding) *

The GCOS will utilize *in situ* and remote sensing techniques to provide high quality, calibrated observations which will be collected and incorporated into data sets and products that are required by various user communities which include *inter alia* scientists, analysts, and forecasters, who will participate in the programme activities. The system will include data and information management to provide an effective distribution and archive for the information to users/participants. The results of their work will be made available, often via other established agencies, to provide information for political and economic decision makers.

To bring the GCOS into being, a framework for planning, and implementation has been established. A Joint Scientific and Technical Committee (JSTC) was established to formulate the overall concept and scope of the GCOS. It will review, assess, and provide oversight of the development and implementation of the various components of the system. A Joint Planning Office (JPO), located at the World Meteorological Organization (WMO) in Geneva, was established to support the JSTC.

The GCOS will provide an international framework to assist nations in obtaining and sharing the fundamental information they need to address climate-related questions now and in the future. The GCOS will provide a strategy for a phased and evolving programme to meet such needs.

With the participation of a broad range of scientists, the JSTC has identified the critical issues requiring observations and recommended priority actions to establish and implement the observing system.

THE WORKSHOP ON LONG-TERM CLIMATE REQUIREMENTS

The workshop reported in this volume has provided an exceptional opportunity for the scientific community to assist in the definition of the global climate observing system. The conference assembled experts from many disciplines – all seeking to describe the observation and information requirements which are needed to

---

* The sponsoring organizations of GCOS are the World Meteorological Organization, the Intergovernmental Oceanographic Commission of UNESCO, the United Nations Environment Programme, and the International Council of Scientific Unions.

answer key questions about the future climate. In addition, representatives of the GCOS JSTC and its panels participated in the development of constructive and implementable recommendations.

The results of this meeting included in the following papers are making significant contributions in defining the scientific issues for GCOS to address. In the articles to follow, a number of significant actions have been identified. Many call for improvements in existing systems, and some for new observational capability. The next steps will be to take the specific recommendations forward to national and international organizations to engage their participation in meeting the requirements.

To implement these recommendations, national and international participation, principally through national operational and research observational programmes as well as data management and distribution activities, must be encouraged. It is anticipated that meetings such as this will forge close links among the various communities to ensure that the appropriate observations and data products are made available for the benefit of the society.

LONG-TERM CLIMATE MONITORING BY THE GLOBAL CLIMATE OBSERVING SYSTEM (GCOS) *

An Editorial

THOMAS KARL[1], FRANCIS BRETHERTON[2], WILLIAM EASTERLING[3],
CHRIS MILLER[4] and KEVIN TRENBERTH[5]

[1] *National Climatic Data Center, 151 Patton Ave., Asheville, NC 20081–5001, U.S.A.*
[2] *University of Wisconsin, Space Science and Engineering , 1225 W. Dayton St., Madison, WI, 53706, U.S.A.*
[3] *University of Nebraska, Dept. of Agricultural Meteorology, L.W. Chase Hall, 33rd Street & Holdrege, Lincoln, Nebraska, 68583–0728, U.S.A.*
[4] *National Environmental Satellite, Data, and Information Service, 1315 East West Highway, Silver Spring, MD 20910, U.S.A.*
[5] *National Center for Atmospheric Research, NCAR sponsored by the National Science Foundation, Climate and Global Dynamics, P.O. Box 3000, Boulder, Co 80307–3000, U.S.A.*

1. Background

The documentation of long-term climate variations and changes is important for several reasons. The early detection of anthropogenically-induced climate change, as well as the sensitivity of the climate system to a variety of human and natural causes, rests upon an observing system capable of delivering adequate long-term data. Additionally, such information is essential to understand impacts on managed and unmanaged social and biophysical systems. So in IPCC (1995) several questions have been posed to the scientific community regarding the present and past states of the climate. However, our present observing system and data management practices have failed to deliver the quality of data required to deduce unequivocal information about the rates and often even the sign of multi-decadal changes and variations. Answers to specific questions such as:

Is the climate warming?
Is the hydrologic cycle changing?
Is the atmospheric/oceanic circulation changing?
Is the climate becoming more variable or extreme?
Is radiative forcing of the climate changing?

are thwarted due to an inadequate or non-existent climate observing system. Each of the above questions is actually quite complex, not only from the standpoint of a multivariate problem, but because of the various aspects of spatial and temporal sampling that must be considered on a global scale. Obviously, without adequate

---

* The U.S. Government right to retain a non-exclusive royalty-free license in and to any copyright is acknowledged.

Climatic Change **31**: 135–147, 1995.
© 1995 *Kluwer Academic Publishers.*

answers to such basic questions, understanding climate change and its predictability is not possible.

The development of a Global Climate Observing System (GCOS) offers the opportunity for scientists to do something about existing observing deficiencies in light of the importance of documenting long-term climate changes that may already be affected by anthropogenic changes of atmospheric composition and land use as well as other naturally occurring changes.

As an important step toward improving the present inadequacies, a workshop was held to help define the long-term monitoring requirements minimally needed to address the five questions posed above, with special emphasis on detecting anthropogenic climate change and its potential impact on managed and unmanaged systems. The workshop focussed on three broad areas related to long-term climate monitoring:

(a) the scientific rationale for the long-term climate products (including their accuracy, resolution, and homogeneity) required from our observing systems as related to climate monitoring and climate change detection and attribution;

(b) the status of long-term climate products and the observing systems from which these data are derived; and

(c) implementation strategies necessary to fulfill item (a) in light of existing systems.

Item (c) was treated more in terms of feasibility rather than as a specific implementation plan.

2. Introduction

NOAA's Earth Systems Data and Information Management Program and the GCOS Joint Planning Office (JPO) helped support a total of 94 participants who met for three days in Asheville, North Carolina, U.S.A. About half the participants were from non-government institutions while others were affiliated with national governments. Nine participants also represented world organizations. Scientists from ten countries participated, Australia, Canada and the U.S.A., Czech Republic, Finland, France, Germany, Japan, Netherlands, and U.K.

There have been many planning meetings for GCOS, but this was one of the first involving a whole community. Given the many previous planning meetings for GCOS, we felt obliged to build on the extensive work by committees already in place and to move ahead from there, rather than reinvent suggestions already being acted upon. Of course, this was not easy as many participants were not familiar with all the GCOS reports. However, the workshop should be viewed as helping to define realistic GCOS goals as related to the consensus of some of the world's leading scientists with special interests in decadal-to-centennial time-scale climate monitoring.

In his opening remarks, John Townshend, speaking on behalf of the Chair of the GCOS Joint Scientific and Technical Commission, Sir John Houghton, reviewed the overall tasks of GCOS and the challenges ahead. He noted that the tasks included:

1. collating the data on observational systems;
2. objectively defining the needs;
3. assessing the capabilities in scientific terms;
4. defining the deficiencies of observational systems, and
5. defining improvements in observing systems.

Clearly, these actions will all lead to improvements in observation, assimilation, and information systems which, in turn, again serve as a basis for re-examining vital GCOS issues. Townshend noted that there will be competition for resources and vigorous questions concerning the value of environmental monitoring and research. There will be changing policy priorities that will need to be addressed and pressures of commercialization. A special challenge will be to balance national policies and needs with international responsibilities.

Additional challenges arise from the inherent complexity of the Earth system, with enormous variation in both time and space. Townshend pointed out the difficulties in obtaining an overall understanding of current and future observing systems, and thus achieving a clear picture of how planned capabilities will be eventually realized. He indicated that it is important to recognize that deficiencies can often be met in a variety of different ways. Planning should recognize that parts of observing systems will fail, but it is impossible to predict which component will fail. Moreover, the impact of either improvements or decay in observing systems is often very difficult to assess. Accordingly, there are many difficulties in prioritizing observing systems.

Townshend went on to discuss the challenges in maintaining and operating GCOS. He recognized the contemporary decay of the *in situ* system, and the problems in ensuring continuity and consistency of observations through time. Often there is a conflict between technological advance and consistency. Maintaining global spatial coverage is a difficult problem and there is always the question of how much redundancy should be built into the system. Continuity of funding is also an issue. The point was made that somehow a balance must be achieved between processing the data we previously gathered and are routinely gathering versus any new observing systems.

Thomas Spence, Director of the Joint Planning Office for GCOS, described the overall framework through which GCOS operates, which is depicted in Figure 1. GCOS was established after the Second World Climate Conference and relies on the overall World Climate Programme infrastructure to help meet its goals. Spence noted that, in addition to climate change detection and response monitoring, the goals of GCOS are to contribute to national economic development and, through routine and systematic measurement systems, contribute toward improved research and toward improved understanding, modeling and prediction of the climate sys-

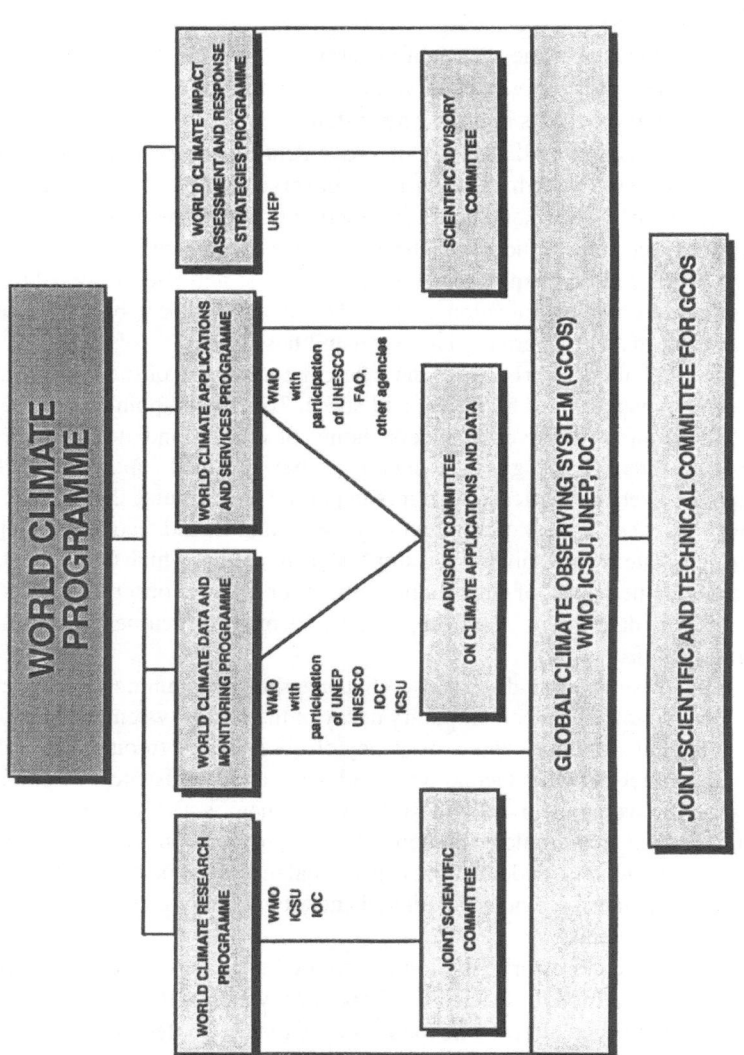

Fig. 1. World Climate Programme Lankages for GCOS.

tem. Spence also described the common interests of GCOS with respect to other global observing systems such as the Global Ocean Observing System (GOOS) and the Global Terrestrial Observing System (GTOS). The climate modules of those observing systems are identical to the ocean and land modules of GCOS, respectively.

Spence indicated that the concept, scope, and scientific and technical guidance of GCOS is the responsibility of the Joint Scientific and Technical Committee (JSTC), established through a Memorandum of Understanding among WMO, IOC, UNEP, and ICSU. The JSTC includes Atmospheric, Land, Ocean, Space Observing, and Data and Information Management Panels. Scientifically and technically sound and economically feasible recommendations are expected from these panels and are a prerequisite to any implementation of changes or additions to existing observing systems. In this regard, Spence noted that the first priority of GCOS is to define the Initial Observing System, which will be based on the current operational components and essential augmentations, including a comprehensive data management system.

Gregory Withee, speaking on behalf of the U.S.A.'s focal point for GCOS, Robert Winokur, who is also Co-Chair of the GCOS JSTC, challenged the participants to address four items of particular concern:

1. What is the science behind monitoring?
2. How can different measurement systems be used together?
3. How can we prioritize our most important global monitoring concerns?
4. How can we best use the measurements we have?

A number of principles of long-term climate monitoring were introduced by the meeting's organizer, Tom Karl, based on the work of Karl *et al.*, (1995), and the U.S.A.'s National Research Council's Climate Research Committee. These principles include:

1. Prior to implementing changes to existing systems, or introducing new observing systems, an assessment of the impacts of such changes on long-term climate monitoring should be standard practice.
2. Routine assessments of the long-term climate monitoring capability of exsisting systems should be standard practice.
3. Observing systems should be complete, possibly including both 'low technology' and 'high technology' components and ground truth validation.
4. The transition from research measurements to operational measurements for long-term climate monitoring must be planned in an orderly and stematic manner.
5. Knowledge of instrument, station, and/or platform history is lessential and should be treated with as much care as the data themselves.
6. *In situ* and other observations with a long uninterrupted record should be given special consideration.
7. Calibration, validation and maintenance are critical to long-term climate monitoring.

8. Processing algerithms and changes in these algorithms must be well documented.

9. Data management, analysis, and diagnostics are a key part of a relevant long-term climate monitoring system.

10. Data management systems must facilitate access (minimum cost and freedom of data availability) and data analysis.

These principles were used as a starting point in the meeting, and to the extent that they are similar or modified in our final recommendations reflects the common thinking of various climate scientists.

Following the introductory lectures and papers, a series of technical papers was presented and discussed during the first two days of the meeting. They have since undergone peer-review and appear in this special issue of *Climatic Change*. They were used as the basis for developing recommendations from the workshop. The authors presented their papers and had an opportunity to discuss them in context with other ideas in three breakout panels that met one evening and the subsequent day. The panels were divided into the following categories:

(a) climate forcings and feedbacks;
(b) climate responses and feedbacks; and
(c) climate impacts.

Some measurements of the climate system can be viewed in more than one of the three categories of response, feedback, or forcing, e.g. precipitation and it's latent heat of condensation, depending on the question posed, so a clean break between panels was not possible. As a result, panels (a) and (b) both considered various aspects.

The task of each break-out group was to develop a set of recommendations that could be used by the GCOS JSTC to help ensure an adequate global long-term climate monitoring capability. It is important to understand that it was not possible to set priorities with respect to the set of specific recommendations developed at this workshop as related to the myriad of needs within each of the three break-out group topics. To set priorities complete information is required i.e., a complete set of issues with a finite number of choices. Obviously, priorities change as assumptions, impacts, and users are varied. Instead, the set of recommendations developed from this meeting should be viewed opportunistically as related to each characteristic of the climate system: forcings, feedbacks, responses, and impacts. All are important and should be implemented to achieve an adequate long-term global climate observing system.

3. Recommendations

In this section a summary of the recommendations that were put forward from each of the working groups is presented. Additional details can be found in each panel report and in the technical papers of this issue of *Climatic Change*.

In developing the recommendations from this workshop, careful attention was paid to the requirements for decadal-to-centennial climate monitoring. The fact is however, as resources are constrained, it will be even more critical in the future to have multi-purpose observing systems. They will have to satisfy a variety of end-users. Moreover, climate is a continuum of time scales. Fortunately, if climate monitoring is done properly for the longest time scales, where detection of change is most demanding, the requirements for high-quality, continuous, and homogeneous data at the shorter time scales will also be fulfilled.

If successful, the GCOS will not be viewed as operational versus research or monitoring versus research, but rather a partnership among observing system operators, data managers, scientists, and other users. Experience with the weather prediction and greenhouse gas flask sampling communities indicates that operations and research can, indeed, be an effective team. This takes time, leadership, and organization. GCOS interests for long-term climate monitoring will have to ensure that such a partnership is forged.

3.1. GENERAL RECOMMENDATIONS

The general recommendations from the workshop cut across disciplines and include:

Recommendation 1:
Prior to implementing changes to existing systems or introducing new observing systems an assessment of the effects on long-term climate monitoring should be standard practice.

Recommendation 2:
Establish the capability for routinely assessing the quality of observations important for long-term climate monitoring, detection, and attribution. This should include the chain of activities involved in the processing of data into useful products.

Recommendation 3:
Continued efforts should be made to improve the quality and volume of the historical data base, including long-term high resolution data capable of resolving important extreme climate events.

Recommendation 4:
Improve existing and develop new information management systems related to the gathering, quality control, assembly, and archiving of climate data; and a system of telemetry, communication and permanent storage dedicated to the objectives of the system.

Recommendation 5:
Ensure that long-term climate assessments that focus on the regular examination and re-examination of the climate data base are an integral part of GCOS. Up-to-date state-of-the-climate assessments should be produced.

Recommendation 6:
Overlapping measurements of both the old and new observing systems for both *in-situ* and satellite data must become standard practice for critical climate variables.

Recommendation 7:
Better satellite data and satellite product calibration is required such as: on-board calibration for all basic measurements of radiance, including the visible channels, and more appropriate 'ground-truth' measurements with the sampling and physical characteristics as comparable as possible to the satellite products.

Recommendation 8:
For polar orbiting satellites minimize the aliasing of the diurnal cycle of the parameter of interest, and most importantly ensure that diurnal sampling biases do not change with time by controlling the flight of the orbiting satellites i.e., eliminate drift.

Recommendation 9:
Pursue alternative means for calibrating stratospheric measurements from satellites. This includes making better use of existing balloon borne systems, such as the ozonesonde network, and microwave measurements.

Recommendation 10:
Ensure that the full and open international exchange of data is maintained at minimal cost of reproduction.

Recommendation 11:
GCOS should encourage the development of metadata information systems that provide, among other types of information, factors that may result in time dependent-biases. The system should be readily accessible. It is further recommended that metadata be given high priority and that GCOS make a similar recommendation to WMO.

Recommendations related to specific climate elements are briefly summarized below. Each of the following three Editorials contain more details.

3.2. CLIMATE FORCINGS AND FEEDBACKS

Greenhouse Gases (CO_2, CH_4, CO, N_2O, CFCs, O_3, H_2O vapor)
Expand the present flask sampling network. More attention must be given to upper tropospheric and stratospheric water vapor measurements. Ozonesondes must be employed in an expanded geographic coverage and improved in quality.

Aerosols
Make better use of satellite-borne aerosol measurements from lidar and passive photopolarimeter instruments, but this is insufficient by itself; so equip a small number of sites with instruments to completely characterize aerosols in areas with key aerosol types.

Solar Radiation
Better accuracy, overlapping satellite measurements, complete spectral resolution, and a ground based program are all critical aspects of a long-term solar monitoring effort.

Clouds
Integrate satellite and surface-based observations. Preserve long-term human-observed *in situ* records, now being replaced by automated measurements, by providing overlapping measurements of adequate duration to establish the cross calibration of the old and new observing systems. Satellite instruments require good in-flight calibration of both longwave and solar channels and overlapping satellite calibration. Stable orbits are also critical as diurnal sampling bias has become a serious satellite problem. The use of three polar orbiters, one precessing and the other two with constant equatorial crossing times can help resolve this problem. *In situ* measurements should be expanded, especially over the Southern Hemisphere, and must include cloud cover, morphological types, and base heights. Cloud base (*in situ*) and cloud top measurements (satellites) are critical cloud properties whose accuracy must be a high priority.

Water Vapor
For radiosondes (and rawinsondes): develop a new cost effective sonde or equivalent technology; standardize algorithms converting relative humidity to dew point temperatures; standardize reporting procedures; develop station histories and an archive of relevant metadata; improve humidity sensors with respect to response time, performance at low temperatures, low humidity, and high humidity; as routine practice, develop a reference radiosonde intercomparison for all operational sondes.

For satellites: implement GEWEX (GVAP) priorities to make existing data useful (Geostationary and Polar Orbiters). More precise pre-launch calibrations of the water vapor channels are required, and higher vertical resolution can and should be achieved using existing technologies.

[13]

Radiation Budget
Sustained long-term monitoring, not necessarily from operational satellites, is critical for understanding climate forcings and evaluating changes in quantities such as clouds and aerosols.

3.3. CLIMATE RESPONSES AND FEEDBACKS

Model Reanalysis
Additional re-analyses should be encouraged that focus on being able to reliably determine decadal-to-centennial climate variability and change.

Climate Data
Upgrade the near real-time data receipt from the Global Telecommunications Network (GTS) by taking advantage of the information superhighway. Countries should be asked to submit more stations per unit area in data sparse regions and more in real time.

Greenhouse Sensitive Variables
Additional efforts are required to determine the most sensitive variables to anthropogenic climate forcing. Long-term measurements of these variables, including measures of atmospheric circulation, must be a high priority.

Reference Stations
Reference stations are needed for surface and ipper air measurements. Adequate resolution is necessary to capture changes and variations of extreme events. Reference stations need to be maintained and operated on century time scales. Data from these stations should be freely, openly, and widely distributed.

Land Surface Air Temperature
Member WMO stations can improve the current state of knowledge by issuing more station data in real time and making more historical time series data available. Existing temperature data sets have the fewest observations at the beginning and end of the period of record, critical times of interest.

Tropospheric and Stratospheric Temperature
New satellite observing systems should be designed to assure backward compatibility with as little complication as possible. Radiosondes require new modes of operation [as described in Section 3.2. (water vapor) and the panel a and b reports) with special attention to radiation induced errors in temperature measurement.

Precipitation
The current *in situ* network must be improved to reduce imeasurement biases, and data inhomogeneities. Modernization of these networks is now creating additional biases which need to be well understood prior to implementing changes. Remote

[14]

sensing of precipitation continues to require attention. The fusion of gauge, satellite, radar, and model generated precipitation estimates should be given a high priority. Streamflow and lake level data should be improved and benchmark data sets developed, free of anthropogenic influences. These data can serve as critical reality checks for precipitation variations.

Surface Marine
More attention to long-term homogeneous measurements are needed for winds and temperature, and the fusion of satellite and *in-situ* observations is critical. Rescue and digitize data that are now in manuscript form only, particularly where these data fill space/time gaps (such as the Kew and Kobe collections).

Sea Level
Develop a subset of the GLOSS sea level network and implement an operational long-term precision altimeter.

Subsurface Ocean
Maintain existing long time series, improve bathymetry data, digitize all upper ocean and interior ocean data before it is 'lost', and foster a global network of subsurface floats, e.g. ALACE .

Cryosphere
There are several specific items: (1) Optimize a procedure for blending satellite and *in situ* observations; (2) Ensure overlap of new automated NOAA/NESDIS snow cover and sea ice products with old methods: (3) Establish an archive for lake ice data; (4) Develop an altimetric baseline for decadal sampling of ice sheets; and (5) Increase the number of shallow cores over ice sheets.

3.4. CLIMATE IMPACTS

Consumers of impact-related climate data from GCOS include resource managers, planners, and governments (local, regional, state, and national), as well as climate impact researchers. So the focus in this area must look beyond the climatological field and consider a community.

Climate Indices
To better assess the change in the earth system global environment integrated information is required such as may be provided through leading environmental indicators.

Other Global Products
More emphasis needs to be placed on the development of long-term global products. This is critically dependent on adequate metadata. GCOS should improve

existing metadata and the initial focus could begin with three critical climate variables: temperature, precipitation, and wind (pressure). The metadata should include past and present information required to interpret the representativeness, accuracy, where and how the data may be accessed, and potential biases of the measurements, such as changes in the time and place of observation, site exposure, instrumentation and methods, data formats, and methodology used to process the data.

Data for Indices and Other Climate Products
GCOS should encourage support for the data needed to construct climate indices and products.

Land Surface Monitoring
Although a consensus could not be reached as to the most appropriate derived product to focus on, there was strong support for long-term monitoring of various land surface characteristics such as vegetation, soil moisture, runoff, skin temperature, etc. Linkages among GCOS, GTOS, and IGBP must continue to be fostered to provide much needed focus for long-term monitoring. Land surface hydrology requires a specific focus, much like GTOS, and we note the current activities focused around the development of a GHyOS (Global Hydrological Observing System).

4. Conclusions

Clearly there are many opportunities for GCOS to improve our long-term climate monitoring capabilities. And clearly monitoring requires the synthesis of observations, analyses, and modeling. As already acknowledged, the recommendations put forth through the experts meeting cannot be easily viewed in terms of priorities, but rather should be used in an opportunistic sense. All the recommendations are important. Despite these words there are two general concepts that stand out among all others. First, adequate long-term climate monitoring will continue to be critically dependent on developing a partnership among network operators, data managers, analysts, and modelers. Multi-purpose observing systems used for operations, research, and monitoring are likely to be the most practical means of achieving an economical long-term climate monitoring system. Second, GCOS should use the recommendations from this meeting of experts in an opportunistic sense, that is they should be implemented as early as possible at appropriate windows of opportunity.

Acknowledgments

This report would not have possible without the support and deliberations of nearly 100 scientists attending the 3-day GCOS long-term climate monitoring meeting.

It reflects to the best of our ability a consensus of their recommendations. Special thanks is also extended to the following who suggested important revisions in earlier versions of this manuscript: Roger Barry, John Christy, Michael Crowe, Bruce Douglas, William Elliott, Joe Elms, Hermann Flohn, Vivien Gornitz, Pavel Groisman, Arnold Gruber, Nathaniel Guttman, James Hansen, Dennis Hartmann, David Hofmann, Judith Lean, Sydney Levitus, Alvin Miller, David Parker, David Robinson, Chester Ropelewski, Neville Smith, Stephen Warren, and Robert Weller. The opportunity to address the global climate monitoring problem was provided through the foresight of Gregory Withee, NOAA's Deputy Assistant Administrator for Environmental Satellites, Data and Information through NOAA's Earth System Data and Information Management (ESDIM) Program Office and J. Michael Hallis Climate and Global Change Program.

LONG-TERM CLIMATE MONITORING BY THE GLOBAL CLIMATE OBSERVING SYSTEM

Report of Breakout Group A – Climate Forcings and Feedbacks

An Editorial

CHRISTOPHER MILLER

National Environmental Satellite, Data and Information Service, 1315 East-West Highway, Silver Spring, MD 20910, U.S.A.

and

FRANCIS BRETHERTON

Space Science and Engineering Center, University of Wisconsin, 1225 W. Dayton Street, Madison, WI 53706, U.S.A.

1. Introduction

The assignment for Breakout Group A was to re-visit and expand upon the plenary session discussion on climate forcings and feedbacks and to develop a set of recommendations for each of the science disciplines or activities covered within this breakout category. Working guidelines for the group included identifying: (1) what has to be done; (2) why it has to be done, i.e., who is the customer? (3) the process for remedying deficiencies and, specifically, how to leverage the activities at operational centers; and (4) priorities (recognizing that it is premature to distinguish between major systems). The science areas addressed included:
- greenhouse gases (GHGs);
- radiation budget;
- water vapor;
- aerosols;
- clouds;
- precipitation;
- tropospheric ozone; and
- solar radiation.

Some of these topics received more attention than others. It was felt that there was neither time nor appropriate participants to adequately address the important topic of land surface processes/land surface characterization and, therefore, there is no discussion reported for this topic. Likewise, the specific recommendations included below are considered important within each science area, but were arrived at without detailed consideration of the associated costs. The meeting began with some general discussion about the customer base for a Global Climate Observing System. Modelers who work within the framework of operational forecast systems

Climatic Change **31**: 149–162, 1995.
© 1995 *Kluwer Academic Publishers.*

(i.e., weather services) are customers because they rely upon timely, high quality, global data sets to update their model forecasts and push the forecast horizon. In turn, people who consume forecasts to support a decision-making process are also customers. As the time scale of the forecasts is extended, policy makers become prime customers since they look to the forecasts to formulate/buttress policy decisions that can have wide-ranging societal consequences. Policy makers are interested in climate change detection but, in addition to evidence of change, they also need to have the observed or projected change interpreted, i.e., attributed to specific causes. The attribution of change is a more demanding challenge than the detection, which itself is a difficult problem. To resolve attribution can impose unique monitoring requirements, e.g., the complex, chemical interactions among atmospheric constituents, like methane and tropospheric ozone. In addition to global-mean climatic changes, possible changes in patterns of variability, e.g., shifts in extreme event statistics, also must be examined and interpreted as part of the detection/attribution process.

GCOS must consider how to achieve an operational system tuned for the longer, climate time scales. Success for GCOS will depend critically on a partnership being forged among the operational observing system, GCM modeling, and climate analysis communities, an arrangement similar to the one that has worked well for the weather community. This is a long-term process that can begin early with the identification of modest additions to the existing operational systems. Operational agencies also will need to consider GCOS requirements when contemplating changes to their systems. This partnership will involve continuous exchange and feedback with modifications being made to operational procedures, as necessary, in the light of experience with modeling requirements and developments in instruments and data interpretation algorithms. However, at the same time, the quality aspects of the existing operational systems would be preserved. This process will depend on identifying key players and key products. Such an arrangement can be successful given leadership, organization, and time.

Group A assumed responsibility to identify the parameters relevant to the decadal-to-centennial time scale, but acknowledged also that a need exists to identify those parameters important to the seasonal-to-interannual time scale since much of the necessary instrumentation, quality control and data assimilation will be driven by such shorter-term requirements. It was recognized by the Group that the decadal-centennial community will have to attend to its own data interests, while simultaneously working with the weather prediction and seasonal-interannual communities. In this regard a guiding principle has been suggested by Tom Karl, i.e., as a matter of standard practice, assessments of the effects on long-term climate monitoring of planned changes to existing systems should be made and submitted to the climate community for comment. For example, the current MSU and future AMSU are projected to have about a one-year overlap. The overlap should not be assumed to be a 'given'; the overlap necessary should be quantified per Tom Karl's proposed Climate Monitoring Impact Reports. The user community should be involved in this evaluation. There has to be assurance that NOAA-14 is not prematurely shut off without adequate consideration of the impacts on all three communities – weather prediction, seasonal-interannual, and decadal-centennial. Proposed system enhancements should, if nothing else, 'do no harm'. If modifying a system will compromise the integrity of an operation that already works well, the temptation to make the change should be resisted.

A paramount objective for GCOS is the creation of quality data sets. To ensure that this development work is done, it is necessary that prospective users *demand* that it be done. For example, the successful NOAA flask sampling network works because it responds to a clearly articulated demand.

The issue of ozone depletion and CFC regulations is a success story because there were good measurements of the forcing agents (polar stratospheric clouds, chlorine-related gases and heterogeneous chemical rates in the laboratory), reasonably good predictive models, and good ozone observations for verification. A similar process for climate change is more demanding. However, the tested recipe remains the same: good measurements of all the forcing elements combined with realistic models. However, the reality is that the forcing and radiation regimes are more complicated; the troposphere is more complex than the stratosphere; the models are more elaborate; and the climate observations for verification are more difficult.

2. Climate Forcings and Feedbacks

2.1. GREENHOUSE GASES (CO_2, CH_4, CO, N_2O, CFCs, O_3, H_2O VAPOR)

The distribution over the globe of sources and sinks for carbon dioxide, as for other long-lived GHGs such as methane and nitrous oxide, is not fully understood and it is important to determine the temporal variability to balance the global carbon cycle. The present NOAA flask sampling network (about 40 sites) shows great promise in studying sources and sinks when the GHG concentrations are combined with isotopic carbon ratios in a two-dimensional transport model. However, to be able to determine temporal variations this system must be expanded. This direct observational approach, when combined with the inverse approach of transport modeling, will create a more robust, integrated strategy. Consequently, another issue is establishing the best transport model within which to maximize the use of these observations. In the following sections recommendations are provided for the monitoring of some of these constituents.

On the Feedbacks side, stratospheric temperature is one example of an important parameter. The stratosphere is expected to show an amplified signal in response to CO_2 buildup. This temperature shift will have an effect on the photochemistry and a possible impact on aerosols. There is also a feedback between temperature in the lower stratosphere/upper troposphere and surface temperature fields. With respect to applicable instrumentation, there are five temperature lidars at the Network for the Detection of Stratospheric Change (NDSC) sites, an array that has implications for future operational satellites.

2.2. RADIATION BUDGET

The history of natural and human-influenced radiative forcing is characterized by a high degree of uncertainty, which, in turn, adversely affects climate modeling. The concept of equivalence for radiative forcing states (postulates) that the effect of radiative forcing is independent of the nature of the forcing agent (e.g., CO_2 solar, etc.) and is simply proportional to the quantitative change in the forcing, i.e., the global-average surface temperature change is related to the global-average radiative forcing (the change in irradiance, in watts/m^2, at the tropopause) through a proportionality factor that is independent of the nature of the forcing. In other words, the climatic response is independent of the exact forcing mechanism. This hypothesis has been called into question especially for spatially-inhomogeneous changes in aerosols and tropospheric ozone. A related issue is the summed effect of forcings that may be of opposite sign. If all the radiative forcings sum to zero, it does not necessarily follow that regional, and possibly even global-scale, climate change is not occurring. The presentation by Wigley discussed this concept of climate sensitivity, which to be operationally valid in a modeling context must be assumed independent of the character of the forcing. Wigley discussed the steps needed to reduce uncertainties (understanding of aerosols through observation and modeling; monitoring of solar variability and volcanic effects; monitoring of global-scale temperature changes; and coupled modeling studies of internally-generated natural variability). The challenges of modeling at the regional scale are illustrated in the paper by Cubasch *et al.*

An appreciation of the complexity of radiative forcing is beginning to develop. Many of the contributions to radiative forcing occur via subtle and inadequately understood mechanisms involving chemistry and/or cloud microphysics. Progress will occur when increased understanding of the individual mechanisms advances (i.e., observed temperature, by itself, does not provide sufficient insight).

There is complementarity between Earth Radiation Budget measurements and other measurements (water vapor, clouds). However, observational support for understanding the role of many of these mechanisms is lacking. Long-term monitoring of the radiation budget needs to be undertaken. Among the requirements is the need for improved surface radiation fields (attributes: gridding, availability, accuracy, UV). Because changes within the troposheric column influence the climate response, radiation at the surface and at the top of the atmosphere is needed.

[22]

2.3.   WATER VAPOR

Water vapor is the most radiatively important GHG and its feedback on climate change has a major influence on climate sensitivity. Water vapor measurements from satellites and radiosondes provide the opportunity for a combined data set to be produced that is of higher quality than either data set by itself. This objective is consistent with the research goals articulated in the international GEWEX Water Vapor Project (GVaP).

No one system is adequate. Satellites provide quasi-global coverage, modest vertical resolution, and suffer from discontinuities among successive satellites and errors introduced by changes in orbits and calibrations. Radiosondes have good vertical resolution, world-wide coverage (but with spatial gaps), a level of quality that diminishes with low temperatures (altitudes greater than 5 km), different humidity sensors and associated data reduction techniques, and changes in instruments over time. Surface observations represent potentially long records for humidity but have not been well studied. Problems include those currently encountered with temperature measurements, changes in the quantity measured, changes in the instrument type, and sensitivity to changes in surroundings.

Stratospheric water vapor at about 20 km has been observed to be increasing at Boulder, Colorado at a rate of about 1% per year over the past 14 years, which is at least twice the rate expected from methane oxidation and increases in methane. If this increase is widespread and represents a signal of global warming, there are important implications for global climate models. The present balloonborne frost-point hygrometer monitoring program should be expanded from a single U.S. site to five global sites, with the additional sites potentially distributed as follows: two in the polar regions and one each in the mid-latitude southern hemisphere and the tropics. There is complementarity of these measurements with satellite infrared and microwave sensors for delineating important vertical profiles. The final design of a monitoring network should be determined after further evaluation. A fuller discussion of the monitoring of water vapor in the troposphere and lower stratosphere can be found in Elliott's paper.

There were several recommendations to bring uniformity and accuracy to upper air radiosonde moisture measurements. The recommendations have been categorized as follows (not in order of priority):

1. Upgrade WWW water vapor measurements
  – Standardize reporting procedures.
  – Continue to respond to WMO requests for radiosonde station histories.
  – Standardize the algorithms that convert relative humidity to dewpoint temperature.
  – Institute higher vertical resolution satellite water vapor observations using proven technologies

– Develop a 'reference' radiosonde system to help evaluate the performance of all radiosondes and promote global change research and assessments. Applications of the radiosonde include: 1) evaluation of *in situ* aircraft sensors (e.g., CASH – Commercial Aviation Sensing of Humidity); (2) air truth for satellite, airborne and ground-based remote sensors (active and passive); and (3) *in situ* testing and evaluation of expendable state-variable and wind sensors and alternative exposure modes.

(note: the latter two recommendations also fall within Category 2 below)

2. Implement GVAP

– Integrate radiosonde and satellite data.
– Implement the GEWEX Water Vapor Project to make existing (15+ years) IR data useable (GOES and POES); utilize the TOVS measurements (note: for the NMC and ECMWF model assimilations no TOVS moisture data were used until 1994).
– Establish a metadata archive.
– Examine the integrity of the radiosonde system in light of: (a) constraints on the frequency band; (b) demise of the omega and Loran-C navigation systems; (c) impending cost increases for expendables and ground equipment
(note: the companion editorial by Trenberth for Breakout Group B, 'Climate Responses and Feedbacks', discusses the threats to the radiosonde/rawinsonde network).
– Establish a small climate change detection network of upper-air stations, i.e., recognize that observing climate is important and not identical to observing weather).

3. Develop Improved Instrumentation

– Improve humidity sensors with respect to response time; performance at low temperatures, low humidity (less than 20%), and high humidity (greater than 80%).
– Develop a new, inexpensive radiosonde, possibly GPS-based.
– Institute accurate pre-launch calibration of water vapor channels, i.e., vacuum testing, and assess the impact of this procedural change.

4. Develop New Measurements

– Institute accurate and precise *in situ* water vapor observations including both 'up' and 'down' sondes. One option is to consider operational dropwindsondes. Commercial and military passenger and cargo aircraft regularly traverse all the world's tropical and temperate oceans at the 200 hPa level and could serve as cost-effective launch and data-telemetry platforms.
– Expand *in situ* measurements in the stratosphere.

Less than 50 of 600 global rawindsonde stations are oceanic. Virtually all ocean weather stations have been decommissioned. Present oceanic windfinding is restricted to omega-based and balloon-tracking systems (radar and theodolite-based), which are expensive to purchase and maintain.

The satellite MSU complements the radiosondes by providing spatial completeness. The MSU (since 1979) can be used in the assessment and removal of radiosonde temperature biases. Temperature, rather than thickness or virtual temperature, should be used for direct comparison with MSU because of moisture biases that affect the radiosonde. Christy *et al.* can be consulted for a description of the MSU measurements and their validation against radiosondes.

2.4. AEROSOLS

There are many types of aerosols; two especially important ones are carbonaceous (organic, black carbon) and water-soluble (sulfates and nitrates) aerosols. Mineral dust, desert dust and sea salt must also be included if human-induced and natural direct effects are to be distinguished (mineral dust, desert dust) and if indirect effects are to be assessed (sea salt). Aerosols apply both direct (scattering and absorption) and indirect (effect on cloud formation) forcing to the climate system. The necessary measurements of aerosol properties are more numerous and difficult than for other climate forcing agents, for example, the aerosol particle size-discriminated chemical composition. While measurement of one extensive variable suffices for GHGs (i.e., concentration), it is necessary to measure both extensive and intensive variables for the aerosol problem. Extensive variables are mass and number concentration and optical depth. Intensive variables are size distribution, chemical composition as a function of size, refractive index, optical scattering efficiencies, CCN spectrum. While satellite-borne lidar will provide information on aerosol vertical distribution and rough estimates of geographic variations in optical thickness, it cannot provide the high-precision optical thicknesses, particle size distributions and chemical composition information that is needed to yield a quantitative connection to temporally or spatially variable aerosol sources or to determine climate forcing. A satellite-borne photopolarimeter offers the hope of measuring aerosol optical depth, refractive index, and particle size, complementing the high vertical resolution of backscattering measured by a lidar.

It is believed that aerosols may mask the greenhouse warming effect on a regional basis. Currently, there is insufficient knowledge of how to measure crucial aerosol forcing operationally. This should be viewed to a large degree as a research problem, for which the research needs to be accelerated. A coordinated attack involving *in situ* observations, satellite observations, and modeling is needed. Appropriate measurement methods must be developed, optimized, and standardized. Concentrated observations should be continued for a few to several years, after which a plan should be developed for monitoring a subset of aerosol characteristics.

Direct forcing is a well-posed problem and the reduction of uncertainties in estimated forcing is realistically achievable. Direct forcings and their uncertainties have been estimated for anthropogenic sulfate, organic and soot aerosols and for volcanic/stratospheric aerosols. Indirect forcing is a very difficult problem and will require substantial research prior to quantification. Indirect forcing via effects of aerosols on clouds has been estimated but all quantitative statements to date are only examples of possible magnitudes based on ad hoc assumptions.

To reduce the uncertainties in a number of aerosol parameters, systematic measurements on key aerosol types combined with model calculations of the regional forcings are necessary. A small number of aerosol observation sites should be established in areas where key aerosol types exist (e.g., industrial haze, biomass combustion smoke). Such sites should be equipped with a complete set of aerosol characterization instruments. Ground-based sites should be supplemented with frequent flights of instrumented aircraft to document horizontal and vertical variability of aerosol properties. Specifically, within the North American continent four aerosol characterization sites, which represent the conditions of 'clean' and 'anthropogenically perturbed', 'continental' and 'marine', are being established by NOAA. This network would be complemented by expansion of the current NOAA five-station spectral optical depth network to include the regional aerosol monitoring sites *plus* a network of approximately 20 optical depth sites in the eastern U.S., where the anthropogenic sulfate concentration is a maximum. Other globally-distributed sites are necessary for full characterization of the aerosol properties. Collaborations are underway with European sites, in particular, the WMO GAW sites. The recent major eruption of Mt. Pinatubo demonstrated the large, transient effects on atmospheric temperature associated with volcanic stratospheric aerosol. Indirect effects on atmospheric chemistry (e.g., ozone reduction causing an increase in UV radiation, causing an OH increase) have consequences for GHG levels. The 20-year record of stratospheric aerosol lidar soundings at Mauna Loa should be continued with improved instrumentation as part of the Network for the Detection of Stratospheric Change (NDSC). The spectral optical depth measurements at five baseline monitoring sites (Barrow, Alaska: Mauna Loa, Hawaii; Samoa; the South Pole; and Cape Grim, Tasmania) and at Boulder should be continued. Also, the 24-year balloonborne aerosol monitoring program by the University of Wyoming at Laramie should continue to be supported.

An international focus is needed to promote efforts among various nations to do exploratory research and sharing of information. From the standpoint of leveraging efforts, there needs to be coordination and merging of efforts between the IGBP IGAC project and the WCRP efforts to achieve efficiencies and a better product. This discussion has already begun. For example, data processing has to be coordinated and integrated, because merging of data afterwards is difficult.

2.5. CLOUDS

Clouds are not being monitored with sufficient accuracy to determine their effects on the radiative forcing of the atmosphere. Maximizing information retrieval on cloud properties from a mixture of satellite and *in situ* measurements presents a challenge. Surface visual observations of clouds are useful because the observer is close to the clouds, clouds can be identified by type (which can be related to cloud dynamics and meteorology), records are available for several decades (which permits interannual and trend analyses), and some clouds are difficult to detect from above (e.g., clouds over snow, low clouds at night). Types of information reported include total cloud cover, base height of lowest cloud, amount of lowest cloud, cloud type at different elevations (low, middle, and high) and present weather (fog, rain, snow). The sampling density over land is 6000 stations every three hours (every six hours in the U.S., Canada, and Australia); over the ocean 1000 ships report every six hours. The average spacing of land stations is 160 km (Europe has many stations; Antarctica too few). An accurate determination of the diurnal cycle from surface observations requires screening the observations for adequate moonlight or twilight.

Ideally, a cloud monitoring system would have uniform coverage globally, a spatial sampling interval less than 50 km, a sampling frequency of greater than six times per day, a record length of greater than 10 years and a calibration precision with overlapping records of 1%. Sampling strategy is very important. Even when the spatial scale is large, the diurnal cycle must be captured, thus imposing spatial/temporal sampling constraints of six times/day and less than 50 km spacing.

In the U.S., cloud type (morphology) is not monitored by the ASOS (Automated Surface Observing System) laser ceilometer which yields base height. ASOS does not report cloud bases if they are more than 12,000 feet above ground level. Previous to the introduction of this replacement system, cloud type was visually observed on a routine basis. An in-depth evaluation of the impact of this loss has not been undertaken and is required. Continuation of visual observations at 100 stations (with a spacing of about 300 km) is planned by the NWS. Changes in total cloud cover and amounts of each cloud type over the oceans can be obtained beginning around 1954. Earlier ship observations are inhomogeneous. Cloud cover appears to have increased during the last 40 years over the ocean by 1% in the Southern Hemisphere and 2% in the Northern Hemisphere. The sampling by ships is random within gridboxes and not regular in time. To reduce the error in the seasonal average cloud cover to 2–3% in an oceanic grid box requires 100 observations. An optimal system would integrate satellite and *in situ* measurements. The International Satellite Cloud Climatology Project (ISCCP) is producing cloud products sufficient for seasonal/interannual requirements but not for the longer term, i.e., the magnitude of the natural changes is smaller than the ability of the satellites to resolve. The data set is global, currently spans more than ten years (beginning in July, 1983), and has a space-time resolution of 30 km and 3 hours. ISCCP recalibrated the satellite data to remove spurious trends and discontinuities. These data allow investigations of cloud variations, e.g., cloud cover, top height, optical thickness.

In a cloud retrieval system, a lot of parameters need to be derived from the satellite radiances. The paper by Rossow and Cairns provides details on the satellite retrievals. Recommendations for GCOS include: (1) continue routine surface and satellite cloud observations and analyses; (2) improve operational humidity profiling (surface and satellite); (3) improve cloud base height measurements from the surface; and (4) augment operational satellite network with specially designed cloud monitoring instruments (spectral resolutions and coverage, high precision calibration).

2.6. PRECIPITATION

For a number of reasons the present network is clearly inadequate to provide the quality precipitation data needed. Problems that confront the precipitation network include: (1) spatial gaps – ocean measurements; (2) presence and removal of biases; (3) discontinuities; (4) adequacy of modernization plans for climate monitoring; and (5) planning for the long-term. A multi-faceted remedial approach is suggested:

– Maintain the current network and facilitate data exchange. The vehicle for realizing this goal is a well-maintained Global Historical Climatology Network (GHCN) producing area-averaged anomalies for climate monitoring. This system will allow a smooth transition to the new generation of instrumentation and provide a reliable product to the modeling community.
– Establish a high quality reference network from the existing set of longest, most complete and highest quality weather stations.
– Consider introducing a new surface instrument, designed specifically for climate monitoring.
– Satellite measurements use 'ground truth' to remove biases but the ground truth itself may be flawed; account for this.
– Use orographic models to segregate orographic effects on precipitation; runoff data can be used as a constraint on the estimates.
– Develop/apply techniques that correct for inhomogeneities and instrument biases.
– Pursue data archaeology (e.g., the colonial archives).

Recent research (see paper by Groisman and Legates) has exposed a serious bias in standard rain gauges, dependent on instrument type and exposure. To adequately study precipitation changes, it is necessary to: (a) use unbiased measurements (unrealistic); or (b) maintain a constant bias (at least on average) in precipitation measurements; select the best stations (attributes: fewer moves, long-term time series, representative of the area).

It must also be recognized that snow is more difficult to deal with because of more extreme and varying biases.

[28]

The need continues to exist to develop new and utilize current precipitation remote sensing algorithms. Remote sensing is based on empirical techniques from IR and microwave measurements. The coverage is quasi-global (IR is most applicable between 40°N and 40°S; microwave is most applicable between 60°N and 60°S, but with poor sampling in low latitudes); the length of record is short; 50 km is an achievable spatial scale; and the frequency of observations is up to eight per day. In the future there will be ocean *in situ* observations using optical rain gauges and acoustical techniques and new satellite instruments like the rain radar TRMM (Tropical Rainfall Measurement Mission). The Global Precipitation Climatology Project (GPCP) is an integration of various precipitation products (gauges, satellites, radars, ECMWF forecasts) and other climate and hydrological variables. GPCP is using gauge and satellite data to produce a gridded (2.5 deg. latitude/longitude), monthly mean, global data set, initially for 1986–1995. Accurate absolute values provided by GPCP will facilitate the calibration of a new generation of instruments.

Some conclusions/recommendations for precipitation measurements are:
– Data availability is still a problem.
– Studies on changes in precipitation require: (a) relatively dense networks; and (b) careful preprocessing of the data.
– Enormous biases and inhomogeneities of unadjusted data may mislead any unaware user who attempts to study precipitation changes over north Eurasia and Canada.
– If homogeneity is not addressed before new measurement systems become operational, the 200–year record of instrumental precipitation observations will be devalued for future climate change analyses.
– No simple solutions exist for the problem of documenting precipitation changes over the ocean.
– No simple solutions exist for the problem of documenting snowfall changes.

2.7. TROPOSPHERIC OZONE

Tropospheric ozone is an important greenhouse gas that has increased in some regions. Ozonesondes are the only currently feasible technique for obtaining ozone profiles in the troposphere; however, the number of profiling stations is entirely inadequate to resolve regional signatures and global trends, e.g., there is a deficiency in the tropics. The present NOAA three-station ozonesonde network is part of a global network that needs to be expanded to improve monitoring of tropospheric ozone trends. International efforts to create a network of this type were endorsed by Group A, which recommended expansion of the network, as appropriate. The proposed International Tropospheric Ozone Year (ITOY) may serve as a catalyst near the end of this century for an expansion of these measurements.

The satellite SBUV, SBUV/2 measurements (from 1979 onward) of ozone provide estimates of total column ozone and ozone profiles in the region 25–55 km above the surface. This means that the lower stratosphere and troposphere are covered only as the residual of these two quantities. The SAGE instruments provide ozone profile data for altitudes 15–50 km, but horizontal sampling is limited and, again, the troposphere is missed. This absence of ozone measurements was noted by Group A.

2.8. SOLAR RADIATION

Changes in solar radiation have the potential to cause natural changes in climate. If there are anthropogenic effects, these will have to be detected within this background of natural variability. Total solar irradiance (TSI) provides direct radiative forcing of surface temperature and other climate parameters. Ultraviolet spectral irradiance affects the direct radiative forcing because it is a significant component of TSI variability. It also contributes to indirect forcing of climate through its influence on ozone. Knowledge of visible and infrared spectral irradiance can complete the picture of how spectral variability affects physical mechanisms related to climate. Climate change simulations require as inputs all climate forcings, including variable solar radiation. The modeling community has utilized estimated TSI forcings in attempting to capture the naturally forced climate change over the past few centuries; direct observations only exist for the past fifteen years. This is an essential element of the climate change detection/attribution exercise. The effects of varying solar radiation must be evaluated if the causes of any future climate change are to be unambiguously identified. The future solar effect may be small (note: there is some evidence that prior to the 19th century variations of solar irradiance contributed to observed temperature changes but since then GHGs have been the principal contributor) but this has to be quantified. Also, future solar changes cannot be predicted; they must be measured. A benchmark for assessing the adequacy of the present measurements would be 20% of the estimated present rate of change of total anthropogenic radiative forcing, equivalent to approximately ±0.04% per decade in TSI. Though individual advanced cavity radiometer instruments may perhaps be achieving this level of precision, uncertainties in the prelaunch calibrations and the observed short-term variations in TSI require extensive overlap between satellites coupled to dense sampling of the measurements to construct valid long-term time series. To date, only one 11-year solar cycle has been measured. Gaps in present day monitoring are likely to occur around the year 2000, compromising the historical record. No solar radiometers are planned for either total or spectral irradiance after EOS; both are required.

Instruments for measuring UV spectral irradiances in the range 200–300 nm should be capable of measuring changes of 0.1% per decade, compared to present day UV radiometers which have relatively high spectral resolution and long-term precision less than 1% per decade. An alternative is broad-band UV monitoring, which has the potential to achieve higher long-term precision for the longer wavelengths and would be less expensive than spectrally dispersing radiometers, but at the expense of spectral information. Monitoring of the entire solar spectrum is ultimately preferred. Additionally, auxiliary ground-based solar monitoring is relevant for modeling to establish proxies of solar variability needed to understand underlying physical mechanisms. This understanding is needed to reconstruct past irradiances using models based on the short, contemporary record. It is recommended that, at a minimum, effective monitoring of the TSI be initiated as soon as possible.

3. Role of Satellites

The role of satellites was a prominent part of the discussion in each of the science areas. Measurements from operational meteorological satellites of vegetation, cloud characteristics, ozone, aerosols, planetary albedo, and surface solar radiation and albedo are based upon solar reflectance observations. As a general principle, although enhancements of the current systems are useful and desirable, correcting problems with the operational satellites and instruments now operating has a priority. The quality of the observations is subject to degradation of the sensors. Although calibrated prior to launch, the visible sensors deteriorate in space so that time series of variables derived from the visible sensors contain spurious trends. Well-calibrated satellite sensors are needed to measure the small signals associated with change on the decadal time scale. Consequently, on-board calibration of solar reflectance measurements should be provided, as is planned for the EOS MODIS instrument. In addition to commencing short wavelength calibration of the operational visible sensors as soon as feasible, the pre-launch calibration procedures for the IR wavelengths need to be improved, and satellite-to-satellite intercomparisons for an adequate overlap period of both temperature and water vapor channels should be made routine. The current IR calibration is not adequate for climate studies.

Current NOAA afternoon polar satellite orbits drift with time. As a result, over a five-year lifetime, a satellite observation time may drift from, say, 2:00 p.m. to 5:00 p.m. This observing time drift introduces artificial signals into the time series measurements. For variables with diurnal variations artificial trends appear. For some variables, for example, vegetation index or ozone, variations in solar zenith over such a large range may also introduce spurious trends because of uncertainty about zenith angle effects. A possible strategy for controlling these spurious trends would be to alter orbit tolerances to include limits on the absolute change as opposed to the rate of change of the equatorial crossing time.

Because no two instruments are exactly alike, instruments on successive satellites must be intercalibrated in orbit to insure continuity and stability of the time series derived from the observations. There should be specific planning for six-month to one-year overlap of successive satellites. Ensuring overlapping periods for instruments is a guiding principle for *in situ* as well as satellite systems.

[31]

The Earth's radiation budget provides a vital diagnostic tool for analyzing long-term climate change. It is valuable for evaluating any changes in cloud, aerosol and surface radiative forcing resulting from natural or anthropogenic climate change. Monitoring of the radiation budget from space on an ongoing basis should be insured.

With respect to international cooperation, the operational environmental satellite agencies should initiate planning to develop an integrated program of long-term climate monitoring. Knowledge gained by global climate monitoring benefits all nations of the world. With the operational satellite agencies sharing the development of such a system, costs for any one participant are substantially reduced.

Hansen *et al.* provide an overview of the climate forcings and feedbacks and argue for long-term monitoring by satellites and complementary *in situ* instruments. They propose a small-satellite mission, called Climsat, designed to exploit information on gases, aerosols, and clouds contained in the satellites' coverage of the thermal and solar spectra. Three instruments on a minimum of two satellites are required. A sun-synchronous near-polar orbiter would give the fixed diurnal reference and a precessing orbiter would capture diurnal variations at latitudes characterized by a significant diurnal signal.

4. Conclusions

In the discussions of Breakout Group A a number of specific recommendations were brought forward in each science area. These recommendations included the enhancement of current measurement systems, the introduction of new measurement systems, and the processing of data to produce quality data sets. As a unit, these recommendations can be considered to be the highest priority activities within each science area. Further prioritization within each science area is possible and probably necessary. This process would require a more thorough delineation of the proposed activities, including estimated costs. Extending this ranking process across the science areas is more difficult. Group A did not address the higher level of prioritization within or across the science areas.

With respect to more general recommendations, there was consensus that an alliance among observing system, climate modeling, and climate analysis communities was key to achieving long term GCOS goals. The weather forecasting community, which is an example of this type of coalition for the shorter time scales, represents a model of success and offers an opportunity for judicious, incremental improvements in the observing systems. It was also agreed that proposed changes to observing systems, either modification of existing systems or introduction of new systems, should be subject to evaluation of their impacts on climate monitoring.

LONG-TERM CLIMATE MONITORING BY THE GLOBAL CLIMATE OBSERVING SYSTEM

Report of Breakout Group B – Climate Responses and Feedbacks

An Editorial

KEVIN E. TRENBERTH

National Center for Atmospheric Research*, P.O. Box 3000, Boulder, CO 80307–3000, U.S.A.

1. Introduction

This report reflects the discussions and recommendations of Breakout Group B. It is recognized that there are gaps and imbalances in the topics covered. These came about from the particular expertise of the collection of scientists who happened to be present and the lack of sufficient time to delve more deeply into several topics. Several fields, notably land surface processes, the middle atmosphere, atmospheric chemistry, and paleoclimate were inadequately represented. As a consequence, the listings of topics below and the recommendations constitute a 'shopping list' which should serve as material for subsequent working groups and committees. In several areas there is overlap with other breakout groups, in particular, discussion of remote sensing occurs mostly in the Group A report.

2. Linkages

Several major concerns arose in the discussions about GCOS. Of greatest importance, probably, is the perceived need for a much stronger link between the observations themselves and the users of the observations and various derived products. Often there are many steps between the eventual user of the information based on the GCOS observations and the basic measurements. In most cases the observations must be quality controlled and analyzed into some kind of product. Most commonly these are maps or gridded fields of one or more variables. There are always assumptions, approximations, extrapolations and/or interpolations involved as well as some kind of model, whether statistical or physically based, to produce these fields.

The most complex physically-based model is a full model of the climate system, or perhaps a component of it such as an atmospheric General Circulation Model (GCM) which is used as an integral part of a four dimensional data assimilation

---

* The National Center for Atmospheric Research is sponsored by the National Science Foundation.

Climatic Change **31**: 163–180, 1995.

(4DDA) process. In 4DDA the model is initiated from a previous analyses and a forecast is made of the state of the fields at the time of interest; the forecast fields are then combined in a statistically optimal way with all the new observations to produce a dynamically and physically consistent set of fields. This means that the error characteristics of the forecast must be known and accommodated, and the analysis is multivariate and takes into account the dynamical relationships among the climate variables and their spatial and temporal scales of variability. Similar methods are now being applied in the ocean.

The analyzed fields and model predictions initialized with the observed fields are the products that are widely used by various communities. It is vital to forge a strong link –much stronger than currently place – among the observations, diagnostic analysis and the modeling communities. For instance, the scientific modeling community uses the observations and analyses to validate and improve models. The review paper by Mahlman provides further background.

A primary constituency is the Intergovernmental Panel on Climate Change (IPCC) who have led the international efforts on assessing the climate changes observed in the past and assessing the utility and uncertainties in predictions of future global warming and climate change associated with anthropogenic effects, most notably, the increases in greenhouse gases.

While GCOS will provide the long-term observations of the climate system, there was recognition of the need for strong linkages to other process oriented programs within the WCRP and IGBP, many of which will also be taking observations for special purposes or for limited periods, and which may be developing prototype operational systems for consideration within GCOS. Of particular note is CLIVAR (Climate Variability and Predictability) in the WCRP which has programs on interannual and decadal-to-century climate variability and anthropogenic climate change.

3. The Process of Building Recommendations

In the discussion group, the workshop participants were asked to think about several things in developing recommendations:
- What is the rationale for the observations?
  - What are the scientific and other needs?
  - Who is the customer?
- What are the critical elements?
- What are the prospects, or what is the doability and feasibility?
- What is the ballpark cost?
- Are there any other special considerations?
- What are the current capabilities?
- What are the deficiencies for decadal-to-century time-scale needs?
- What is the relative scientific importance?

– In assessing the importance, what are the assumptions?
– Can we set priorities, at least in a limited way? Note that to properly set priorities, complete information on the choices must be available, and the choices must be limited. It is therefore reasonable to set priorities within a fairly narrow framework such as for a single climate variable or limited domain, but to set priorities among different variables and constituencies requires considerable education of all those participating. There simply was not sufficient time to achieve this at the workshop.
– What are the consequences of not doing the observations, and what are the benefits?
– What are the recommendations?
– How might they be implemented, or what are the issues in doing so?

There was also the suggestion that we examine what could be done with existing systems. In particular, what needs to be done to make observing systems which are there for other reasons more suitable for climate, and especially the decadal time scales? Factors in observing systems developed for weather forecasting, for example, that do not meet the needs of the decadal climate requirements include:

– Changes in instrumentation, calibration, and measurement techniques
– Changes in time of observation
– Changes in the exposure of instruments, station environment, and effects of urbanization (i.e., building a city around the location where the measurements are made)
– Changes in spatial distribution and areal coverage
– Changes in methods of processing and analyzing observations
– Inadequate documentation of the above changes.

In considering a new system that might be started, an important consideration therefore is that it must be sustainable, calibrated, and global so that a consistent record is obtained. It was noted that there are many suggested needs and a lot of suggested new measurement systems, but that how to implement a system that fulfills the above requirements requires careful consideration. Observing system components from satellite have been especially prone to limited lifetimes, lack of continuity, changing instruments, changing orbits or other characteristics of the spacecraft from one mission to the next, and drifts in orbit during a mission so that the diurnal cycle can be aliased onto the long-term changes which are of interest.

Consequently there was discussion of strategies for dealing with these realities in the absence of the solution where the same system is sustained with constant calibration. To create decadal to century time series, as desired, it is probably necessary to combine many segments, taking into account any offsets and altered characteristics of the data from one segment to another. These can only be assessed by a period of overlap and, because of the spatial variability over the globe and changes in climate with the annual cycle, in general the adjustments required to piece together a continuous record are a function of the annual cycle. Accordingly, the overlap should exist for at least one year. An example of the successful implementation of

[35]

this strategy is the temperature records derived from the Microwave Sounder Units (MSUs) over a series of satellite missions from 1979 to the present.

Following the general discussion, the participants were asked to specifically address:

- Definition of needs, issues and rationale
- Assessment of capabilities
- Deficiencies
- Recommendations for improvements, both what and how
- Consequences or impacts

Subgroups consisting of the lead authors and the panelists, plus one or two others then convened to develop the recommendations, which were the focus of the remainder of the group session. It is recognized that the recommendations are incomplete. In particular, there was some discussion of land surface processes in Group B, but as most of the participants with expertise in that area were distributed among the other groups, no recommendations were endorsed in this area. In addition, discussion in several other areas should be regarded as preliminary.

4. Past Data and Data Archaeology

A major theme of many of the papers presented at the workshop was various analyses of the past data already available. A common thread was the need to correct and adjust the data in various ways to accommodate changes in the way the measurements were made, changes in locations and temporal sampling, changes in spatial coverage with time and so on. These data are used to determine the past climate record and thus to provide estimates of the variability on decadal and longer time scales, and trends. Such variability induces both the natural and anthropogenic components.

These analyses are vitally important for providing a baseline for GCOS, especially for issues on decadal to century time scales. The future observations build upon this base and therefore, it is desirable to utilize the past record in making decisions about which observations should be continued. Continuity and homogeneity of the record are extremely important considerations. This does not mean that advantage should not be taken of new technology and measurement systems. On the contrary, more efficient and cost effective observations can only help the cause. But it does mean that special steps are needed to ensure continuity of the record and that adjustments, if needed, are well determined, for instance by a period of overlap of measurements usually for at least one annual cycle.

Because of the importance of the past record, there were many strong recommendations of the need to continue to recover past data, make it available in computer-compatible form, and to evaluate, correct and adjust the data to reconstruct homogeneous time series with as complete coverage as possible. *The workshop endorses these activities.* In particular:

Recommendation: Continued efforts should be made to improve the quality and volume of the historical data base. This includes the need for data 'archaeology'.

Examples include the projects known as Comprehensive Ocean-Atmosphere DataSet (COADS) for surface marine data, the Comprehensive Aerological Reference Data Set (CARDS) for data in the free atmosphere, Data Rescue (DARE), the World Climate Data and Monitoring Programme's Climate Computing (CLICOM), Global Oceanographic Data Archaeology and Rescue (GORDA), and the rescue of cryospheric data from the Former Soviet Union. Where different compilations of historical data have been prepared, such as for COADS and the U.K. Meteorological Office marine data base, efforts should be made to merge them. Also the data from these efforts must be made available. Such efforts can help future reanalyses but are also useful for documenting long-term circulation changes. As a particular example, there is a need for much more back data of maximum and minimum temperatures at many locations to provide the needed analyses of these quantities and the diurnal range in temperature.

Nevertheless, in the discussion group B, the focus was on recommendations concerning future observations and products.

5. Data Processing

5.1. OPERATIONAL CLIMATE MONITORING

All observations made should be communicated, processed and archived with the expectation that they will be reused multiple times as part of reanalysis of the past record. This provides a basis for designing a data processing system and archive that facilitates reuse of the data. However, there is concern in the scientific community about the gap between the users of the data and the observers. It is inevitable for global data that most of the observations are made by someone else. So the issue arises about how to ensure that the quality concerns of the decadal community are addressed, and how the data are adequately monitored so that retroactive adjustments to the data are minimized. This is especially a concern when many of the observations are made for weather forecasting purposes.

One means of addressing this issue would be to establish one or more groups who monitor the data as they are taken. This already happens to a very limited extent for certain time series. Other related activities are those like CDAS – the Climate Data Assimilation System – in the United States (see the paper by Trenberth). However, there is currently no operational entity that has as a task the comprehensive monitoring of the atmospheric data for climate purposes and thus forestalling the many problems experienced in the past. At present, most data are reviewed only much later by the data archival centers, but this tends to perpetuate the system in place and will ensure the need to continue to adjust time series after the fact.

[37]

In general the data serve multiple purposes and, in particular, will also be used for seasonal-to-interannual monitoring and prediction. It seems most likely that operational climate centers will be established for this time scale. An example may be the proposed International Research Institute. Therefore an option may be to append to those activities a monitoring capability that looks after the needs of the long-term climate monitoring.

Recommendation: Establish the capability of monitoring the quality of important observational data bases important for long-term climate monitoring, detection and attribution.

Closely related to the above recommendation are several concerns expressed by the ocean community, especially the recommendation for an 'Evaluation Unit', as given in section 10, below.

5.2. REANALYSIS

A comprehensive view of the observed climate change must be pursued by analyzing all climate variables and accounting for relationships among them wherever possible. The atmospheric and oceanic circulation provide vital links in building this understanding. Previous operational analyses have been optimized to produce the best forecast, rather than the best analysis, and there are examples where these lead to conflicts (such as the need to reduce moisture in the analyses so that the prediction model does not rain excessively at the start of a forecast). In the past, inhomogeneities have limited the utility of the operational global atmospheric analyses for climate purposes. Because of the spurious changes that disrupt the climate record, a strong case has been made for *reanalysis* of the observations using a constant state-of-the-art analysis system. Such retrospective analysis raises the prospects for a number of improvements, as summarized in the paper by Trenberth.

An important basic concept inherent in reanalysis is that it should not be done just once. Having a system frozen in time means that any bad points or shortcomings are also frozen in time. Several problems are apparent already in current atmospheric reanalysis efforts: some data are not properly assimilated, treatment of future data is rudimentary, treatment of the earth's surface is unrefined, the 4DDA will improve, and so on. Therefore, it is expected that reanalysis should become a routine activity that would start over again at regular intervals, perhaps every 5 to 10 years, and in this way could take advantage of the the latest state-of-the-art system.

Recommendation: Reanalysis activities should be encouraged and should occur at regular intervals, every 5 to 10 years, as data assimilation systems improve.

Current reanalysis activities (see paper by Trenberth) were initiated by the TOGA progam for the purposes of understanding and predicting seasonal to interannual climate variations. They do not adequately address the needs of the decadal-to-century community. Changes in the data base will still be present, so it is essential for numerical experiments to be carried out to assess their impact. For example, satellite data have had a substantial positive impact on Southern Hemisphere analyses, so it is important to assess whether analyses are viable in the absence of space-based observations, or what is the impact on the quality of the analyses.

Recommendation: Additional reanalyses should be undertaken that address the needs of the decadal-to-century time scales and the problem of the continually changing data base.

One way to achieve this may be to carry out reanalysis in a series of segments with at least one year of overlap. Within each segment of, say, 5 or 6 years, the database is constrained to be reasonably homogeneous by discarding some data. The database is permitted to change between segments, such as when a new satellite observing system is launched and comes on line, and the overlap allows differences in statistics to be assessed and allowed for in constructing a continuous record. However, even with such a strategy, it must be recognized that the error levels attached to each analysis will change as the database changes.

An alternative suggestion is to degenerate the database to include only the rawinsonde network. It is possible, however, that the changes in the rawinsonde database alone are too great with time, and so would still allow too many prospects for spurious change to creep in. This would need to be assessed. Another consideration may be what to use as the 'first guess' for the analyses. Possibly a climatology should be used rather than a short-term forecast which is model and data dependent.

Other reasons for pursuing further reanalyses are to address concerns of communities who have not been adequately consulted in the design of the present activities. For instance, the Arctic community is very interested but concerned about the quality of the analyses in the Arctic that are likely to result in the absence of resolution of certain problems. Because the needs of several communities may not be addressed by present reanalyses, it may be desirable to have a workshop to address these issues.

Currently, 'reanalysis' refers to the 4DDA atmospheric activity, but the concept can and should be generalized to include ocean data sets, surface fields (such as SST, sea ice, snow cover), and so on.

5.3. CLIMAT DATA

The flow of monthly climate summaries is of vital importance for operational climate monitoring and in building databases. This has been carried out with the CLIMAT network, but the network does not operate as well as it might and the network is degrading. Few areas have greater than 2 stations per 5° box.

Recommendation: Upgrade the network by taking advantage of the information superhighway.

Recommendation: Countries should be asked to submit more stations per unit area and more in real time.

6. Which Physical Variables?

A key issue for detection of anthropogenic climate change is which climate variables are best suited in the sense that the signal expected for the anthropogenic forcing emerges most clearly from the noise of natural variability. Closely related to this question is whether there may be an optimal combination of variables that forms the most distinctive 'fingerprint'. A need therefore exists to identify the atmospheric, oceanic, land surface and cryosphere variables with high signal to noise ratios.

Recommendation: Perform detailed signal-to-noise analysis of global climate model experiments forced with realistic 'transient' anthropogenic forcings of the increases in greenhouse gases and changes in atmospheric aerosol.

These analyses need to be performed with several models to determine the robustness of the results. The results should be assessed bearing in mind the imperfections and possible biases of the models, implying that *a priori* validation of the models is essential, with special attention given to their variability and sensitivity. The feasibility is high and the cost low.

In designing or ranking options for GCOS, the signal-to-noise ratio is but one factor to be considered. To go further, as is required, and understand and *attribute* an observed change to a particular forcing or cause, much more complete information is needed. For example, experiments with one model have indicated that global mean surface atmospheric temperature is indeed the most important variable while the sea level pressure field is not helpful for detection. But analyses of the three dimensional temperature field, the atmospheric circulation and sea level pressure field are essential for attribution because of the need to gain an appreciation of the processes involved. So GCOS should give close attention to the full three dimensional structure of the atmosphere and oceans.

7. Atmospheric Monitoring

7.1. THREATS TO THE RAWINSONDE

There has been a decline of the World Weather Watch, in particular a reduction in the number of rawinsonde observations taken since the Global Weather Experiment in 1979. Cost of expendables is a factor in this decline. On the other hand, there has been a significant improvement in the quality of the rawinsonde observations since 1979 in most regions, associated with new rawinsonde designs and greater levels of automated processing.

There is an imminent threat to some of the conventional sounding data as the phase out of Omega and Loran-C navigation systems proceed. Several manufacturers are already working on replacements for windfinding by switching to a GPS-based system. Other threats to the rawinsonde network come from expected narrowing of radio-frequencies available for use, potentially substantially increasing the cost of each unit. It is also desirable to continue the process of improving the quality and stability of the rawinsonde measurements. Therefore for monitoring climate changes in the atmospheric circulation it is vital that:

Recommendation: Resources and facilities must be provided to update, adapt or improve rawinsonde designs to cope with expected operational changes, and more exacting user requirements for measurement accuracy and stability, in a cost-effective way.

In the last decade, large-scale testing of the major operational radiosondes has been performed. Results indicate that it is desirable to improve the quality and stability of the rawinsonde measurements and establish performance characteristics and calibration of the different rawinsondes, by intercomparing with a reference rawinsonde.

Recommendation: Develop one or more reference rawinsondes and continue to develop programs for intercomparison and calibration of all operational sondes.

If or when changes are made in a sonde, it is important to assess the impact on measurements. Improved procedures are needed for archiving and obtaining metadata (information about the instrument and its operation). The feasibility of sustaining high quality rawinsonde observations and associated adequate archiving is high. This can be accomplished without excessive additional costs.

7.2. REFERENCE STATION NETWORK

In order to provide a greatly improved measure of how the climate has changed, to determine trends reliably, and to enable detection, a reliable subset of observations must be established. Accordingly, as already recommended by the GCOS Atmospheric Observation Panel:

Recommendation: Networks of reference stations should be established both at the surface and for the upper air network.

The upper air reference network consists of selected rawinsonde stations taking into account spatial distribution and past record length and performance. At the surface, a full complement of climate statistics is desirable, and should include surface pressure, present weather, maximum, minimum and mean temperatures, numbers of ground frosts, precipitation amounts, raindays, sunshine, moisture variables (dew point, relative humidity), clouds, snowfall and depth, and wind. The number of reference stations should be sufficient to support climate change detection activities. The highest priority should be given to stations that are spatially well distributed and which retain the homogeneity of the climate records.

Nations should be encouraged to make special efforts to ensure the integrity and continuity of these designated stations so that a skeletal network will accurately depict the changes occurring. Such a network also provides needed ground truth for satellite-based observations. For these stations it is highly desirable to have a period of overlap when changes are introduced so that the impact on the climate record can be assessed. A high quality metadata record at each station is essential. There is unlikely to be a good match between the needs in this area for the surface versus the free atmosphere.

In the stratosphere, the main source of information nowadays is the data from various satellites. Conventional soundings terminate at 10 mb. In previous years, a spartan network of rocketsondes was used to calibrate the remotely sensed data. This network has decayed and is unlikely to be restored.
Recommendation: Pursue alternative means for calibrating the stratospheric measurements.

One option may be to utilize the ozonesonde network. Soundings of ozone include temperature measurements and they sometimes go to much higher elevations, to less than 5 mb pressures (above 35 km).

8. Satellite Temperature Data

Satellite data have been notorious for the lack of continuity from one instrument and/or platform to another. An exception is the series of microwave sounder unit observations from two channels, channel 2 in the lower troposphere and channel 4 in the lower stratosphere. These have proven useful in providing continuous records from 1979 to 1994, but this is entirely from luck rather than design.
Recommendation: Design a satellite-based system with stable instruments that are capable of producing continuous time series of temperatures.

The next set of NOAA satellites will no longer contain instruments that exactly match the previous MSU channels, although the latter may be reconstructed by a combination of other channels. There is, however, increased risk of failure or drift by involving more channels, and there is an accumulation of errors. The risk of loss of continuity is lessened by the strategy of continuing at least two polar orbiters with the same monitoring capabilities, thereby preserving the much needed overlapping series between satellites that allows the continuity to be maintained.

The MSU data are useful for evaluating the rawinsonde data and for evaluating the reanalysis results.

9. Surface Marine Observations

In the following, there are a number of suggestions concerning the need for improved marine observations. As with all improved instrumentation, increased sensitivity and shorter response times are apt to alter frequency distributions, especially the extremes and tails of the distributions.

Bulk SST observalions from ships come from insulated bucket samples as well as measurements from engine intakes. The later are most common but vary depending on the draft of the ship and the configuration of the engine room.

The two kinds of measurements differ typically by several tenths °C. Root mean square errors in ship measurements overall are typically 1 °C. Accordingly there is a need for improved technology and more representative measurements. The *in situ* observations need to be blended with satellite observations of SST using AVHRR, which have their own error characteristics, including the fact that the remotely-sensed quantity is a skin rather than bulk temperature. Sensors mounted on moored and drifting buoys have proven invaluable for calibrating and quality controlling SST observations from space and should continue. Improved surface marine air temperatures are also needed. Somewhat lower in priority than the above for detection, but vital for improving understanding of atmosphere-ocean coupling, are the needs to improve measurements of surface pressure and near-surface winds. The need here is to obtain better observations of the surface atmospheric circulation and surface fluxes, of which the wind is a key component, yet poorly observed. In addition, improved measurements of marine variables that relate to the fresh water flux are needed routinely.

Recommendation: Improve *in situ* SST by converting to ship hull sensor thermometers.

This is potentially expensive if retrofitting is required for existing ships, but the recommendation can be much more readily carried out on all future ships.

Recommendation: Improve SST analyses by blending *in situ* data with precision satellite data. See also the Group A report.

In addition, air temperatures from ships are notorious for the contamination by the heating during the day of the ship's hull, and by effects of the ship itself which depend upon the location of the instrument relative to the superstructure of the ship.

Recommendation: Improve air temperature measurements from ships, such as by using electronic sensing and/or better placed screens.

In this case new technology exists to place sensors well up on the mast remote from any influence of the ship itself, but the screening and ventilation of the sensors must be properly implemented.

Recommendation: Improve *in situ* surface pressures, such as by using carefully-sited precision electronic barometers.

Recommendation: Improve near-surface winds, such as by using adequately exposed automated instruments that measure true wind, and thus remove the need for using the Beaufort wind scale.

Recommendation: Blend *in situ* observations of wind with those from remote sensing, such as from scatterometers and altimeters.

More ambitious products may indude a blending of both *in situ* and satellite data in a 4DDA framework. Further needs are for better measurements of surface humidity, especially as it is important for surface fluxes of moisture.

Recommendation: Improve measurements of surface humidity, such as by using electronic humidity sensors with salt protection.

Recommendation: Increase the number and improve the assessment of surface wave measurements.

These may be combined with validated wave models to create a climate record useful for assessing changes in waves and their impacts on coastal regions.

Recommendation: Pursue the prospects for improved estimates of precipitation over the ocean using several technologies.

10. Ocean Observations

10.1. WORKSHOP DISCUSSION AND OOSDP

This part of the workshop agenda focussed on the interior ocean and those ocean measurements which were considered important for monitoring, describing, detecting and understanding climate change at decadal and longer time scales. The sessions on subsurface ocean circulation and sea level provided the background for the recommendations discussed below. Models were presented as the means of unifying the system, providing the means to exploit observed information in many different ways as well as for processing complicated and diverse data into a form which has several practical applications.

Over the last four years the Ocean Observing System Development Panel (OOS-DP) has been drafting the conceptual design for an ocean observing system for climate, partly to satisfy the requirements of GCOS. The OOSDP (1995) report has been subject to wide scientific review, and benefitted from advice from the ocean and climate communities. It was not possible to review all the material dealt with in that report and endorse the recommendations at the Workshop. The goals are prioritized using assessments of the feasibility (which includes cost) and impact. It is suggested that an ordered implementation of the elements supporting these goals will lead to a sensible, staged implementation of the observing system.

The terms of reference of OOSDP called for the conceptual design of a system 'to monitor, describe and understand the physical and biogeochemical processes that determine ocean circulation and the effects of the ocean on seasonal to decadal climate change, and to provide the observations necessary for climate prediction'. Clearly the domain of this panel covers the principal oceanographic interests of the Workshop. The discussion here focusses on the interior ocean and sea level though, as will become clear, to meet these goals many other activities must also be undertaken.

The OOSDP design is constructed around a set of goals and subgoals, the details of which will not be discussed here. However, since the recommendations depend to some extent on the outcomes from other goals, some of which are also covered in other parts of this Workshop report (SST, marine fields and sea ice), a brief background is included here. The goals are:

1. Surface fields and surface fluxes (e.g. SST, sea ice, wind stress, heat and moisture flux);

2. Upper ocean monitoring and prediction;
3. The ocean interior (long time scales); and
4. Processing and synthesis.

A full suite of recommendations has been developed for these goals. The first goal overlaps with the surface marine fields discussed in section 9 though from a somewhat different perspective. The last goal targeted several areas which, while not discussed in detail at the Workshop, are of some relevance.

10.2. APPLICATIONS AND EVALUATION

The OOSDP made specific recommendations for the establishment of processing and data management activities suited to long-term climate change applications. The Workshop did not discuss these in detail but they are broadly consistent with the themes from other sessions, see sections 4 and 5.

Recommendation: Establish a routine, long-term climate assessment center whose tasks would include regular examination and re-examination of the ocean climate data base in order to produce up-to-date state-of-the-ocean-climate assessments.

Recommendation: Develop an information management system dedicated to the gathering, quality control, assembly and archiving of ocean climate data and a system of telemetry, communication and permanent storage dedicated to the objectives of the ocean climate observing system.

Recommendation: Establish an 'Evaluation Unit' to monitor the chain of activities involved in the processing of data into useful products.

In particular, the Evaluation Unit would provide oversight of the information flow from raw data through to practical, useful products and ensure that the specified quality and efficiency is maintained. This would ensure that the goals of the ocean observing system for climate are met with the accuracy and quality required.

10.3. PARALLEL ACTIVITIES

In framing the recommendations for particular elements, it has been assumed that implementation has proceeded for the following fields: sea surface temperature, surface wind stress, sea surface salinity, sea ice extent (concentration and volume), and fluxes of heat, moisture and carbon across the air-sea interface. The recommendations for these products have been framed in such a way as to ensure that the quality standards necessary for application to long-term decadal time scale climate change can be met. The Levitus paper and the OOSDP recognized the important contribution of relentless monitoring of the upper ocean with, for example, XBTs (and perhaps XCTDs) from Volunteer Observing Ships. These were given second level priority in the OOSDP Report. The majority of these activities were given higher initial priority than the interior ocean and so would be in place as the following recommendations for the interior ocean were implemented.

10.4. THE INITIAL OBSERVING SYSTEM

The initial observing system comprises some elements which already exist and elements which are recommended for implementation now. The OOSDP report describes the minimal requirements for satisfying the particular objectives of the interior ocean.

There are few elements for the interior ocean which can be classed as existing, one exception being sea level records.

Recommendation: A subset of the GLOSS sea level network, preferably those with long consistent records, should be supported. These stations should be geocentrically located.

Gornitz's paper discussed the effectiveness of *in situ* sea level data for detecting changes in mean sea level due to an enhanced greenhouse gas warming. The cost-effectiveness of this option compared with other systems, and the high profile of sea level in terms of impact, provide strong motivation for this recommendation. Note that the Workshop did not agree on what the total number of gauges should be though about 60 were suggested by the OOSDP.

There are several potential systems, mostly operating within research programs, which should be added to complete the initial observing system. The interior ocean buffers exchanges of heat, water and carbon with other components of global climate. The aim of the observing network, in concert with measurements of surface fluxes and upper ocean change on shorter time scales, is to gather sufficient data to to construct inventories of the key physical and chemical quantities for the full depth of the ocean. The inventories would be redone at regular intervals, say, 5 to 10 years. At somewhat lower priority, but nevertheless still important for long time scale change, are measures to determine the ocean circulation and its transport of these same physical and chemical quantities. The altimeter is given particular emphasis because it offers one of the few avenues for truly global coverage, in this case of the integrated effect of thermal expansion and freshwater storage.

Recommendation: Implement a fully operational long-term precision altimeter satellite mission, accompanied by a small number of precision, satellite reporting sea level gauges.

Recommendation: A global network of subsurface autonomous floats (e.g., ALACE) measuring Lagrangian velocity at intermediate depths and profiles of temperature should be implemented subject to successful demonstrations by WOCE.

The OOSDP, and the Levitus paper, stressed the tremendous value derived in the past from time series stations, principally the Ocean Weather Ships, many of which are now defunct. New technology means such stations can now be operated more cost-effectively providing the opportunity to reoccupy some of these sites, subject to scientific assessment, and perhaps start new ones.

Recommendation: Existing long-time series stations measuring temperature, salinity, etc. should be maintained and, on the basis of further scientific assessment and improving technology, new time series stations should be established at old sites or at carefully selected new sites.

Repeat sections (hydrography) should also be used to construct time series over the full depth for decadal and longer time scale applications. The details of the repeat sections and, to some extent, the time series, would depend on on-going research programs like WOCE and IGOFS.

10.5. ENHANCEMENTS, RESEARCH AND DEVELOPMENT

The OOSDP suggests that the initial observing system could be enhanced by trans-ocean hydrographic sections, development of salinity sensors for autonomous profilers, monitoring of exchange through key straits and of transport in certain locations (e.g., using submarine cables), and a specific satellite mission to determine the earth's gravitational field (geoid). This latter enhancement would assist in making the altimeter useful for absolute current determinations rather than relative flow.

In addition to these specific recommendations there are several research and development activities which were identified by the OOSDP as being important for the implementation of an effective and efficient system. These include the development of an autonomous profiling CTD, e.g., for application at deep water monitoring sites. A specific recommendation for improved bathymetry is worth-while emphasising here:

Recommendation: Many bathymetry data sets are only marginally adequate for modern climate studies. It is recommended that these data sets should be enhanced.

The Workshop emphasized the valuable information that already exists but is not yet available, either because it has not been digitized or because it is not part of the ocean data exchange and archive system. The OOSDP discussed the importance of data archeology as part of the information management system but, since it is an activity that is so important for monitoring and understanding decadal and interannual variability, we include a further specific recommendation here.

Recommendation: A concerted effort should be mounted to digitize all upper ocean and interior ocean data that presently exist on fragile, non-permanent media and that efforts be continued to trace and make part of the permanent archive temporarily 'lost' data (e.g., as in the IOC GORDA data archeology project).

The OOSDP has provided prioritization of all goals and subgoals, and of individual elements recommended for those subgoals, some of which were presented at the Workshop. There has been no attempt, as yet, to examine trade-offs and priorities across GCOS components or across GOOS and GTOS activities.

11. Cryosphere

There are a number of possibilities for improved monitoring of the cryosphere and the list below is prioritized to include only those considered most important and practicable.

11.1. SNOW COVER

Changes in snow cover have already been shown to have amplified the recent warming over northern land areas. Continued monitoring of snow in similar formats to those produced previously is high priority for diagnosis of climatic change in these areas, which are locations where permafrost is susceptible to thaw. The first problem in the monitoring of snow is the lack of time series of fields (grids) of snow cover and water equivalent.

Recommendation: Optimize the procedure for blending satellite data and *in situ* measurements of snow depth to produce historical time series of gridded snow depth and water equivalent.

Changes in snowfall are likely to accompany greenhouse warming. Aside from direct impacts (requirements for snow removal, recreation, etc.), the hydrologic regime will be affected through changes in runoff and soil moisture. Air temperatures will also be modified by changes in snow depth through timing of snow disappearance. A historical database depicting snow depth, and preferably snow water equivalent, is required to quantify and provide a basis to predict these impacts.

11.2. SEA ICE

The second problem is the potential discontinuity of weekly charts/grids of snow cover (NOAA/NESDIS) and sea ice (Navy/NOAA Joint Ice Center).

Recommendation: Ensure a period of overlap ($\sim$ 2 years) of new automated products and current products; assess and demonstrate continuity before discontinuing the older product.

Continuity of product is required for detection of model-projected greenhouse signals in the polar regions. Models project strongest warming over sea ice in the Arctic; *in situ* drifting buoy observations of air temperature over sea ice have been subject to considerable uncertainty (although improved sensors are now being deployed), so sea ice cover will likely be the best indicator of warming over polar oceans.

11.3. ICE SHEETS AND GLACIERS

Changes in ice sheet volume have direct consequences for global sea level. Ice sheet volumetric assessments are essential to diagnosis and prediction of changes in sea level. Accurate mass balances would remove a source of non-uniqueness in the interpretation of Earth polar wander, rate-of-rotation, time-variant gravity and isostatic rebound. Changes in ice sheet accumulation and ablation in the event of global warming will be greatest at the margins. Without special attention to the margins, and complete coverage of the West Antarctic Ice Streams, an observing system may fail to detect warming-related changes in accumulation or ablation, or fluctuating discharge due to changing bed conditions in West Antarctica.

There is a lack of an altimetric baseline for long-term ice sheet mass balance monitoring for the climatically sensitive ice sheet margins, and for Antarctica south of 82°S.

Recommendation: Establish an altimetric baseline for decadal sampling of the ice sheets.

Priority should be given to evaluating technical solutions for a baseline for the ice sheet margins. The impact of cloud on laser solutions should be carefully examined; alternative strategies using aircraft platforms, particularly in Greenland, and novel microwave technology, should be evaluated.

The largest uncertainty in components of the mass balance of ice sheets arises from the extrapolation of accumulation data that are sparse in time and space. Volumetric changes from altimetry require better understanding of near-surface densification to convert these changes to mass changes.

Recommendation: Increase the spatial density of shallow cores, particularly in the interior of the ice sheets, with a view to determining the temporal and spatial correlation of accumulation, and their dependence on geographical location, and direct measurement of firn densification.

Better understanding of spatial and temporal variations of accumulation will permit a considerable decrease in current uncertainties deduced from historical observations. Shallow cores may also provide records of atmospheric constituents over the past few decades and centuries. Without a better knowledge of accumulation variations, improvement in balance estimates from historical data will be limited.

Glaciers provide useful indicators of climate change as well as being an important part of the climate system that can feedback and reinforce changes. Hence there is also a need for improved monitoring of glaciers, particularly glacier volume and mass balance.

11.4. LAKE ICE

The absence of a baseline network for the monitoring of lake ice in the era of future satellite missions, such as the Earth Observing System (EOS), needs to be addressed.

Recommendation: Establish a central archive for historical lake ice data now held by various national agencies and other sources. Implement data quality control and reformatting, and use the existing data to define an optimal network for monitoring by satellite in the EOS era.

The baseline network will permit systematic monitoring of a quantity that integrates variability of key climatic quantities, thereby providing a measure of climatic change over high-latitude land areas. Costs will be relatively low, as raw data now exist and high-resolution satellite data will be available for other purposes. Deduced changes will likely be relevant to ecosystem changes over northern land areas.

11.5. PERMAFROST AND SEASONALLY FROZEN GROUND

The absence of a baseline network for monitoring the depth of the active layer overlying the permanently frozen ground (permafrost) and the extent, thickness and vertical temperature profile of the permafrost itself, needs to be addressed. Currently there are but a few effective archives of such data at national levels. Yet these data are increasingly required for climate system model validation and improvement. There are no suitable remote sensing techniques.

Recommendation: A central archive should be established to assemble, quality control and archive historical data on permafrost and seasonally frozen ground.

12. Paleoclimate

The short time span of available instrumental records limits the prospects for determining whether the observed environmental changes are caused by human activities. This problem is compounded by the nature of natural variations and inadequate knowledge of how the climate has varied in the past. Therefore a complementary activity to GCOS is the collection and analysis of paleoenvironmental data from centuries-long high temporal resolution records.

References

The Ocean Observing System Development Panel: 1995, *Scientific Design for the Common Module of the Global Ocean Observing System and the Global Climate Observing System: An Ocean Observing System for Climate*, Dept. Oceanogr., Texas A&M Univ. College Station, Texas, 265 pp.

LONG-TERM CLIMATE MONITORING BY THE GLOBAL CLIMATE OBSERVING SYSTEM (GCOS)

Report of Breakout Group C – Climate Impacts

An Editorial

WILLIAM E. EASTERLING

Department of Agricultural Meteorology, University of Nebraska-Lincoln, Lincoln, NE 68583–0728, U.S.A.

1. Introduction

Historically, decisions shaping the essential characteristics of climate observing systems – including the spatial distribution of observing sites and areas, the choice of which climate elements to observe, the frequency of such observations, the numerical expression of the observations, and the formatting, archiving and distribution of resulting climate data sets – were conditioned by the need to know how climate and its determinants vary through time and space. Heretofore, such systems were constructed primarily to benefit the research community interested in understanding the functioning of the climate system. The needs of users of climate data for impact assessment, including researchers, came last. Times have changed.

At the World Meteorological Organization's (WMO) Global Climate Observing System (GCOS) Workshop on Long-Range Climate Data Needs (see lead editorial, this issue, for details), one of three working groups was asked to identify the highest priority climate data needs for impact assessment. The following is a summary of such needs as identified by the impacts working group. The needs come under the rubrics of *usable climate knowledge, leading climate indicators of impacts, global climate-related data, metadata for impact assessments and climate data in support of indicators and climate-related data.*

2. Usable Climate Knowledge

The development of climate data and information product streams that are useful to a wide spectrum of value-added consumers, including resource managers (e.g., foresters, agricultural extension educators, water reservoir operators), planners (e.g., developers, civil engineers, city and regional planners), government analysts (e.g., local, regional, state, and national), technical and news media (e.g., trade publications, newsletters, newspapers and electronic broadcast) and climate impact researchers (e.g., climatologists, geographers, economists, ecologists, hydrologists)

Climatic Change **31**: 181–184, 1995.
© 1995 *Kluwer Academic Publishers.*

is vital. The key point here is that there is a shortage of usable climate knowledge
– defined as climate information that has social utility beyond its value to climate
system research. Products of climate observing systems, even those intended for
consumers, suffer deficiencies with respect to social utility. GCOS must evenly
balance the needs of consumers of usable climate knowledge with the needs of
the climate research community in devising a strategy for future climate observing
systems.

What is needed specifically? Climate data and information should be tailored
to match the temporal and spatial scale resolution, and types of climate elements
required for impact assessment. Climate data and information products need to be
aggregated (or disaggregated) to spatial and temporal scales that mesh with vul-
nerable biophysical processes and human activities. Those who track agricultural
production, for example, must integrate climate information into the management
of annually renewing biological systems that are best represented at scales of a
few tens of kilometers while dam operators must integrate climate information
into the management of reservoirs that, in some cases, fill and empty over time
periods measurable in decades, with catchments spanning thousands of kilometers
and many different geophysical domains. Such tailoring needs to select critical
climate elements that elicit impacts in a particular sector or ecosystem. It should
also distinguish regions by their vulnerability to certain kinds of climate variation.
In many cases, indices need to be constructed that combine climate and climate-
related information in order best to capture vulnerability (the need for indices is
discussed fully below). GCOS should develop a mechanism to identify and catalog
the temporal and spatial scale, and climate element requirements of broad classes
of consumers of climate data and information, and such requirements should be
incorporated into the planning of the architecture and attributes of future climate
observing systems.

3. Leading Climate Indicators of Impacts

Indexes which integrate several leading climate indicators of impacts are needed to
anticipate and assess better the impacts of long-term climate variability and change
(see Easterling and Kates, this issue, for an expanded discussion of such indexes).
Like the U.S. Index of Leading Economic Indicators anticipates business cycles,
broad classes of indexes of leading climate indicators are needed to anticipate
climate impacts. Leading climatic indicators are defined here as empirical time
series that usually experience the onset of a fluctuation before some aspect of
society or the environment is affected but rarely experience one if no such effects
are imminent (i.e., few false signals). Such indexes will help key GCOS to a
minimum global standard of climate information needed to assess impacts.

Initial emphasis should be placed on the development of indexes of leading
climate indicators related to: (1) *climate extremes and increased variability*, includ-

ing characterization of trends and patterns of droughts, floods and severe storms; (2) *greenhouse climate response*, including long-term trends in minimum temperatures, precipitation intensity and mid-latitude summer drought, and including climate-related trends, such as alpine glacial retreats and sea level rise – all of which help detect greenhouse climate changes that are relevant to impacts; (3) *ecosystem health*, including climate-related changes in biodiversity and other ecosystem properties in sensitive ecotones; (4) *energy and renewable natural resources*, including trends in climate-related demand for and supplies of commodities and services related to energy, agriculture, water, forestry, fisheries and recreation; and (5) *early warning of major climate events*, including utilization of ENSO events, drought persistence, snow cover anomalies and other precursors of impacts. To this end, GCOS should consider establishing a working group on climate indices to assess the current suite of indices, maintain coherence among regions and provide much needed focus for this topic.

4. Global Climate-Related Data

Climate-related geophysical, ecological and socioeconomic data are required for impact analyses. More emphasis needs to be given to the development of long-term global products.

What is needed specifically? Satellite and *in situ* data are needed for a range of phenomena which reflect long-term trends in climate. Very high-resolution data (e.g., Landsat and SPOT) on land cover and land use should be updated at regular intervals (every 5 years is recommended). Such data are crucial to understanding the role of terrestrial ecosystems in the global carbon cycle and how such is influenced by humans. More frequently updated medium-resolution data (e.g., AVHRR) on, for example, snow cover (weekly), land use and land cover (bi-annual), spring onset of green-up (semi-annual) and NDVI-based drought indices (weekly) are needed. Other climate-related data needed *in situ* include trends in surface hydrologic characteristics (e.g., streamflow, lake levels and ice-out, runoff, groundwater levels, glacial terminus), marine characteristics (e.g., sea level, sea ice, wave height) and socioeconomic characteristics (e.g., population, natural resources, nutritional status, per capita income). Digital terrain data are needed in concert with climate data in certain types of impact analyses. It is recommended that the GCOS Space Panel consider our endorsement of global coverage of very high-resolution satellite data on land cover and land use that are updated every five years in its own recommendations.

5. Metadata for Climate Impact Assessment

Metadata – descriptive information about data that is critical to the appropriate use and interpretation of the data – must include information such as the time and place of observation, the instrumentation and methods of measurement used to obtain the data, the format of the data and the methodology used to process the data for archiving. The availability of such metadata should be widely advertised to consumers.

What is needed specifically? It is recommended that the GCOS Data and Information Panel give high priority to the development of metadata information systems in which users can readily access and make inquiries. We suggest, as an initial focus, that WMO request procedures from its member states to document and preserve, as part of their data management procedures, metadata on surface temperature, precipitation and wind (including estimates made from surface pressure measurements). Eventually, metadata should be available on the full suite of climate and climate-related data required for impact assessment. WMO should provide information on the availability of such metadata for current and future observations, and for historical records.

6. Data in Support of Indices and Climate-Related Products

GCOS should encourage support for the data needed to construct climate indices and climate-related products.

What is needed specifically? Long-term and near-real time data required to update the indices and products are needed. An integral part of the construction of such indices and products is a steady stream of quality-assured data with adequate coverage of key regions and elements. It is recommended that WMO assist member states in the establishment of a minimum standard of data needed to construct such indices, to maintain climate-related products for impact assessment and, where necessary, to assist member states to revitalize their observation networks (in Africa especially).

7. Conclusion

Implementation of the aforementiond recommendations will be a positive start to improving the utility of information generated by climate observing systems to consumers. Impact researchers are served notice that efforts are needed to develop and test indexes of leading climate indicators of impacts. From these efforts will emerge the detailed data requirements for future indexes. Directories of climate-related data needs also should be compiled. Such information will provide critical guidance to the design of future global climate observing systems.

CRITICAL ISSUES FOR LONG-TERM CLIMATE MONITORING *

THOMAS R. KARL[1], VERNON E. DERR[2], DAVID R. EASTERLING[1],
CHRIS K. FOLLAND[3], DAVID J. HOFMANN[2], SYDNEY LEVITUS[4],
NEVILLE NICHOLLS[5], DAVID E. PARKER[3] and GREGORY W. WITHEE[6]

[1]*NOAA/NESDIS/NCDC, Federal Building, Asheville, NC 28801, U.S.A.*
[2]*NOAA/OAR/ERL, 325 Broadway, Boulder, CO 80303, U.S.A.*
[3]*Hadley Centre/UKMO, London Rd., Bracknell, Berkshire RG12 2SY, U.K.*
[4]*NOAA/NESDIS/NODC, 1825 Connecticut Ave., NW, Washington D.C., U.S.A.*
[5]*BMRC, 180 Lonsdale, Melbourne, Australia*
[6]*NOAA/NESDIS, Federal Building #4, Suitland, MD 20233, U.S.A.*

Abstract. Even after extensive re-working of past data, in many instances we are incapable of resolving important aspects concerning climate change and variability. Virtually every monitoring system and data set requires better data quality, continuity, and homogeneity** if we expect to conclusively answer questions of interest to both scientists and policy-makers. This is a result of the fact that long-term meteorological data, (both satellite and conventional) both now and in the past, are and have been collected primarily for weather prediction, and only in some cases, to describe the current climate. Long-term climate monitoring, capable of resolving decade-to-century scale changes in climate, requires different strategies of operation. Furthermore, the continued degradation of conventional surface-based observing systems in many countries (both developed and developing) is an ominous sign with respect to sustaining present capabilities into the future. Satellite-based observing platforms alone will not, and cannot, provide all the necessary measurements.

Moreover, it is clear that for satellite measurements to be useful in long-term climate monitoring much wiser implementation and monitoring practices must be undertaken to avoid problems of data inhomogeneity that currently plague space-based measurements. Continued investment in data analyses to minimize time-varying biases and other data quality problems from historical data are essential if we are to adequately understand climate change, but they will never replace foresight with respect to ongoing and planned observing systems required for climate monitoring. Fortunately, serious planning for a Global Climate Observing System (GCOS) is now underway that provides an opportunity to rectify the current crisis.

1. Introduction

Long-term climate monitoring is the process of delivering and transforming data and information to describe the state and the changing state of climate. Long-term climate monitoring requires observing and data management programs that provide observations and data bases of sufficient quality and sensitivity to address questions of interest to both policy-makers and scientists. Examples of important questions include:

---

* The U.S. Government and the British Crown right to retain a non-exclusive royalty-free license in and to any copyright is acknowledged.
** Data homogeneity requires ensuring that data represent changes and variations of a specific aspect of the climate system unaffected by the measuring device, processing system, or local changes in instrument exposure or other man-made modifications of the environment in the proximity of the instrument.

Climatic Change **31**: 185–221, 1995.
© 1995 *Kluwer Academic Publishers.*

- Is the climate warming?
- Is the hydrologic cycle changing?
- Is the atmospheric or oceanic circulation changing?
- Is the climate becoming more variable or extreme?
- Can we detect anthropogenic effects in the climate record?
- What does the climate record reveal regarding the sensitivity of the climate system to various climate forcings?

Answers to these questions are critically dependent upon the selection of the most appropriate variables to be monitored and their respective temporal and spatial sampling frequency. This includes consideration of the processes determined by atmospheric and oceanic chemistry and biogeochemistry, atmospheric and oceanic dynamics, thermodynamics, land surface hydrology, ecological systems and dynamics, solid earth processes, solar influences, and human interactions. The attributes of a successful long-term climate monitoring system, however, also require timely delivery of data and information with large signals relative to time-dependent data biases. A successful monitoring program must be able to evolve with changes in technology and funding such that there are minimal impacts on data quality and homogeneity of past, present, and future measurements. Sound data management systems are an integral part of a successful climate monitoring program.

The purpose of this article is to focus on some of the more important issues related to the operation and maintenance of a long-term global climate monitoring program. This is highlighted by examining both the inadequacies of past and present observing and data management systems and offering potential solutions to alleviate these problems (section 2). Various new systems have been proposed in recent years for improved long-term monitoring of climate, many aimed at providing global observing systems. Satellite systems have been in operation for over the last 2 decades and are already proving to be invaluable for long-term climate monitoring, but inadequacies are widespread. In section 3 we focus on these and other new approaches to climate monitoring. In section 4 we focus on strategies for ensuring that the variety of observing and data management systems, conventional and new, can be effectively used for climate monitoring, and finally in Section 5 we offer a set of guiding principles for long-term climate monitoring.

The context of this article is framed by the emerging Global Climate Observing System (GCOS). The Conference Statement of the 1990 Second World Climate Conference asserted that there was an urgent need to create a GCOS which would include both space-based and surface-based observing components. Among other GCOS goals are the needs for a long-term climate monitoring system for detection of long-term climate change. GCOS is being planned around an improved World Weather Watch program, a Global Atmospheric Watch Program, the climate components of a Global Ocean Observing System (GOOS) and a Global Terrestrial Observing System (GTOS), and the maintenance and enhancement of monitoring programs of other key components of the climate system. GCOS will encompass all components of the climate system including the atmosphere, biosphere, cryosphere, hydrospere, and land surface over the full global domain. Thus it will go beyond the scope of presently established observational programs. GCOS can provde a means to develop a rational and systematic overview of the requirements and priorities for the best use of monitoring resources. Because the resources for GCOS must come from national efforts, its success depends on our ability to avoid past mistakes and promulgate successes. This article highlights some of these lessons.

2. Status of Long-Term Climate Monitoring

Examples of climate monitoring are examined within three broad categories: agents of climate change, feedbacks, and responses. Special attention is given to operational systems. Systems that may become operational are discussed in the next section. By no means are the examples exhaustive, but they serve to illustrate the types of problems evident in long-term climate monitoring, and provide a basis for developing solutions to minimize future problems.

2.1. FORCING FACTORS ASSOCIATED WITH CHANGE

2.1.1. *Variations of Solar Irradiance*
The most fundamental quantity affecting the entire Earth-system is solar irradiance at the top of the atmosphere. Measurements of solar irradiance have been taken by balloons and rockets, but continuous measurements of 'top-of-the-atmosphere' solar irradiance did not begin until the late 1970s with the Nimbus 7 and the Solar Maximum Mission (SSM). These measurements were made by NASA, and were justified on the basis of research and technology demonstrations in contrast to a continuous operational monitoring system. The ACRIM-1 (Active Cavity Radiometer Irradiance Monitor) instrumed used on the SSM measured substantially more solar irradiance than was measured by Nimbus 7 (Figure 1). Although both satellites depict a 1 to 2 Watts/m^2 11-year oscillation associated with the sunspot cycle, the intersatellite differences overwhelm the oscillations. The 1–2 W/m^2 change in solar irradiance (Figure 1) is not equivalent to a 1–2 W/m^2 change in greenhouse radiative forcing because the earth's averaged solar irradiance (which is equivalent) is the solar constant divided by 4. NOAA satellites also measure solar irradiance as part of a calibration procedure. However, these measurements were made only once every two weeks, and are consistently 5 to 8 Watts/m^2 less than those from Nimbus 7.

The solar irradiance measurements clearly present a challenge with respect to documenting long-term climate variations. New data are now becoming available from the Upper Atmospheric Research Satellite (UARS) using an ACRIM-2 instrument. Indications are that these measurements are within the range of values

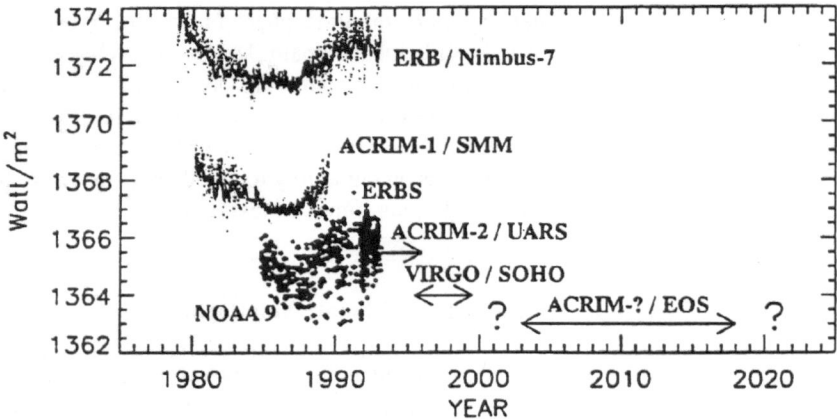

Fig. 1.   Changes in total solar irradiance as measured from different satellites and instruments (adapted from NRC, 1994).

measured by the NOAA satellites, but about 2 Watts/m^2 lower than the ACRIM-1 instrument. If the stability and precision of the ACRIM instrument stands the test of time, an operational version of the ACRIM-2 instrument aboard polar or geostationary satellites will provide the long-awaited commitment to long-term stable monitoring of the solar irradiance, but as evident from Figure 1, only if overlapping measurements between satellites are maintained. An uninterrupted overlapping series of spacecraft radiometers employing in-flight sensitivity tracking is also the primary recommendation of a multi-year study on solar influences on global change completed by the U.S. National Research Council (1994).

2.1.2. *Trace Greenhouse Gases*

Perhaps the best example of a high-quality, sustained, long-term monitoring program is the record of CO_2 concentrations at Mauna Loa Observatory on the island of Hawaii. Measurements were begun by Keeling in 1957, with the support of the NOAA Mauna Loa Observatory staff, and continue to this day. In 1973, CO_2 measurements and measurements of other greenhouse gases were instituted at all NOAA baseline stations, including Mauna Loa. The early record of Keeling exists because of the active engagement of a scientist who championed a long-term monitoring program and overcame obstacles inherent in funding such programs. Such successes in other monitoring programs cannot be relied upon. Atmospheric CO_2 and other greenhouse gases such as methane are now being operationally measured at over thirty sites globally plus numerous shipboard measurements. NOAA operates a number of these sites and provides calibrations and flask sample analyses in cooperative programs at the others. NOAA has been actively engaged in the processing, calibration, and archiving of these data, and now has numerous stations with multi-decadal measurements of radiatively active trace gases. Fortunately, the institutionalization of these atmospheric observing systems have been carried out in such a manner that they have continued to fulfill requirements for long-term climate monitoring.

Based on present understanding, the carbon budget cannot be balanced and the uptake of carbon over the oceans and land remains very uncertain. Part of this can be attributed to inadequate information regarding the role of the oceans and the terrestrial biosphere in the carbon cycle. For example, oceanic total dissolved inorganic carbon (DIC) has in the past been routinely measured, but the accuracy of the measurements was poor (Atwood *et al.*, 1993). When recent measurements from the Ocean-Atmosphere Carbon Exchange Study (OACES) cruise were compared with data derived from a 1972 GEOSECS program an expected increase in DIC was not observed in the upper water column. Figure 2 provides an illustration of the types of differences obtained. Clearly, extreme care is required to measure DIC if environmental change is to be properly monitored. Fortunately, today's measurements are being calibrated by repeated oceanic transects and the use of reference materials, whose carbon content is well known. Efforts such as these must continue over many years.

Simply measuring CO_2 concentrations will not substantially advance our present knowledge of the global carbon cycle. New diagnostic techniques are being used, for example, combining global atmospheric and surface ocean water CO_2 measurements with general circulation models (GCM), to provide information on CO_2 sources and sinks. To date, this work suggests that large amounts of CO_2 are absorbed on the continents by terrestrial ecosystems (Tans *et al.*, 1990). Isotopic ratios of carbon and oxygen in CO_2 are new tools which will prove to be invaluable in studying CO_2 sources and sinks. For example, fossil fuel CO_2 emissions, having a lower $^{13}C/^{12}C$ ratio, tend to decrease the isotopic carbon ratio. The variation of this isotopic anomaly in the atmosphere and oceans can provide new constraints on the global carbon budget (Tans *et al.*, 1993).

Keeling *et al.* (1993) and Keeling and Shertz (1992) make the case for measuring O_2 as a function of latitude, similar to the CO_2 measurements. Fossil fuel burning and deforestation reduces the amount of O_2 in the atmosphere. This downward trend and a seasonal cycle, related to photosynthesis and biological productivity of the global oceans, can now be measured. The O_2 monitoring data, in conjunction with CO_2 and carbon isotope data (C^{12}/C^{13} and C^{14}), will provide the information required to more confidently identify changes in the sources and sinks of atmospheric CO_2. The active engagement of scientists is required to improve the precision, calibration, and standards used to document changes in atmospheric O_2. Finally, models will be required in conjunction with the data to derive carbon flux estimates.

Total ozone has been measured by several different observing systems. Figure 3 reveals some of the inherent calibration problems associated with detecting long-term trends of total ozone from ground-based Dobson spectrophotometers. These

THOMAS R. KARL ET AL.

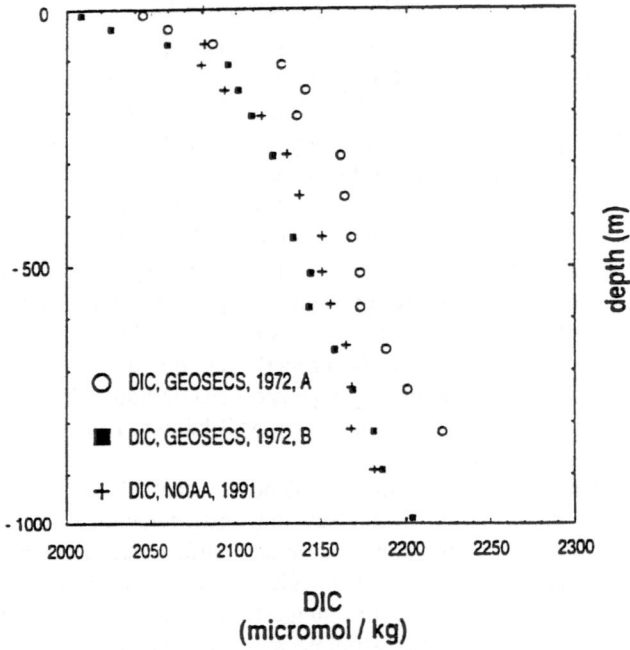

Fig. 2.   Dissolved inorganic carbon measured in 1972 using gas chromatography (circles) and potentiometric titration (squares), and in 1991.

stations are a critical component of the long-term ozone monitoring strategy, yet calibration is a critical issue (Komhyr *et al.*, 1989). Fortunately, NOAA maintained the world standard Dobson instrument (#83), which is used to calibrate most of the world's Dobson instruments, to ±1% over the past 30 years. Figure 4 shows a systematic difference of the ozone derived from the Dobson network and that derived from the NOAA 11 SBUV2 instrument. Of particular interest is the drift in the SBUV2 data when using processing algorithm version 5.5 versus the small drift for version 6.0. The use of both *in-situ* and space-based measurements increases the confidence of both data sets. Measuring the same variable with two different observing systems is sometimes the only way to minimize the adverse impacts of time-varying biases.

Recent studies (Wang *et al.*, 1993; Mohnen *et al.*, 1993) point to the sensitivity of the climate system to changes in the vertical distribution of tropospheric and strato-spheric ozone. Ozone profile information, in particular long-term trends, is required to conduct modeling studies of climate change. Ozonesondes, lidars, microwave

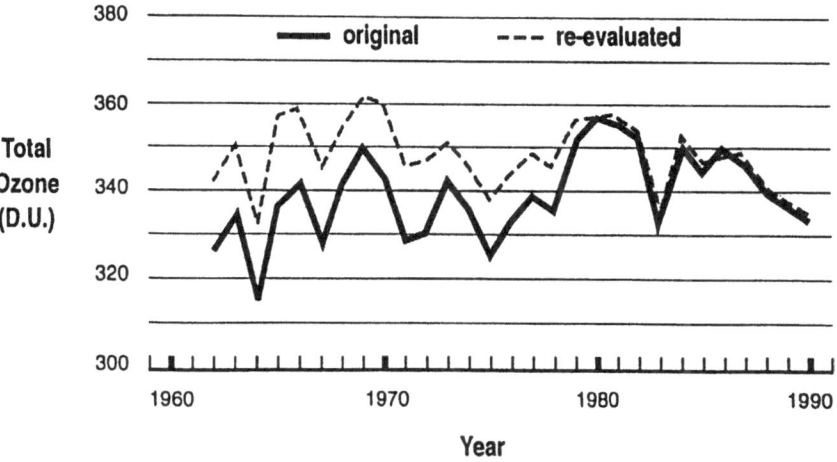

Fig. 3. Changes in total ozone, Hradrec Kralove, prior and subsequent to re-evaluation of calibration (Figure courtesy of W. Planet, NOAA).

SBUV/2 minus Dobson (BP)/Dobson
NOAA-11

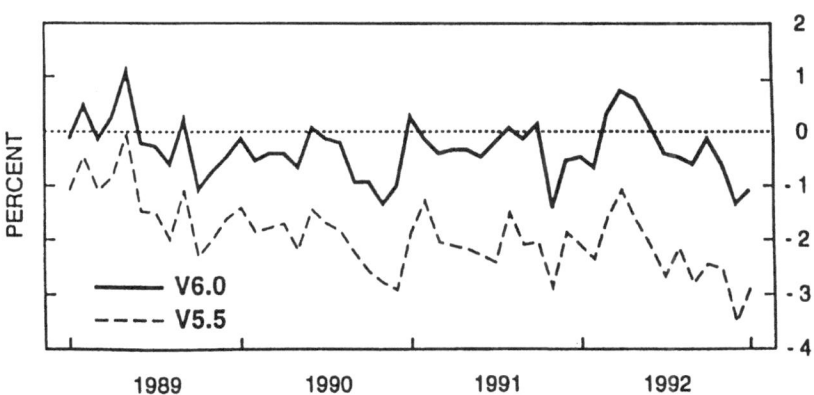

Fig. 4. Systematic differences of total ozone as measured by satellite SBUV2 and *in-situ* Dobson measurements. SBUV2 data was processed by algorithms version 5.5 (NOAA-9) and version 6.0 (NOAA-11) (Figure courtesy of K. Rao, NOAA).

instruments and inversion of spectral UV data (for example, the 'Umkehr' technique applied to Dobson data) provide vertical profiles of ozone; however, these surface-based techniques provide inadequate spatial and temporal resolution for determining long-term geographic and vertical changes of ozone, particularly in the troposphere. In addition, lidar and Umkehr data are subject to contamination by aerosols, especially following major volcanic eruptions. Satellite measurements of scattered UV, extinction in the visible and microwave emissions are being used to retrieve ozone profiles. While they are able to obtain near-global coverage, there are complications due to aerosol scattering, pressure broadening of microwave lines, etc., not to mention the difficulty in maintaining long-term calibration in some of the instruments. The effect of tropospheric ozone on surface tempera- ture has only recently been emphasized (Wang *et al.*, 1993). Other examples will certainly emerge as time progresses, and this implies that it will not be possible for any scientific committee or expert panel to identify all the forcings required for long-term climate monitoring. This suggests that scientists must continue to be opportunists, using operational measurements from multiple use observing systems to help understand the climate system.

2.1.3. *Aerosols*

Over the past few years research has suggested that historical changes of anthro- pogenic aerosols cause a climatic forcing of comparable magnitude, but with opposite sign to greenhouse gas forcing (Charlson *et al.*, 1990, 1991, 1992; Penner *et al.*, 1992; IPCC, 1992; Taylor and Penner, 1994; Karl *et al.*, 1994; Santer *et al.*, 1995). Penner *et al.* (1993) have outlined a plan to narrow the uncertainties regard- ing anthropogenic aerosol forcing that includes space-based, *in-situ*, and aircraft monitoring of aerosol type and quantity as well as precursor gas sources and source fields. In addition, a number of observational strategies have been articulated to determine and understand the processes which control the size, refractive index, hygroscopic growth, production, and removal of anthropogenic aerosols. Such an observing system would alleviate existing gaps in measurements of atmospheric anthropogenic aerosols. In this respect it is worthwhile to mention the history of the U.S. turbidity monitoring network. As knowledge about the role of aerosols increased rapidly, the inadequacy of this observing program became more apparent; however, instead of evolving and improving the program stagnated. This points out that when observing programs become institutionalized, scientific users of the data and information must continue to be closely linked to the program to ensure the observations are adequate to monitor the most important characteristics of climate system.

Measuring the atmospheric characteristics of anthropogenic aerosols is only one aspect of understanding aerosol forcing. The determination of the source terms through the use of emission inventories and budgets is also critical. For exam- ple, existing space-based measurements can be used to routinely estimate global biomass burning from fires which act as an important source of carbonaceous aerosols. Clearly, the success of many of these efforts will be tied to multi-purpose satellite observing systems, and a long-term monitoring program free of undesirable time-varying biases. This will require a strategy of operation that can accommo- date recommendations for improved measurements without diversely affecting the homogeneity of the data bases.

Stratospheric aerosols have been given increased attention mainly because measurements after the eruption of Mt. Pinatubo have shown conclusively that such volcanic eruptions can cause a large negative radiative forcing on climate even if only for a relatively short time (Minnis *et al.*, 1993). Stratospheric aerosols have been monitored by the NASA Stratospheric Aerosol and Gas Experiment (SAGE) and similar experiments such as the Halogen Occultation Experiment (HALOE) on UARS. These measure vertical profiles of aerosol-related radiative extinction in the atmosphere by viewing the sun during sunrise and sunset as observed from the satellite (Minnis *et al.*, 1993). NOAA satellites are capable of providing estimates of aerosol optical depth by using visible channels on its polar orbitors, but only over areas of constant surface albedo, e.g. the oceans, and only for optical depths in excess of about 0.05. Of course, the long-term monitoring of the atmospheric composition, size distribution, and vertical distribution of tropospheric and stratospheric aerosols is critical to understanding climate variations. Hansen *et al.* (1995) outline a strategy to monitor these quantities, which entails a number of small satellites. Given the small size of the satellite payloads they argue that the observing system would be flexible with a small planning horizon. This could help ensure overlapping coverage between new satellites and instruments.

2.2. CLIMATE FEEDBACKS

The separation of climate-related quantities into forcing (agents of change) and responses is easily accomplished, but there exists another class of climate-related quantities best described as feedbacks. It is not always possible to uniquely ascribe climate quantities to this class. For example, some climate feedbacks can be viewed as climate responses depending on the question of interest. Similarly some climate feedbacks can just as easily be thought of as climate forcings. Nonetheless, as related to the climate forcings already described, it is convenient to discuss a number of climate quantities in terms of their climate feedbacks.

2.2.1. *Clouds and Related Measurements*
Few would argue that clouds and their changes in composition, amount, level, and distribution are critical to understanding future and past climate changes (Balling, 1993; Lindzen, 1993; Schneider, 1993; Hansen *et al.*, 1993; Michaels, 1993; Schlesinger, 1993; Karl, 1993). Despite the evidence presented from conventional observations regarding changes in cloud amount, (Figure 5) our confidence is not high in our ability to monitor even basic changes in total cloud amount. Karl and Steurer (1990) point out these difficulties, even when two related measurements are compared, both with long records (Figure 6). With the advent of automated cloud observing networks in countries such as the U.S.A., new inhomogeneities in the cloud records are already occurring, which are mostly incorrigible. Unfortunately, turning to the satellite record for global cloud monitoring is not much help. Even the short data sets from projects such as the International Satellite Cloud Climatol-

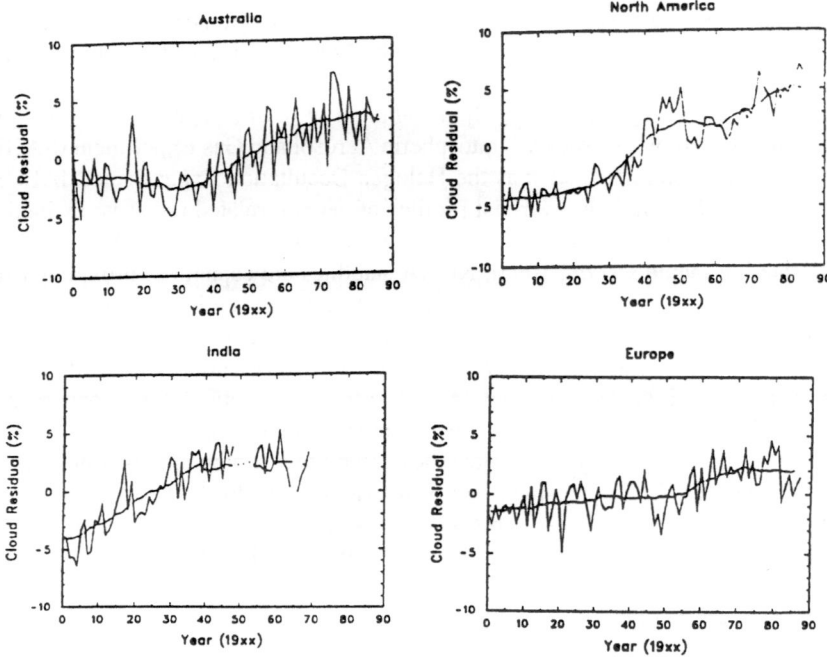

Fig. 5.   Changes in anomalies of observed cloud cover over Australia, Europe, India, and North
America (from McGuffie and Henderson-Sellers, 1993).

ogy Project (ISCCP) have serious inhomogeneities (Figure 7), despite scientists's
struggles to improve the cloud algorithms. The problem stems from inadequqate
overlap between space-based observing systems.

Homogeneous measurements of solar radiation at the earth's surface could be
another means to help clarify long-term trends of cloud cover, but the impact of
natural and anthropogenic aerosols hinders direct comparison between trends in
these two data sets. Moreover the solar measurements have been plagued by major
data inhomogeneities and data quality problems over the past few decades. The
history of NOAA's U.S. surface solar radiation network (SOLRAD) is perhaps a
good example of what can happen to operational observing systems. Over the past
two decades the data from the instruments within the SOLRAD monitoring system
were often beset with unwanted drift and poor data continuity. Funds were provided
to calibrate the measurements for a short time, but it was too late to salvage a multi-
decadal record of global surface radiation. The SOLRAD network is now virtually
dismantled in favor of a new U.S. surface radiation network (SURFRAD) being
developed by NOAA. SURFRAD will have fewer sites, but combined with sites
of other federal agencies, could provide an adequate solar radiation monitoring
network in the U.S. Nonetheless, if the observations and data from the SURFRAD

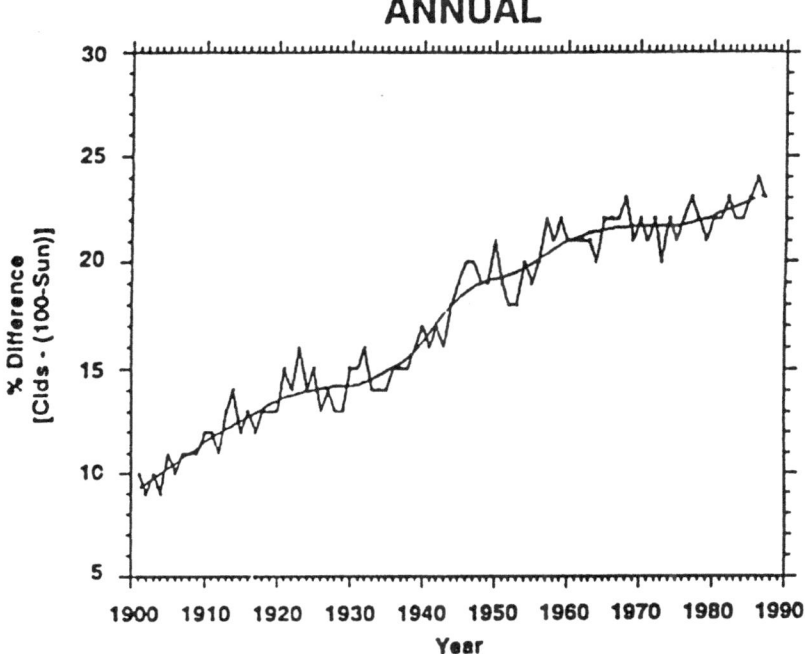

Fig. 6.   Difference between the observed cloud cover and 100% minus the percent of possible sunshine (actual sunshine is subtracted from 100% to covary in the same sense as cloud amount) over the contiguous United States (from Karl and Steurer, 1990).

monitoring system are not routinely assessed for trends and time-varying biases it is quite conceivable that it will also fail the test of time as the network matures and stations are relocated or instruments changed. Overlapping measurements can help overcome these problems should stations have to be relocated or instruments changed. On a global basis, NOAA has been monitoring solar and terrestrial radiation at its baseline sites (Barrow, Alaska; Mauna Loa, Hawaii; American Samoa; South Pole) for several decades. These sites are supplemented by several island sites and the Boulder 300 m tower. These are long-term, high quality sites operated in a research mode and when combined with satellite radiation data, have revealed important information on, for example, anomalous cloud absorption of solar radiation (Cess et al., 1995). The WMO Global Atmosphere Watch (GAW) program is instituting six new monitoring stations in Algeria, Argentina, Brazil, China, Indonesia, and Kenya. Among the measurements are high quality solar radiation data, being implemented with the advise and collaboration of NOAA personnel. Three of these stations were operating at the end of 1994. New efforts by WMO's Global Atmospheric Watch program to periodically check calibrations

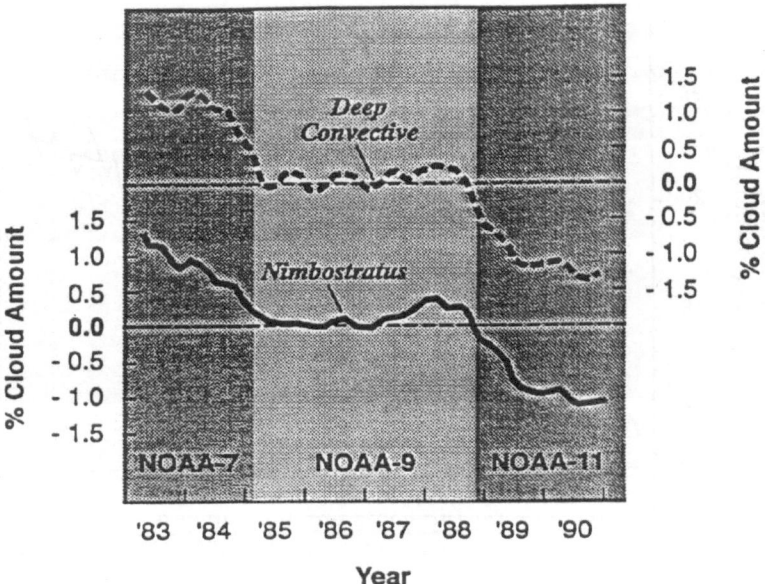

Fig. 7.   Changes in monthly anomalies of mean cloud cover based on the ISCCP data base (from Klein and Hartmann, 1993).

and perform intercomparisons will be critical to the long-term success of multi-decadal surface radiation monitoring.

An international surface-based UVB monitoring network is now being set up through WMO's Global Atmospheric Watch program. Periodic calibrations and intercomparisons will be critical to the success of this program since there are various competing instruments that can be used to measure spectrally-resolved UVB. Because these measurements are also likely to have a high degree of spatial variability, e.g., urban versus rural, cloudy versus clear skies, it will be essential to have active scientific users, data managers, and network managers working toward the maintenance of a long-term reliable network.

2.2.2. *Cryosphere*

Year-to-year variations of sea-ice and snow cover can amount to 10 million square kilometers. Long-term changes of these quantities have been reported (Chapman and Walsh, 1993; Robinson *et al.*, 1993; IPCC, 1990, 1992), and indeed, may have played an important role in the magnitude of springtime warming (Groisman *et al.*, 1994a, b). Nonetheless, existing data bases of snow cover, snow depth and

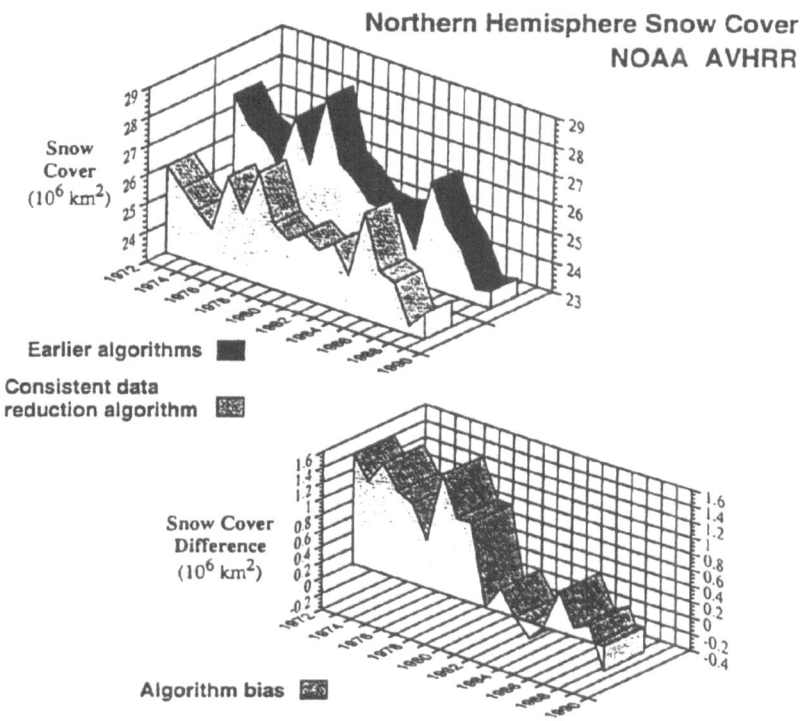

Fig. 8.   Changes in Northern Hemisphere snow cover as reported in IPCC (1990) and IPCC (1992) and the difference between the two time series.

its water equivalent, and sea ice require much more attention to data quality and long-term continuity. For example, simple changes in a land-ocean mask and an undocumented change in the post-processing of the snow cover data set caused IPCC (1990) to exaggerate the retreat of satellite-derived snow cover by nearly 1.5 million square kilometers (Figure 8).

Satellite data of sea-ice extent are important because changes in sea-ice extent have large regional effects on temperature. The historical record of sea ice extent (IPCC, 1992) reveals anomalously high Southern Hemisphere sea ice extent during the period 1973–76, but doubt has been cast on the homogeneity of the record during these years because data before 1976 were largely based on a microwave radiometer which differed from later instruments. Sea ice mapping using passive microwave data, moored buoys, visible and near-infrared spectral radiances, and *in-situ* measurements are now being integrated, but operational products that are later used to assess multi-decadal trends do not have routine checks for long-term data homogeneity, so problems can go unnoticed for long periods of time.

2.2.3. *Surface Hydrology*

At the present time there is no federal commitment to long-term soil moisture measurements, and information on long-term variations of soil moisture in the United States is unknown. The value of these measurements has long been recognized (Shukla, 1993), and observations have been recorded in the former U.S.S.R. for many decades. Space-based measurements of soil moisture are still not technically feasible. It is unclear why routine soil moisture measurements, such as occurred in Russia and in some states (Illinois), in the U.S.A. are not being made. It may be related to a perceived lack of importance, or an inability to get the attention of operational multiple use network managers to consider these measurements as part of their routine operations.

Streamflow measurements in the United States are affected by local anthropogenic changes (diversions, withdrawals, discharges, etc.) that often make it impossible to distinguish between local and global changes. Slack and Landwehr (1992) have worked to identify a list of long-term stations with minimal local impacts. It is unclear what mechanisms exist within each country to ensure that the intrinsically valuable monitoring stations, such as these, are protected in future years. Minimally, adequate information about how the measurements have been affected by local changes must be archived with the data to enable separation of local and global changes.

2.2.4. *Subsurface Ocean Heat Flux and Circulation*

Uncertainty regarding the amount of heat that can be stored or given up by the oceans can prevent an adequate separation of natural versus anthropogenic climate forcing. From a climate change detection perspective, there are two main interests in sub-surface ocean data. The first lies in the possibility that global warming could be detected with strong statistical significance in the deeper oceans, beneath the reach of the seasonal cycle, because the interannual variability of temperature is thought to be very low there. Even very small changes of a couple of tenths of degrees Celsius may be detectable as a significant trend. The second interest lies in the possibility of significant changes in ocean circulation e.g., a slowing of the Atlantic thermohaline circulation.

The major data required include temperature, salinity, and some current velocity data. In the top several hundred meters of the ocean expendable bathythermographs (XBTs) are often used to measure vertical profiles of temperature. XBT's can be deployed from merchant ships and are therefore relatively abundant. Intermediate and deep waters require other instruments such as Conductivity-Temperature-Depth whose deployment requires research vessels and highly trained scientists and technicians. The type of long-term thermometric measurements available are depicted in Figure 9. It depicts the increase in the annual mean temperature observed at 1750 m depth at Station 'S' in the subtropical gyre near Bermuda for the 1955–1990 period. The linear warming trend has a magnitude of 1.0 °C per century for the 1960–90 period (Levitus *et al.*, 1994). In the subarctic gyre of the North Atlantic,

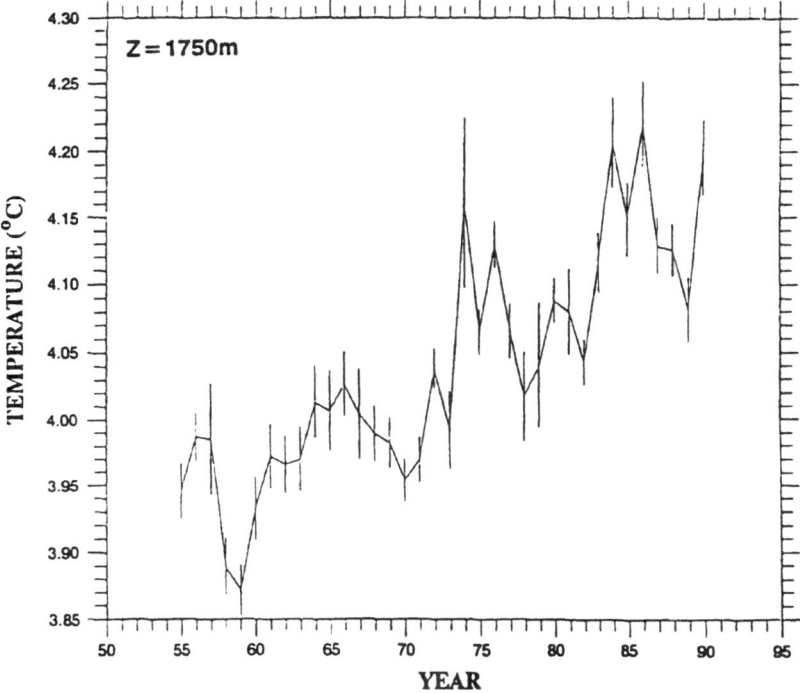

Fig. 9. Time series of annual mean temperature (°C) at 1750 m depth at station 'S' (32.16° N, 64.5° W). The vertical lines about each annual mean temperature represents ±1 standard error of estimate based on the twelve monthly mean temperatures about the annual mean.

deep ocean temperatures have been cooling. Gyre and basin scale changes over the past few decades in the thermal structure of the upper Pacific Ocean are also being documented (Levitus *et al.*, 1994; Levitus and Boyer, 1994). Little change has been noted in the deep North Pacific, but little change would be expected on this time scale given the differences in the time-scale of mixing between deep and intermediate waters in the North Atlantic versus the North Pacific (Decadal versus Century time scales). Measurements of both temperature and salinity however, are required to infer circulation and vertical mixing.

Other attempts to detect warming in restricted regions of the ocean have been made (IPCC, 1990; Bindoff and Church, 1992; Ellett and Blindheim, 1992; Antonov, 1993), but all efforts suffer because of the paucity of observations. For this reason it is essential to continue developing data bases and observing networks. This includes XBT's and hydrographic data. In view of the sparsity of observations, much of the data will have to be derived from ocean model assimilations. Just taking more observations, however, will not be adequate. An example of how monitoring using XBTs can go astray is illustrated by current concerns over systematic errors that may have arisen because of incorrect fall rate calculations used with Sipican XBTs (Wright, 1993). A task team from the International Global Ocean Service System (IGOSS) has investigated this problem. New fall rate coefficients are expected to be incorporated into routine processing in 1995. Other potential problems affecting the continuity of long-term data include the effect of changes in manufacturer and controller boards used for data retrievals.

2.2.5. *Water Vapor*

The most important greenhouse gas on the planet is also a poorly measured element. United States radiosondes are unable to measure very high and low relative humidities. Moreover, changes in the methods of measurement have occurred in ways that have seriously affected the homogeneity of the water vapor record. For example, in October of 1993, the calculation of relative humidity was changed for all VIZ sondes, which is the bulk of the U.S.A. network (Elliott and Gaffen, 1993). Relative humidities will now be calculated below 20%, and a correction to the resistance value used in the algorithm to calculate relative humidity will increase relative humidities above 60% (Figure 10). The dispersion of the relative humidity and water vapor calculation will increase due to the changes and the average value will change in an unpredictable manner. At least researchers are being warned about the change. These changes are occurring without any change in instrumentation. It is crucial that changes in operational processing and instruments in multipurpose networks receive adequate discussion prior to implementing changes.

Water vapor in the stratosphere is now being monitored by space-based instruments such as SAGE, HALOE and the Microwave Limb Sounder (MLS) on UARS. However, intercomparisons with *in-situ* methods will be required to assure the validity of trend studies. Only one stratospheric and upper tropospheric water vapor monitoring record is of adequate length and accuracy for trend studies. This is the NOAA record of balloonborne frost-point hygrometers which have been flown on a monthly basis at Boulder since 1981. The instrument is a more accurate, higher resolution descendant of the Mastenbrook instrument which was flown prior to this period at Washignton, DC. Recent analysis of the 14-year Boulder record (Oltmans and Hofmann, 1995) indicates that a significant positive water vapor trend of about 8% per decade exists in the lower stratosphere (19 km). This value, representative of northern mid-latitudes, is nearly a factor of two larger than expected from the oxidation of methane in the stratosphere and the observed increases in methane, and may be related to global warming.

2.3. CLIMATE RESPONSES

2.3.1. *Surface*

2.3.1.1. *Temperature.*   Probably no single climate element has been studied more than near-surface temperature over land. Unfortunately, if the recent rate of

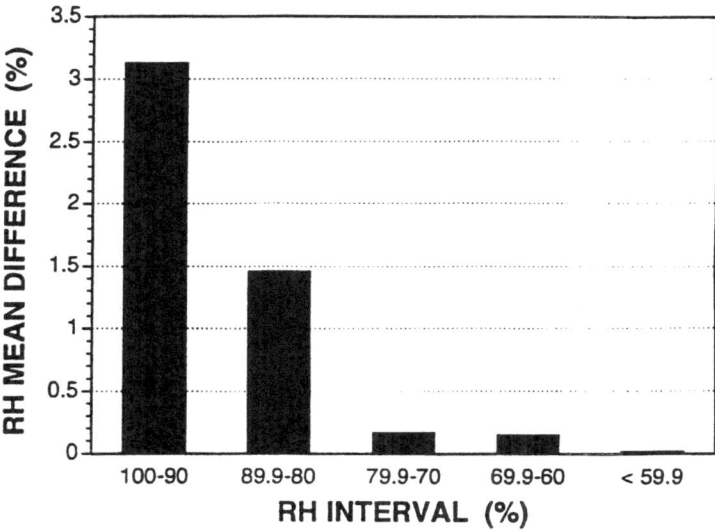

Fig. 10. The effect of a change in the resistance value used in processing relative humidity data from radiosondes based on test results from the National Weather Service Testing and Evaluation Branch (from Elliott and Gaffen, 1993).

decline of global data acquisition across the Global Telecommunications System and international data exchange continues, there will soon be inadequate spatial coverage even to attempt an estimate of near-surface global temperature change (Figure 11). The inadequacy arises not so much from the total number of stations available, but because stations are being dropped in areas that already have sparse networks, e.g. Africa, South America, Central America, etc. In addition, historical temperature data are beset by inhomogeneities due to changes in instrumentation, exposure, site-changes, and time-of-observation bias. Changes in instrumentation in recent times continue to pose problems (Figure 12). Some of these problems can be overcome by thorough comparisons of data between stations with the aid of metadata (documentation regarding site and instrumentation changes etc.). Differences of temperature depited in Figure 12 can only be detected by hundreds of comparisons with neighboring stations and the exact bias of any one instrument is unknown. Moreover, the accuracy of newly introduced modified hygrothermometers in the U.S.A.'s modernized automated first-order weather observing network is ±1 °C. This is inadequate for many long-term climate monitoring applications. Quayle et al. (1991) note that in the Cooperative Observing network the introduction of an electronic maximum-minimum thermistor housed in a plastic instrument shelter has introduced a systematic bias by reducing the maximum temperature by 0.4 °C and increasing minimum temperatures by 0.3 °C (Figure 13). Little consideration was given to overlapping measurements between old and new

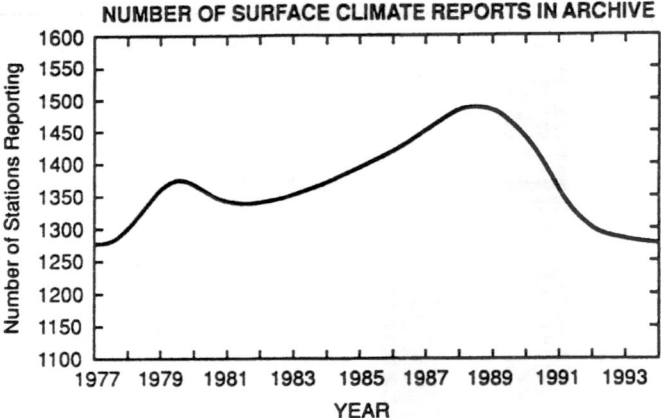

Fig. 11. Number of 'CLIMAT' messages (containing monthly temperature summaries) received at the U.K. Meteorological Office through the Global Telecommunications Systems.

instruments. Unfortunately this is standard practice across the world, rather than the exception.

The primary source of data from rural stations that are free from urbanization throughout the world is often non-automated stations. Systematic biases are introduced when observers change the time of the daily observation (they have systematically tended to switch from evening to morning observations in the U.S.A.). Biases of the order over 1 °C (Figure 14) can enter into the temperature record. In the age of space-based remote sensing all of these problems could be solved with very modest investments. At this time we are solely dependent on conventional measurements to provide us information about changes in the near-surface (1.5 to 2 m) tempeature since satellites are incapable of resolving temperature in the surface boundary layer.

Sea surface temperatures from ships of opportunity and drifting and moored buoys are used to complement the land-based surface temperature measurements. Original data are so poor, however, that data adjustments of a magnitude similar to the apparent climate-change signal are necessary to remove inhomogeneities (Figure 15). For instance, in the mid-nineteenth century, wooden buckets were often used to sample water for measuring sea surface temperature, but canvas buckets (which lose heat more rapidly) became common by the early twentieth century. The buckets were placed on deck while the temperature was measured (and sometimes left on deck for considerable periods prior to measurement), so changes in the insulation properties of the buckets would lead to biases. Around the middle of the twentieth century a rapid switch took place to mainly engine-intake data, producing a sudden, artificial warming in the data, because the water sample could no longer cool (through evaporation, convection, and conduction) on deck

MAXIMUM TEMPERATURE

Average difference:
+ 0.50 °C

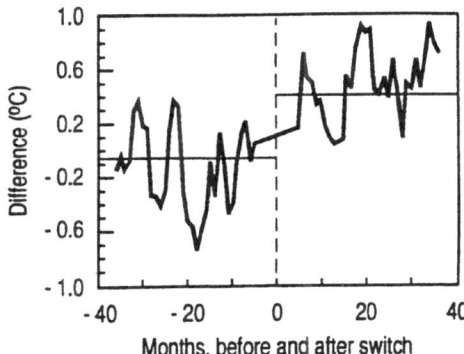

MINIMUM TEMPERATURE

Average difference:
maybe + 0.10 °C

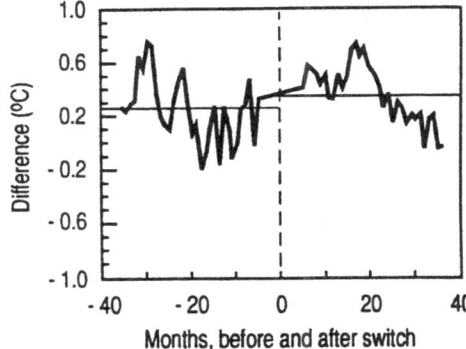

Fig. 12. Average differences of temperature between stations prior and subsequent to the switch from the HO60 to the HO83 series hygrothermometers.

prior to measurement. Even though methods to correct for these time-varying biases have been developed (Figure 15), information on the proportion and distribution of the switch to engine intake temperatures, and between wooden and canvas buckets, is incomplete. So some doubts remain about the adequacy of the corrections to the historical data, even if the methods used to adjust these data were perfect (Folland *et al.*, 1993; Folland and Parker, 1995). For example, a progressive increase in the fraction of engine intake data since the 1950s may have caused a small artificial warming trend (Folland and Parker, 1994). Information about the measuring technique used has often been a low priority, since it has not been a critical factor related to weather forecasting, but ship engine intake, hull contact, and insulated buckets are all potential kinds of observing devices in use today.

Satellite-based SSTs have the virtue of more completely covering the ocean. For a variety of reasons, however, including the interference of episodic volcanic aerosols, the AVHRR temperatures had to be adjusted to remove time-varying and mean biases while retaining (as much as possible) the geographical patterns of

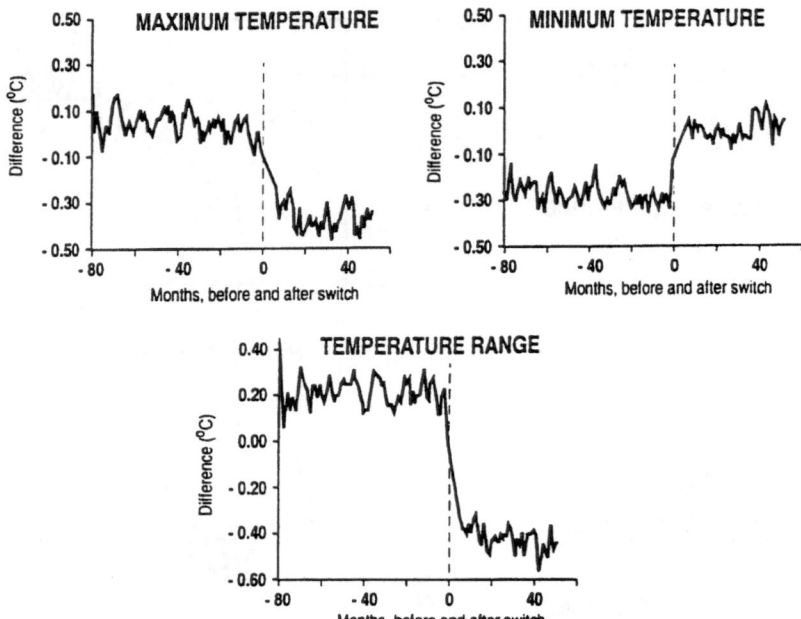

Fig. 13. Same as Figure 12 except for the switch from an alcohol maximum-minimum temperature in a wooden Cotton Region Shelter to a thermistor in a plastic 'Beehive' shelter (from Quayle *et al.*, 1991).

these gradients (Figure 16). The biases were extensive and large after the recent Mt. Pinatubo eruption. However, even 'perfect' satellite SSTs, or much-improved data anticipated from the Along-Track Scanning Radiometer (ATSR), measure the ocean surface 'skin' rather than bulk temperature, so this will not provide homogeneous data with the historical conventional marine observations. Genuine differences between these two temperatures can be substantial (Schluessel *et al.*, 1990). Thus adjustments to satellite SST data will be required for climate change studies for the foreseable future and methods have been devised to do this (Reynolds, 1993). This however, will still be unsatisfactory if the mix of *in-situ* observations also has time-varying biases.

2.3.1.2. *Precipitation.*   Precipitation is a critical climate element affecting all forms of life, yet we cannot routinely measure it, let alone monitor multi-decadal changes (IPCC, 1990, 1992). Precipitation gauge design has changed over time, affecting wind-induced turbulence over precipitation gauges leading to time-varying biases in the under-catch of solid, and to a lesser extend liquid, precipitation (Sevruk, 1982; Folland, 1988; Karl *et al.*, 1993b) cf., Figure 17. In the high latitudes,

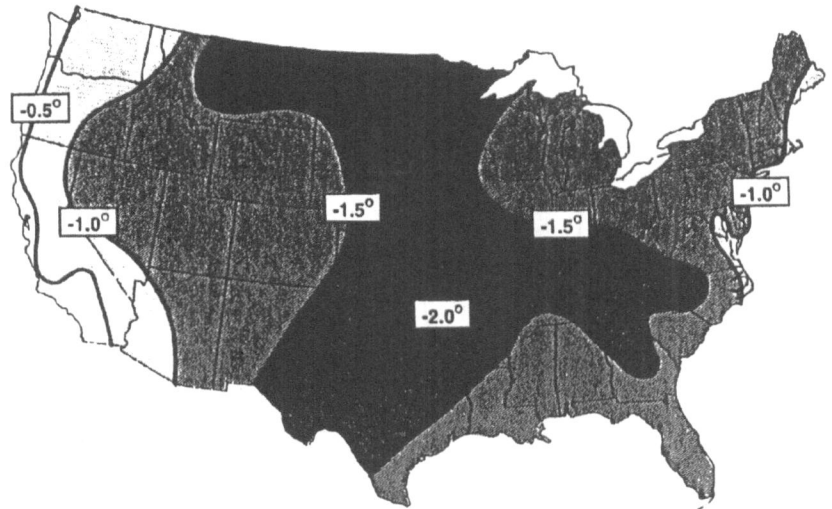

Fig. 14. The effect on the average monthly mean temperature of a change in the observation time using the 24-hour maximum and minimum temperatures ending at 5 PM and 7 AM local time (from Karl *et al.*, 1986).

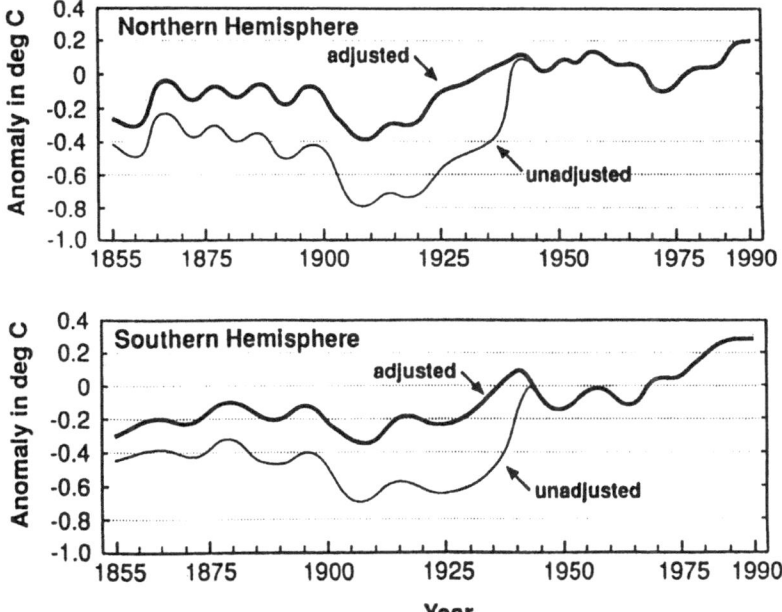

Fig. 15. Variations of adjusted and unadjusted SST anomalies (1951–80) for the northern and southern hemisphere. Annual anomalies have been filtered using a 21-point low pass filter.

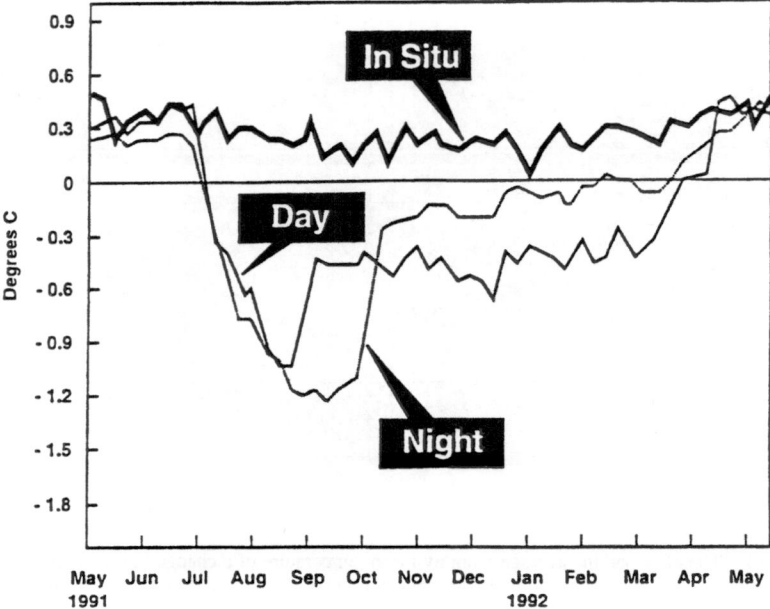

Fig. 16.   Global ocean *in-situ* and AVHRR-derived anomalies of SST prior and subsequent to the eruption of Mt. Pinatubo (from Reynolds, 1993).

measurements have underestimated the true precipitation by 50% or more, but substantial absolute biases also exist in mid-latitudes (Figure 18).

More long-term data are needed as many rainfall records remain inaccessible. This is particularly critical because the spatial and temporal variability is significantly greater for precipitation compared to temperature. Local time-varying biases in precipitation data are very difficult to correct. In some countries well-located stations, free of urban-induced and other local biases, have closed in recent years due to economic pressures. Too often, precipitation is accumulated over weekends, rather than being measured daily at these non-automated stations. This confounds the determination of changes in precipitation frequency distributions. All these factors will complicate the future use of ground-based observations of precipitation on a large scale. Meanwhile many countries are now automating their networks, introducing new instruments. If the characteristics of these new instruments relative to the true precipitation and to the measurements from previous instruments are not well identified prior to their introduction, the long-term climate record is likely to be seriously degraded, rather than improved as might otherwise be anticipated with the introduction of new technology.

Satellites provide much needed near-global indirect measures of precipitation. Yet even as near-global coverage of space-based rainfall measurements are becom-

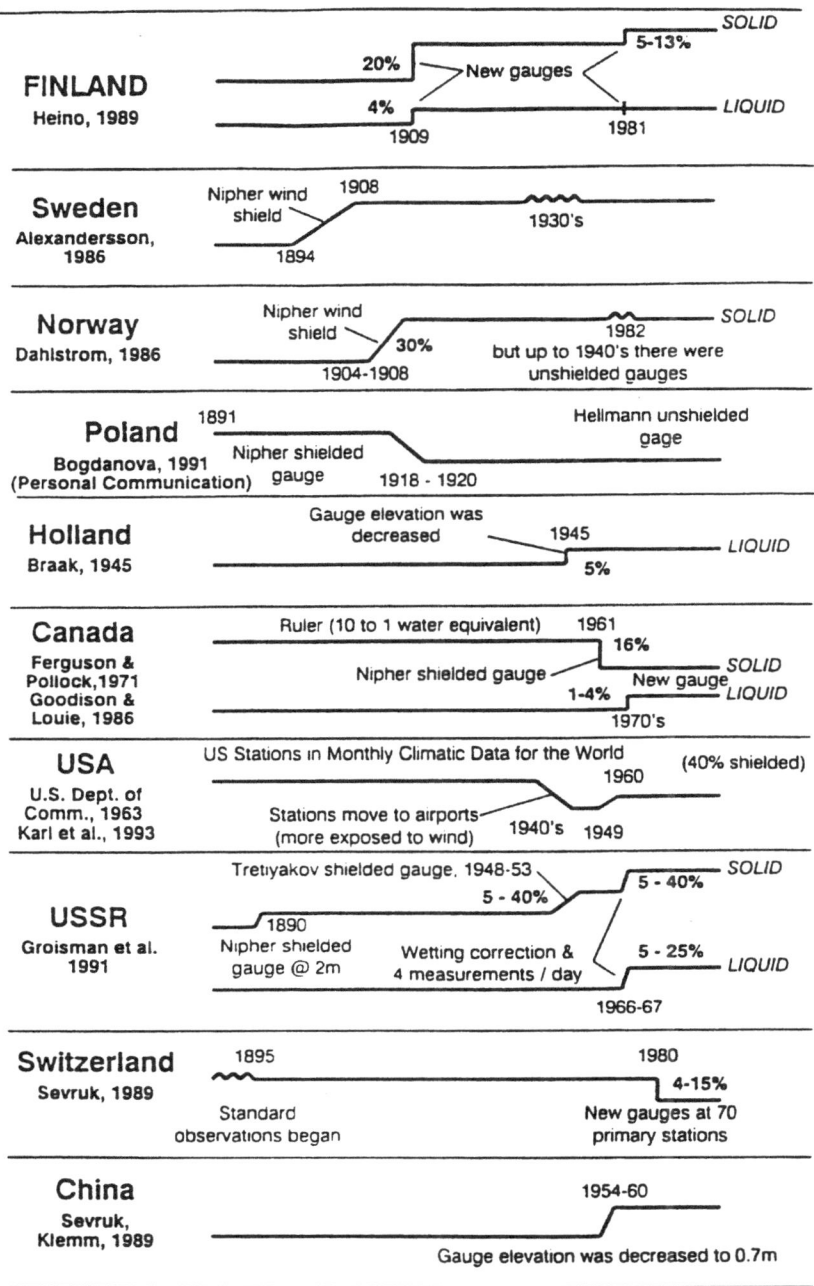

Fig. 17. Time-varying biases of precipitation measurement for various countries (from Karl *et al.*, 1993b).

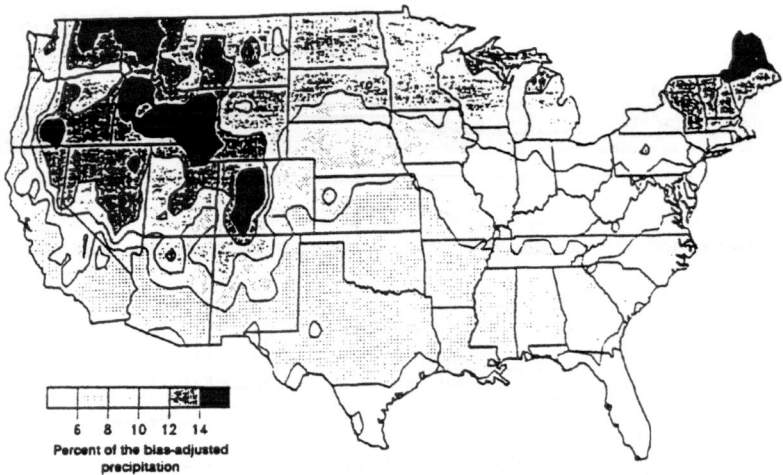

Fig. 18.   Estimated biases in annual precipitation using the NOAA first order and cooperative observing network (from Legates and DeLiberty, 1993).

ing available (Spencer, 1993) they are not adequately resolving the diurnal cycle for long-term continuous climate monitoring. Moreover, surface-based precipitation measurements are required for ground-truthing, but if these measurements seriously underestimate precipitation, *in-situ* biases will be reflected in space-based measurements as well. This includes data collected from satellite-based precipitation radars such as those being deployed in the Tropical Rainfall Measuring Mission (TRMM), as described by Thiele (1993) as well as those derived from Doppler radar. Optical precipitation devices are being tested in some areas and may eventually provide yet another routine means to measure surface precipitation.

2.3.1.3. *Winds and Pressure.*   Changes in wind speed affect evaporative fluxes over the ocean and have the ability to enhance or diminish the hydrological cycle. Unfortunately, biases in observed wind speeds over the ocean in existing archives weaken potentially important analyses such as those depicted in Figure 19 where the change in evaporation attributed to changes in wind speed remain uncertain. Programs such as GEOSAT have been far too short to provide any useful information about long-term changes of winds. Planned satellite missions using a scatterometer offer the opportunity to help document long-term changes, but only if there is a strong commitment to data continuity.

2.3.1.4. *Glaciers.*   There is a continuing need to monitor the length and mass-balance of glaciers worldwide. The 1990 IPCC Report demonstrated the ubiquitous retreat of glaciers throughout the 20th century. This evidence was used to confirm the existence of global warming independently of the various surface temperature

[78]

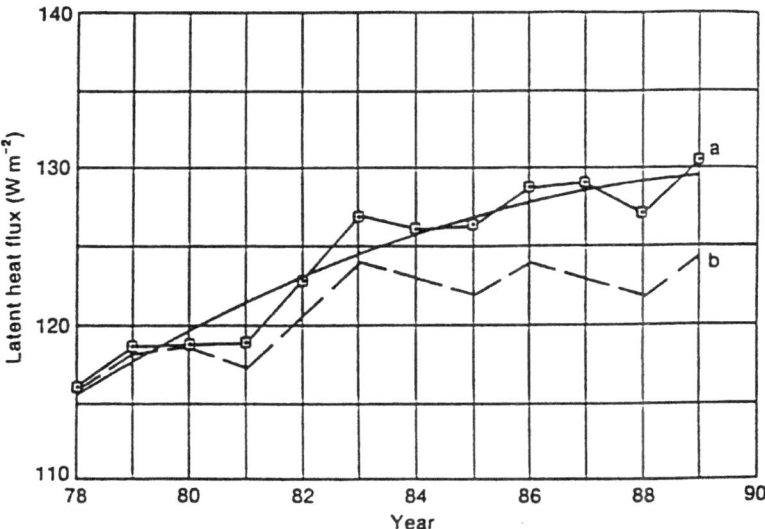

Fig. 19. Trends in global annual oceanic latent heat flux 62° N to 42° S from IPCC (1992). Curve a uses unadjusted wind speeds and curbe b uses reduced wind speeds following Cardone *et al.* (1990).

data sets. It was felt that this was valid because a worldwide retreat is unlikely to be related to a reduction in global mountain rainfall as this is not expected on this scale. However, individual glaciers do respond strongly to rainfall changes so that a few, mostly in coastal areas, are currently strongly advancing. The importance of worldwide glacier retreat also lies in the fact that melting of mountain glaciers is expected to substantially contribute (up to 40%) to sea level rise over the next century (IPCC, 1990). However, glacier data for some regions are unavailable and need to be incorporated into global assessments: for example, the Pyrenees where glacial retreat during the 20th century has been substantial (Parker, 1993a; Tihay, 1990). Aerial photography (Hambrey and Alean, 1992) and satellite-based altimetry (ISCU/UNEP/UNESCO/WMO, 1993) are useful tools for monitoring of mountai glaciers and ice sheets. Very recent evidence (IAHS/UNEP/UNESCO, 1993) shows that retreat in 1985–1990 was at least as great as in the previous 5 year period, and many more glaciers are now being monitored, especially for their mass balance changes. Such data need to be brought together and compared carefully with worldwide regional air temperature data.

2.3.1.5. *Tropospheric and Stratospheric Temperature.*   The global radiosonde data base is affected by instrumental and other changes and requires homogenization (Elliott and Gaffen, 1991; Gaffen, 1994; Parker and Cox, 1994). Biases have been introduced by changes in observing times, the introduction of new or modified radiosondes, and station relocations. For example, changes in housing of the radiosonde instrument can introduce biases, as the dissipation of solar heating by the housing can affect reported daytime humidities. In order to homogenize the radiosonde data base, historical 'metadata' on instrumentation, observing hours, radiation corrections, location and elevation of balloon launch sites, etc., needs to be accurate and brought up to date. International comparisons of instruments are essential, along with statistical and thermodynamic models of the radiosondes, to adjust the data. For climate monitoring purposes an international reference radiosonde should be an integral tool for assessing the impact new or modified radiosondes introduced into the global network. Past and ongoing changes in the processing of the upper air data have also lead to discontinuities in the climate record. In October, 1993 the United States changed the gravitational constant used in the hypsometric equation to derive geopotential heights. This has been shown by Elliott and Gaffen (1993) to result in nearly a 0.2 °C artificial decrease in layer-thickness derived middle troposphere temperatures. This is greater than the magnitude of estimated decadal changes of global temperature at this height. Gaffen (1993) describes in detail the great number of inhomogeneities that have occurred in the upper tropospheric and stratospheric temperature record that severely degrades our ability to reconstruct changes. At these heights direct and indirect radiation can lead to substantial biases of the measurement of ambient air temperature (Figure 20). Over the years there has been a gradual improvement in minimizing these effects. These improvements, however, have been poorly documented (inadequate metadata). As a result, the rate of temperature change remains uncertain.

A homogenous radiosonde data base is essential input into climate data assimulation system. The reanalysis effort of the U.S.A. and Europe cannot hope to produce comprehensive climate data capable of adequately resolving multi-decadal climate variations (see Section 4.2) without such data.

Recently Spencer and Christy (1990) have shown how the satellite-based Microwave Sounding Unit (MSU) can be used to derive a data set of temperatures above the surface layer. These data are generally agreed to be one of the best examples of space-based measurements providing decadal (1979-present) estimates of climate change and variability. Nonetheless, even these data are not free from data inhomogeneities. The equatorial crossing time of NOAA-9 drifted over two hours in four years, but because of overlapping satellite coverage it was possible to remove this observation time bias in the data set. More recently, NOAA-11 underwent severe orbital drift that required a 0.03 °C per decade year adjustment to the temperature trend from the data set (Figure 21), an adjustment of comparable magnitude to the estimated trend. Since overlapping coverage was not available during recent years, NOAA-11 data are no longer used. In these instances, long-term solutions include the installation of small rocket engines that allow the satellite to be periodically repositioned.

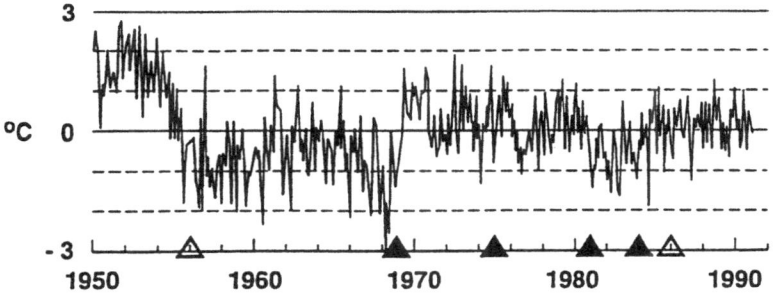

Fig. 20.   Changes in upper troposphere (200 hPa) temperatures and the associated changes on balloon-borne instrument hardware (from Gaffen, 1994).

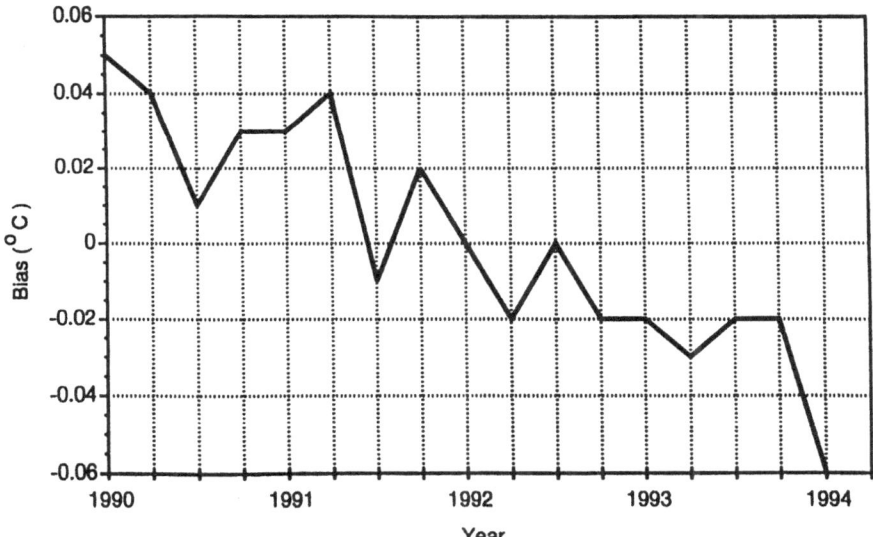

Fig. 21.   MSU2R tropospheric bias attributed to diurnal sampling changes due to orbital satellite drift (Christy *et al.*, 1995).

2.3.1.6. *Extreme Weather Events.*   Many extreme weather events (e.g., extreme maximum and minimum temperatures; intense rainfalls) suffer even greater data problems than do means (e.g., temperature and total precipitation). For example, during times of severe weather, equipment malfunctions are much more common, but vary with the type and exposure of the equipment. Changes in observing schedules, correctable at the monthly or seasonal time scale are virtually impossible to deal with at daily time scales. The requirement for high resolution data is a necessity, but many countries do not have such data readily available, or if they do, it is not accessible for redistribution throughout the scientific community. Thus there is no established mechanism for the exchange of these data, often voluminous, between nations, such as exists for monthly mean data.

For other events (e.g., storms, including tropical cyclones), the problems are compounded by changes in observing systems and coverage, and by analysis system changes. It has become virtually impossible to adequately address questions related to changes in frequency of tropical or extratropical storm frequency. As described in Section 4.2, changes in the frequency of extratropical storms will be facilitated by the reanalysis efforts of the U.S.A. and European communities but may not be very useful for changes in tropical storm frequencies. Often with little additional effort when changes in operational classification systems occur, such as algorithms to calculate tropical storm strength, it would be useful to retain with the classification the minimum set of observed variables required to calculate storm intensity.

3. New Systems to Monitor Climate Change

3.1. SPACE-BASED OBSERVATIONS

Satellite observations are certainly not new. By comparison with conventional *in-situ* measurements however, and for all practical purposes they can be considered as new instruments each time a satellite is launched with new orbits, instruments, and measured radiances. Numerous instruments, useful for observing global climate, have been flown on operational satellites. Some of the data from such instruments have already proved to be invaluable in monitoring climate (e.g., Spencer and Christy, 1990; Spencer, 1993). As demonstrated in several examples in the previous section however, more often than not, the past performance of space-based instruments in monitoring multi-decadal climate variations and changes has been fraught with discontinuities and inhomogeneities that have not been correctable. Too little attention has been given to the need for overlapping measurements of similar spectral bandwidths, in-flight calibration of instruments, observation sampling time changes, and to the general removal of other time-varying biases that may be achieved through integration or blending with conventional measurements. Until these issues are addressed, space-based observations will continue to be the largest portion of the observing system that remains under-utilized in long-term climate monitoring.

3.2. ATMOSPHERIC ELECTRICAL CIRCUIT

Scientists have long sought after more convenient methods of measuring global temperatures. One recent idea is to infer global mean temperature changes through the measurement of the atmospheric electrical circuit. Global thunderstorm activity leads to an electrification of the fair-weather atmosphere. Measurements of the atmospheric electrical circuit provide information on global thunderstorm and lightning activity based on one reliable measurement. It is known that thunderstorm activity is non-linearly related to temperature. Measurements in the tropics have shown that lightning frequencies in thunderstorms are extremely sensitive to small changes in global surface air temperature (Williams, 1992). So, the global electrical circuit has been suggested as a natural global thermometer (Williams, 1992; Price, 1993). Price (1993) suggested that a 1% increase in global temperature may lead to a 20% increase in the ionospheric potential. However, this measurement may only reflect regional temperatures when and where thunderstorms are common (warm seasons of the mid-latitudes and in tropical climates). Furthermore, regional temperatures can increase while thunderstorms decrease. This often occurs when changes in atmospheric thermodynamics inhibit thunderstorms, for example within subtropical high pressure systems or during major droughts. So, Williams' result may be confounded by changes in atmosphere dynamics e.g., in El Niño years, which are not only globally warmer but also have an increased temperature gradient between the tropics and mid-latitudes. Moreover, changes in atmospheric composition of aerosols, ions, and water vapor would also influence the ionospheric potential. Further study is required to identify the sensitivity of the global electrical circuit to these and other confounding influences and ambiguities.

3.3. BOREHOLE TEMPERATURES

The thermal regime at shallow depths within the earth is controlled in part by the conditions at the surface, and therefore is closely tied to the atmospheric temperature (Huang *et al.*, 1994). Past changes of temperature at the surface have penetrated into the Earth's subsurface and have been recorded as transient perturbations to the steady state temperature field. Recent reconstructions of ground surface temperatures have demonstrated that borehole temperatures provide valuable information about climate change over the past few centuries. Low frequency surface temperature variations occurring in the past 300 years can be detected in a borehole over 600 meters deep. The geothermal approach provides a direct recovery of temperature, and therefore the results are free of many of the uncertainties associated with other proxy records (e.g., tree rings). Thousands of available borehole temperature data sets provide the potential for increased spatial coverage of the surface temperature history of the past few centuries. Since boreholes are often located in remote areas they are usually free of urbanization effects. The interpretation of these records however, are not without difficulty. Many boreholes must be analyzed, to filter out noise introduced by unwanted local anthropogenic and natural changes around the borehole, e.g. changes in snow cover, tree growth, etc. Some analyses have found unrealistically large century-scale gradients of temperature change from nearby boreholes.

3.4. OCEAN ACOUSTIC TRANSMISSIONS

Munk (1989) noted the possibility that ocean warming could be detected by long-path acoustic travel through the oceans. Munk and Forbes (1989) proposed measuring the changes in travel times of long-distance acoustic transmissions from Heard Island in the southern Indian Ocean to receivers scattered around other ocean basins. These sound waves would be trapped within the sound channel which is normally found in the deep ocean 1000 m below the surface. The change in travel times over a period of years can provide a measure of the average temperature change along the acoustic ray paths (the sound speed depends also, to a lesser degree, on salinity and pressure). Mikolajewicz *et al.* (1993) suggest that use of several paths (i.e., siting detectors at several locations) would provide a valuable indicator of climate change, and would increase the probability of detecting an enhanced greenhouse effect. This measurement offers the possibility of measuring layer temperatures in the deep ocean somewhat analogous to the layer temperatures observed through the satellite-derived Micro-wave Sounding Unit for the atmosphere.

4. Strategies for Improved Climate Monitoring

4.1. DATA RECOVERY AND REHABILITATION

Enormous amounts of conventional data have been collected and archived in meteorological services. Many of these data are not digitized, and most suffer from time-varying biases, as discussed in section 2. Careful analysis of these data and metadata (documentary records about site changes, exposure details etc. regarding specific sites) can result in data largely free of artificial biases, at least for some types of data. Increased emphasis on data recovery and rehabilitation will enhance our ability to monitor climate change. Better coordination among data processing centers, both nationally and internationally, must occur if irrecoverable data processing biases such as depicted in Figure 22 are to be minimized. The change in precipitation frequency in Figure 22 is attributed to a change in coding practices unknown to the processing center.

Careful analysis of conventional meteorological data will provide an important tool for diagnosing climate change for the foreseeable future. New monitoring systems (section 3) can greatly enhance the conventional data. Identification and protection of high-quality sites (such as the reference Climate Stations program being coordinated by the WMO) must be an integral part of a monitoring strategy, and programs to ensure that historical data are preserved (such as the Data Rescue,

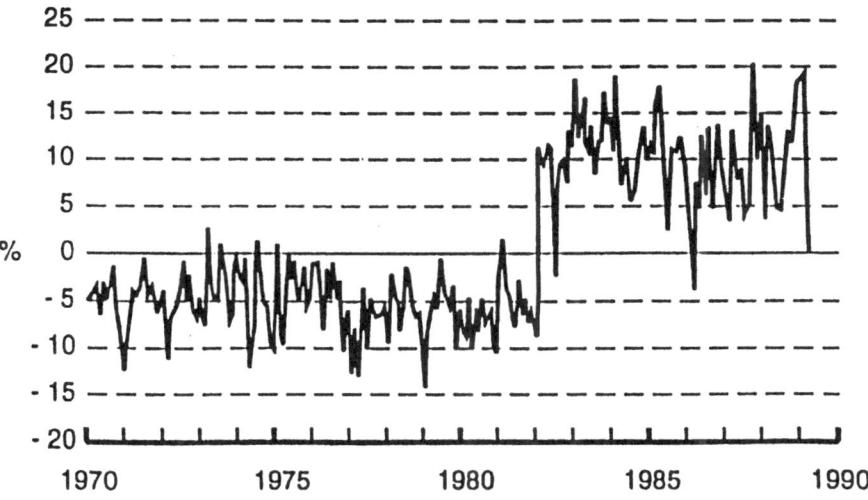

Fig. 22. Precipitation frequency anomalies (60° N–70° N).

DARE, program also coordinated by WMO) are essential for the current and future monitoring of climate.

4.2. REANALYSIS

Modern meteorological computer analysis techniques produce globally-complete datasets of the ensemble of meteorological parameters that can be used to define climate. The models and general circulation techniques used in such analyses, however, have changed with time, thereby introducing a further set of biases. Several National Meteorological Agencies are creating historical grid-point data-bases by reanalysis of recent historical data using fixed numerical models and analysis techniques.

Analyses from models run in weather forecasting mode can add further informa-tion about climate, supplementing data, especially of variables difficult to observe directly. Rainfall derived from the European Centre for Medium-range Weather Forecasts model is already being used experimentally as a major input in the new Global Precipitation Climatology Project data set (Rudolf, 1993). Other quantities only reliably calculable, on a large-scale, from analysis/model systems include heat fluxes, moisture divergence, diabatic heating, and soil moisture. Models can also be used, when forced with observed sea surface temperatures, to estimate evap-oration over the tropical oceans and land (once the current severe problems with land surface processes are reduced), for comparison with meager observations. In the future, models and analysis systems will provide considerable amounts of 'data' useful for climate monitoring, as long as biases introduced by changes in the observing systems do not confound model and analysis systems. Techniques for producing globally-complete fields have not explicitly treated time-varying biases (as opposed to random errors). So these must eventually be removed if the reanalysis projects are to produce high-quality data needed for long-term climate monitoring.

4.3. EXPANSION OF PRESENT IN-SITU EFFORTS

Major improvements in climate monitoring could be achieved with relatively modest increases in on-going activity involving *in-situ* measurements. Some examples include greenhouse gas flask sampling, aerosol chemical and physical characteristic monitoring and balloonborne ozone and water vapor measurements. For example, expansion of the NOAA cooperative flask sampling network in 1992 allowed cleaner separation of terrestrial and oceanic CO_2 sources and sinks through isotopic carbon measurements. These measurements, before expansion, were not adequately accurate and consistent for such interpretations. Similarly, a stratospheric water vapor monitoring program such as is conducted at Boulder, with five additional sites (mid-latitude southern hemisphere, tropical and two polar), could be instituted with relatively little cost and would serve as validation points for future satellite measurements, much as the Dobson network does for ozone today.

4.4. PRINCIPLES OF LONG-TERM CLIMATE MONITORING

Long-term climate monitoring that will be capable of providing robust information about decadal-to-century-scalechanges and variations of climate will require adherence to a number of guiding precepts. Similar to concepts articulated by the U.S.A. National Research Council (1993), these include:

1. The effects on the climate record of changes in instruments, observing practices, observation locations, sampling rates etc. must be known prior to implementing such changes. This can be ascertained through a period of overlapping measurements between old and new observing systems or sometimes by comparison of the old and new observing systems with a reference standard. Site stability for *in-situ* measurements, both in terms of physical location and changes in the nearby environment, should also be a key criterion in site selection. Thus, many synoptic network stations, primarily used in weather forecasting but which provide valuable climate data, and all dedicated climatological stations intended to be operational for extended periods, must be subject to such a policy.
2. The processing algorithms and changes in these algorithms must be well documented. Documentation of these changes should be carried along with the data throughout the data archiving process.
3. Knowledge of instrument, station and/or platform history is essential for data interpretation and use. Changes in instrument sampling time, local environ-

mental conditions for *in-situ* measurements, and any other factors pertinent to the interpretation of the observations and measurements should be recorded as a mandatory part of the observing routine and be archived with the original data.

4. *In-situ* and other observations with a long uninterrupted record should be maintained. Every effort should be applied to protect the data sets that have provided long-term homogeneous observations. Long-term for space-based measurements is measured in decades, but for more conventional measurements long-term may be a century or more. Each element of the observation system should develop a list of prioritized sites or observations based on their contribution to long-term climate monitoring.

5. Calibration, validation and maintenance facilities are a critical requirement for long-term climatic data sets. Climate record homogeneity must be routinely assessed, and corrective action must become part of the archived record.

6. Wherever feasible, some level of 'low-technology' backup to 'high-technology' observing systems should be developed to safeguard against unexpected operational failures.

7. Data poor regions, variables and regions sensitive to change, and key measurements with inadequate spatial and temporal resolution should be given the highest priority in the design and implementation of new climate observing systems.

8. Network designers and instrument engineers must be provided long-term climate requirements at the outset of network design. This is particularly important because most observing systems have been designed for purposes other than long-term climate monitoring. Instruments must have adequate accuracy with biases small enough to document climate variations and changes.

9. Much of the development of new observation capabilities and much of the evidence supporting the value of these observation stem from research-oriented needs or programs. A lack of stable, long-term commitment to these observations, and lack of a clear transition plan from research to operations, are two frequent limitations in the development of adequate long-term monitoring capabilities. The difficulties of securing a long-term commitment must be overcome if the climate observing system is to be improved in a timely manner with minimum interruption.

10. Data management systems that facilitate access, use, and interpretation are essential. Freedom of access, low cost, mechanisms which facilitate use (directories, catalogs, browse capabilities, availability of metadata on station histories, algorithm accessibility and documentation, etc.) and quality control should guide data management. International cooperation is critical for successful management of data used to monitor long-term climate change and variability.

5. Conclusions

Enormous amounts of meteorological data have been collected and archived over the past century. Even greater amounts will be collected, using new observing systems, in the future. The old and new observations will need to be combined carefully, and comprehensive efforts made to reduce the influence of time-varying biases in all the data. This must occur if the incipient Global Climate Observing System is to provide a vehicle to deliver much needed information about long-term climate change and variability.

Acknowledgements

This work was partially supported by the Office of Global Programs, Climate Change Date and Detection Program Element, and a NOAA/DOE Interagency Agreement. Scott Miller and Michael Burgin assisted with the graphics used in this report.

References

Antonov, J. I.: 1993, 'Linear Trends of Temperature at Intermediate and Deep Layer of the North Atlantic and North Pacific Oceans: 1957–1981', *J. Clim.* **6**, 1928–1942.

Atwood, D., Freely, R., and Wanninkhof, R.: 1993, 'The Role of the Ocean in Modulating Atmospheric CO_2 Increases: Long Term Accuracy in Oceanic Inorganic Carbon Measurements', *NOAA Data Quality and Continuity Program: Issues and Status* **V. 1**, No. 3.

Balling, R. C.: 1993, 'Global Temperature Data', Research and Exploration, Publication of the *National Geographic Society* **9**, 201–207.

Bindoff, N. L. and Church, J. A.: 1992, 'Warming of the Water Column in the Southwest Pacific Ocean', *Nature* **357**, 59–62.

Chapman, W. L. and Walsh, J. E.: 1993, 'Recent Variations of Sea Ice and Air Temperature in High Latitudes', *Bull. Amer. Met. Soc.* **74**, 33–46.

Cardone, V. J., Greenwood, J. G., and Cane, M. A.: 1990, 'On Trends in Historical Marine Data', *J. Clim.* **3**, 113–127.

Cess, R. D., Zhang, M. H., Minnis, P., Corsetti, L., Dutton, E. G., Forgan, B. W., Garber, D. P., Gates, W. L., Hack, J. J., Harrison, E. F., Jing, X., Kiehl, J. T., Long, C. N., Morcrette, J.-J., Potter, G. L., Ramanathan, V., Subasilar, B., Whitlock, C. H., Young, D. F., and Zhou, Y.: 1995, 'Absorption of Solar Radiation by Clouds: Observations versus Models', *Science* **267**, 496–499.

Charlson, R. J., Langner, J., and Rodhe, H.: 1990, 'Sulphate Aerosol and Climate', *Nature* **348**, 22.

Charlson, R. J., Langner, J., Rodhe, H., Leovy, C. B., and Warren, S. G.: 1991, 'Perturbation of the Northern Hemisphere Radiative Balance by Backscattering from Anthropogenic Sulfate Aerosols', *Tellus* **43AB**, 152–163.

Charlson, R. J., Schwartz, S. E., Hales, J. M., Cess, R. D., Coakley, J. A. Jr., Hansen, J. E., and Hofmann, D. J.: 1992, 'Climate Forcing by Anthropogenic Aerosols', *Science* **255**, 423–430.

Christy, J. R., Spencer, R. W., and McNider, R. T.: 1995, 'Reducing Noise in the MSU Daily Lower Tropospheric Global Temperature Data Set', *J. Clim.*, (in press).

Ellett, D. J. and Blindheim, J.: 1992, 'Climate and Hydrographic Variability in the ICES Area During the 1980's, *Int. Counc. Explor. Sea* **195**, 11–31.

Elliott, W. P. and Gaffen, D. J.: 1991, 'On the Utility of Radiosonde Humidity Archives for Climate Studies', *Bull. Amer. Meteor. Soc.* **72**, 1507–1520.

Elliott, W. P. and Gaffen, D. J.: 1993, 'Upper Air Processing Changes', *NOAA Data Quality and Continuity Program: Issues and Status* **V. 1**, No. 2.

Folland, C. K.: 1988, 'Numerical Models of the Raingauge Exposure Problem, Field Experiments and an Improved Collector Design', *Q. J. Roy. Meteorol. Soc.* **114**, 1485–1516.

Folland, C. K. and Parker, D. E.: 1995, 'Correction of Instrumental Biases in Historical Sea Surface Temperature Data Using a Physical Approach', *Q. J. Roy. Meteorol. Soc.* **121**, 319–367.

Folland, C. K., Reynolds, R. W., Gordon, M., and Parker, D. E.: 1993, 'A Study of Six Operational Sea Surface Temperature Analyses', *J. Clim.* **6**, 96–113.

Gaffen, D.: 1994, 'Temporal Inhomogeneities in Radiosonde Temperature Records', *J. Geophys. Res. Atmos.* **99**, 3667–3676.

Groisman, P. Ya. and Easterling, D. R.: 1994, 'Variability and Trends of Total Precipitation and Snowfall over the United States and Canada', *J. Clim.* **7**, 184–205.

Groisman, P. Ya., Karl, T. R., and Knight, R. W.: 1994, 'Observed Impact of Snow Cover on the Heat Balance and the Rise of Continental Spring Temperatures', *Science* **263**, 198–200.

Hambrey, M. and Alean, J.: 1992, *Glaciers*, Cambridge University Press, p. 208.

Hansen, J., Rossow, W., Carlson, B., Lacis, A., Travis, L., DelGenio, A., Fung, I., Cairns, B., Mishchenko, M., and Sato, M.: 1995, 'Low-cost Long-term Monitoring of Global Climate Forcings and Feedbacks', *Clim. Change* **31**, 247–271 (this issue).

Hansen, J., Lacis, A., Ruedy, R., Sato, M., and Wilson, H.: 1993, 'How Sensitive is the World's Climate?', Research and Exploration, Publication of the *National Geographic Society* **9**, 142–158.

Huang, S., Pollock, H. N., and Shen, P. Y.: 1994, 'Advantages and Limitations of Reconstructing a Ground Surface Temperature History from Borehole Temperatures', *Abstracts*, AGU Western Pacific Geophysics Meeting, EOS, 1994.

IPCC: 1990, *Climate Change: The IPCC Scientific Assessment*, Houghton, J. T., Jenkins, G. L., and Ephraums, J. J. (eds.), Cambridge University Press, Cambridge, U.K., 365 pp.

IPCC: 1992, *Climate Change, 1992, Supplementary Report*, WMO/UNEP, Houghton, J. T., Callander, B. A., and Varney, S. K. (eds.), Cambridge University Press, pp. 62–64.

Karl, T. R.: 1993, 'Missing Pieces of the Puzzle', Research and Exploration, Publication of the *National Geographic Society* **9**, 234–249.

Karl, T. R., Jones, P. D., Kukla, G. *et al.*: 1993a, 'A New Perspective on Global Warming: Asymmetric Increases of Day and Night Temperatures', *Bull. Amer. Meteor. Soc.* **74**, 1007–1023.

Karl, T. R., Knight, R. W., Kukla, G., and Gavin, J.: 1995, 'Evidence for Radiative Effects of Anthropogenic Sulfate Aerosols in the Observed Climate Record', Dahlem Konferenzen, *Aerosol Forcing of Climate*, John Wiley & sons, Ltd., pp. 363–382.

Karl, T. R., Quayle, R. G., and Groisman, P. Y.: 1993b, 'Detecting Climate Variations and Change: New Challenges for Observing and Data Management Systems', *J. Clim.* **6**, 1481–1494.

Karl, T. R. and Steurer, P. M.: 1990, 'Increased Cloudiness in the United States During the First Half of the Twentieth Century: Fact or Fiction?', *Geophys. Res. Lett.* **17**, 1925–1928.

Karl, T. R., Williams, C. N. Jr., and Young, P. J.: 1986, 'A Model to Estimate the Time of Observation Bias Associated with Monthly Mean Maximum, Minimum, and Mean Temperatures for the United States', *J. Clim. Appl. Met.* **25**, 145–160.

Keeling, R., Najjar, R., Bender, M. L., and Tans, P. P.: 1993, 'What Atmospheric Oxygen Measurements Can Tell Us about the Global Carbon Cycle', *Global Biogeochem. Cycles* **7**, 37–67.

Keeling, R. R. and Shertz: 1992, 'Seasonal and Interannual Variations in Atmospheric Oxygen and Implications of the Global Carbon Cycle', *Nature* **358**,

Klein, S. A. and Hartmann, D. L.: 1993, 'Spurious Changes in the ISCCP Data', *Geophys. Res. Lett.* **20**, 455–458.

Komhyr, W. D., Grass, R. D., and Leonard, R. K.: 1989, 'Dobson Spectrophotometer 83: A Standard for Total Ozone Measurements, 1962–1987', *J. Geophys. Res.* **94**, 9847–9861.

Legates, D. R. and DeLiberty, T. L.: 1993, 'Measurement Biases in the United States Raingage Network', *Proc. Eighth Symposium on Meteorological Observation and Instrumentation*, Anaheim, California, Amer. Meteor. Soc., J48–J51.

Levitus, S., Antonov, J., Zhou, X., Dooley, H., Selemenov, K., and Tereschenkov, V.: 1994, 'Decadal-Scale Variability of the North Atlantic Ocean', in *Natural Climate Variability on Decade-to-Century Time Scales*, Press National Academy Press.

Levitus, S. and Boyer, T.: 1994, *World Ocean Atlas, 1993*, **5**, *Interannual Variability of the Upper Ocean Structure, NOAA Atlas Series*, (in preparation).

Lindzen, R.: 1993, 'Absence of Scientific Basis', Research and Exploration, Publication of the *National Geographic Society* **9**, 191–200.

McGuffie, K. and Henderson-Sellers, A.: 1993, 'Cloudiness Trends This Century from Surface Observations', Presented at the NOAA/DOE MINIMAX Workshop, Sep 27–30, 1993, College Park Md, and in Review *Atmos. Res.*

Michaels, P.: 1993, 'Benign Greenhouse', Research and Exploration, Publication of the *National Geographic Society* **9**, 222–233.

Mikolajewicz, U., Maier-Reimer, E., and Barnett, T. P.: 1993, 'Acoustic Detection of Greenhouse-Induced Climate Changes in the Presence of Slow Fluctuations of the Thermohaline Circulation', *J. Phys. Oceanogr.* **23**, 1099–1109.

Minnis, P., Harrison, E. F., Stowe, L. L., Gibson, G. G., Denn, F. M., Doelling, D. R., and Smith, W. L. Jr.: 1993, 'Radiative Forcing by the Mount Pinatubo Eruption', *Science* **259**, 1411–1414.

Mohnen, V. A., Goldstein, W., and Wang, W.-C.: 1993, 'Tropospheric Ozone and Climate Change', *J. Air and Waste Managem. Assoc.* **43**, 1332–1344.

Munk, W. H.: 1989, 'Global Warming: Detection by Long-Path Acoustic Travel Times', *Oceanography* **2**, 40–41.

Munk, W. H. and Forbes, A. M. G.: 1989, 'Global Ocean Warming: An Acoustic Measure?', *J. Phys. Oceanogr.* **19**, 1765–1778.

National Research Council: 1993, *Ocean-Atmosphere Observations Supporting Short-Term Climate Predictions*, National Academy Press, Washington D.C., (in press).

National Research Council: 1994, *Solar Influences on Global Change*, National Academy Press, Washington D.C.

Oltmans, S. J. and Hofmann, D. J.: 1995, 'Increase in Lower-Stratospheric Water Vapour over Boulder, Colorado from 1981 to 1994', *Nature* March 9, 1995.

Parker, D. E.: 1993a, 'Glacial Retreat in the Pyrenees', *Weather* **48**, 116–117.

Parker, D. E. and Cox, D. I.: 1994, 'Towards a Consistent Global Climatological Rawinsonde Data-Base', Accepted by *Internat. J. Climatol.*

Penner, J. E., Dickinson, R. E., and O'Neill, C. A.: 1992, 'Effects of Aerosol from Biomass Burning on the Global Radiation Budget', *Science* **256**, 1432–1433.

Penner, J. E., Eddleman, H., and Novakov, T.: 1993, 'Towards the Development of a Global Inventory of Black Carbon Emissions', *Atmos. Environ.*, (in press).

Price, C.: 1993, 'Global Surface Temperatures and the Atmospheric Electrical Circuit', *Geophys. Res. Lett.* **20**, 1363–1366.

Quayle, R. G., Easterling, D. R., Karl, T. R., and Hughes, P. Y.: 1991, 'Effects of Recent Thermometer Changes in the Cooperative Station Network', *Bull. Amer. Meteor. Soc.* **72**, 1718–1723.

Reynolds, R. W.: 1993, 'An Improved Real-Time Global Sea Surface Temperature Analysis', *J. Clim.* **6**, 114–119.

Robinson, D. A., Dewey, K. F., and Heim, R. R. Jr.: 1993, 'Global Snow Cover Monitoring: An Update', *Bull. Amer. Met. Soc.* **74**, 1689–1696.

Rudolf, B.: 1993, 'Management and Analysis of Precipitation Data on a Routine Basis', *Contribution to Symposium on Precipitation and Evaporation*, Bratislava, 8 pp.

Santer, B. D., Taylor, K. E., Wigley, T. M. L., Penner, J. E., Jones, P. D., and Cubasch, V.: 1995, 'Towards the Detection and Attribution of an Anthropogenic Effect on Climate', *Clim. Dyn.*, in press.

Schlesinger, M. E.: 1993, 'Greenouse Policy', Research and Exploration, a Publication of the *National Geographic Scoiety* **9**, 159–172.

Schluessel, P., Emery, W. J., Grassl, H., and Mannen, t.: 1990, 'On the Bulk-Skin Temperature Difference and Its Impact on Satellite Remote Sensing of Sea Surface Temperatures', *J. Geophys. Res.* **95**, 13341–13356.

Schneider, S. H.: 1993, 'Degree of Certainty', Research and Exploration, Publication on the *National Geographic Society* **9**, 173–190.

Sevruk, B.: 1982, 'Methods of Correcting for Systematic Error in Point Precipitation Measurements for Operational Use', *Hydrology Rep.* **21**, WMO 589.

Shukla, J.: 1993, 'On the Initiation and Persistence of the Sahel Drought', *Decadal-to-Century-Scale Climate Variability*, National Academy Press, (in press).

Slack, J. R. and Landwehr, J. C.: 1992, 'Hydro-Climatic Data Network (HCDN): A.U.S. Geological Survey Streamflow Data Set for the United States for the Study of Climate Variations, 1874–1988', OFR 92–129, USGS Water Supply Paper No. 2406.

Spencer, R. W.: 1993, 'Global Oceanic Precipitation from the MSU During 1979–91 and Comparisons to Other Climatologies', *J. Clim.* **6**, 1301–1326.

Spencer, R. W. and Christy, J. R.: 1990, 'Precise Monitoring of Global Temperature Trends from Satellites', *Science* **247**, 1558–1562.

Tans, P. P., Fung, I. Y., and Takahashi, T.: 1990, 'Observational Constraints on the Global Atmospheric CO_2 Budget', *Science* **247**, 1431–1438.

Tans, P. P., Berry, J. A., and Keeling, R. F.: 1993, 'Oceanic $^{13}C/^{12}C$ Observations: A New Window on Ocean CO_2 Uptake', *Global Biogeochem. Cycles* **7**, 353–368.

Taylor, K. E. and Penner, J. E.: 1994, 'Response of the Climate System to Atmospheric Aerosols and Greenhouse Gases', *Nature* **369**, 734–737.

Thiele, O. W.: 1993, 'Tropical Rainfall Measuring Mission (TRMM) and Rainfall Validation', in *Analysis Methods of Precipitation on a Global Scale*, Report of a GEWEX Workshop, Koblenz, Germany, organized by the *Global Precipitation Climatology Centre*, WCRP-81, WMO-TD-No. **558**, ICSU/IOC/WMO, A/56.

Tihay, J. P.: 1990, 'Glaciers: le Lent Recul [Glaciers: The Slow Retreat]', *Pyrénées Magazine* **8**, 24–31, (in French), English translation available in National Meteorological Library, Bracknell, U.K.

Wang, W.-C., Zhuang, Y.-C., and Bojkov, R. D.: 1993, 'Climate Implications of Observed Changes in Ozone Vertical Distributions at Middle and High Latitudes of the Northern Hemisphere', *Geophys. Res. Lett.* **20**, 1567–1570.

Williams, E. R.: 1992, 'The Schumann Resonance: A Global Tropical Thermometer', *Science* **256**, 1184–1187.

Wright, D.: 1993, *4,000,000 XBTs: Can They Be Wrong?*, Report of the 2nd NOAA Data Quality and Continuity Workshop, April, 1993, Rockville, MD.

(Received 28 February, 1994; in revised form 9 May, 1995)

TOWARD A SCIENTIFIC CENTERED CLIMATE MONITORING SYSTEM *

J. D. MAHLMAN

Geopysical Fluid Dynamics Laboratory/NOAA, Princeton, NJ 08542, U.S.A.

1. Introduction

Recognition that earth's climate and biogeophysical conditions are likely changing due to human activities has led to a heightened awareness of the need for improved long-term global monitoring. The present long-term measurement efforts tend to be spotty in space, inadequately calibrated in time, and internally inconsistent with respect to other instruments and measured quantities. In some cases, such as most of the biosphere, most chemicals, and much of the ocean, even a minimal monitoring program is not available. Monitoring is defined here as long-term measurements of key climate variables with continuous coverage in time, maintenance of required instrument calibration accuracy, and empowered by careful diagnostic analysis of the data.

Recently, it has become evident that emerging global change issues demand information and insights that the present global monitoring system simply cannot supply. This is because a monitoring system must provide much more than a statement of change at a given level of statistical confidence. It must describe changes in diverse parts of the entire earth system on regional to global scales. It must be able to provide enough information to allow an integrated physical characterization of the changes that have occurred. Most importantly, it must allow a separation of the observed changes into their natural and anthropogenic parts. This separation can only be accomplished by combining theoretical and modeling insights with reality, as obtained from the monitoring data. The enormous policy significance of global change virtually guarantees an unprecedented level of scrutiny of the changes in the earth system and why they are happening.

These pressures create a number of emerging challenges and opportunities. For example, they will require a growing partnership between the observational programs and the theory/modeling community. Without this partnership, the scientific community will likely fall short in the monitoring effort.

The monitoring challenge before us is not to solve the problem now, but rather to set appropriate actions in motion so as to create the required framework for solution. Each individual instrument system needs to establish its role in the large problem and how the required interactions are to take place. Below, we emphasize

* The U.S. Government right to retain a non-exclusive royalty-free license in and to any copyright is acknowledged.

some of the needs and opportunities that could and should be addressed through participation by theoreticians and modelers in the global change monitoring effort.

2. Requirements for Theory/Modeling Support for Monitoring

CONTEXT

All observing systems are incomplete in the sense that they will never be able to measure everything, everywhere, all of the time with perfect accuracy and sustained calibrations. Moreover, even if this impossible goal could be achieved, the changes recorded by the 'perfect' measurements would still need to be interpreted in the context of previous predictions and to be explained scientifically. Thus, the challenge before us is to seek the mechanisms by which models and theoretical analysis can be used in cooperation with observational systems to yield the maximum information and to produce the required synthesis.

INFORMATION CONTENT OF OBSERVATIONAL NETWORKS

One of the most straightforward ways to utilize models in a monitoring context is in the evaluation of existing or hypothetical networks. For the atmosphere, successful studies conducted at GFDL have included evaluations of the global radiosonde network (Oort, 1978), the Dobson total ozone network (Moxim and Mahlman, 1980), and satellite temperature soundings (Graves, 1986). In such approaches, time-depenoent, three-dimensional model output statistics are sampled in ways identical or similar to that of a given network. The advantage of using the model is that the 'right' answers in this context are readily available for comparison against the answers inferred from the network subsample. Such research has revealed a number of significant deficiencies in the existing networks, as well as in the models used in the evaluations.

A frequent objection to using models for research in this context is that the models are seriously incomplete depictions of reality. True enough. However, models do have the virtue of constituting a self-consistent global dataset. Moreover, models produce only a restricted version of the much richer spatial and temporal structure found in nature. Thus, model diagnoses of network information tend to err on the conservative side; problems identified in networks through use of models are likely to be even worse in the real world.

EVALUATION OF MODELS FROM SPARSE OBSERVATIONAL DATA

The other side of the coin is that even the current monitoring networks can be very powerful tools for evaluating strengths and weaknesses of models. Surprisingly, this is still true even for seriously undersampled quantities such as tropospheric ozone or oceanic salinity. It is a common misconception that 3-D global models can only be tested through use of complete 3-D global datasets. Just the opposite is true. Even individual local time series can (and often do) demonstrate that a global model is deficient in certain respects. This is because a global 3-D model attempts to capture both regional and global structures. If a global model exhibits some local temporal variations quite unlike the real world, the model has already been determined to be deficient. Thus, observed data properly gathered at local sites can provide a powerful tool for model evaluation. In turn, improved models can provide a means for filling in the inevitable gaps in monitoring systems. We shall return to this theme below.

DESIGN OF OBSERVATIONAL NETWORKS

A particularly attractive possibility is to use models to design more effective networks at the outset. This concept is almost irresistible because of the prodigious expense of constructing and maintaining dense sampling networks. In principle, models can provide perspective and predictions on the value of data at various accuracies and sampling densities. In practice, this approach will be somewhat limited by the accuracy and credibility of the model employed. Models themselves undersample the environment because their data densities are also limited by costs, in this case computational. Moreover, model-based network designs must also include separate information about data errors, calibration drifts, and likelihood of data losses.

It is becoming increasingly common to hear that a new proposed monitoring network can be designed in advance using model-based insights. In principle, this is true; in practice, serious barriers remain. The most serious barrier seems to be the lack of properly focussed human talent. Each potential network design problem represents a major research problem that typically requires several years of concentrated research to provide targeted, useful answers. Currently, there is a major deficiency of properly trained and focussed researchers, backed by serious commitment, both personal and institutional, to solve such problems. The design of observational networks has the potential to become a significant new priority area in the context of global change monitoring and assessment.

MODEL IDENTIFICATION OF GLOBAL CHANGE 'FINGERPRINTS'

Questions what the monitoring networks are capable of measuring are strongly influenced by the presence of an evolving theoretical/modeling perspective on what the expected changes should look like. Unfortunately, the issue is clouded by the presence of significant uncertainty in the model predictions. Even though they are uncertain, the model predictions still can provide major guidance to the kinds of signals that a network needs to be able to detect in order to test those predictions.

As examples, can the network detect a global warming signal in the ocean? Change in cloud-radiation feedbacks? How about CO_2 uptake changes? How will

Surface Air Temperature (N.H.)

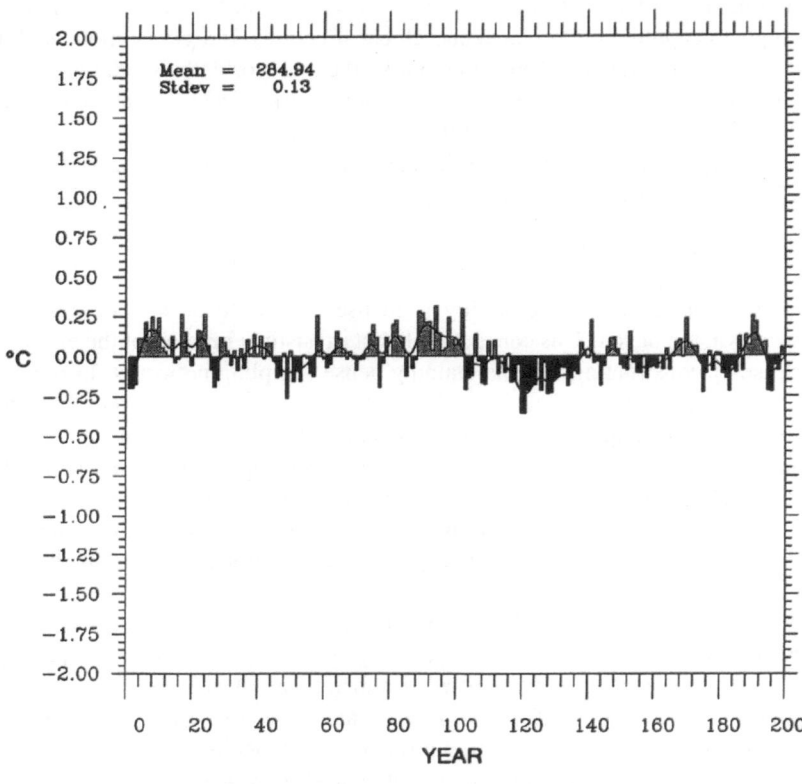

Fig. 1a.

Figs. 1(a)–(c). Annual-average surface air temperature (°C) from a 200-year integration of low resolution (≈ 500 km grid spacing) coupled ocean-atmosphere GDFL climate model. This is a model in statistical equilibrium in which no trends in climate forcing are applied. (Courtesy S. Manabe and T. Delworth). Part (a) is for the Northern Hemisphere; (b) is for the contiguous U.S.; (c) is for the grid box encompassing Washington, DC.

the warming signal differ from the expected low frequency variability operating on time scales similar to the expected anthropogenic climate signal? Can the signals be separated and understood independently?

An instructive example of the role of modeling in interpreting climate change can be seen in Figure 1, taken from a 200 year integration of the low-resolution coupled ocean-atmosphere GFDL climate model (Manabe and Stouffer, 1993). This is an integration which is in near-perfect long-term statistical equilibrium and which incorporates no trends in climate forcing. Figure 1a for the Northern

Surface Air Temperature (U.S.)

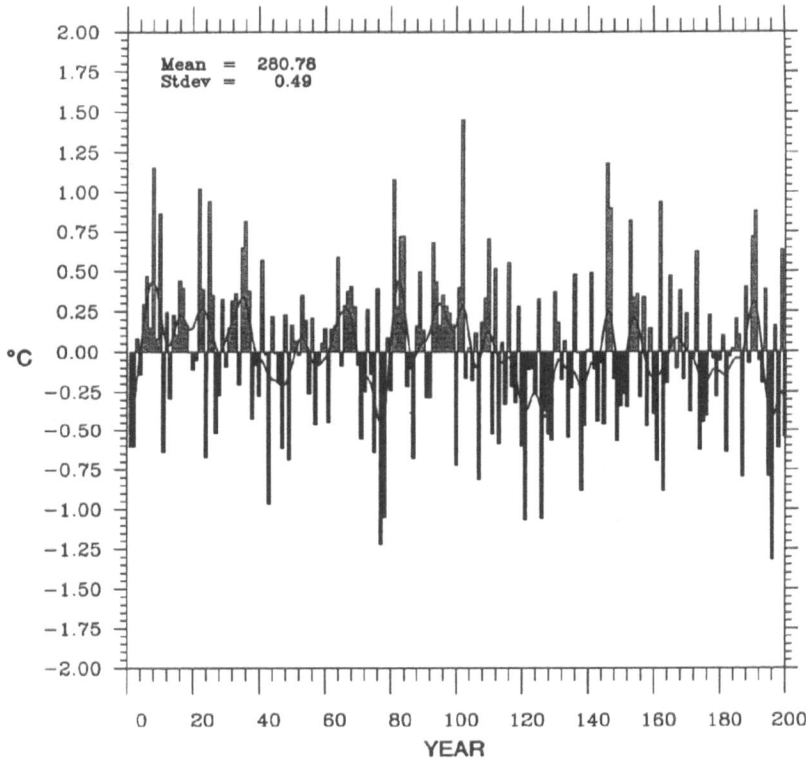

Fig. 1b.

Hemisphere annual-mean surface air temperature shows modeled trough-to-peak swings of nearly 0. °5C over time intervals to 40–60 years. These changes are of comparable magnitude to the observed changes in this century. Natural variability can thus appear to either amplify or damp anthropogenically induced climate warming signals. Figures 1b and 1c show the same quantities but for the contiguous U.S. and for the 'Washington, DC' grid box (roughly 500 km on a side). These model results show how the natural variability increases dramatically as the region size decreases. A well planned monitoring system must take such variability under careful consideration, particularly on time scales longer than a decade or so.

There are many important questions that we cannot answer about climate change at this time. However, it is a safe prediction that we will have to deal with them through the signals from the global monitoring system. At the minimum we must design our systems so that we at least deal with the difficult interpretative questions

Surface Air Temperature (Washington, D.C.)

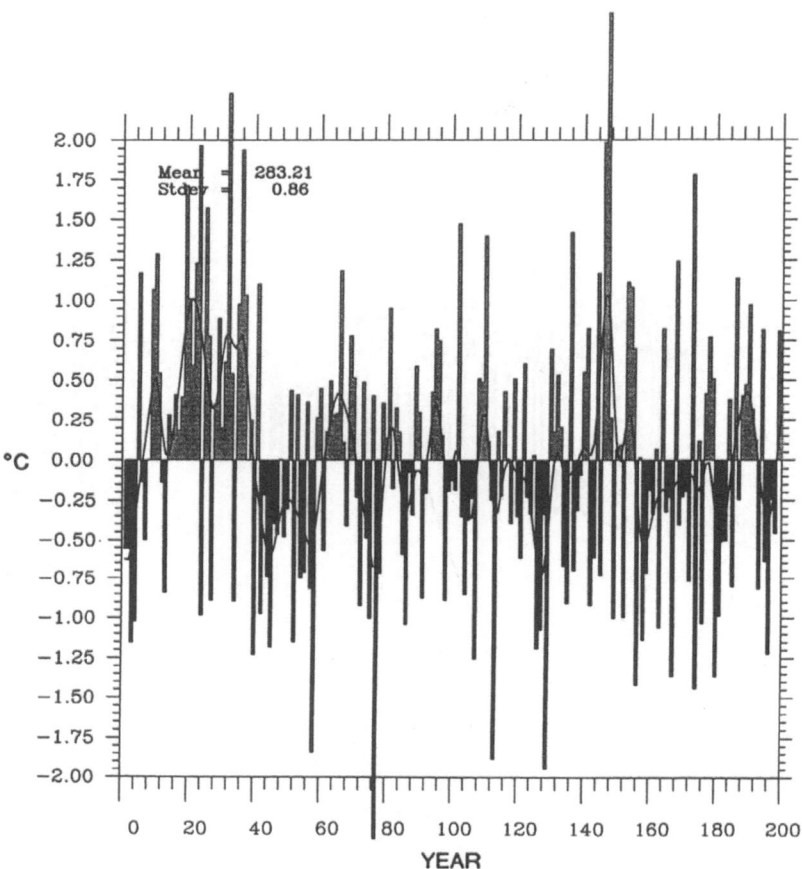

Fig. 1c.

that are already before us. We must take on the natural variability question head on as a concomitant part of global change. We also must address the global sampling and long-term calibration question with sufficient skill to address adequately the proper monitoring identification of the regional climate change signals that are already predicted. In many cases, the currently predicted regional changes, if correct, may have major societal impact (WMO/UNEP, 1990).

3. Model Assimilation of Data in the Context of Climate Change

One of the inevitable aspects of an expanded global monitoring system is that it will be composed of data from diverse sources. The data will be heterogenous in terms of types of instruments, the nature of the data obtained, the variables being measured, and their error characteristics. The sampling will frequently be spotty in space and sporadic in time. The systems will be dynamically incomplete; temperature may be globally available, but winds and tropospheric ozone amounts likely will not be. Much of the data will be in the form of extended time series that contain gaps, errors, and calibration problems.

All of these data inconsistencies create the need for a unified approach for combining and synthesizing the data. Fortunately, over the past decade or so, viable approaches for accomplishing this have been developed for both the atmosphere and the ocean. Particularly powerful is the so-called four-dimensional data assimilation approach (4DDA).

The 4DDA approach uses comprehensive numerical models to provide a phys-ically consistent synthesis and global analysis. In effect, 4DDA uses a general circulation model to accept input data in a dynamically consistent manner. The model serves a 'traffic cop' determining which data in which forms are acceptable for inclusion. The data are incorporated into the model in such a way as to yield the most self-consistent analysis of the data. In this context, the model serves as a nonlinear interpolator to fill in missing spatial and temporal information as well as missing variables (such as winds, ocean salinity, or trace constituents). For a more in-depth discussion of 4DDA, see Trenberth (this issue).

A great strength of this approach is the production of a self-consistent final analysis. A great weakness is that the quality of the analysis can be quite sensitive to the quality of the model used. This is a particular concern for regions where the data coverage is extremely coarse and model quality remains relatively low. However, the insightful use of 4DDA techniques hold great promise to help improve the models as well as the data analyses.

In the monitoring context, perhaps the most promising use of 4DDA is in the retrospective analysis of historical datasets, such as is now in preparation at NOAA's National Meteorological Center. This approach may eventually be able to yield analyses over decades that are appropriately time-calibrated for monitoring use and evaluation. An unsolved problem with this approach is the limited ability of today's data checking procedures to filter out small apparent 'trends' due to calibration drift or instrument changes. For a specific analysis, this is a small effect; for climate change analysis, it can be as large as the signal itself. However, the advantage of the reanalysis procedure is that it can be redone as many times as necessary to glean the maximum information from the dataset. A major hurdle in reanalysis (and rereanalysis) is that it is computationally burdensome and labor intensive. Obviously, there will be tradeoffs between the quality of the analyses and resources available, just as in the monitoring networks themselves.

4. Final Comments

It is clear that success in the monitoring problem will require a growing partnership between theory/modeling researchers and the observational data analysts. It is equally clear that the task will be extraordinarily difficult. It will take a long time, perhaps decades, and will require a new generation of scientific talent, institutional resolve, and financial resources. At the very minimum, we must begin today to improve the monitoring viability of our current climate measuring systems.

Finally, some will argue that the monitoring problem is too difficult and too unglamorous to command the sustained resources and commitment required. When such arguments are advanced, it will be important to remember the challenge facing us all:

> We are faced with nothing less that the need to identify how the earth system is changing over the next century, explain why the changes are occurring, separate natural from anthropogenic change, and learn if our predictions were correct or incorrect.

If we in the scientific community cannot step up to this challenge, it is a safe prediction that all of us will be held accountable.

Acknowledgements

The author is indebted to the small band of data diagnosticians who have worked tirelessly to extract useful information from various climate measurement systems. Their work has shown us the enormous potential value of an improved climate monitoring system. Much of this essay has been extracted from Mahlman (1993), part of a key NASA workshop on improved climate monitoring.

References

Graves, D. S.: 1986, *Evaluation of Satellite Sampling of the Middel Atmosphere Using the GFDL 'SKY-HI' general Circulation Model*, Princeton University, Program in Geophysical Fluid Dynamics, Ph.D. Dissertation, 314 pp.

Mahlman, J. D.: 1993, 'Monitoring Issues from a Modeling Perspective', in *Long-Term Monitoring of Global Climate Forcings and Feedbacks*, NASA Conference Publication 3234, pp. 1–5.

Manabe, S. and Stouffer, R. J.: 1993, 'Century-Scale Effects of Increased Atmospheric CO_2 in the Ocean-Atmosphere System', *Nature* **364(6434)**, 215–218.

Moxim, W. J. and Mahlman, J. D.: 1980, 'Evaluation of Various Total Ozone Sampling Networks Using the GFDL Tracer Model', *J. Geophys. Res.* **85**, 4527–4539.

Oort, A. H.: 1978, 'Adequacy of the Rawinsonde Network for Global Circulation Studies Tested Through Numerical Model Output', *Mon. Wea. Rev.* **106**, 174–195.

Trenberth, K. E.: 1995, 'Atmospheric Circulation Climate Changes', *Clim. Change* **31**, 264.

WMO/UNEP: 1990, *Climate Change: The IPCC Scientific Assessment Intergovernmental Panel on Climate Change*, Cambridge University Press (Library of Congress ISDN 0 521 40720 6), 365 pp.

(Received 23 January, 1995; in revised form 11 July, 1995)

LONG-TERM CLIMATE MONITORING AND EXTREME EVENTS

NEVILLE NICHOLLS

BMRC, PO Box 1289K, Melbourne Vic 3001, Australia

Abstract. Problems with long-term monitoring of various extreme meteorological events (including tropical and extratropical cyclones, extreme winds, temperatures and precipitation, and mesoscale events) are examined. For many types of extreme events, the maintenance of long-term homogeneity of observations is more difficult than is the case for means of variables. In some cases, however, a strategy of using more than a single variable to define an event, along with the careful elimination of biases in the data, can provide quantitative information about trends. Special care needs to be taken with extreme events deduced from meteorological analyses, because changes in analysis and observation systems are certain to have affected extremes. Also, compositing of observations from more than one station, using differences in means (of temperature for instance) to produce a single long-term site, may not remove the biases in the extremes. These problems, along with ambiguities in defining extreme events, and difficulties in combining different analyses from different sites, complicate (and perhaps invalidate) attempts to determine whether extreme weather is becoming more frequent. The best that is likely to be achieved, even with increased emphasis on attaining the high-level of homogeneity necessary in the observations, is to monitor long-term variations in certain important extreme events, in select locations with high-quality data. Regional indices of important extreme events, selected on the basis of their damage potential and capable of adequate monitoring, may be established. If, in the future, we are to answer the question "Are extreme weather events becoming more frequent?", we must establish and protect high-quality stations capable of monitoring the most important extreme events (perhaps with such regional indices), and ensure that changes affecting the recording of extreme events (e.g., changes in exposure) are meticulously documented.

Introduction

Many of the deleterious impacts of a global climate change may result from changes in frequency or intensity of extreme weather events such as tropical cyclones, wind storms, heavy rainfall, and extreme temperatures, rather than from changes in mean values of atmospheric variables such as temperature. Already, these extreme events cause considerable damage and loss of life. Numerical model simulations of the climatic effects of an enhanced greenhouse effect suggest that the frequency and intensity of some extreme events may change. It is important, therefore, that there exist effective methods for the monitoring of changes in extreme events. Only then can we answer the frequently-posed question "Are extreme weather events becoming more frequent or severe?".

Karl *et al.* (1995a) note the difficulties involved in monitoring most aspects of the climate system. Monitoring extreme events is even more difficult than monitoring changes in climate means, in many cases. Serious problems with past and present monitoring systems prevent or complicate the credible estimation of changes or variations of many types of extreme events. In this paper the problems

Climatic Change **31**: 231–245, 1995.

of monitoring extreme events are discussed, for an illustrative range of extreme events. The problems are not uniform across all types of extreme events. Nor are the potential solutions. For some types of events the difficulties and the potential solutions are rather simple, or are similar to the solutions necessary to ensure adequate monitoring of changes in means of variables. For others, inadequacies in the monitoring system appear to be more severe, and less amenable to ready solution. Some possible methods for overcoming the potential problems, and for long-term continuous monitoring of extreme events, are considered.

For the purposes of this paper an extreme weather event is defined as a short-lived meteorological phenomenon capable of causing loss of life or substantial damage. 'Short-lived' is assumed to include time-scales up to a few days, thereby encompassing tropical cyclones and mid-latitude wind storms. The definition excludes very long events which could still result in severe economic or social damage, such as droughts.

The various extreme meteorological events can be categorised in a variety of ways. In this paper the following types of extreme events are considered separately:
 − Tropical cyclones;
 − Extratropical cyclones;
 − Strong winds;
 − Heavy precipitation;
 − Extreme temperatures;
 − Mesoscale disturbances (eg., tornadoes, thunderstorms);
These categories involve some overlap: tropical cyclones and tornadoes are usually associated with extreme winds and sometimes with heavy rains. The methods of monitoring the various types noted above are, however, different, and have their own range of problems. Monitoring the number of tropical cyclones as well as the number of extreme wind events, therefore, can provide extra useful information, despite their overlap. The types of extreme events selected for consideration here include the best-documented, and the most damaging. The discussion here of these 'representative' extreme events also indicates the type and seriousness of the difficulties involved in monitoring almost all types of extreme weather events.

Tropical Cyclones

Intense Atlantic hurricane activity over the period 1970–87 was less than half that in the period 1947–69 (Gray *et al.*, 1992). A similar quiet period had occurred in the western North Pacific, suggestive of a decrease in the number of very intense tropical cyclones. Bouchard (1990) and Black (1992), however, demonstrated that this apparent change in intensity in the western North Pacific was an artefact, due to a change around 1970 in the method used to derive wind estimates from pressure estimates. Chen (1990) independently determined that the historical typhoon data sets were biased and required correction prior to 1970, if they were to be compred with recent behaviour. Black found that when a consistent method for determining wind estimates was used throughout the period of record, then the pre-1970 data was statistically indistinguishable from the post-1970 data. Bouchard used estimates of surface pressure to determine the number of very intense typhoons (defined as a cyclone with a central pressure below 910 mb), and found no evidence of any strong trend in the number of very intense typhoons, although the 1980s had, on average, more very intense typhoons than earlier decades. Landsea (1993) suggested that Atlantic hurricane intensity record was probably also biased. Winds were five kt higher before 1970, compared with hurricanes with the same minimum pressure after 1970. This suggests that the apparent drop in Atlantic hurricane activity may, at least partially, also be an artefact, as in the northwest Pacific. Landsea found, however, that after adjusting for this bias a substantial downward trend in intense hurricane activity is still apparent. There remains a possibility that not all the bias in the Atlantic records has been removed, and that this drop in activity is artificial. However, a similar drop in activity around 1970 was also observed in the frequency of storms hitting the U.S.A. These storms were categorised by using minimum sea level pressure recorded at landfall. Such observations should not be as suspect as observations over the ocean. Also, the decrease in hurricanes appears to reflect a relationship between hurricane activity and Sahel rainfall (Landsea and Gray, 1992), which has also been low since about 1970. Finally, hurricane activity is also weaker during El Niño episodes. The tendency for the El Niño-Southern Oscillation to remain more frequently in the El Niño mode since the mid-1970s (Trenberth, 1995) would, therefore, have led to a tendency for weak hurricane activity. There are strong grounds, therefore, for concluding that much of the decrease in intense Atlantic hurricane activity is real (Landsea *et al.*, 1995). However, it is worth pointing out that our confidence in this conclusion rests partly on empirical relationships between hurricane activity and other meteorological variables.

Apparent changes in cyclone frequency elsewhere may be artefacts. The Southern Oscillation Index (SOI) is a good predictor of tropical cyclone numbers in the Australian region. Nicholls (1992) demonstrated that the relationship between the SOI and cyclone numbers apparently changed around 1986. After this year the numbers of tropical cyclones appeared to be consistently lower than would be expected from the relationship with the SOI derived only with data beforethis date, although the interannual variations were still closely related. The sudden change in this cyclone-SOI relationship is suspicious, and may be due to changes in the analysis of cyclones. Figure 1 shows a time series of the number of named tropical cyclones in the Australian region (105° E–165° E) between 1949 and 1994. The influence of the El Niño–Southern Oscillation, which affects interannual variations in cyclone numbers, has been removed from these data, to facilitate identification of trends. The steady upward trend in cyclone numbers from 1949 to the mid 1980s seems likely to be at least partly artificial, reflecting improvements in observing systems (e.g., the progressive introduction of improved meteorological satellites over this period). On the other hand, tropical cyclone activity around Australia is also related to sea surface temperatures (Nicholls, 1984), and these increased over this period (Figure 2). So perhaps part of the apparent upward trend in cyclone activity is real. The 'spike' in 1962 is the result of some weak storms being identified as tropical cyclones. In previous and later years such systems would not have been named as cyclones. The sudden drop in cyclone numbers in 1986 is probably at least partly artificial, reflecting a change in policy of which tropical storms will be named as cyclones. However, there is some evidence of a real drop in activity, again from other data. Tropical cyclones often bring good rains to northern Australia, and summer rainfall was low through the mid to late-1980s. This may reflect a decrease in cyclone frequency, supporting the apparent decrease seen in Figure 1. The gradual increase in cyclone numbers in the most recent years may also be real, based on higher rainfalls after the late-1980s. This example demonstrates the difficulties in even establishing numbers of tropical cyclones in recent years, let alone examining changes over decades. This applies even to a good observing network, on the Australian coast and surrounding islands. Obstacles in establishing whether observed trends in cyclone numbers in, for instance, the south Indian Ocean and the south central Pacific, are even more severe. However, the use of proxy variables, such as rainfall and station pressure, may provide evidence in support of apparent changes in cyclone activity determined from analyses.

As was the case with the monitoring of the numbers of cyclones, there are grounds for believing that the changes in observing and analysis techniques over the past few decades may have biased the trends in cyclone intensity. Figure 3 shows the maximum sustained wind speed attained each year in Atlantic hurricanes (C. W. Landsea, pers. comm., 1994). There is no obvious trend. After 1970, however, the spread of maximum wind speeds increased: in some years the maximum wind speed was relatively weak, while in others very high maximum wind speeds (higher than had been observed prior to 1970) were reached. The suddenness of the change around 1970 suggests that it might be the result of changes in the observing techniques, specifically the introduction of satellite monitoring. A similar change in spread is not apparent in the *mean* maximum sustained wind speed attained each year in Atlantic hurricanes. This suggests that the changes in the observing systems may be causing problems for an occasional very weak or very strong storm.

Observing and analysis techniques will continue to change in the future. A fresh approach to monitoring cyclone numbers and intensity is therefore necessary, if we are to be able to credibly claim that we have a clear picture of variability and change. One possible approach is to parameterise tropical cyclone numbers in terms of a few large-scale variables, i.e. extending the approach of Gray and his colleagues. If such a parameterisation was successful, the changes in the large scale variables could then be used to deduce the changes in cyclone numbers, or to verify such changes. Changes in these variables could then be used to deduce changes in tropical cyclone activity.

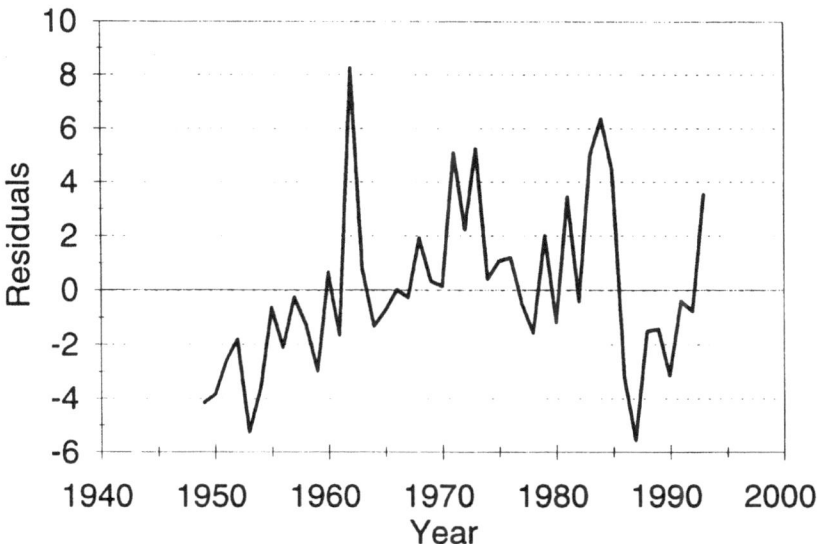

Fig. 1. Time series of the number of named tropical cyclones in the Australian region (105° E–165° E) between 1949 and 1994. The influence of the El Niño–Southern Oscillation, which affects interannual variations in cyclone numbers, has been removed from these data by linear regression, to facilitate identification of trends.

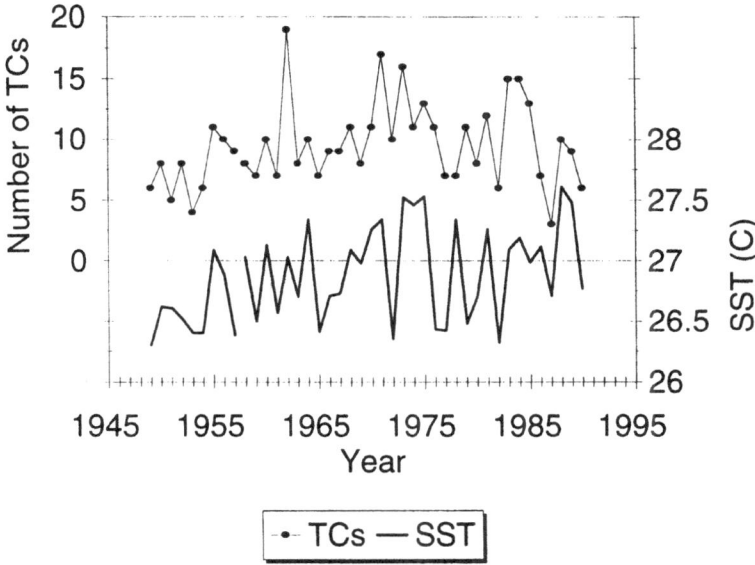

Fig. 2. Time series of the number of named tropical cyclones in the Australian region (105° E–165° E) between 1949 and 1991, and September sea surface temperatures in the box 120–160° E, 5–15° S.

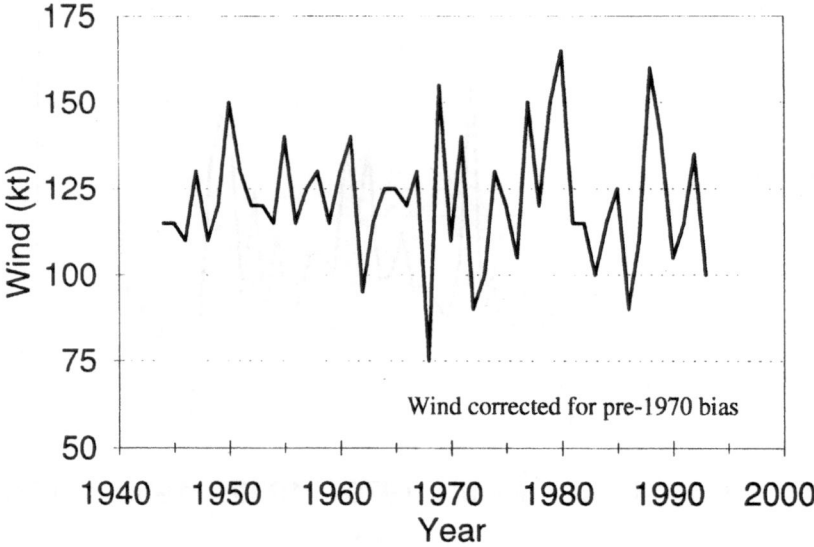

Fig. 3. The maximum sustained wind speed attained each year in Atlantic hurricanes (C. W. Landsea, per. comm.).

A different approach is to rely on deductions about numbers or intensity of cyclones from stable, land-based observations. As noted earlier, the sharp decrease in North Atlantic hurricane activity after 1970 was also observed in the frequency of storms hitting the U.S.A. Such landfall observations should not be as suspect as observations over the ocean, and they therefore provide partial confirmation of the reality of the changes discussed above. Another approach (suggested by F. Woodcock) is to define systems as tropical cyclones if a satellite image indicates a clear 'eye', and to base long-term monitoring of numbers of cyclones on this simple criterion.

Extratropical Cyclones

Cyclonic depressions in higher latitudes rarely cause as much damage as a severe tropical cyclone. However, they can still be extremely destructive. The most obvious method for monitoring extratropical cyclone occurrence is to use historical weather maps. Schinke (1993) used once-daily analyses (U.S.A. analyses (1939–1964; thereafter German analyses) to count the number of storms below certain thresholds. Agee (1991) combined data from three previous studies of cyclone and anticyclone frequency around North America to examine trends.

Schmidt and von Storch (1993) note, however, that the use of operational analyses for the studies cited above may be complicated by the possibility of improvements in the analyses resulting from more and better observations and other improvements in monitoring the state of the troposphere. Trenberth (1995) notes some of the changes in analysis systems in recent decades. The use of analysis systems over longer periods is fraught with dangers, especially for deducing changes in frequency or intensity of extreme events such as extratropical depressions. More detailed mapping, for instance, would tend to result in cyclones with lower central pressures. Such a change might arise because of improved observational systems, or because of improvements in the analysis systems. Because of the doubts caused by changes in observing/analysis techniques, other data are needed to confirm any deduced changes. Changnon and Changnon (1992) used insurance data and found that temporal variations in winter storm disasters closely matched the trends in cyclone numbers for North America (Agee, 1991). Insurance data, however, are themselves inhomogeneous, because of changes in insurance coverage, as well as other factors reflecting social rather than weather changes.

As with tropical cyclones, parameterisation of cyclone activity in terms of variations of other variables, with stable observing characteristics, might provide more credible estimates of changes and variations. Hopkins and Holland (1995) determined the numbers of cyclones affecting the Pacific coast of Australia with an objective method of specification based on a consistent set of observing stations spread along the coast. They found a clear upward trend in the numbers of cyclones between 1958 and 1992. Such a trend, if deduced from analyses, would be suspect; the use of the consistent synoptic observations, however, lends credibility to the trend.

Reanalysis projects provide some hope for improved determination of trends in extratropical circulation systems. Modern meteorological computer analysis techniques produce globally-complete data-sets. The techniques used in such analysis have, however, changed with time, thereby introducing a further set of biases. Several groups plan to create historical grid-point data-bases by reanalysis of recent historical data using fixed numerical models and analysis techniques (cf., Kalnay and Jenne, 1991). Trenberth (1995) describes these reanalysis projects in some detail. The reanalysis projects will remove the biases associated with the analysis techniques, but will inevitably still contain time-varying biases because some temporally heterogeneous data, for example from radiosondes, will be supplied to the models. Techniques for producing globally-complete fields have not so far treated biases explicitly (as opposed to random errors), so these must be removed first, if the reanalysis projects are to produce data free of time-varying biases. But even with these time-varying instrumental problems, the reanalyses should improve confidence in our ability to determine trends or variations in extreme events, such as extratropical depressions. Doubts about the homogeneity of the analyses due to possible changes in at least some observing system could be addressed by examining how well the analyses can reproduce those data known to be homogenous over time, e.g., global lower tropospheric temperatures derived from microwave sensor unit (MSU) observations since 1979.

Strong Winds

Wind records are amongst the worst in terms of consistency over long periods. Fairly small changes in exposure would significantly affect wind speeds. As well, during times of strong winds, weather equipment malfunctions are more likely, but vary with the type and exposure of the equipment. So, progressive strengthening of equipment might lead to artificial biases in strong winds. Schmidt and von Storch (1993) suggest that local studies with homogenous air pressure data may provide more definite answers, at least for specific regions, because of the uncertainties with direct observations of winds. They used daily air pressure observations at three stations in the southeast North Sea to calculate the annual distributions of daily geostrophic wind speeds and concluded that the frequency of extreme storms in this area has not changed in the past 100 years. There are numerous areas of the world where long-term surface pressure observations could be used in this way, to assess the possibility of changes in extreme wind speeds.

Again, other variables (which might themselves be more robust than wind speeds) might provide confirmation of changes, or lack of changes, in wind speeds. Von Storch et al. (1993) also examined high water levels at Hoek van Holland, after removing the effects of tides and sea level changes. The resultant time series should reveal storm-related surge heights. No trend was found in the frequency of extreme surge heights. Of course, high water levels may also be affected by inhomogeneities, thereby invalidating their use to monitor trends in extreme wind speeds. Comparison of extreme wind records and nearby storm-surge records goes, however, provide an avenue for checking the consistency of both forms of data.

Heavy Precipitation

Karl et al. (1995c) analysed the trends in the percentage of total seasonal and annual precipitation occurring in heavy daily rainfall events (days with rainfall exceeding 50.8 mm/day) over the U.S.A., the former Soviet Union, and China. A significant trend to increased percentages of rainfall falling in heavy events was evident in the U.S.A., largely due to a strong increase in extreme rainfall events in summer. However, the accumulation of rainfall totals over more than one day can complicate such studies. For instance, in some countries rainfall reports at some stations are not taken on weekends. As a result the Monday report is an accumulation of rainfall since Friday. In some countries this tendency has occurred more frequently in recent years than in the past, at some stations. This will bias the estimation of trends in extreme rainfalls in a complicated fashion. Careful checks of documentation regarding station observing/reporting schedules, conducted as part of a general check on station quality and consistency (eg., Lavery et al., 1992) should reveal whether such accumulations were likely to bias the results of any apparent trend in rainfall occurrence. It may be possible to model the rainfall distribution statistically, and use this model to calculate the likely influence of the increased prevalence of accumulation on extreme event frequency.

Karl and Knight (1995) used a different approach to check whether the accumulations were biasing their results. They repeated their calculations with 3-day rainfalls to check the possibility that the 1-day rainfall results simply reflected a trend to increased numbers of 'accumulated' totals, i.e., totals reflecting more than a single day of rainfall. A similar trend was apparent in heavy rainfall events calculated from 3-day total rainfalls, to that from the 1-day totals. They concluded, on this basis, that accumulations were not substantially affecting their results. Note that this accumulation problem is unlikely to affect determination of trends in mean rainfall, except insofar as evaporation from the gauge over several days reduces observed precipitation.

Changes in instrumentation may also bias the results. IPCC (1990, 1992) note that precipitation is generally underestimated by conventional measuring devices, typically by 10 to 15%, and that progressive improvements in instrumentation have introduced artificial, systematic increases in estimates of precipitation, especially in areas where the proportion of solid to liquid precipitation is relatively high. Karl *et al.* (1995a) discuss the effects of changes in instrumentation, and instrument bias, in precipitation measurements. Very important for the measurement of heavy rainfalls would be biases due to changes in wind protection. Strong winds will often accompany heavy rainfall and snowfall. Any changes to minimise losses due to wind-induced turbulence over the orifice of a raingauge would be likely to result in artificial trends in extreme precipitation. So, in some areas, determination of trends in heavy rainfalls might be more difficult than is the case for total rainfall.

Some changes in instrumentation might affect the apparent frequency of heavy rains, relative to days with light falls. For instance, Australian rain gauges were changed in 1974 to metric units. There are grounds for concern (Nicholls and Kariko, 1993) that this change may have led to an increase in the numbers of days with very light rainfall. So, even though this change would not have led to an artificial change in the absolute numbers of heavy rainfalls (or of the total rainfall recorded in some extreme events), it may have led to an artificial change in the relative frequency of heavy events compared to light rainfall days. Nicholls and Kariko calculated an average intensity of rainfall events, by simply dividing the total rainfall by the number of rain days. If the number of light rainfall days had increased artificially this might have led to an artificial decrease in average intensity, which might be assessed, incorrectly as evidence of decreasing relative frequency of heavy rainfall.

Some supporting evidence regarding heavy rainfalls may be obtained from flood heights of extreme streamflow records. Care would need to be taken, of course, to ensure that such records are not affected by changes in rivers, dams, or vegetation and building changes in catchments.

Extreme Temperatures

Stone *et al.* (1995) examined daily temperature series for several stations in inland eastern Australia and found a significant decrease in the numbers of days with minimum temperatures below 0 °C, and the date of last frost, over the 20th century. They also found similar decreases in frequency for other low temperature thresholds. Similar changes in frost frequency have been noted in several other studies.

There are concerns, however, with such studies. Increased urbanisation may result in increased minimum temperatures, including a reduction in the frequency of very low minimums. Other effects, e.g., increased spray irrigation, or increased vegetation round observing sites, might also reduce the number of frost observed. These effects might be even more prevalent in rural areas than urban areas (despite the widespread concern about the effects of urbanisation). For instance, many rural observations in Australia are from farms, where the thermometer screen is sited nearby the farm buildings. Even if the site is well separated from buildings, this area often features more large trees and more complete vegetation cover than the surroudning areas. The trees will have grown over long periods and their influence on local winds may have influenced the temperature recordings. So, the increasing height and cover in the area around the screen might be expected to bias the extreme temperatures. Note that this may be a problem even if correct standards are applied for siting of screens, especially for extreme temperatures can be accompanied by strong winds. It may be that long-term phenomenological reports, e.g. visual reports of heavy frosts on fields, may provide at least confirmatory evidence of changes in extreme events. Such reports, in some cases, may be more reliable than temperature recordings from screens, as they are more likely to represent conditions some distance from the screen, and therefore may be less-affected by such factors as the growth of trees near the instrument enclosure.

A common problem with long-term station records is that a high-quality station may have closed at some time, and have been replaced by another station nearby. Overlapping periods can be used to compare the two stations, and to produce a composite station, with the data at one station adjusted to take into account the differences between the sites. This is commonly done with monthly or annual mean data, and can produce good composite sites for estimation of trends in mean temperatures. However, inter-station differences in extreme temperatures can be very different to the differences in the means and this may confound the use of composite sites in estimating trends in extreme temperatures (Trewin and Trevitt, 1995). Since even a change in exposure, or a slight shift in location, at a specific station may result in a need for adjustments, this problem will affect many, if not most, sites used for long-term monitoring of temperatures. For example, sheltering by a tree might affect extreme maximum temperatures more than mean temperatures. So, correcting the 'sheltered' part of the record by adjusting mean temperatures may not result in adequate corrections of the extreme temperatures.

Trewin and Trevitt (1995) have proposed an approach to compositing stations likely to overcome such problems.

Changing the exposure of the thermometers (e.g., from the Glaisher stands common in the 19th century to Sevenson screens) introduces different biases to the mean, maximum, and minimum temperatures (Parker, 1994). Such changes could have affected the extremes differently to the mean temperatures. For instance, Laing (1977) found differences of around 2 °C between Stevenson and Glaisher maxima on some very warm days, even though the difference in the annual mean temperatures was statistically insignificant (Parker, 1994). Laing's results suggest that correction of historical records to ensure homogeneity of extreme temperatures, despite changes in screens, may be more difficult than for mean temperatures. Again, as long as overlapping records from the various exposures are available, the data can be combined as suggested by Trewin and Trevitt (1995).

Mesoscale Disturbances

The final type of extreme weather events considered here consists of events normally subject only to visual reports, rather than resulting from quantitative readings on conventional meteorological observing equipment. The apparent prevalence and intensity of such events depend crucially on the objectivity, availability and consistency of the observers. Identification of trends in such data is likely to be problematic, because of doubts about consistency of observer's behaviour and reactions over very long periods such as several decades.

Ostby (1993) examined the U.S.A. data base of reports of tornadoes from 1953–1992, for evidence of changes in frequency or intensity. He noted that tornado reports are probably the most 'noisy' and biased of all meteorological data. There appears to have been an increase in the reporting of weak tornadoes, perhaps the result of increased population, eagerness in reporting, or improved reporting procedures.

Similar problems plague determination of changes or variations in the frequency of thunderstorms, dust storms, and other mesoscale meteorological events. For at least some such events, truly global monitoring systems may soon be available. For instance, monitoring lightning strikes from space appears to be feasible. Alternatively, tropical thunderstorm activity leads to electrification of the fair-weather atmosphere through what is known as the atmospheric electrical circuit. So measurements of the atmospheric electrical circuit may be able to provide information on thunderstorm and lightning activity (e.g., Williams, 1992; Price, 1993).

Some General Problems

One problem with assessing possible changes in extreme events, to answer such questions as "Are extreme weather events becoming more frequent", is that such questions must be addressed at numerous individual locations. It is difficult to establish protocols for combining information regarding extreme events at the various locations. The behaviour may be very different at various locations. Thus, Karl *et al.* (1995c) found a significant trend to increased percentages of rainfall falling in heavy events in the U.S.A., but not in the former Soviet Union nor in China, making it difficult to conclude whether there has been an overall (global?) increase in extreme rainfall events.

Investigators may have used very different criteria in their considerations of extreme events. For instance, a variety of approaches to the study of intense rainfalls have been taken in Australia in recent years. Yu and Neil (1991) examined whether daily rainfalls above a certain intensity threshold were likely to behave in the same way as rainfall totals (e.g., seasonal or annual rainfall), in a changing climate. Cumulative rainfalls in excess of 40 m/day were calculated for stations in southeast Australia. In this area there has been an increase in rainfall totals through the 20th century (e.g., Nicholls and Lavery, 1992). However, Yu and Neil found no trend in the amount of rainfall received in intense rainfall events. Nicholls and Kariko (1993) calculated an average rainfall intensity (total rainfall divided by number of rain days) for stations in eastern Australia and also found no trend in intensity through the 20th century. They noted that the changes in rainfall totals reflected increases in the number of rain events, rather than a change in intensity. Yu and Neil compared the rainfall in excess of various intensity thresholds in this region during the 1920s, with the situation in the 1950s. The 1950s had higher total rainfall, but the 1920s had more falls of high intensity (at all thresholds above 30 mm/day). Yu and Neil (1993) examined the relationship between total rainfall and high intensity falls for the southwest of Australia, an area where rainfall has been decreasing in recent decades (Nicholls and Lavery, 1992). They did not find a concurrent decrease in high intensity rainfall. In fact high intensity rainfall increased in summer, offsetting a decrease in winter. Lough (1993) however found that for the northeast Australia, interannual variations in rainfall totals are closely related to variations in rainfall intensity, as well as to the number of events. This area is tropical and received much of its rainfall in only a relatively few events. Lough (1993) found no evidence of a trend to more intense rainfall, or for greater numbers of heavy rain days between 1921 and 1987. However, Suppiah and Hennessy (1995) examined trends between 1910 and 1989 across tropical Australia and found that most stations revealed positive trends in the 90th percentile rainfall intensity and frequency, although few were statistically significant. It is difficult to decide how much of the difference between the conclusions of these studies represent real geographical variations, and how much is due simply to the approaches used in the various studies.

Different criteria (what is an extreme event?), when used on the same data, may lead to different conclusions, reflecting the fact that changes in extremes and variability on different time scales do not necessarily have to be identical. It might be, for instance, that intensities of say, 10-minute ranfalls, might increase without a concomitant increase in intensity measured with 24-hour rainfall totals. Yu and

Neil (1991) analysed 6-minute rainfall data from Canberra Airport, Australia to test this possibility. The rainfall volume occurring above a given intensity threshold in the 6-minute data was compared with that above the same threshold from daily data. Annual high-intensity rainfall volumes, derived from the two data sets, with an intensity above a certain threshold are positively correlated but the correlation decreases as the threshold is increased. For a threshold of 20 mm/day the r^2 is 0.92, but it decreases to 0.76 for intensity greater than 40 mm/day. It is feasible, therefore, that results of studies of trends in intense rainfalls using daily data may not reflect trends in shorter period intense rainfalls. Such problems might also arise at longer time-scales. For instance, a trend in one-day extreme rainfalls might not be reflected by a similar trend in 3- or 30-day extreme rainfalls. Similarly, temperature extremes might be defined as the number of days exceeding a certain threshold, or alternatively as the highest temperature recorded during a given period (e.g., a year). Trends in such different 'extremes' might not be identical.

Care is also required in extrapolating observed trends in extremes even to nearby regions. Grace and Curran (1993) showed that for many southern Australian coastal sites, the frequency distribution of daily maximum temperatures was a combination of maritime and continental airstreams, with each distinct airstream having a separate normal distribution. A change in the intensity or frequency of one of these airstreams (e.g., more frequent maritime stream) might result in a change in extreme temperatures at the coastal sites, but perhaps not further inland (where the maritime stream might be less influential). So, a change in extreme temperatures at the coastal sites could be very different to the situation at a nearby inland site.

These problems, and those discussed earlier for the specific types of extreme events, mean that it is unlikely that we will be ever be able to answer, on a global basis, the question "Are extreme weather events becoming more frequent?". The difficulty of combining such disparate information as trends in tropical cyclone activity, trends in extreme temperatures (e.g., frosts), etc., probably renders the question meaningless in any case. Even if one form of extreme weather event changes in frequency, it is likely that this will be offset be an opposite change, either in a different sort of event at the same location or elsewhere. The only tactic that seems likely to realise useful results is to concentrate on important extreme events and to determine for specific regions (e.g. an ocean basin), whether this subset of extreme events is canging in frequency or intensity. The array of problems discussed earlier clearly demonstrates that even such a limited, location-specific and variable-specific approach is difficult. With care, however, and the use of supporting evidence from a variety of data types, this does seem feasible, at least for some types of extreme events, in some areas.

It may then be possible to select a subset of important, but well-monitored, extreme weather events from a specific region. These could then be used to form an index of extreme weather for that region, and be routinely monitored for future changes. Karl *et al.* (1995b) used this approach to develop an index of extreme weather for the U.S.A. They included the area with very high or very low maximum and minimum temperatures, the area with very dry or very wet conditions, the area with a large proportion of rainfall falling in extreme rainfalls, and the area with very large or very small number of rainfall days. These were combined into an index which was calculated for every year on record. In the same way an index of extreme weather events could be developed for other countries or regions or even ocean basins (e.g., the Atlantic). The specific extreme events to be included in such indices, and their relative weighting, should be developed a priori, based on an estimate of their relative importance or destructiveness.

Improved consistency of routine meteorological observations, and intensified efforts to remove past inconsistencies and biases in these data, will however be needed, even for such studies. The prevalence of these problems in historical data, however, does not preclude the development of such indices for monitoring *future* changes in extreme weather events. Care will need to be taken in the development of the indices, to ensure that a consistent and reliable data base for calculation of the indices is available in the future. This will require that sufficient high-quality observing stations are either established or protected from future changes liable to introduce artificial changes in the time-series. If this is done we may, in the future, have more evidence of whether or not extreme weather events have been changing, possibly even on a quasi-global scale.

References

Agee, E. M.: 1991, 'Trends in Cyclone and Anticyclone Frequency and Comparison with Periods of Warming and Cooling over the Northern Hemisphere', *J. Clim.* **4**, 263–267.

Black, P. G.: 1992, 'Evolution of Maximum Wind Estimates in Typhoons', *ICSU/WMO International Symposium on Tropical Cyclone Disasters, October 12–16, 1992, Beijing.*

Bouchard, R. H.: 1990, 'A Climatology of Very Intense Typhoons: Or Where Have All the Super Typhoons Gone?', *1990 Annual Tropical Cyclone Report*, Joint Typhoon Warning Center, Guam, 266–269.

Changnon, S. A. and Changnon, J. M.: 1992, 'Temporal Fluctuations in Weather Disasters: 1950–1989', *Clim. Change* **22**, 191–208.

Chen, X-Z.: 1990, *Correction of the Historical Data of the Maximum Wind Speed around Tropical Cyclone Centers of the Western North Pacific Ocean*, Shanghai Typhoon Institute. National Typhoon Conference Research and Technological Reports, 19 pp.

Grace, W. and Curran, E.: 1993, 'A Binormal Model of Frequency Distributions of Daily Maximum Temperature', *Aust. Meteorol. Mag.* **42**, 151–161.

Gray, W. M., Landsea, C. W., Mielke, P. W., and Berry, K. J.: 1992, 'Predicting Atlantic Seasonal Hurricane Activity 6–11 Months in Advance', *Wea. Forecasting* **7**, 440–455.

Hopkins, L. C. and Holland, G. J.: 1995, 'Australian East-Coast Cyclones and Heavy Rain Days: 1958–1992', *J. Clim.* (submitted).

Intergovernmental Panel on Climate Change (IPCC): 1990, *Climate Change, The IPCC Scientific Assessment*, Houghton, J. T., Jenkins, C. J., and Ephraums, J. J. (eds.), Cambridge University Press, 365 pp.

Intergovernmental Panel on Climate Change (IPCC): 1992, *Climate Change 1992, The Supplementary Report to the IPCC Scientific Assessment*, Houghton, J. T., Callandar, B. A., and Varney, S. K. (eds.), Cambridge University Press, 198 pp.

Kalnay, E. and Jenne, R.: 1991, 'Summary of the NMC/NCAR Reanalysis Workshop', *Bull. Amer. Met. Soc.* **72**, 1897–1904.

Karl, T. R., Derr, V. E., Easterling, D. R., Folland, C., Levitus, S., Nicholls, N., Parker, D., and Withee, G. W.: 1995a, 'Critical Issues for Long-Term Climate Monitoring', *Clim. Change*, (in press).

Karl, T. R., Knight, R. W., and Plummer, N.: 1995c, 'Trends in High-Frequency Climate Variability in the Twentieth Century', *Nature*, (in press).

Karl, T. R., Knight, R. W., Easterling, D. R., and Quayle, R. G.: 1995b, 'Trends in U.S. Climate During the Twentieth Century', in *Consequences*, USGCRP.

Laing, J.: 1977, 'Maximum Summer Temperatures Recorded in Glaisher Stands and Stevenson Screens', *Meteorol. Mag.* **106**, 220–228.

Landsea, C. W.: 1993, 'A Climatology of Intense (or Major) Atlantic Hurricanes', *Mon. Wea. Rev.* **121**, 1703–1713.

Landsea, C. W. and Gray, W. M.: 1992, 'The Strong Association between Western Sahelian Monsoon Rainfall and Intense Atlantic Hurricanes', *J. Clim.* **5**, 435–453.

Landsea, C. W., Nicholls, N., Gray, W. M., and Avila, L. A.: 1995, 'Quiet Early 1990s Continues Trend of Fewer Intense Atlantic Hurricanes', *Geophys. Res. Letts.*, (submitted).

Lavery, B., Kariko, A. P., and Nicholls, N.: 1992, 'A High-Quality Historical Rainfall Data Set for Australia', *Aust. Met. Mag.* **40**, 33–39.

Lough, J. M.: 1993, 'Variations of Some Seasonal Rainfall Characteristics in Queensland, Australia: 1921–1987', *Int. J. Climatol.* **13**, 391–409.

Nicholls, N.: 1984, 'The Southern Oscillation, Sea Surface Temperature, and Interannual Fluctuations in Australian Tropical Cyclone Activity', *J. Climatol.* **4**, 661–670.

Nicholls, N.: 1992, 'Recent Performance of a Method for Forecasting Australian Seasonal Tropical Cyclone Activity', *Aust. Met. Mag.* **40**, 105–110.

Nicholls, N. and Kariko, A.: 1993, 'East Australian Rainfall Events: Interannual Variations, Trends, and Relationships with the Southern Oscillation', *J. Clim.* **6**, 1141–1152.

Nicholls, N. and Lavery, B.: 1992, 'Australian Rainfall Trends During the Twentieth Century', *Int. J. Climatol.* **12**, 153–163.

Ostby, F. P.: 1993, 'The Changing Nature of Tornado Climatology', *Preprints, 17th Conference on Severe Local Storms, October 4–8, 1993, St. Louis, Missouri*, pp. 1–5.

Parker, D. E.: 1994, 'Effects of Changing Exposure of Thermometers at Land Stations', *Int. J. Climatol.* **14**, 1–32.

Price, C.: 1993, 'Global Surface Temperatures and the Atmospheric Electrical Circuit', *Geophys. Res. Letts.* **20**, 1363–1366.

Schmidt, H. and von Storch, H.: 1993, 'German Bight Storms Analysed', *Nature* **365**, 791.

Schinke, H.: 1993, 'On the Occurrence of Deep Cyclones over Europe and the North Atlantic in the Period 1930–1991', *Beitr. Phys. Atmosph.* **66**, 223–237.

Stone, R., Nicholls, N., and Hammer, G.: 1995, 'Frost in NE Australia: Trends and Influence of Phases of the Southern Oscillation', *J. Clim.*, (submitted).

Suppiah, R. and Hennessy, K. J.: 1995, 'Trends in the Intensity and Frequency of Heavy Rainfall in Tropical Australia and Links with the Southern Oscillation', *Aust. Meteorol. Mag.*, (in press).

Trenberth, K. E.: 1995, 'Atmospheric Circulation Climate Changes', *Clim. Change* **31**, 427–453 (this issue).

Trewin, B. C. and Trevitt, A. C. F.: 1995, 'The Development of Composite Temperature Records', *Int. J. Climatol.*, (in press).

von Storch, H., Guddak, J., Iden, K. A., Jónson, T., Perlwitz, J., Reistad, M., de Ronde, J., Schmidt, H., and Zorita, E.: 1993, *Changing Statistics of Storms in the North Atlantic?*, Report No. 116, Max-Planck-Institut für Meteorologie, Hamburg, 18 pp.

Williams, E. R.: 1992, 'The Schumann Resonance: A Global Tropical Thermometer', *Science* **256**, 1184–1187.

Yu, B. and Neil, D. T.: 1991, 'Global Warming and Regional Rainfall: The Difference between Average and High Intensity Rainfalls', *Int. J. Climatol.* **11**, 653–661.

Yu, B. and Neil, D. T.: 1993, 'Long-Term Variations in Regional Rainfall in the South-West of Western Australia and the Difference between Average and High Intensity Rainfalls', *Int. J. Climatol.* **13**, 77–88.

(Received 23 January, 1995; in revised form 26 June, 1995)

LOW-COST LONG-TERM MONITORING OF GLOBAL CLIMATE FORCINGS AND FEEDBACKS

J. HANSEN, W. ROSSOW, B. CARLSON, A. LACIS, L. TRAVIS, A. DEL GENIO,
I. FUNG, B. CAIRNS, M. MISHCHENKO and M. SATO

NASA Goddard Institute for Space Studies, New York, NY 10025, U.S.A.

Abstract. We describe the rationale for long-term monitoring of global climate forcings and radiative feedbacks as a contribution to interpretation of long-term global temperature change. Our discussion is based on a more detailed study and workshop report (Hansen *et al.*, 1993b). We focus on the potential contribution of a proposed series of inexpensive small satellites, but we discuss also the need for complementary climate process studies and ground-based measurements. Some of these measurements could be made inexpensively by students, providing both valuable climate data and science educational experience.

Introduction

Climate varies on all time scales, and there are many aspects of climate change with practical importance. Without prejudice to other issues, this paper focuses on global temperature change on time scales from a year to several decades, a topic of societal concern because of the suspected role of human-made greenhouse gases in causing long-term change. We call attention to the need for high precision monitoring of all significant global climate forcings as an essential requirement for interpretation of global temperature changes. Without such data, which are not covered adequately by current monitoring plans, uncertainty about the causes and implications of observed climate change will persist indefinitely, and it will be much harder to decide on a prudent environmental policy.

Quantitative knowledge of all global climate forcings, natural and human-made, is essential to national and global policymakers. Environmental and energy policies, for example, will be influenced by the degree to which greenhouse gases and fine particles in the lower atmosphere, produced by use of fossil fuels and by biomass burning, are judged to influence global climate. Even after policy decisions are made, monitoring of the climate forcings to a precision that accurately defines their changes is necessary in order to judge the effectiveness of the policies.

In this paper we discuss the principal global climate forcings and radiative feedbacks. We show that many of these quantities could be observed with the required high precision, global coverage, and time-space sampling by a pair of small, inexpensive satellites. We also underline the need to observe other climate forcings and radiative feedbacks not included on the specific small-satellite we propose, called Climsat, as well as the need for complementary field projects and measurements from surface stations. Together with existing observations, these measurements will

Climatic Change **31**: 247–271, 1995.

strongly constrain interpretation of observed global temperature change and permit quantitative comparison of climate forcing mechanisms which presently involve substantial uncertainty. We recognize that the proposed Climsat measurements also have application to regional climate phenomena of much practical interest, but our strategy is to focus on global climate change, using the needs for interpretation of global temperature change to define measurement requirements.

We assume that Climsat would be carried out within the context of, and as one contribution to, a comprehensive global observing system. Climsat thus is meant to be a key addition to existing and planned observing systems, including the international Global Climate Observing System and the Global Ocean Observing system. Other fundamental climate diagnostics, such as precipitation and ocean parameters, need to be measured by existing and new experimental systems, and by their follow-on programs. It is also important that the meteorological observing system of NOAA be upgraded to enhance its effectiveness for climate monitoring. NASA's Earth Observing System can provide detailed measurements important to the study of many climate processes. DOE's existing and planned Atmospheric Radiation Measurements ground sites could effectively complement Climsat observations, as could appropriate observations by high school students in the planned GLOBE program.

Forcings and Feedbacks

Global temperature has increased significantly during the past century (IPCC, 1990; Hansen and Lebedeff, 1987; Jones *et al.*, 1986). Understanding the causes of observed global temperature change is impossible in the absence of adequate monitoring of changes in global climate forcings and radiative feedbacks. We define climate forcings as changes *imposed* on the Earth's energy balance which work to alter global temperature, for example, a change of incoming solar radiation or a man-made change of atmospheric composition. Radiative feedbacks are responses to climate change, such as altered cloud properties or sea ice cover, which may magnify or diminish the initial climate change.

Monitoring of global climate forcings and feedbacks, if sufficiently precise and long-term, can provide a *very strong constraint* on interpretation of observed temperature change. Such monitoring is essential to eliminate uncertainties about the relative importance of various climate change mechanisms including tropospheric sulfate aerosols from burning of coal and oil (Charlson *et al.*, 1992), smoke from slash and burn agriculture (Penner *et al.*, 1992), changes of solar irradiance (Friis-Christensen and Lassen, 1991), changes of several greenhouse gases, and many other mechanisms.

The considerable variability of observed temperature, together with evidence that a substantial portion of this variability is unforced (Barnett *et al.*, 1992; Manabe *et al.*, 1990; Hansen *et al.*, 1988; Lorenz, 1963), indicates that observations of climate forcings and feedbacks must be continued for decades. Since the climate system responds to the time integral of the forcing, a further requirement is that the observations be carried out continuously.

However, precise observations of forcings and feedbacks will also be able to provide valuable conclusions on shorter time scales. For example, knowledge of the climate forcing by increasing CFCs relative to the forcing by changing ozone is important to policymakers, as is information on the forcing by CO_2 relative to the forcing by sulfate aerosols. It will also be possible to obtain valuable tests of climate models on short time scales, if there is precise monitoring of all forcings and feedbacks during and after events such as large volcanic eruption or an El Niño.

Greenhouse gases. The measured increase of homogeneously mixed green-house gases since the beginning of the industrial revolution causes a climate forcing of about 2 W/m^2 (IPCC, 1992; Hansen and Lacis, 1990; Dickinson and Cicerone, 1986; Ramanathan et al., 1985; Wang et al., 1976), as illustrated in Figure 1. Although a host of chemical species, e.g., NO_x, CO and OH, requires monitoring for the purpose of understanding greenhouse gas trends and predicting future changes, the well mixed greenhouse gases are measured accurately already. However, there is major uncertainty about the total anthropogenic greenhouse forcing, especially because of uncertain changes of the ozone profile (IPCC, 1992; Ramaswamy et al., 1992; Lacis et al., 1990). Stratospheric water vapor also may be increasing not just because of oxidation of increasing methane (Ellsaesser, 1983; Le Texier et al., 1988), but also other as-yet-not-understood mechanisms (Oltmans and Hofmann, 1995), so, in the absence of adequate monitoring, its net climate forcing is very uncertain.

Climate forcing due to ozone change is complicated because ozone influences both solar heating of the Earth's surface and the greenhouse effect. Both of these mechanisms influence surface temperature, but their relative importance depends on the altitude of the ozone change. Figure 2 illustrates the equilibrium response of a GCM to specified ozone changes: (a) Ozone loss in the upper stratosphere warms the Earth's surface, because of increased ultraviolet heating of the troposphere; (b) Added ozone in the troposphere warms the surface moderately; (c) Ozone loss in the tropopause region causes a strong cooling because the low temperature at the tropopause maximizes the ozone's greenhouse effect; (d) Coincidentally, removal of all ozone causes only a moderate surface cooling.

The ozone changes that had been predicted for many years on the basis of homogeneous (gas phase) chemistry models included upper stratospheric ozone loss and tropospheric ozone increase. Both of those ozone changes would cause surface heating. But limited ozone measurements in the 1970s (Tiao et al., 1986; Reinsel et al., 1984) suggested the possibility that upper tropospheric ozone and lower stratospheric ozone may be decreasing. Discovery of the Antarctic ozone hole in the 1980s (Farman et al., 1985) and analysis of the mechanisms involved in the ozone depletion led to the realization of the effectiveness of heterogeneous loss

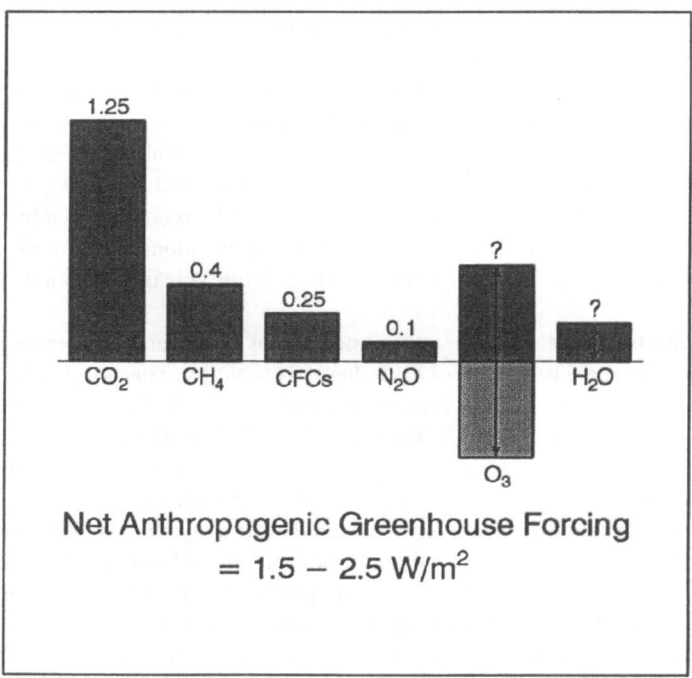

Fig. 1. Anthropogenic greenhouse climate forcings (W/m^2) due to measured or estimated trace gas changes between 1850 and 1990. The forcing is calculated as the change in net radiative flux at the tropopause caused by the change in atmospheric composition (Hansen and Lacis, 1990).

processes in the 15–25 km region (WMO, 1990). Satellite data for the 1980s (Stolarski *et al.*, 1991; McCormick *et al.*, 1992) have shown that the lower stratospheric ozone loss is not confined to the Antarctic.

It will not be possible to accurately evaluate the total anthropogenic greenhouse effect unless ozone change is monitored as a function of altitude, latitude and season. Useful stratospheric ozone profile data are presently supplied by the SAGE II instrument on the ERBS satellite, which is over 10 years old. As discussed below, this data record could be extended and enhanced by flight of a proposed improved version of the instrument (SAGE III) with greater sensitivity, higher spectral resolution, and increased spatial sampling, as one contribution to more comprehensive ozone monitoring.

Aerosols. Perhaps the greatest uncertainty in climate forcing is that due to tropospheric aerosols (Charlson *et al.*, 1992). Aerosols cause a direct climate forcing, by reflecting sunlight to space and absorbing solar and terrestrial radiation in the atmospheric column, and an indirect climate forcing, by altering cloud properties. Existence of the latter effect is supported by satellite observations of increased

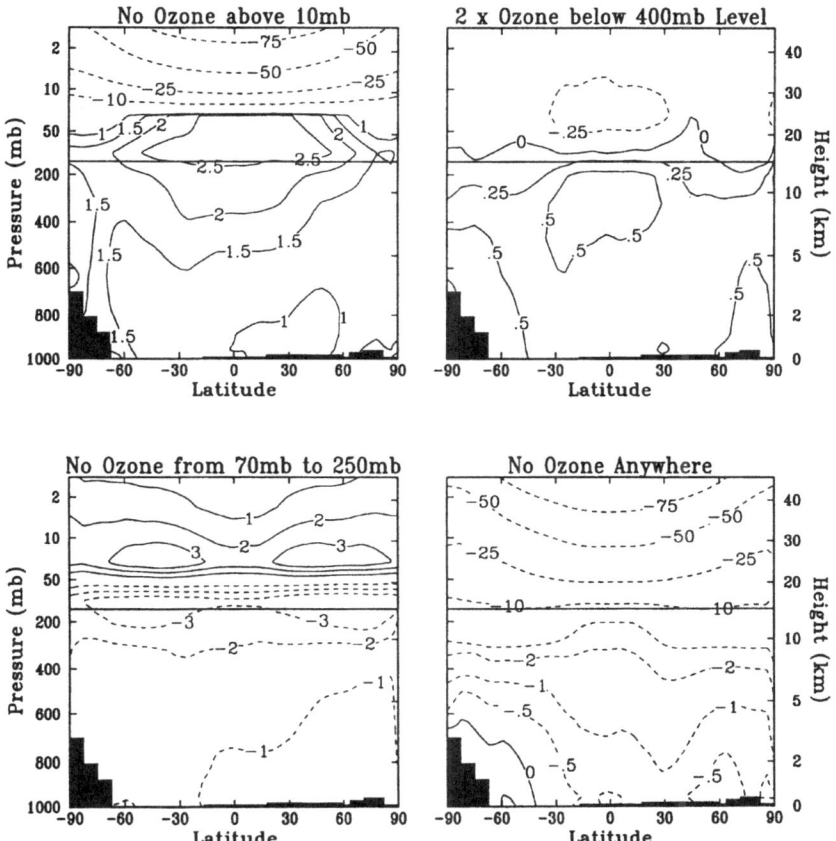

Fig. 2. Zonal mean equilibrium temperature change (°C) estimated for several arbitrary changes of the ozone distribution. Results were obtained from 100 year runs of the GISS GCM.

cloud brightness in ship wakes (Coakley *et al.*, 1987), satellite observations of land-ocean and hemispheric contrasts of cloud droplet sizes (Han *et al.*, 1994), and *in situ* data concerning the influence of aerosol condensation nuclei on clouds (Radke *et al.*, 1989). Sulfate aerosols originating in fossil fuel burning may produce a global climate forcing of order –1 W/m² (Charlson *et al.*, 1991), and aerosols from biomass burning could conceivably produce a comparable forcing (Penner *et al.*, 1992). Wind-blown dust, influenced by anthropogenic activities, has long been suspected of being an important forcing on some regional climates (Tanre *et al.*, 1984; Joseph, 1984; Coakley and Cess, 1985; Tegen and Fung, 1994). It has also been suggested (Jensen and Toon, 1992; Sassen, 1992) that volcanic aerosols sedimenting into the upper troposphere may alter cirrus cloud microphysics, thus producing a possibly significant climate forcing, which is uncertain even as to sign.

Unfortunately, no global data exist that are adequate to define any of these aerosol climate forcings.

Aerosols can be seen against the dark ocean surface by the imaging instruments on present operational meteorological satellites (Rao *et al.*, 1988; Jankowiak and Tanre, 1992). For example, these images clearly show Sahara/Sahelian dust spreading westward from Africa, summertime sulfate aerosols moving eastward from the United States, and aerosols from seasonal biomass burning in the tropics. However, the nature and accuracy of these data are inadequate to define the climate forcing, and, indeed, the observed optical depths may be in part thin cirrus clouds. The climate forcing issue requires aerosol data of much higher precision, including information on aerosol altitude and aerosol physical properties such as size and refractive index. Cloud properties, including optical depth, particle size, and phase, must be monitored simultaneously to very high precision, so that the temporal and spatial variations of aerosols and clouds can be used to help define the indirect aerosol climate forcing.

Stratospheric aerosol optical depth was monitored in the polar regions from late 1978 until 1994 by a solar occultation instrument (SAGE I) on the Nimbus-7 spacecraft (McCormick *et al.*, 1979). The record reveals seasonal polar stratospheric (condensation) clouds, especially in Antarctica, as well as the influence of aperiodic volcanic sulfuric acid aerosols, especially the El Chichon eruption in 1982 and the Mt. Hudson and Mt. Pinatubo eruptions in 1991. An approximate 50% increase of 'background' aerosol optical depth between 1979 and 1990 is thought by some (e.g., Hofmann, 1990) to be a result of anthropogenic impact on the sulfur cycle, perhaps due to aircraft emissions.

The global radiative forcing of the El Chichon aerosols reached a maximum of about -2 W/m^2 (Hansen and Lacis, 1990) and the forcing by Pinatubo aerosols was even larger (Hansen *et al.*, 1992, 1993a; Minnis *et al.*, 1993), thus exceeding the magnitude of the forcing by all anthropogenic greenhouse gases, though opposite in sign. Although the aerosol forcing from individual volcanic eruptions is more short-lived, it must be monitored if the global temperature record is to be interpreted. The Nimbus-7 spacecraft measuring polar stratospheric aerosols ceased operations recently. SAGE II has been obtaining data at low and middle latitudes from the ERB spacecraft since 1984, but that spacecraft is showing signs of age and is already well beyond its design life.

Solar irradiance. Another potentially important climate forcing is change of solar irradiance. The spectrally integrated irradiance has been monitored for about 15 years, showing a decline of about 0.1% between 1979 and 1986, followed by at least a partial recovery. If this measured variability were spectrally uniform, it would imply a climate forcing of about 0.3 W/m^2 of absorbed solar energy. Solar variability of a few tenths of a percent could cause a global temperature change of the magnitude of the observed cooling between 1940 and 1970, and there have been suggestions that the sun may be responsible for the warming trend of the past century (Friis-Christensen and Lassen, 1991). Thus we need to monitor solar irradiance on longer time scales, including the spectral distribution of changes, because the nature of the solar forcing varies strongly depending on the altitude where the radiation is absorbed. There are offsets of the absolute irradiance even among the best calibrated instruments (Lean, 1991), which implies the necessity of overlapping coverage by successive instruments for successful monitoring.

Surface reflectivity. Perhaps the next climate forcing mechanism to be redis-covered as a competitor to increasing greenhouse gases is change of the Earth's surface reflectivity. Sagan *et al.* (1979) argued that anthropogenic deforestation and desertification could have increased the planetary albedo sufficiently to cause a cooling of about 1 °C over the past few millennia, and may have been responsible for the observed global cooling after 1940. Potter *et al.* (1981) calculated a smaller global cooling, 0.2 °C, with a two-dimensional climate model, but nevertheless surface albedo change is a potentially significant climate forcing. For example, a change of mean land albedo from 0.15 to 0.16 would cause a global climate forcing of about 0.5 W/m^2, comparable in magnitude to the forcing due to expect-ed increases of anthropogenic greenhouse gases during the next two decades. Although a global mean change that large may be unlikely, regional effects could be substantial and the global effects need to be quantified.

Operational meteorological satellites currently measure the Earth's surface reflectivity at one or two wavelengths, but the instruments are not calibrated well enough to provide reliable long-term data (Brest and Rossow, 1992). However, it is not difficult to obtain both higher accuracy and precision than that of the meteorological instruments, which were not designed for long-term climate moni-toring. More complete spectral coverage also is needed because vegetated regions in particular have a strong, seasonally variable, spectral dependence of reflected radiation.

Radiative feedbacks. There are many feedback processes, some known and others yet to be discovered, which alter the climate system's ultimate response to a climate forcing. In studies with current GCMs, it has been found that the net response of global temperature to a forcing such as doubled carbon dioxide can be separated quantitatively into contributions arising from the forcing plus three major radiative feedbacks: (1) changes of atmospheric water vapor, and its vertical distribution; (2) changes of cloud cover, optical depth, and altitude; and (3) changes of the duration and area of ice and snow cover (Cess *et al.*, 1989, 1990, 1991; Schlesinger and Mitchell, 1987; Hansen *et al.*, 1984). For example, for doubled CO_2 the no-feedback climate sensitivity of 1.2–1.3 °C is increased to about 2–5 °C in the GCM simulations, with the latter value depending upon the strength of these three feedbacks in each global model.

As indicated by the schematic Figure 3, the largest feedback in the GCMS is caused by water vapor. Lindzen (1990) maintains that the models exaggerate the water vapor feedback and has argued that the feedback could be negative. Although there is theoretical and empirical evidence against Lindzen's hypothesis of a negative feedback (Betts, 1991; DelGenio *et al.*, 1991; Rind *et al.*, 1991; Raval

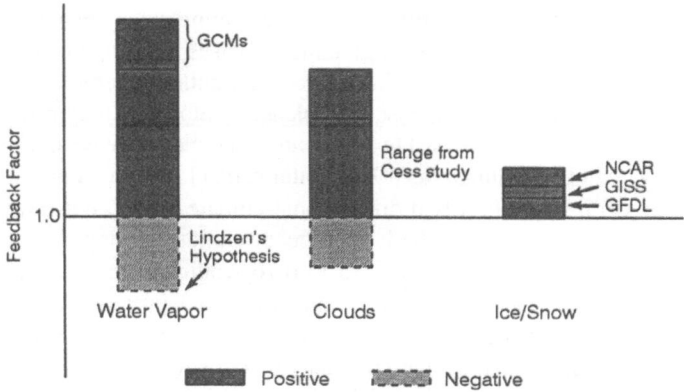

Fig. 3. Schematic indication of the radiatiative feedback factors which have been found to determine the global sensitivity of the general circulation models to climate forcings such as doubled atmospheric CO_2 (Hansen *et al.*, 1993b).

and Ramanathan, 1989), this does not diminish the importance of changes of the water vapor profile in determining the magnitude of the water vapor feedback. Cloud feedbacks are probably the most uncertain, with the range from GCMs including negative as well as positive feedbacks (Cess *et al.*, 1989, 1990). The ice/snow feedback also shows a wide variation among models (Cess *et al.*, 1991; Randall *et al.*, 1994).

Climate feedbacks are the cause of large uncertainty about climate sensitivity to a specified forcing. Continued efforts to improve the representation of the feedback processes in climate models are important and are receiving much attention, but it is unlikely that general agreement on the magnitude of global feedbacks on longer time scales can be obtained on the basis of only process studies and models. Thus it is crucial that observations of current and future climate change be accompanied by measurements of the feedbacks to an accuracy sufficient to define their contribution to observed climate change. As we demonstrate below, it is possible to obtain the required accuracies with existing technology.

It is appropriate to ask whether there are other important climate forcings or feedbacks, in addition to those wich the scientific community has already identified. Although the processes that have been considered account for all the major mechanisms for exchange of energy with space, it is likely that there will be future surprises in our understanding of both climate forcings and feedbacks. Therefore, it is important that a monitoring strategy include measurements covering practically the entire spectra of both the solar and thermal radiation emerging from the Earth, because all radiative forcings and feedbacks operate by altering these spectra. Although efforts to measure integrated reflected solar and emitted thermal fluxes are underway (Kandel, 1990), measurement of changes in the spectral distribution of the radiation are required to provide diagnostic information about causes of flux changes.

Monitoring Rationale

Ambiguity would persist in interpretation of observed temperature changes even if all global climate forcings and feedbacks were measured accurately, because of possible but unmeasured changes of atmospheric and oceanic energy transports. Nevertheless, a long-term record of the forcings and feedbacks will provide a very strong constraint on interpretation of future global temperature change.

Our rationale is inspired by Keeling's CO_2 record, which is a prototype of high-precision long-term monitoring of a climate parameter. The CO_2 record cannot by itself provide an understanding of either the global carbon cycle or the global thermal energy cycle, but it provides a very strong constraint. Also, the CO_2 monitoring is not competitive with detailed observations required to understand carbon and thermal energy processes. On the contrary, it inspires and helps guide such studies. Note that CO_2 monitoring, after being proven as a research product, became an operational activity of NOAA. Similarly, monitoring of other climate forcings and feedbacks could become an operational activity once a measurement approach is demonstrated.

Many of the missing climate parameters can be measured by three small instruments which we describe below. These instruments measure with high precision the spectra of reflected solar radiation and emitted thermal radiation. We have used observed global datasets and global climate models to determine the minimum measurements needed to define changes of the climate forcings and feedbacks with the required accuracies on seasonal and longer time scales, as described in the Climsat report (Hansen *et al.*, 1993b). A more detailed consideration of cloud monitoring is presented by Rossow and Cairns (1995). The fewest number of satellites required is two. A sun-synchronous near-polar orbiter provides a fixed diurnal reference. A precessing orbiter inclined 50–60 degrees to the equator provides a statistical sample of diurnal variations at latitudes with significant diurnal change. The need to determine diurnal changes has been highlighted by the discovery that global warming of the past several decades has a strong day/night asymmetry (Karl *et al.*, 1993). The two orbits together provide good global observing conditions for all three instruments and reduce sampling errors below the level required for detecting expected decadal time-scale change. Two satellites are also required to allow satellite-to-satellite transfer of calibration when one satellite fails and must be replaced.

Climsat is low-cost. The Climsat instruments have well-proven long-lived predecessors; thus their production can take advantage of, but does not depend upon, costly technological advances. They are light-weight (about 25 kg without SAGE, or 60 kg with SAGE), permitting launch by a Pegasus-class vehicle. These characteristics make Climsat economically feasible for repeated missions over decades.

The three instruments we propose would provide many of the missing climate forcings and feedbacks, but certain complementary monitoring is required to complete the full set of data requirements. One particular need is long-term satellite monitoring of both the total and spectral solar irradiance; this data is presently being obtained by the UARS mission, but there is urgent need of real plans for continued monitoring, which could be effectively carried out by a small satellite. Also satellite monitoring of parameters such as tropospheric aerosols and the ozone profile must be supplemented by ground-based monitoring networks to assure acquisition of complete climate forcing and feedback information. This is discussed further below.

Proposed Climsat Measurements

Measurements by the three proposed Climsat instruments cover practically the entire thermal and solar spectra, as summarized in Figure 4, and are designed to exploit the information on gases, aerosols and clouds contained in these spectra. In the thermal wavelength region information is contained primarily in the high resolution spectral variations of the radiance (Conrath et al., 1970; Hanel et al., 1972; Kunde et al., 1974; Clough et al., 1989). On the other hand, because incident sunlight is unidirectional, reflected solar radiation is in general strongly polarized, and the polarization is highly diagnostic of aerosol and cloud properties (Hansen and Travis, 1974; Coffeen and Hansen, 1974). The full spectral coverage of the Climsat instruments is a crucial characteristic of the proposed measurements, because it means they should be capable of providing information on climate 'surprises', as well as the climate forcings and feedbacks about which we already know, because all radiative forcings and feedbacks operate by altering the solar or thermal spectrum in some way.

SAGE III (Stratospheric Aerosol and Gas Experiment III) observes the sun and moon through the Earth's atmosphere, obtaining extinction profiles with very high vertical resolution. SAGE III uses the same grating spectrometer as its immediate predecessors, but, unlike them, it records the spectrum on a continuous linear array of detectors, yielding a spectral resolution of 10 Å (10^{-3} μm) from 0.29 μm to 1.02 μm. It also adds a detector at 1.55 μm to improve determination of aerosol sizes and provide discrimination between aerosols and thin cirrus clouds. SAGE III will provide absolutely calibrated profiles of stratospheric aerosols, stratospheric water

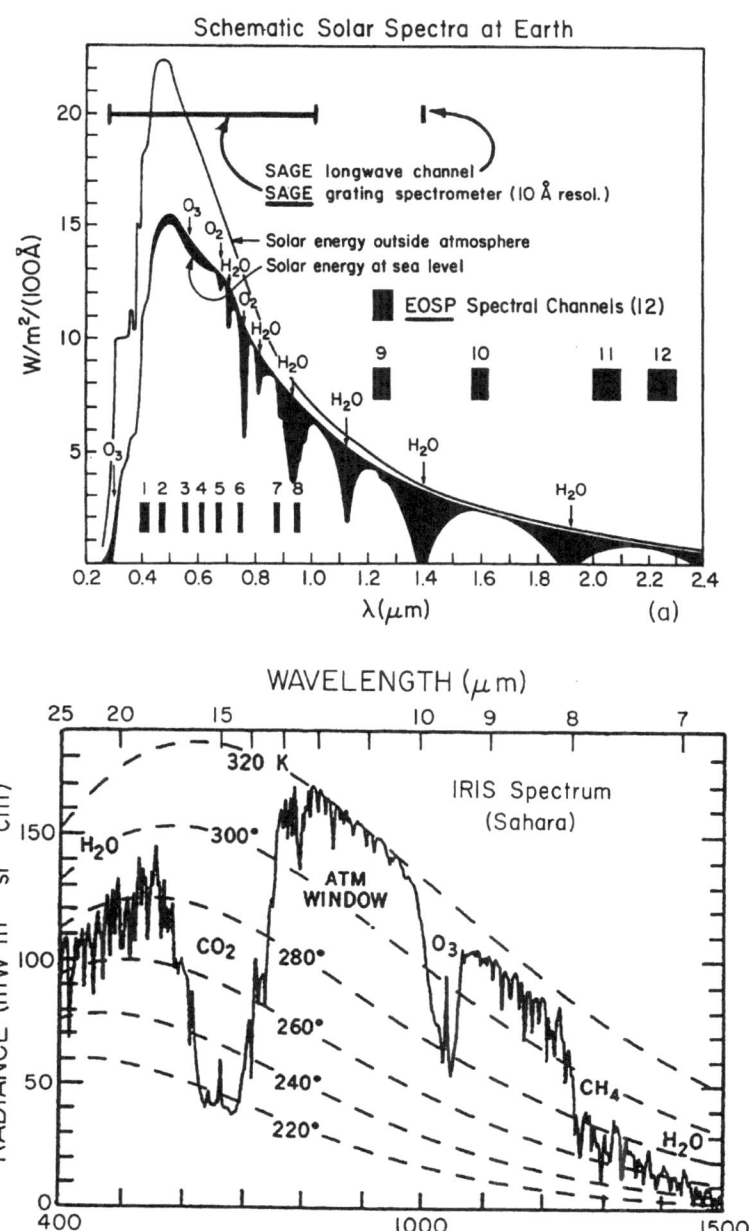

Fig. 4. Approximate spectral regions covered by proposed instruments: (a) Location of the EOSP and SAGE III spectral channels, relative to a typical spectrum of solar radiation; (b) Example of terrestrial thermal spectrum, obtained by the Nimbus-3 IRIS instrument over the Sahara desert. MINT will have a somewhat broader spectral coverage, 250–1700 cm^{-1}, and higher resolution (2 cm^{-1}).

vapor, and ozone, extending and improving upon predecessor data (McCormick, 1993).

MINT (Michelson Interferometer) covers the spectral range 5–50 μm, the longer wavelengths being important for defining the water vapor distribution. Its high spectral resolution and high wavelength-to-wavelength precision provide the essential ingredients for accurate long-term monitoring of cloud properties (cloud cover, effective temperature, optical thickness, ice/water phase and effective particle size) day and night in six 8 km fields of view. MINT simultaneously monitors tropospheric water vapor, ozone and temperature (Lacis and Carlson, 1993).

EOSP (Earth Observing Scanning Polarimeter) covers the solar spectrum from the near ultraviolet (0.4 μm) to the near infrared (2.25 μm) in 12 spectral bands, obtaining global maps of the radiance and polarization with a spatial resolution of 8 km at the subsatellite point. Its unique contributions are accurate global distribution and physical properties of tropospheric aerosols (optical thickness, particle size and refractive index) and precisely calibrated surface reflectance. In addition, EOSP yields detailed cloud properties, complementing and refining cloud information from MINT (Travis, 1993).

Table I summarizes specific technical data on each of the three instruments. A more detailed description of the instruments is given in the workshop report (Hansen *et al.*, 1993b) and references provided there.

Perhaps the most crucial characteristic of the Climsat instruments is that they are all self-calibrating to high precision. The SAGE calibration is obtained by viewing the sun (or moon) just before or after every occultation. MINT records its interferogram on a single detector, thus obtaining high wavelength-to-wavelength precision. Absolute calibration of the MINT measurements is achieved by periodic viewing of space (cold reference) and an internal black body (warm reference). EOSP interchanges the roles of its detector pairs periodically by using a stepping half-wave retarder plate, calibrating polarization to 0.2% absolute accuracy. The EOSP radiance calibration is based primarily on internal lamps with a demonstrated stability of better than 2% per decade, implying a decadal precision for surface reflectivity of better than 0.002 for a surface reflectivity of 0.1. This radiance calibration stability exceeds that of operational satellites by a factor of about five (Brest and Rossow, 1992).

All three Climsat instruments are based on space-proven predecessors, with incremental but significant enhancements in capability, incorporating recent advances in detector and electronic technology. Each of the three instruments has a predecessor with a lifetime in space exceeding 10 years.

The original Climsat concept had these three instruments together on a small satellite. Recently funding has been approved for three SAGE III instruments, with the first two flights expected to be on a Russian polar orbiting satellite and the inclined orbiting space station. Thus one strategy for completing the proposed set of Climsat measurements would be to fly the MINT and EOSP instruments on a smaller spacecraft. These two instruments need to be on the same spacecraft, with

TABLE I

Climsat sensors

SAGE III	EOSP	MINT
Earth-limb scanning grating spectrometer, UV to near IR, 10 Anstrom resolution.	Along-track scans of radiance and polarization, 12 bands near UV to near IR.	Michelson interferometer, 2 cm^{-1} resolution from 5 μm to 50 μm; nadir viewing by 2 × 3 array of detectors.
IFOV = 30 arcsec ($\sim$0.5 km); inversion resolution 1–2 km.	IFOV = 12 mrad (8 km at nadir).	IFOV = 12 mrad (8 km from 650 km altitude).
Yields profiles of T, aerosols, O_3, H_2O, NO_2, NO_3, OClO – most down to cloud tops.	Yields aerosol optical depths, particle size and refractive index, cloud optical depth and particle size, and surface reflectance and polarization.	Yields cloud temperature, optical depth, particle size and phase, temperature, water vapor and ozone profiles and surface emissivity.
Mass: 35 kg	Mass: 16 kg	Mass: 10 kg
Power (mean/peak) 10/45 W	Power (mean/peak): 15/22 W	Power (mean/peak): 14/22 W
Mean Data Rate: 0.45 Tbpy*	Mean Data Rate: 0.8 Tbpy*	Mean Data Rate: 0.7 Tbpy*
Cost: About $20M for first copy, about $10M each additional copy	Cost: About $20M for first copy, about $10M each additional copy	Cost: About $20M for first copy, about $10M each additional copy

* Tbpy = Terabits/year, Mission comparison: ISCCP = 0.2 Tbpy; CLIMSAT = 4 Tbpy; EOS = 2500 Tbpy [one Terabit is approximately 1000 tapes (6250 bpi) per year].

their identical fields of view boresighted at nadir, so that the information obtained on clouds, atmosphere and surface is enhanced.

Estimated Measurement Accuracies

Specification of measurement accuracy requirements and instrument capabilities is difficult and somewhat subjective. Nevertheless, it is important to provide plausible estimates of needs and capabilities. As summarized below, we considered two independent criteria for specifying accuracy requirements (Chapters 3 and 7 in Hansen *et al.*, 1993b). The estimates of instrument capabilities were based on consideration of predecessor instrument performance, planned instrument improvements, and Climsat sampling characteristics. Actual capabilities can only be determined after the fact, but we note that it is not unusual for ultimate capabilities to exceed expectations in the case of stable calibrated instruments.

The first criterion for accuracies was based on the desire to simply detect plausible changes of the forcings and feedbacks estimated to be possible during the next 20 years, as discussed by DelGenio (1993). The second criterion was the more demanding desire to determine quantitatively the contribution of every significant forcing and feedback to the planetary energy balance. We defined a significant global mean flux change as 0.25 W/m^2 or greater, based on the consideration that anticipated increases of greenhouse gases during the next 20 years will cause a forcing of about 1 W/m^2. The accuracy requirements resulting from these two criteria are listed in Table II. Although the precise magnitudes are uncertain, the indicated values are plausible and consistent with current understanding of the climate system.

The capabilities of the proposed Climsat mission depend on the instrumental accuracies and precisions, and also on the spatial and temporal sampling from the Climsat orbits. The instrumental capabilities are discussed in Sections 8–10 and the sampling in Sections 11–12 of the Climsat workshop report (Hansen *et al.*, 1993b). Predecessor SAGE instruments provide the basis for expected accuracies of the improved SAGE III. Estimates of retrieval precisions for MINT are based mainly on analysis of data from the Infrared Interferometer Spectrometer (IRIS) instrument, accounting for expected decreases in instrument noise levels. EOSP has an inherent advantage over instruments measuring only reflected radiance, because its high accuracy measurement of linear polarization is very sensitive to particle microphysics (Hansen and Travis, 1974). The nonuniqueness in retrievals based on the radiance can be overcome using the polarization, as demonstrated by Pioneer Venus retrieval of the properties of haze particles located above and within a lower cloud deck (Kawabata *et al.*, 1980). Comparable complications exist for aerosols over land-vegetation surfaces on Earth, but present evidence that the polarization from vegetated surfaces derives mainly from specular reflection at leaf surfaces (Vanderbilt *et al.*, 1985), together with the broad EOSP wavelength coverage, suggests that the surface contributions can be accurately accounted for in the EOSP retrievals.

Reliable determination of the ultimate retrieval capabilities prior to flight of the instruments is extremely difficult, but further simulations of instrument performance, data inversion techniques, and sampling studies are being pursued. Sampling studies for the stratospheric quantities, for example, are hindered by inadequate knowledge of small scale spatial variability of the parameters being measured. Our present estimates of potential Climsat capabilities are given in the fourth column of Table II for regional (1000 km by 1000 km), seasonal (3 month) averages and in the fifth column for global decadal change. Generally the sampling is not a factor in determining the global decadal change, but it does influence the ability to determine regional seasonal change.

It appears that, in general, these instruments are capable of measuring the changes of climate forcings and feedbacks projected as being plausible during the next 20 years. The more difficult criterion, quantifying the flux changes to 0.25 W/m^2, can also be achieved for all the climate forcings except aerosol induced cloud changes. But this latter forcing may be measurable in the regions of largest (measured) aerosol changes, which may allow an inference of the corresponding

TABLE II

Comparison of estimated Climsat measurement accuracies with changes of forcing and feedback parameters anticipated on a 20 year time span (A. DelGenio, Chapter 3 in Hansen *et al.*, 1993b) and with the parameter changes required to yield a flux change of 0.25 W/m^2

Forcing or feedback	Plausible 20 year change			Global change required to yield $\Delta Flux = 0.25$ W/m^2	Climsat accuracy estimated for regional seasonal mean	Climsat accuracy estimated for global decadal change
Ozone	Altitude and height dependent			10% of O_3 at 15–20 km	10%	3%
Stratospheric H_2O	$\dfrac{\Delta q}{q}$	=	0.3	0.25	0.10	0.03
Stratospheric aerosol	$\Delta\tau$	=	0.04	0.01	0.02	0.002
Tropospheric aerosol	$\Delta\tau$	=	0.04	0.01	0.02	0.005
Total solar irradiance	0.1–0.3%			0.1%	not on Climsat, but ACRIM, if flown *continuously*, could readily achieve the needed accuracy	
Surface (land) reflectivity	0.01 (land)			0.006 (land)	0.01	0.003
Tropospheric H_2O upper lower	$\dfrac{\Delta q}{q}$	=	0.1 0.04	0.02 0.02	0.05 0.03	0.03 0.02
Cloud cover cirrus stratus	ΔC	=	0.03 (regional) 0.03 (regional)	0.004 0.003	0.02 0.02	0.004 0.004
Cloud top temperature pressure	ΔT Δp	= =	1 K 12 mb	0.4 K 5 mb	1 K 15 mb	0.3 K 5 mb
Cloud optical depth cirrus stratus	$\Delta\tau$	=	0.1 1	0.02 0.07	0.1 0.5	0.05 0.2
Cloud particle size (water)	Δr	=	1 μm	0.2 μm	0.5 μm	0.2 μm

global forcing. It appears that these instruments may be just marginally capable of measuring most of the feedbacks, mainly cloud parameter changes, to the 0.25 W/m^2 criterion. Direct measurement of cloud optical thickness change to this accuracy does not seem to be achievable. The alternative of measuring the corresponding cloud albedo changes over decades is also estimated to be just outside the capability which is proven for the EOSP calibration lamps on the basis of planetary flight experience. However, we emphasize that the accuracies considered here are several times better than those of current meteorological satellites, which are already capable of detecting some interannual changes (Ardanuy *et al.*, 1992).

In summary, these instruments are capable of detecting plausible decadal changes of many climate forcings and feedbacks. In most cases, the forcings and feedbacks can be quantified to the high precision (0.25 W/m^2) desired to help interpret global climate change. The merits of precise long-term monitoring of the fundamental climate data represented by the solar and thermal spectra also can be argued simply on the basis of analogous examples such as Keeling's CO_2 monitoring. However, the fact that it appears feasible to derive climate parameters with an accuracy at least approaching that needed to interpret flux changes of 0.25 W/m^2 makes the case even stronger.

Complementary Monitoring Requirements

Although the proposed instruments can provide with the required accuracies many of the climate forcings and feedbacks missing from the existing observational system, other monitoring is needed to complete the full set of data requirements. We summarize the key requirements here.

Perhaps the most crucial need is long-term monitoring of the sun, which provides the ultimate drive for the Earth's climate. A plausible case has been made that solar irradiance changes might be responsible for climate changes such as those characterized by the Little Ice Age (Eddy, 1976), which may only require solar changes of several tenths of a percent (Wigley, 1988; Wigley and Kelly, 1990). Precise monitoring of the total solar irradiance during the past decade (Willson and Hudson, 1991; Hoyt *et al.*, 1992) confirmed the existence of significant variations of solar irradiance, of the order of 0.1% over the last 11 year solar cycle. It is important that this fundamental measurement be continued. There must be an overlap of the successive monitoring instruments, because it is not possible to obtain sufficient absolute accuracy of the irradiance (Lean, 1991). The UARS mission (Reber, 1990) includes ACRIM II, which precisely monitors total solar irradiance, but prompt flight of another ACRIM or its equivalent is needed.

It is also important to monitor the spectrum of the solar irradiance. The climate forcing due to solar change is entirely different if the change occurs at wavelengths absorbed in the upper atmosphere, as opposed to wavelengths which reach the troposphere. Furthermore changes in ultraviolet irradiance may cause an indirect climate forcing by altering the abundances of greenhouse gases such as ozone (Chandra, 1991; Stolarski *et al.*, 1991). The UARS mission includes two instruments which monitor the solar spectral irradiance in the ultraviolet region, where large variability is known to occur (Rottman, 1988), but plans for a follow-up are needed. Total and spectral irradiance monitors both appear to be prime candidates for flight on small satellites.

Several of the parameters which our proposed instruments can monitor require complementary detailed measurements from ground stations, specifically ozone, tropospheric aerosols and tropospheric water vapor. The change of the ozone profile in the upper troposphere and lower stratosphere is difficult to measure accurately from space, because that region lies below the bulk of the ozone. A crude measure of total tropospheric ozone change can be obtained by combining SAGE and TOMS satellite data (Fishman *et al.*, 1990), but this does not provide an accurate measure of ozone climate forcing. Although the increased sensitivity of the SAGE III instrument should increase the accuracy of the ozone profile in the tropopause region, it is also important to have monitoring from a number of well placed ground stations. If the plans for the Network for Detection of Stratospheric Change (Kurylo and Solomon, 1990) and plans for tropospheric monitoring (Prinn, 1988) are implemented, and if SAGE III flies in both polar and inclined orbits, monitoring of the ozone profile will probably be adequate for the purpose of defining ozone climate forcing.

Similarly, monitoring of tropospheric aerosols from space with the required high precision is difficult and the capabilities remain to be proven. It will be important to have detailed aerosol monitoring at a number of continental and marine stations and periods of special detailed study to verify changes detected from satellites and obtain additional aerosol properties. Regional ground-based aerosol monitoring networks need to be supported and strengthened. In the United States the Department of Energy ARM sites are expected to make aerosol measurements which are hopefully the beginning of long-term monitoring. As inexpensive sunphotometers are capable of accurate local aerosol measurements, it may be feasible to involve schools in maintaining instrument sites and working with the data, thus yielding useful climate data while providing valuable science educational experience.

Several satellite instruments can measure upper tropospheric water vapor, a key climate feedback parameter influencing climate sensitivity. But such measurements need to be supplemented by a reference network of radiosondes with improved calibration aided by use of a traveling calibration standard.

Relation to Climate Process and Diagnostic Studies

Long-term monitoring of global climate forcings and radiative feedbacks is, of course, only a portion of required global climate measurements (cf., USGCRP, 1993). There is also need for monitoring of climate diagnostics and for detailed measurement and analysis of a number of climate processes, especially relating to the oceans, clouds, precipitation, and fluxes between the surface and the atmosphere. It is important that measurements of these climate diagnostics and processes proceed apace with the long-term climate monitoring of climate forcings and radiative feedbacks. The combination of improved knowledge of changing climate forcings and radiative feedbacks, together with improved understanding and modeling of climate processes, is required to obtain predictive capability for future climate.

The rate at which the climate system responds to a change of climate forcing depends upon how rapidly a heat perturbation mixes into the ocean, which requires appropriate knowledge of ocean circulation. Moreover, it is essential to understand how ocean circulation may change in response to atmospheric changes (Broecker, 1987). The WOCE (World Ocean Circulation Experiment) program (WCRP, 1986), especially if it is continued and expanded, promises to improve our understanding of ocean circulation and its relation to atmospheric climate change. Acoustic tomography, in particular the proposed near-global expansion of the Heard Island experiment (Munk and Forbes, 1989), appears to have exciting potential for monitoring heat uptake by the ocean on decadal time scales. This needs to be complemented by a continuing series of altimetry and scatterometer space missions, together with a reference network of moored buoys, to measure surface winds and ocean currents.

Clouds are probably the most uncertain climate feedback. In addition to monitoring of possibly small decadal cloud changes, it is important to make detailed observations which allow us to understand and model cloud processes better. The EOS mission, which includes the CERES instrument and relatively high resolution imaging, should provide an improved ability to analyze the relation of clouds and the earth's radiation budget, as well as other cloud studies, because almost all of the EOS instruments have some cloud measurement objectives.

Precipitation is a climate diagnostic of great practical importance. Moreover, changes of precipitation can complicate attempts to interpret long-term temperature changes, because of the latent heat associated with evaporation and precipitation. Although there is no expectation that rain rates will be monitored with a precision comparable to that attainable for the radiative forcings and feedbacks, it is important that rainfall monitoring be advanced as much as practical, to improve the simulation and prediction capability of climate models. Thus the TRMM mission (Simpson *et al.*, 1988) planned for 1997 should be just the beginning of a rainfall monitoring satellite series, with measurement capabilities and coverage that improve with time.

Fluxes between the atmosphere and the earth's surface of energy, momentum, water, carbon, and other substances are intimately involved in the functioning of the earth's climate. Many measurements related to these fluxes will be obtained by EOS, and these data should contribute toward improved modeling of climate processes. Many of these data would be more valuable if they were accompanied by accurate measurements of near surface winds; this requires advances in instrument technology and may be a good candidate for a focused small satellite mission. Regional ground-based and ocean field studies are also essential for improved understanding of surface fluxes.

We emphasize that the EOS and other planned experimental measurements are complementary to the monitoring of the proposed Climsat instruments, but the existence of these projects does not negate the need for the proposed Climsat monitoring. For example, EOS does not measure all climate forcings and is not well suited for long-term monitoring. Because of its high cost and the absence of 'hot spares' to replace a failed instrument or spacecraft, EOS is not likely to provide continuous multidecadal monitoring. EOS does not provide the required space-time sampling and coverage as Climsat, which proposes two identical satellites, one with a precessing inclined orbit. The Climsat approach also allows instrument cross-calibration when one must be replaced, which is critical to long-term data precision. EOS does not adequately sample diurnal variations, which are particularly important in defining cloud forcings and feedbacks. EOS does not plan to address the greatest uncertainty in human-made climate forcings, lower atmospheric fine particles, with the required precision until the second AM spacecraft in 2004. The very high wavelength-to-wavelength precision of the Michelson Interferometer proposed for Climsat, which uses a single, passively cooled detector without scanning, is crucial for obtaining the required accuracy. In contrast, the infrared spectrometer on EOS uses separate detectors for each eavelength, requiring individual calibrations, is actively cooled and does not cover the thermal spectrum either continuously or as extensively as MINT.

Discussion

The concept of monitoring missing climate forcings and feedbacks with small inexpensive satellites was discussed by one of us (JEH) in December 1989 at a round-table meeting chaired by Senator Albert Gore, in response to Senator Gore's request for suggestions to reduce uncertainties about global change. Senator Gore asked for a written description of the small satellite proposal, which led to a publication in Issues in Science and Technology (Hansen *et al.*, 1990). On the basis of that article, Congress allocated funds in 1991 to initiate Climsat, but the funding was rescinded by Congress when the program was not initiated. Subsequently, formal proposal of this small satellite concept has awaited the possibility of open competition via an Earth Probe budget line. Such a small satellite budget line has been discussed in connection with recent NASA budgets, but it has not received a priority sufficient to obtain funding.

Independently of Climsat, funding has been approved to build three copies of the SAGE III instrument with the first two expected to be placed on already planned polar and inclined orbit missions. As the limb-viewing SAGE III does not view the same region simultaneously with the downward-looking instruments, there is no scientific requirement that it be on the same spacecraft as the other Climsat instruments. MINT and EOSP, on the other hand, have identical fields-of-view at nadir and need to be on the same satellite to enhance information on cloud, atmospheric and surface properties. In view of the new plans for SAGE, this suggests the possibility of including MINT and EOSP, whose combined mass is less than 40 kg, on an even smaller spacecraft. Such a satellite could also be used for later follow-on flights of SAGE III as a single instrument, since the SAGE III mass is about 40 kg. These alternatives should be explored if a small satellite budget line with open competition becomes a reality.

A satellite-borne lidar could provide valuable data on tropospheric aerosols and clouds, complementary to that of EOSP and MINT. A lidar measures backscatter-ing, which is not simply related to optical depth, but it provides precise determina-tion of vertical layering. It would be particularly useful to have lidar measurements nearly simultaneous with those of Climsat. A lidar on a small satellite launched to fly 'in formation' with Climsat would provide an important test of this approach for Earth monitoring.

Perhaps the most difficult challenge for long-term monitoring of climate forcings and feedbacks is the tropospheric fine particles (aerosols) and associated cloud changes. Although the proposed Climsat instrument EOSP has better capabilities for measuring aerosols than existing instruments, it would need to be coordinated with ground based measurements at a number of globally distributed sites. We believe it would be possible for schools, supported by scientists in their regions, provided with training, and connected by computer network to a coordinating center, to make valuable contributions to this climate monitoring. Specifically, a multichannel sunphotometer can be used to measure aerosol amount (optical depth and size), as well as other relevant quantities such as ozone, water vapor and cloud cover. In conjunction with appropriate curricula, these measurements of the solar spectra could provide valuable science learning experience. It is notable that the concept of involving school students in significant scientific measurements was introduced by Senator Gore at the same 1989 round-table discussion mentioned above.

We note that the Climsat data and data products would be well suited to contribute to teaching Earth sciences in schools because the Climsat data would provide a low volume, comprehensive, on-going description of important climate parameters. Global maps of monthly-averaged distributions of surface and atmo-spheric properties, routinely distributed over Internet, will reveal seasonal and interannual changes which can be used to illustrate global change topics to students. The measurements of both solar and thermal spectra are fundamental quantities which can be related to curricula topics. Curricula could be designed not only to give the typical student some understanding of global change and a taste of how scientific research is performed, but to challenge the precocious ones to get involved in actual global change research.

Acknowledgements

We thank Peter Stone, Robert Charlson and an anonymous reviewer for valuable comments on our manuscript. This research has been supported by the NASA EOS, Climate Modeling and Tropospheric Aerosol research programs.

Acronyms

ACRIM:	Active Cavity Radiometer Irradiance Monitor
ARM:	Atmospheric Radiation Measurements (DOE program)
AVHRR:	Advanced Very High Resolution Radiometer (flown on NOAA satellites)
CFCs:	Chlorofluorocarbons
CERES:	Clouds and the Earth's Radiant Energy System (EOS instrument)
DOE:	U.S. Department of Energy
EOS:	Earth Observing System
EOSP:	Earth Observing Scanning Polarimeter (EOS instrument)
ERBS:	Earth Radiation Budget Satellite
GCM:	Global Climate Model or General Circulation Model
GISS:	Goddard Institute for Space Studies (NASA)
GLOBE:	Global Learning and Observations to Benefit the Environment
IPCC:	Intergovernmental Panel on Climate Change
ISCCP:	International Satellite Cloud Climatology Project
MINT:	Michelson Interferometer
NASA:	U.S. National Aeronautics and Space Administration
NOAA:	U.S. National Oceanic and Atmospheric Administration
SAGE:	Stratospheric Aerosol and Gas Experiment (EOS instrument)
TOMS:	Total Ozone Mapping Spectrometer
TRMM:	Tropical Rainfall Measuring Mission (Japan-U.S. satellite mission)
UARS:	Upper Atmospheric Research Satellite
USGCRP:	United States Global Change Program
WCRP:	World Climate Research Program
WMO:	World Meteorological Organization
WOCE:	World Ocean Circulation Experiment

References

Ardanuy, P. E., Kyle, H. L., and Hoyt, D.: 1992, 'Global Relationships among the Earth's Radiation Budget, Cloudiness, Volcanic Aerosols, and Surface Temperature', *J. Clim.* 5, 1120–1139.

Barnett, T. P., DelGenio, A. D., and Ruedy, R. A.: 1992, 'Unforced Decadal Fluctuations in a Coupled Model of the Atmosphere and Ocean Mised Layer', *J. Geophys. Res.* **97**, 7341–7354.

Betts, A. K.: 1991, *Global Warming and the Tropical Water Budget*, Testimony to the United States Senate Commerce Committee, October 7, Washington, D.C.

Brest, C. L. and Rossow, W. B.: 1992, 'Radiometric Calibration and Monitoring of NOAA AVHRR Data for ISCCP', *Int. J. Remote Sensing* **13**, 235–273.

Broecker, W. S.: 1987, 'The Biggest Chill', *Nat. Hist.* **96**, 74–82.

Cess, R. D. and 19 co-authors: 1989, 'Interpretation of Cloud-Climate Feedback as Produced by 14 Atmospheric General Circulation Models', *Science* **245**, 513–516.

Cess, R. D. and 31 co-authors: 1990, 'Intercomparison and Interpretation of Climate Feedback Processes in Nineteen Atmospheric General Circulation Models', *J. Geophys. Res.* **95**, 16,601–16,615.

Cess, R. D. and 32 co-authors: 1991, 'Interpretation of Snow-Climate Feedback as Produced by 17 General Circulation Models', *Science* **253**, 888–891.

Chandra, S.: 1991, 'The Solar UV Related Changes in Total Ozone from a Solar Rotation to a Solar Cycle', *Geophys. Res. Lett.* **18**, 837–840.

Charlson, R. J., Langner, J., Rodhe, H., Leovy, C. B., and Warren, S. G.: 1991, 'Perturbation of the Northern Hemisphere Radiative Balance by Backscattering from Anthropogenic Sulfate Aerosols', *Tellus* **43AB**, 152–163.

Charlson, R. J., Schwartz, S. E., Hales, J. M., Cess, R. D., Coakley, J. A., Hansen, J. E., and Hofmann, D. J.: 1992, 'Climate Forcing by Anthropogenic Aerosols', *Science* **255**, 423–430.

Clough, S. A., Worsham, R. D., Smith, W. L., Revercomb, H. E., Knuteson, R. O., Woolf, H. W., Anderson, G. P., Hoke, M. L., and Kneizys, F. X.: 1989, 'Validation of FASCODE Calculations with HIS Spectral Radiance Measurements', in Lenoble, J. and Geleyn, J. F. (eds.), *IRS '88: Current Problems in Atmospheric Radiation*, Deepak Publishing, Hampton, Va., pp. 376–379.

Coakley, J. A., Bernstein, R. L., and Durkee, P. A.: 1987, 'Effect of Ship-Stack Effluents on Cloud Reflectivity', *Science* **237**, 1020–1022.

Coakley, J. A. and Cess, R. D.: 1985, 'Response of the NCAR Community Climate Model to the Radiative Forcing of Naturally Occurring Tropospheric Aerosol', *J. Atmos. Sci.* **42**, 1677–1692.

Coffeen, D. L. and Hansen, J. E.: 1974, 'Polarization Studies of Planetary Atmosphere', in *Planets, Stars and Nebulae*, T. Gehrels (ed.), Univ. Arizona Press, Tucson, 1133 pp.

Conrath, B. J., Hanel, R. A., Kunde, V. G., and Prabhakara, C.: 1970, 'The Infrared Interferometer Experiment on Nimbus 3', *J. Geophys. Res.* **75**, 5831–5857.

DelGenio, A. D.: 1993, 'Accuracy Requirements', in Hansen, J., Rossow, W., and Fung, I. (eds.), *Long-Term Monitoring of Global Climate Forcings and Feedbacks*, NASA Conference Publication 3234, 91 pp., (available Goddard Institute for Space Studies, New York).

DelGenio, A. D., Lacis, A. A., and Ruedy, R. A.: 1991, 'Simulations of the Effect of a Warmer Climate on Atmospheric Humidity', *Nature* **251**, 382–385.

Dickinson, R. E. and Cicerone, R. J.: 1986, 'Future global Warming from Atmospheric Trace Gases', *Nature* **319**, 109–115.

Eddy, J. A.: 1976, 'The Maunder Minimum', *Science* **192**, 1189–1202.

Ellsaesser, H. W.: 1983, 'Stratospheric Water Vapor', *J. Geophys. Res.* **88**, 3897–3906.

Farman, J. C., Gardiner, B. G., and Shanklin, J. D.: 1985, 'Large Losses of Ozone in Antarctica Reveal Seasonal ClO_x/NO_x Interaction', *Nature* **315**, 207–210.

Fishman, J., Watson, C. E., Larsen, J. C., and Logan, J. A.: 1990, 'Distribution of Tropospheric Ozone Determined from Satellite Data', *J. Geophys. Res.* **95**, 3599–3617.

Friis-Christensen, E. and Lassen, K.: 1991, 'Length of the Solar Cycle: An Indicator of Solar Activity Closely Associated with Climate', *Science* **254**, 698–700.

Han, Q., Rossow, W. B., and Lacis, A. A.: 1994, 'Near-Global Survey of Effective Droplet Radii in Liquid Water Clouds Using ISCCP Data', *J. Clim.* **7**, 465–497.

Hanel, R., Conrath, B., Hovis, W., Kunde, V., Lowman, P., Maguire, W., Pearl, J., Pirraglia, J., Prabhakara, C., and Schlachman, B.: 1972, 'Investigation of the Martian Environment by Infrared Spectroscopy on Mariner 9', *Icarus* **17**, 423–442.

Hansen, J., Fung, I., Lacis, A., Rind, D., Lebedeff, S., Ruedy, R., Russell, G., and Stone, P.: 1988, 'Global Climate Changes as Forecast by Goddard Institute for Space Studies Three-Dimensional Model', *J. Geophys. Res.* **93**, 9341–9364.

Hansen, J. E. and Lacis, A. A.: 1990, 'Sun and Dust versus Greenhouse Gases: An Assessment of Their Relative Roles in Global Climate Change', *Nature* **346**, 713–719.

Hansen, J., Lacis, A., Rind, D., Russell, G., Stone, P., Fung, I., Ruedy, R., and Lerner, J.: 1984, 'Climate Sensitivity: Analysis of Feedback Mechanisms', in Hansen, J. E. and Takahashi, T. (eds.), *Climate Processes and Climate Sensitivity*, Geophys. Monogr. Ser. 29, AGU, Washington, D.C., pp. 130–163.

Hansen, J. E., Lacis, A., Ruedy, R., and Sato, M.: 1992, 'Potential Climate Impact of Mount Pinatubo Eruption', *Geophys. Res. Lett.* **19**, 215–218.

Hansen, J., Lacis, A., Ruedy, R., Sato, M., and Wilson, H.: 1993a, 'Global Climate Change', *Nat. Geogr. Res. Explor.* **9**, 142–158.

Hansen, J. and Lebedeff, S.: 1987, 'Global Trends of Measured Surface Air Temperature', *J. Geophys. Res.* **92**, 13, 345–13, 372.

Hansen, J., Rossow, W., and Fung, I.: 1990, 'The Missing Data on Global Climate Change', *Issues Sci. Tech.* **7**, 62–69.

Hansen, J., Rossow, W., and Fung, I.: 1993b, *Long-Term Monitoring of Global Climate Forcings and Feedbacks*, NASA Conference Publication 3234, 91 pp., (available Goddard Institute for Space Studies, New York).

Hansen, J. E. and Travis, L. D.: 1974, 'Light Scattering in Planetary Atmospheres', *Space Sci. Rev.* **16**, 527–610.

Hofmann, D. J.: 1990, 'Increase in the Stratospheric Background Sulfuric Acid Aerosol Mass in the Past 10 Years', *Science* **248**, 996–1000.

Hoyt, D. V., Kyle, H. L., Hickey, J. R., and Maschhoff, R. H.: 1992, 'The Nimbus-7 Solar Total Irradiance: A New Algorithm for Its Derivation', *J. Geophys. Res.* **97**, 51–63.

IPCC (Intergovernmental Panel on Climate Change): 1990, 'Climate Change', in Houghton, J. T., Jenkins, G. J., and Ephraums, J. J. (eds.), *WMO/UNEP*, Cambridge (U.K.), Cambridge University Press, 365 pp.

IPCC (Intergovernmental Panel on Climate Change): 1992, 'Climate Change 1992', in Houghton, J. T., Callander, B. A., and Varney, S. K. (eds.), *The Supplementary Report to the IPCC Scientific Assessment*, Cambridge University Press, 200 pp.

Jankowiak, I. and Tanre, D.: 1992, 'Satellite Climatology of Saharan Dust Outbreaks: Method and Preliminary Results', *J. Clim.* **5**, 646–656.

Jensen, E. J. and Toon, O. B.: 1992, 'The Potential Effects of Volcanic Aerosols on Cirrus Cloud Microphysics', *Geophys. Res. Lett.* **19**, 1759–1762.

Jones, P. D., Wigley, T. M. L., and Wright, P. B.: 1986, 'Global Temperature Variations between 1861 and 1984', *Nature* **322**, 430–434.

Joseph, J. H.: 1984, 'The Sensitivity of a Numerical Model of the Global Atmosphere to the Presence of Desert Aerosol', in Gerber, H. E. and Deepak, A. (eds.), *Aerosols and Their Climatic Effects*, Deepak Publ., Hampton, Va., pp. 215–226.

Kandel, R.: 1990, 'Satellite Observation of the Earth Radiation Budget and Clouds', *Space Sci. Rev.* **52**, 1–32.

Karl, T. R., Jones, P. D., Knight, R. W., Kukla, G., Plummer, N., Razuvayev, V., Gallo, K. P., Lindseay, J., Charlson, R. J., and Peterson, T. C.: 1993, 'A New Perspective on Recent Global Warming', *Bull. Amer. Meteorol. Soc.* **74**, 1007–1023.

Kawabata, K., Coffeen, D. L., Hansen, J. E., Lane, W. A., Sato, M., and Travis, L. D.: 1980, 'Cloud and Haze Properties from Pioneer Venus Polarimetry', *J. Geophys. Res.* **85**, 8129–8140.

Kunde, V. G., Conrath, B. J., Hanel, R. A., Maguire, W. C., Prabhakara, C., and Salomonson, V. V.: 1974, 'The Nimbus-4 Infrared Spectroscopy Experiment, 2. Comparison of Observed and Theoretical Radiances from 425–1450 cm^{-1}', *J. Geophys. Res.* **79**, 777–784.

Kurylo, M. J. and Solomon, S.: 1990, *Network for the Detection of Stratospheric Change: A Status and Implementation Report*, NASA/NOAA Joint Report, Code EEU, NASA Headquarters, NOAA Aeronomy Laboratory, Boulder, 71 pp.

Lacis, A. and Carlson, B.: 1993, 'Michelson Interferometer (MINT)', in Hansen, J., Rossow, W., and Fung, I. (eds.), *Long-Term Monitoring of Global Climate Forcings and Feedbacks*, NASA Conference Publication 3234, 91 pp., (available Goddard Institute for Space Studies, New York).

Lacis, A. A., Wuebbles, D. J., and Logan, J. A.: 1990, 'Radiative Forcing of Climate by Changes of the Vertical Distribution of Ozone', *J. Geophys. Res.* **95**, 9971–9981.

Le Texier, H., Solomon, S., and Garcia, R. R.: 1988, 'The Role of Molecular Hydrogen and Methane Oxidation in the Wter Vapour Budget of the Stratosphere', *Quart. J. Roy. Met. Soc.* **114**, 281–295.

Lean, J.: 1991, 'Variations in the Sun's Radiative Output', *Rev. Geophys.* **29**, 505–535.

Lindzen, R. S.: 1990, 'Some Coolness Concerning Global Warming', *Bull. Amer. Meteorol. Soc.* **71**, 288–299.

Lorenz, E.: 1963, 'Deterministic Non-Periodic Flow', *J. Atmos. Sci.* **20**, 130–141.

Manabe, S., Bryan, K., and Spelman, M. J.: 1990, 'Transient Response of a Global Ocean-Atmosphere Model to a Doubling of Atmospheric Carbon Dioxide', *J. Phys. Oceanog.* **20**, 722–749.

McCormick, M. P.: 1993, 'Stratospheric Aerosol and Gas Experiment (SAGE III)', in Hansen, J., Rossow, W., and Fung, I. (eds.), *Long-Term Monitoring of Global Climate Forcings and Feedbacks*, NASA Conference Publication 3234, 91 pp., (available Goddard Institute for Space Studies, New York).

McCormick, M. P., Hamill, P., Pepin, T. J., Chu, W. P., Swissler, T. J., and McMaster, L. R.: 1979, 'Satellite Studies of the Stratospheric Aerosol', *Bull. Amer. Meteorol. Soc.* **60**, 1038–1046.

McCormick, M. P., Veiga, R. E., and Chu, W. P.: 1992, 'Stratospheric Ozone Profile and Total Ozone Trends Derived from the SAGE I and SAGE II Data', *Geophys. Res. Lett.* **19**, 269–272.

Minnis, P., Harrison, E. F., Stowe, L. L., Gibson, G. G., Denn, F. M., Doelling, D. R., and Smith, W. L.: 1993, 'Radiative Climate Forcing by the Mount Pinatubo Eruption', *Science* **259**, 1411–1415.

Munk, W. H. and Forbes, A. M. G.: 1989, 'Global Ocean Warming: An Acoustic Measure?', *J. Phys. Oceanogr.* **19**, 1765–1778.

Oltmans, S. J. and Hofmann, D. J.: 1995, 'Increase in Lower-Stratospheric Water Vapor at a Mid-Latitude Northern Hemisphere Site from 1981 to 1989', *Nature* **374**, 146–149.

Penner, J. E., Dickinson, R. E., and O'Neill, C. A.: 1992, 'Effects of Aerosol from Biomass Burning on the Global Radiation Budget', *Science* **256**, 1432–1434.

Potter, G. L., Ellsaesser, H. W., MacCracken, M. C., and Ellis, J. S.: 1981, 'Albedo Change by Man: Test of Climatic Effects', *Nature* **291**, 47–49.

Prinn, R. G.: 1988, 'Toward an Improved Global Network for Determination of Tropospheric Ozone Climatology Trends', *J. Atmos. Chem.* **6**, 281–298.

Radke, L. F., Coakley, J. A., and King, M. D.: 1989, 'Direct and Remote Sensing Observations of the Effects of Ships on Clouds', *Science* **246**, 1146–1149.

Ramanathan, V., Cicerone, R. J., Singh, H. B., and Kiehl, J. T.: 1985, 'Trace Gas Trends and Their Potential Role in Climate Change', *J. Geophys. Res.* **90**, 5547–5557.

Ramaswamy, V., Schwartzkopf, M. D., and Shine, K. P.: 1992, 'Radiative Forcing of Climate from Halocarbon-Induced Stratospheric Ozone Loss', *Nature* **355**, 810–812.

Randall, D. A. and 26 co-authors: 1994, 'Analysis of Snow Feedbacks in 14 General Circulation Models', *J. Geophys. Res.* **99**, 20,757–20,771.

Rao, C. R. N., Stowe, L. L., McClain, E. P., Sapper, J., and McCormick, M. P.: 1988, 'Development and Application of Aerosol Remote Sensing with AVHRR Data from the NOAA Satellites', in Hobbs, P. V. (ed.), *Aerosols and Climate*, A. Deepak Publ., Hampton, Va., 486 pp.

Raval, A. and Ramanathan, V.: 1989, 'Observational Determination of the Greenhouse Effect', *Nature* **342**, 758–761.

Reber, C. A.: 1990, 'The Upper Atmosphere Research Satellite', *Eos* **71**, 1867–1868, 1873–1874, 1878.

Reinsel, G. C., Tiao, G. C., DeLuisi, J. J., Mateer, C. L., Miller, A. J., and Frederick, J. E.: 1984, 'Analysis of Upper Stratospheric Umkehr Ozone Profile Data for Trends and the Effect of Stratospheric Aerosols', *J. Geophys. Res.* **89**, 4833–4840.

Rind, D., Chiou, E. W., Chu, W., Larsen, J., Oltmans, S., Lerner, J., McCormick, M. P., and McMaster, L.: 1991, 'Positive Water Vapor Feedback in Climate Models Confirmed by Satellite Data', *Nature* **349**, 500–503.

Rossow, W. B. and Cairns, B.: 1995, 'Monitoring Changes of Clouds', *Clim. Dynam.*, (in press).

Rottman, G. J.: 1988, 'Observations of Solar UV and EUV Variability', *Adv. Space REs.* **7**, 53–66.

Sagan, C., Toon, O. B., and Pollack, J. B.: 1979, 'Anthropogenic Albedo Changes and the Earth's Climate', *Science* **206**, 1363–1368.

Sassen, K.: 1992, 'Evidence for Liquid-Phase Cirrus Cloud Formation from Volcanic Aerosols: Climatic Implications', *Science* **257**, 516–519.

Schlesinger, M. E. and Mitchell, J. F. B.: 1987, 'Climate Model Simulations of the Equilibrium Climatic Response to Increased Carbon Dioxide', *Rev. Geophys.* **25**, 760–798.

Simpson, J., Adler, R. F., and North, G. R.: 1988, 'A Proposed Tropical Rainfall Measuring Mission (TRMM) Satellite', *Bull. Amer. Meteor. Soc.* **69**, 278–295.

Stolarski, R. S., Bloomfield, P., McPeters, R. D., and Herman, J. R.: 1991, 'Total Ozone Trends Deduced from Nimbus 7 TOMS Data', *Geophys. Res. Lett.* **18**, 1015–1018.

Tanre, D., Geleyn, J. F., and Slingo, J.: 1984, 'First Results of the Introduction of an Advanced Aerosol-Radiation Interaction in the ECMWF Low Resolution Global Model', in Gerber, H. E. and Deepak, A. (eds.), *Aerosols and Their Climatic Effects*, Deepak Publ., Hampton, Va., pp. 133–177.

Tegen, I. and Fung, I.: 1994, 'Modeling of Mineral Dust in the Atmosphere: Sources, Transport and Optical Thickness', *J. Geophys. Res.* **99**, 22,897–22,914.

Tiao, G. C., Reinsel, G. C., Pedrick, J. H., Allenby, G. M., Mateer, C. L., Miller, A. J., and DeLuisi, J. J.: 1986, 'A Statistical Trend Analysis of Ozonesonde Data', *J. Geophys. Res.* **91**, 13,121–13,136.

Travis, L.: 1993, 'Earth Observing Scanning Polarimeter', in Hansen, J., Rossow, W., and Fung, I. (eds.), *Long-Term Monitoring of Global Climate Forcings and Feedbacks*, NASA Conference Publication 3234, 91 pp., (available Goddard Institute for Space Studies, New York).

USGCRP: 1993, *Our Changing Planet: The FY 1993 U.S. Global Change Research Program*, Committee on Earth and Environmental Sciences, National Science Foundation, Washington, D.C., 79 pp.

Vanderbilt, V. C., Grant, L., Biehl, L. L., and Robinson, B. F.: 1985, 'Specular, Diffuse, and Polarized Light Scattered by Two Wheat Canopies', *Appl. Optics* **24**, 2408–2418.

Wang, W. C., Yung, Y. L., Lacis, A. A., Mo, T., and Hansen, J. E.: 1976, 'Greenhouse Effects Due to Man-Made Perturbations of Trace Gases', *Science* **194**, 685–690.

WCRP: 1986, *Scientific Plan for the World Ocean Circulation Experiment*, WCRP Series No. **6**, WMO/TD no. 122.

Wigley, T. M. L.: 1988, 'The Climate of the Past 10,000 Years and the Role of the Sun', in Stephenson, F. R. and Wolfendale, A. W. (eds.), *Secular Solar and Geomagnetic Variations in the Last 10,000 Years*, Kluwer Publ., Dordrecht, pp. 209–224.

Wigley, T. M. L. and Kelly, P. M.: 1990, 'Holocene Climatic Change, [14]C Wiggles and Variations in the Solar Irradiance', *Phil. Trans. R. Soc. Lond. A* **330**, 547–560.

Willson, R. C. and Hudson, H. S.: 1991, 'The Sun's Luminosity over a Complete Solar Cycle', *Nature* **351**, 42–44.

WMO: 1990, *Scientific Assessment of Stratospheric Ozone: 1989*, World Meteorological Organization Global Ozone Research and Monitoring Project – Report, No. 20, WMO, Geneva.

(Received 23 January, 1995; in revised form 29 June, 1995)

REGIONAL CLIMATE CHANGES AS SIMULATED IN TIME-SLICE EXPERIMENTS

U. CUBASCH[1], J. WASZKEWITZ[1], G. HEGERL[2] and J. PERLWITZ[2]

[1] *Deutsches Klimarechenzentrum GmbH, Bundesstr. 55, 20146 Hamburg, Germany*
[2] *Max-Planck-Institut für Meteorologie, Bundesstr. 55, 20146 Hamburg, Germany*

Abstract. Three 30 year long simulations have been performed with a T42 atmosphere model, in which the sea-surface temperature (SST) and sea-ice distribution have been taken from a transient climate change experiment with a T21 global coupled ocean-atmosphere model. In this so-called time-slice experiment, the SST values (and the greenhouse gas concentration) were taken at present time CO_2 level, at the time of CO_2 doubling and tripling.

The annual cycle of temperature and precipitation has been studied over the IPCC regions and has been compared with observations. Additionally the combination of temperature and precipitation change has been analysed. Further parameters investigated include the difference between daily minimum and maximum temperature, the rainfall intensity and the length of droughts.

While the regional simulation of the annual cycle of the near surface temperature is quite realistic with deviations rarely exceeding 3 K, the precipitation is reproduced to a much smaller degree of accuracy.

The changes in temperature at the time of CO_2 doubling amount to only 30–40% of those at the 3 * CO_2 level and show hardly any seasonal variation, contrary to the 3 * CO_2 experiment. The comparatively small response to the CO_2 doubling can be attributed to the cold-start of the simulation, from which the SST has been extracted. The strong change in the seasonality cannot be explained by internal fluctuations and cold start alone, but has to be caused by feedback mechanisms. Due to the delay in warming caused by the transient experiment, from which the SST has been derived, the 3 * CO_2 experiment can be compared to the CO_2 doubling studies performed with mixed-layer models.

The precipitation change does not display a clear signal. However, an increase of the rain intensity and of longer dry periods is simulated in many regions of the globe.

The changes in these parameters as well as the combination of temperature- and precipitation change and the changes in the daily temperature range give valuable hints, in which regions observational studies should be intensified and under which aspects the observational data should be evaluated.

1. Introduction

Even though global warming has been much in debate during the last decade, there is still a difference in the quantities the modelers analyse (mainly those they know they can trust most in their models) and those, which effect the daily life. The perception of the public suggests that the climate change has a visible impact on a number of meteorological parameters like temperature, storms, droughts, precipitation and so on. However, this perception can rarely be verified with observational data, since only few data exist in a coherent form and only for a limited number of regions.

A number of studies have been carried out to predict regional climate changes. However, these studies have attracted a lot of criticisms, since it was felt that the model resolution was too coarse and the model performance was too poor to

Climatic Change **31**: 273–304, 1995.
© 1995 *Kluwer Academic Publishers.*

allow for a regional interpretation of the results (von Storch *et al.*, 1993 Grotch and McCracken, 1991; Wigley *et al.*, 1990; Karl *et al.*, 1990; Bardossy and Plate, 1992). Various techniques are currently been employed to overcome this problem. Besides statistical methods (Karl *et al.*, 1990; Bardossy and Plate, 1992; von Storch *et al.*, 1993), dynamical regional high resolution models have been nested into global models (for an overview see: Giorgi and Mearns, 1991). These regional models have the advantage that affordable very high resolution simulations can be performed for a certain region of interest for a particular time, but have the disadvantage that they encounter severe problems at the boundaries and that they are currently coupled only one way, i.e. an interaction of the regional scale with the global scale is not possible. An alternative strategy using dynamical models is the so called 'time slice' method using a global atmosphere model.

The present study investigates regional climate changes with a (for climate models) high resolution global general circulation model (T42, i.e. a Gaussian and of ca. 2.8°) using the time-slice method. In this technique the atmosphere model is forced by the sea surface temperature (SST) and sea ice distribution is taken from a transient simulation with a coarse resolution (T21, i.e. a Gaussian grid of 5.6°) globally coupled ocean-atmosphere model (Cubasch *et al.*, 1992) at the point of CO_2 doubling and tripling, and by the corresponding changed greenhouse gas concentration. (Perlwitz *et al.*, 1994; Mahfouf *et al.*, 1994). This method has the advantage that the model can be integrated for several decades around the time of interest with a high resolution, and that it gives a large statistical sample of the changed climate similar to equilibrium experiments with mixed-layer models (IPCC, 1990). Additionally it has a more credible distribution of the SST than the mixed layer models.

Currently there is a growing interest to identify variables suitable for the detection of the climate change. These variables should have a high signal to noise ratio (Santer *et al.*, 1993), should be easy to measure, should have been recorded in the past and should be reliably simulated by the models. As longer records only exist on a regional scale, it is of particular interest to identify not only the parameters but also the regions which are optimal for a detection.

At first it will be analysed how well the model simulates the seasonal cycle of basic weather parameters and whether the seasonal cycle is changed (section 3.1) under the increased greenhouse gas concentration (section 3.1). The analysis will focus on the near surface temperature (section 3.1.1) and the precipitation (section 3.1.2). A number of studies have dealt with the question how the climate change will impact the plant growth and how vegetation regions are shifted (Lohmann *et al.*, 1993; Claussen and Esch, 1992). While these vegetation models mainly use a combination of precipitation and temperature, a shift in vegetation is a quantity difficult to observe. Some observational studies suggest that the combination of precipitation- and temperature change is a better indicator for the climate change than either variable on its own (Karl, 1995, pers. com.). Such a combination is analysed in section 3.1.3.

Studies in North America suggest that the observed global warming is in some regions partially due to a larger rise in the minimum temperature than in the maximum temperature (for an overview see: Kukla *et al.*, 1994). In section 3.2 we will investigate whether this effect can be found in climate models as well, and whether it can be found globally or only in specific regions.

Observational findings from North America (Karl *et al.*, 1995) indicate that there is a tendency of an increase of the 'strong' rainfall. This supposition will be tested with the model results as well (section 3.3).

Finally, during the severe drought in Central U.S. in the year 1988 there have been speculations by some scientists and the press that it is the first sign of the anthropogenic greenhouse effect. While it is impossible to verify this statement with these particular model results, it is, however, possible to investigate, if there is a tendency towards longer drought periods in a changed climate (section 3.5).

2. The Model and the Experiment

The model employed is the T42 version of the ECHAM3 model (DKRZ, 1993). It is a spectral transform model with a triangular truncation and has a vertical resolution of 19 levels. The parametrization of sub-grid scale processes, which is performed on a Gaussian grid of ca. 2.8°, includes a radiation scheme with a broad-band formulation of the radiative transfer equation with six spectral intervals in the infrared and four intervals in the solar part of the spectrum (Hense *et al.*, 1982; Rockel *et al.*, 1991). Gaseous absorption due to water vapor, carbon dioxide and ozone is taken ito account as well as scattering and absorption due to aerosols and clouds. The cloud optical properties are parameterized in terms of cloud water content which is an explicit variable of the model.

The vertical turbulent transfer of momentum, heat, water vapor and cloud water is based upon the Monin-Obukov similarity theory for the surface layer and the eddy diffusivity approach above the surface layer (Louis, 1979). The drag and heat transfer coefficients depend on roughness length and Richardson number, and the eddy diffusion coefficients depend on wind stress, mixing length and Richardson number, which has been reformulated in terms of cloud-conservative variables (Brinkop, 1991).

The effect of orographically excited gravity waves on the momentum budget is parameterized on the basis of linear theory and dimensional considerations (Miller *et al.*, 1989). The vertical structure of the momentum flux induced by the gravity waves is calculated from a local Richardson number which describes the onset of turbulence due to convective instability.

The parametrization of cumulus convection is based on the concept of mass flux and comprises the effect of deep, shallow and mid-level convection on the budget of heat, water vapor and momentum (Tiedtke, 1989). Cumulus clouds are represented by a bulk model including the effect of entrainment and detrainment on the updraft and downdraft convective mass fluxes. Mixing due to shallow stratocumulus convection is considered as a vertical diffusion process with the eddy diffusion coefficients depending on the cloud water content, cloud fraction and the gradient of relative humidity at the top of the cloud.

Stratiform clouds are predicted in accordance with a cloud water equation including sources and sinks due to condensation/evaporation and precipitation formation both by coalescence of cloud droplets and sedimentation of ice crystals. Sub-grid scale condensation and cloud formation is taken into account by specifying appropriate thresholds of relative humidity depending on height and static stability.

The land-surface model considers the budget of heat and water in the soil, snow over land and the heat budget of permanent land and sea ice (Dümenil and Todini, 1992). The heat transfer equation is solved in a five-layer model assuming vanishing heat flux at the bottom. Vegetation effects such as interception of rain and snow in the canopy and the stomatal control of evapo-transpiration are grossly simplified.

A detailed description of the model and a documentation of its performance has been presented by Roeckner et al. (1992), Geckler et al. (1994) and Arpe et al. (1994).

For the experiments presented here, the model has been run in the so called 'time-slice' mode: First a control simulation is carried out with prescribed climatological SST and present day CO_2 concentration ($1 * CO_2$). Then the CO_2 cconcentration is doubled and the change of the SST at the time of doubling of CO_2 of a transient experiment done with a low resolution version of the model coupled to a realistic ocean model (Cubasch et al., 1992) is added to the climatological SST ($2 * CO_2$). In a third experiment the CO_2 concentration has been set to three times its present day level, and the change of the SST at the time of the tripling of CO_2 in the transient experiment has been added to the climatological SST ($3 * CO_2$). All experiments have been run for 30 years and provide therefore an adequate sample for subsequent studies.

This time-slice method, as well as the same SST change data and the changes in the CO_2 concentration has been employed by Mahfouf et al. (1994) and by Parey (1994). The dynamical aspects of the time-slice simulations with the ECHAM3 model and a comparison to the low resolution globally coupled ocean-atmosphere model (ECHAM1 + LSG) can be found in Perlwitz et al. (1994).

3. Results

The model data are compared with the rainfall and surface temperature data after Legates and Willmott(1990a).

Beside the regions defined in the IPCC 1990 report, Central and Northern Europe have been analysed (Figure 1). We only considered only the land points. These

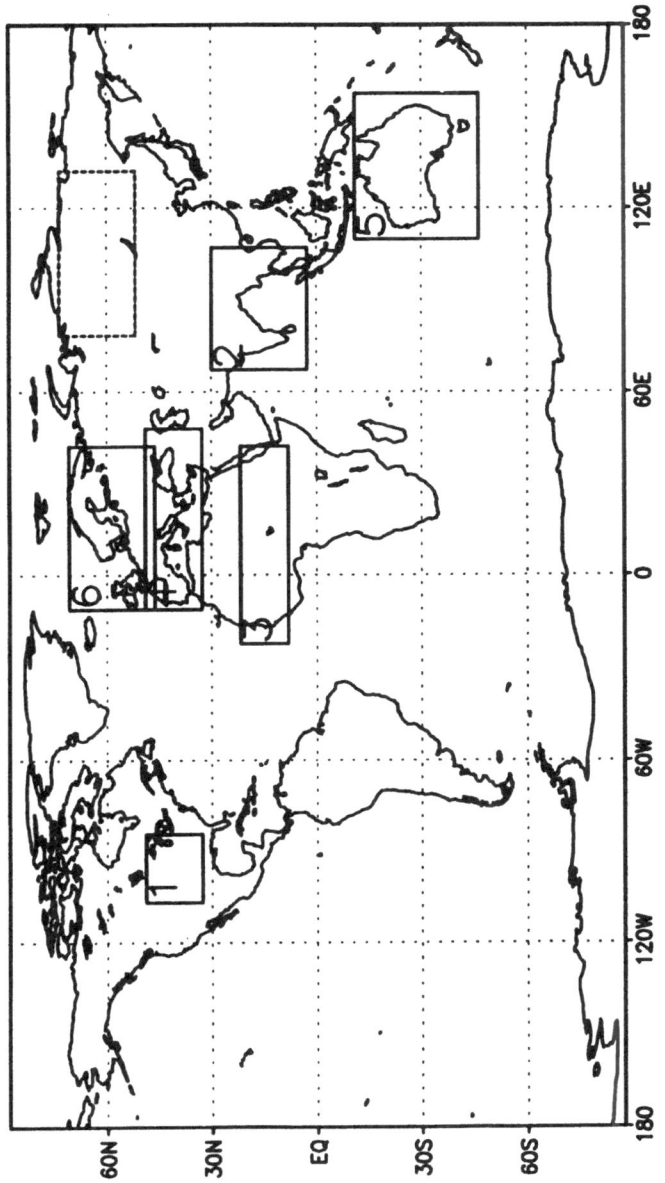

Fig. 1. The six regions discussed in this study. Regions 1–5 represent the regions proposed by IPCC, 1990. (Northern Asia – indicated by dashed lines – is referred to in section 3.1.3). Additionally region 6 (Central and Northern Europe) has been analysed.

TABLE I

The regions selected for this study

No.	Boundaries	Location	Number of land-points in area	Remarks
1	85°–105° W, 35°–50° N	Central North America	46	IPCC '90
2	70°–105° E, 5°–30° N	Southern Asia	69	IPCC '90
3	20° W–40° E, 10°–20° N	Sahel	97	IPCC '90
4	10° W–45° E, 35°–50° N	Southern Europe	76	IPCC '90
5	110°–155° E, 12°–45° S	Australia	84	IPCC '90
6	10° W–40° E, 50°–70° N	Central and Northern Europe	91	

regions represent only a subset of the regions actually analysed. To discuss all of them would exceed the sensible size of a paper. We therefore restrict ourselves to the regions of general interest and we only refer to other regions to deepen the insight into the problems.

3.1. CHANGES IN THE SEASONAL CYCLE

Two quantities are of primary concern for mankind and are among the most commonly measured, at least over land, i. e. precipitation and near surface temperature.

3.2. CHANGES IN SURFACE TEMPERATURE

The annual cycle of the surface temperature (Figure 2(a)) is simulated for Central North America in the $1 * CO_2$ simulation with the right amplitude, but with a warm bias of around 5 K in summer and winter. The temperature change is only marginal by going from $1 * CO_2$ to $2 * CO_2$ and exceeds only in the autumn and winter season 1 K. In the $3 * CO_2$ experiment, a clear signal is established with an average above 3 K. In this experiment the annual cycle is altered with a minimum in spring and a maximum in late fall (longer Indian summer). The maximum in autumn is connected with an increased cloud cover, thus trapping the infrared radiation near the surface. In spring the cloud cover is reduced, allowing for more infrared cooling of the surface. The temperature changes for both (Figure 3(a)), the $2 * CO_2$ and $3 * CO_2$ experiments, are significant. However, they are still

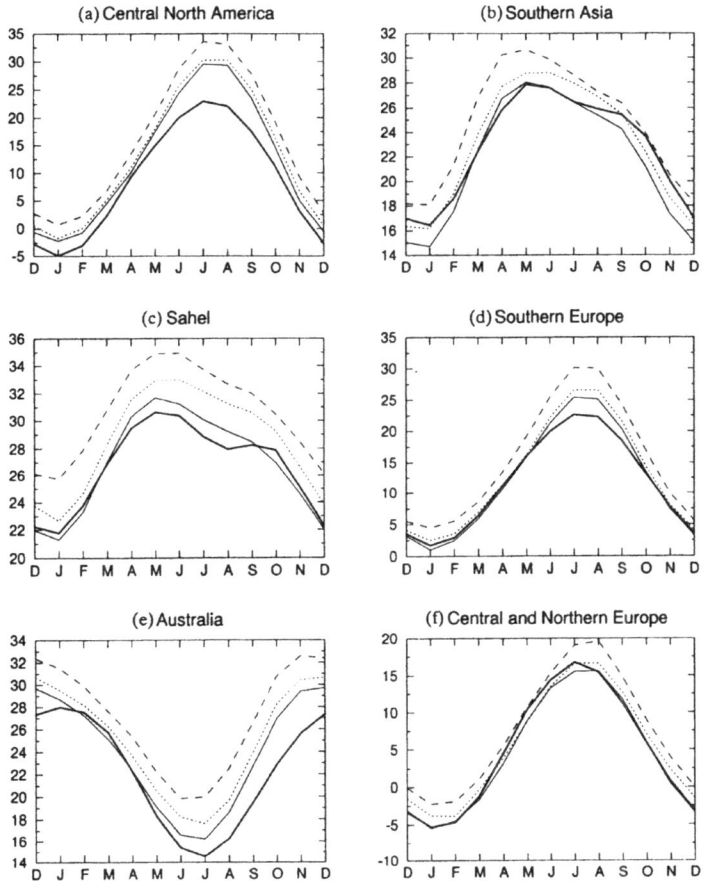

Fig. 2. The annual cycle of the near surface temperature (°C) for the observation (bold solid) and the $1 * CO_2$ (thin solid), $2 * CO_2$ (dotted) and $3 * CO_2$ (dashed) integrations for Central North America (a); Southern Asia (b); the Sahel region (c); Southern Europe (d); Australia (e); and Central and Northern Europe (f).

smaller than the difference to the observation. The year to year variability does not increase significantly in either experiment.

In Southern Asia the observed annual cycle of the near surface temperature (Figure 2(b)) is during spring and summer comparable with observations, but during the other seasons too cold (ca. 2 K in December). The temperature change (Figure 3(b)) for the $2 * CO_2$, experiment over Southern Asia is almost the same in every season, while in the $3 * CO_2$ simulation it has a marked intraseasonal variability with a maximum of 4.5 K in March and a minimum of only 2 K in autumn. The temperature change is correlated with the change in the solar radiation at the

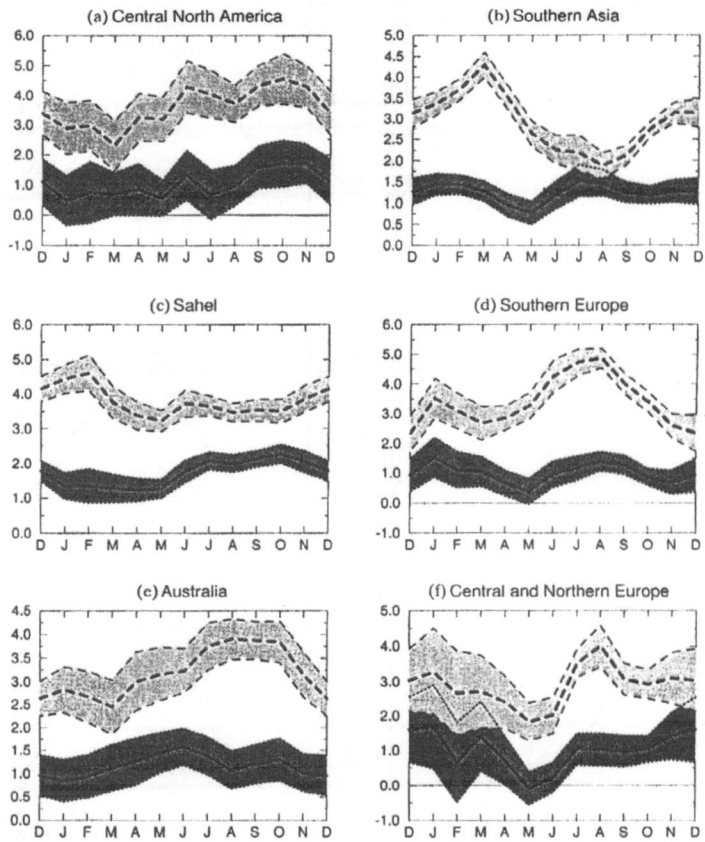

Fig. 3. The annual cycle of the near surface temperature change (°C) for the 2 * CO$_2$ (dotted, dark shading) and 3 * CO$_2$ (dashed, light shading) integrations relative to the 1 * CO$_2$ experiment for Central North America (a); Southern Asia (b); the Sahel region (c); Southern Europe (d); Australia (e); and Central and Northern Europe (f). The shading indicates the confidence limits.

top of the atmosphere with 74% and with the solar radiation reaching the surface with 77%. The key to this phenomenon is the monsoon circulation. In March the conditions in this region are determined by flow from the north giving dry and clear sky conditions. In this case solar radiation passes almost unhindered to the surface, while the enhanced greenhouse effect directly warms the surface. During the pre-monsoon season (March) the radiative input at the surface is increased by up to 6 W/m^2 for the 3 * CO$_2$ simulation. In the summer, more and thicker clouds due to the enhanced hydrological cycle reflect more solar radiation at the top of the atmosphere and absorb more solar radiation on their way to the surface. During the summer month, an average of about 14 W/m^2 less solar radiation reaches the ground in the 3 * CO_2 simulation compared to the control simulation. Even though the thermal radiation is enhanced due to the greenhouse effect, the net radiation at the surface is reduced by about 2.5 W/m^2 during the summer month. The interannual variability does not change with increased greenhouse gas concentration.

In the Sahel region, the observed seasonal change of the near surface temperature (Figure 2(c)) has about the same amplitude as in the model with deviations to observed below 2 K. A real jump in the temperature occurs in the Sahel region only by the 3 * CO_2 experiment with more than 3.5 K on average (Figure 3(c)), while with 2 * CO_2 the warming is just between 1.5 and 2 K. The change in the season cycle is comparatively small and not consistent between the 2 * CO_2 and the 3 * CO_2 experiment. Again, the interannual variability is not significantly altered.

In Southern Europe (Figure 2(d)), the observed seasonal cycle is simulated well in the model for all seasons with the exception of summer, where the simulated temperature is about 3 K too warm. Like in Central Northern America, the temperature change over Southern Europe in the 2 * CO_2 experiment is only marginal and displays no distinct annual cycle. In the 2 * CO_2 experiment, however, the temperature difference (Figure 3(d)) rises from 2.5 K in winter to a maximum of 5 K in late summer. The temporary minimum in May is caused by a decrease in the cloud cover and a decrease in the soil moisture. The gain of solar radiation at the surface is overcompensated by thermal radiation, thus in the mean the temperature is raised less than the annual average. In the subsequent month the solar radiation dominates the budget, while the greenhouse effect allows less thermal radiation to leave the atmosphere, thus leading to a temperature maximum, since the soil-moisture and the cloud cover conditions are almost unchanged compared to the control simulation. Again, the interannual variability is not significantly altered.

As in Southern Europe, the annual cycle of simulated temperature in Australia (Figure 2(e)) has the minimum and maximum in about the right month as observed. The model is too warm in all seasons with the exception of southern hemisphere autumn. The largest deviations occur in southern hemisphere summer with differences of up to 4 K. The temperature increase (Figure 3(e)) has only a marked seasonal cycle in the 3 * CO_2 experiment, where it accelerates the Austral spring. Under 2 * CO_2 condition the temperature rise is seasonally almost independent and only about 30% to 40% of the 3 * CO_2 value. Again, the interannual variability is not significantly altered.

In Central and Northern Europe, the seasonal temperature cycle is simulated almost perfectly (Figure 2(f)). In the 2 * CO_2 as well as in the 3 * CO_2 experiment the change of the temperature is significant and counteracts the seasonal cycle, i. e. the temperature change is much larger in winter than in summer, where in the 2 * CO_2 simulation hardly any temperature change can be found. In this simulation, only in the winter season the temperature change exceeds 1 K. The variability is not significantly altered.

In some regions the temperature change is for both climate change simulations still smaller than the deviation of the simulation to observations. The predicted climate change for the 2 * CO_2 experiment generally only amounts to about 30% to 40% of the one obtained by the 2 * CO_2 experiment instead of the expected 50%. This can be attributed to the cold start phenomenon (Cubasch et al., 1995; Hasselmann et al., 1992). It is caused by the fact that the start of the transient experiment was 1985, not at the beginning of the industrial burning of fossil fuel ca. 1750. Therefore all the warming between 1750 and 1985 has been neglected, which causes an underestimation of the warming in the experiment, from which the SST change had been calculated (Cubasch et al., 1992) by about 15–20% (Fichefet and Tricot, 1992). Additionally the thermal inertia of the ocean delays the temperature response. Transient simulations have at the time of CO_2 doubling only achieved about 60% of their equilibrium warming (Cao et al., 1992; IPCC, 1992). Assuming a response time of the atmospheric part of the Hamburg coupled model of about 30 years (Cubasch et al., 1994), the simulation will have obtained the amplitude of the CO_2 equilibrium response at around 100 years, which coincides with the time, when its transient forcing has reached the 3 * CO_2 mark. Therefore it is not surprising that the 3 * CO_2 time-slice experiments resemble the CO_2 doubling equilibrium simulations with mixed-layer models (see Schlesinger and Mitchell, 1997) more than the 2 * CO_2 time-slice experiment.

However, not all the differences in the response between 2 * CO_2 and 3 * CO_2 can be explained by the cold start and random fluctuations: While the climate change simulated by tbe 2 * CO_2 experiment hardly displays much seasonal variability, the 3 * CO_2 run exhibits in many regions a clear annual signal, which is too strong to be explained by cold start and random fluctuations alone. The forcing of the 3 * CO_2 simulation appears to be strong enough to trigger off a number of positive feedback mechanisms, (for example soil-drying – surface temperature rise) which in some regions reinforce the signal and alter the seasonal cycle.

A change in the variability can generally not be detected. As has been stated by Bengtsson et al. (1994) the model underestimates the natural variability in the tropics when driven by climatological sea-surface temperatures. Its sensitivity towards changes in the variability might therefore not addressable.

3.2.1. Changes in Precipitation

A thorough investigation of the hydrological cycle in the ECHAM3 model has been carried out by Arpe et al. (1994). It indicates that the uncertainties in the observations cause severe problems in the validation of the model results. The model simulates the precipitation mainly within the brackets of uncertainty of the observations. Instead of going through the exercise of comparing the model results to all available observations again, we restrict ourselves to the climatology of Legates and Willmott (1990b) bearing in mind the large uncertainty inherent in this dataset.

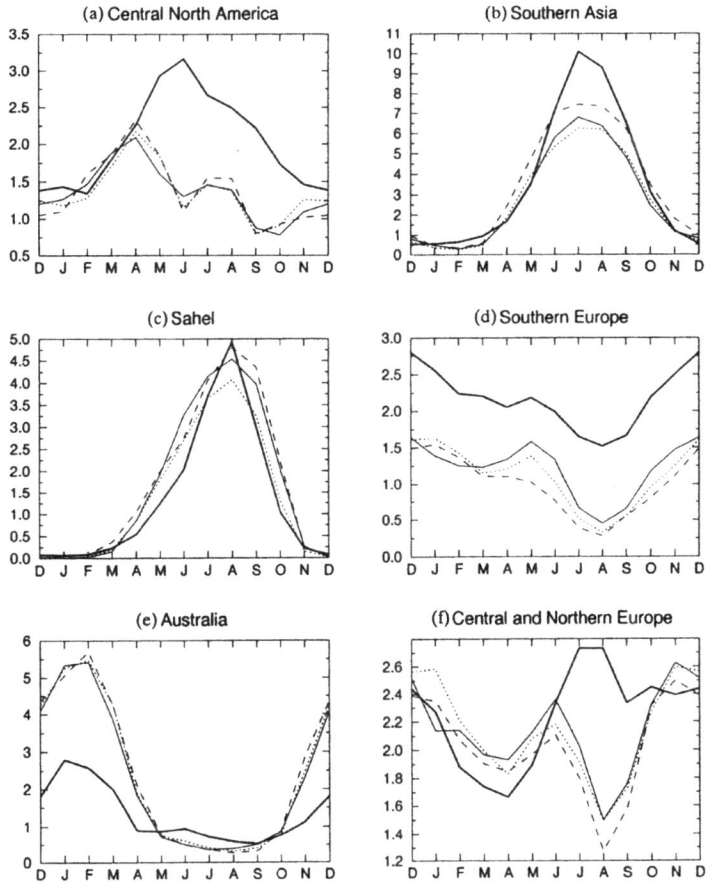

Fig. 4. The annual cycle of the precipitation (mm/d) for the observation (bold solid) and the $1 * CO_2$ (thin solid), $2 * CO_2$ (dotted) and $3 * CO_2$ (dashed) integrations for Central North America (a); Southern Asia (b); the Sahel region (c); Southern Europe (d); Australia (e); and Central and Northern Europe (f).

The simulated precipitation over Central North America completely misses the peak in late spring and early summer (Figure 4(a)). The mean value is only about 50% of the observed amount. The precipitation does not significantly change in Central North America in any of the climate change simulations. Differences of a maximum of 20% still fall within the interannual variability of the control simulation (Figure 5(a)). The interannual variability is not significantly altered either.

The precipitation over Southern Asia (Figure 4(b)) has a marked seasonal cycle with a minimum in early spring (the winter monsoon) and a maximum in late

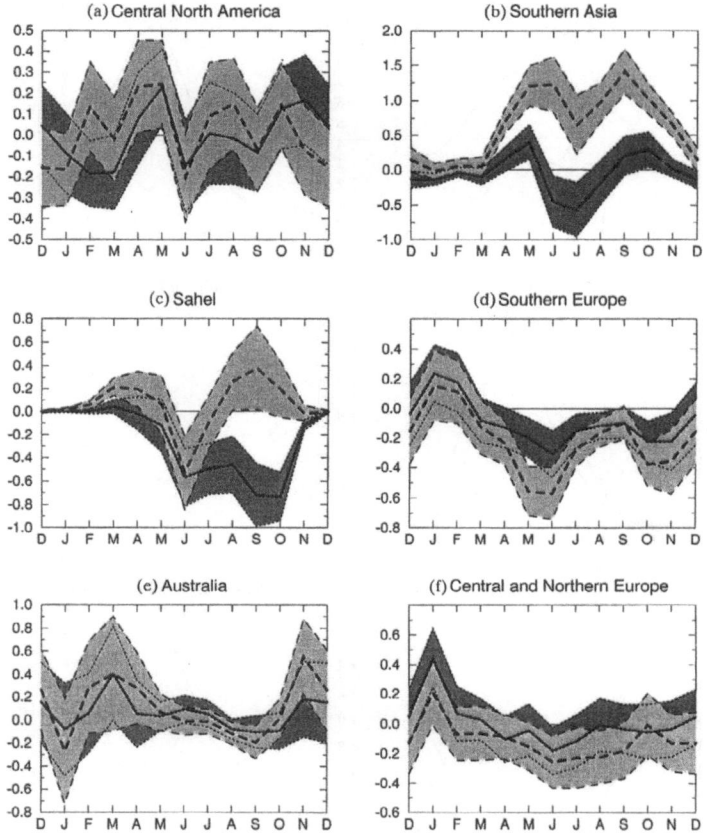

Fig. 5.   The annual cycle of the precipitation (mm/d) for the 2 * CO_2 (dotted, dark shading) and 3 * CO_2 (dashed, ligh shading) integrations relative to the 1 * CO_2 experiment for Central North America (a); Southern Asia (b); the Sahel region (c); Southern Europe (d); Australia (e); and Central and Northern Europe (f). The shading indicates the confidence limits.

summer. The annual cycle of precipitation is simulated quite well over Southern Asia, even though the maximum during the summer monsoon season is underestimated about 30%. This, of course, also drags the mean simulated value below the observed mean.

Under the 2 * CO_2 conditions this situation is hardly altered (Figure 5(b)), and changes are barely significant. Under the 3 * CO_2 conditions, the precipitation in the monsoon season increases by 10%. It has already been stressed by Lal *et al.* (1994) that the enhanced hydrological cycle results in an increase of the monsoon precipitation over India. The interannual variability is not significantly influenced by the greenhouse gas concentration.

The annual cycle of precipitation, its mean and its amplitude, are well simulated in the Sahel region (Figure 4(c)). The change of precipitation is very inconsistent among the experiments. While the 2 * CO_2 experiment predicts a decrease of 40% during summer and autumn, the 3 * CO_2 experiment predicts a 20% increase (Figure 5(c)). The variability increases during the dry season (December to March) in the 3 * CO_2 experiment. Even though the statistical test suggests that this increase is significant, the low absolute amount during this season casts some doubts about the sense of this test under these circumstances.

The annual cycle of the precipitation over Southern Europe (Figure 4(d)) is simulated well, however the absolute amount is underestimated by a factor of two. The precipitation is increased during the winter season in both climate change experiments, but it is decreased during the summer season (Figure 5(d)). The seasonal cycle of precipitation is therefore enhanced.

There are indications that the variability during the summer season is enhanced in both climate change simulations.

While the winter precipitation over Australia (Figure 4(e)) has been simulated well, the steep increase in spring and autumn simulated by the model cannot be found in the observations. The annual mean is therefore overestimated. No clear signal can be identified for the precipitation change (Figure 5(e)). It is as well as under 2 * CO_2 as well as 3 * CO_2 conditions not significant and displays no seasonal cycle. No unequivocal significant change in the variability of precipitation can be found under the changed climate conditions.

The observed summer precipitation of Central and Northern Europe shows a relative maximum while the model simulates a minimum. During the other seasons the observed and simulated values agree quite well. The mean value is underestimated. Under climate change conditions, the model predicts on average a drying, which reaches its maximum during the (already unreasonably dry) summer season, but which is not statistically significant. This decrease of precipitation is slightly higher in the 3 * CO_2 compared to the 2 * CO_2 experiment. The variability is not influenced by the climate change conditions.

The analysis of the seasonal precipitation shows that this parameter is only poorly simulated. The change of precipitation confirms the statements made by Santer et al. (1994) that precipitation has a low signal-to-noise ratio, because it does not show an unequivocal sign. Even the tripling of CO_2 which has a large impact on the near surface temperature, did not influence precipitation in a distinct way, with the exception of Southern Asia and the Sahel. Generally the deviation in the precipitation simulation compared to observations is larger than the predicted climate change.

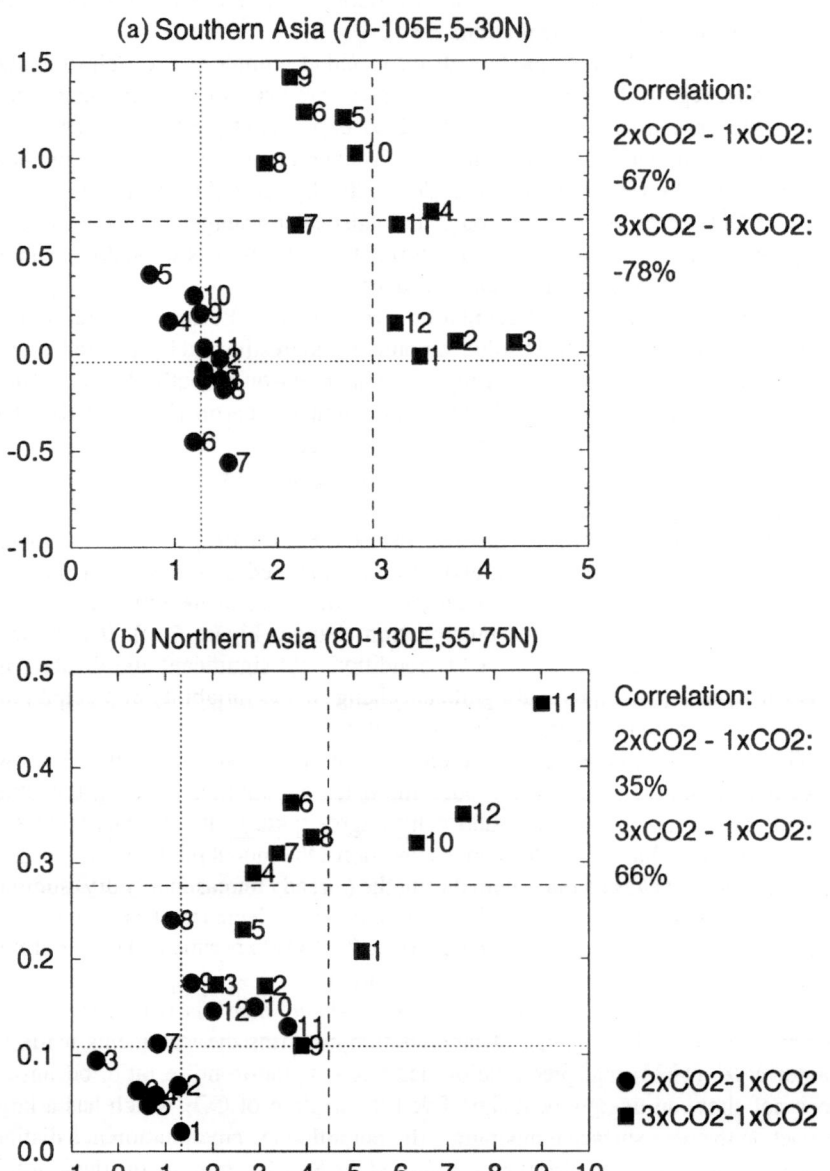

Fig. 6. The monthly ratio between temperature change (*x*-axis (°C)) and the precipitation change
(*y*-axis (mm/d)) for the 2 * CO₂ (circles) and the 3 * CO₂ (squares) experiment for Southern Asia (a)
and Northern Asia (b). The numbers beside the circles and squares are representing the months.

3.2.2. *The Combined Change of Precipitation and Surface Temperature*

To investigate the combination of changes of temperature and changes of precipitation in a region, we have plotted for all regions the changes of precipitation versus the changes of temperature of every month. Figure 6(a) shows the results for Northern Asia, which has been selected, because it has a strong positive correlation between precipitation and temperature change (i.e. in the month with the largest temperature increase the precipitation has gained its maximum rise as well), and for Southern Asia (Figure 6(b)), which has a strong negative correlation. These two regions are prime examples of the two possible correlations between changes of temperature and precipitation. It is not very distinct in the experiments with the $2 * CO_2$, but can clearly be seen in the experiments with the $3 * CO_2$ concentration.

The mechanisms of this different behavior can be explained as follows. In northern Asia (not shown) the temperature increase is largest in late fall and early winter, since the global warming delays the onset of snowfall. At the same time the precipitation increase is at a relative maximum, because the increased temperature allows more humidity to be transported northwards. This rainfall is connected with a substantial increase of the cloud cover, trapping the longwave radiation at the surface, while reflecting little shortwave radiation at the top of the atmosphere, since the radiative input in winter is low. The net radiative gain at the surface is $3 \ W/m^2$ for the winter season. In spring and summer the higher snow depth due to the increased winter precipitation takes more time to melt, and more snow is removed, which damps the warming, and increases of the hydrological cycle.

In southern Asia the temperature increases predominately in the seasons with the minimum percentage of the annual rainfall directly due to the greenhouse effect (cf. section 3.1.1). In the monsoon season the rainfall increases due to the increased hydrological cycle (Cubasch *et al.*, 1994). This is connected with an increased cloudiness. The increased albedo allows less solar radiation to reach the ground. Furthermore, the increased soil wetness counteracts any direct warming. The combined precipitation-temperature change in the other regions can be explained by similar mechanisms, but they rarely are as distinct as in Southern and Northern Asia.

The global correlation between temperature- and precipitation change has been displayed in Figure 7 for the $3 * CO_2$ experiment. A long term observation is only sensible in those regions, where the correlation between temperature- and precipitation change has the same sign with increasing CO_2 concentration. The correlation as simulated by the model therefore has only been displayed for those points, where it has the same sign for both, the $2 * CO_2$ and $3 * CO_2$ experiments, and where it is statistically significant at the 95% level. Polar regions clearly belong to the category where the seasonal temperature change is positively correlated with an increase in precipitation, while tropical regions and part of the subtropics show a negative correlation. For large areas, particularly in the mid-latitudes, no unequivocal correlation between the $2 * CO_2$ and $3 * CO_2$ experiment can be

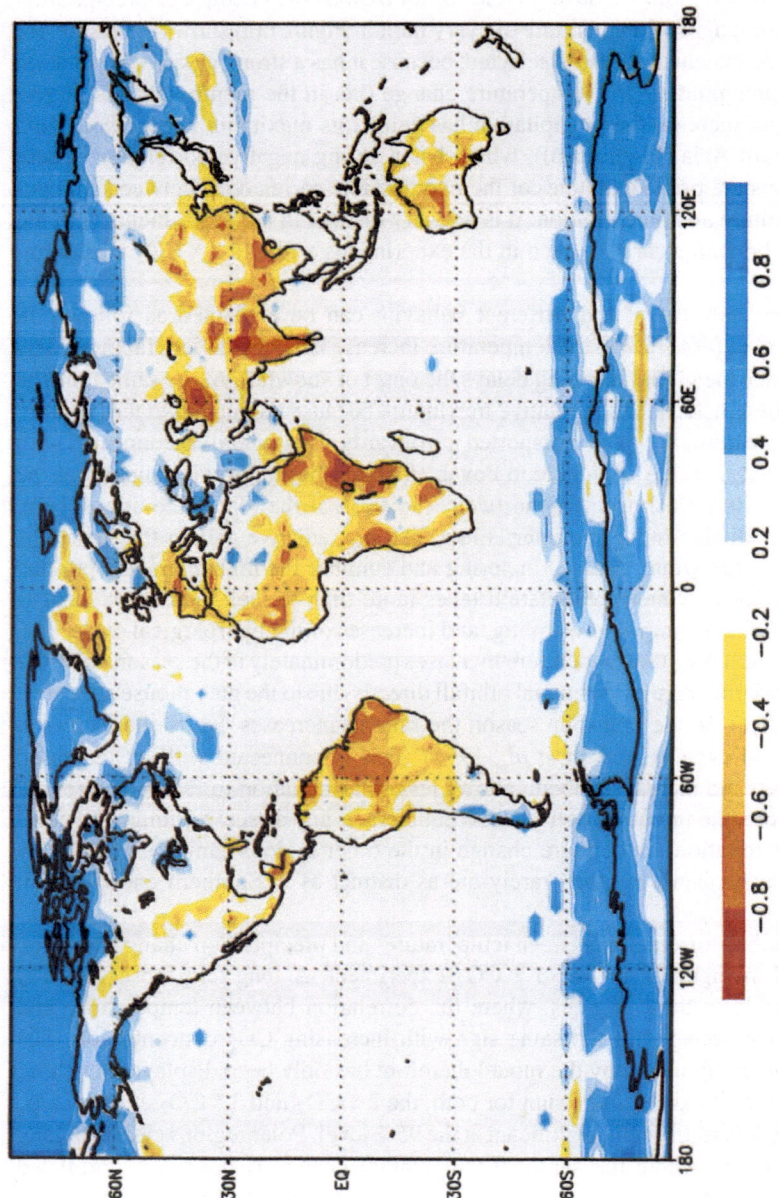

Fig. 7. The latitudinal distribution of the correlation between temperature change and precipitation change (annual mean) for the 3 * CO₂ experiment. Only those points have been plotted, where the correlation of the 2 * CO₂ and 3 * CO₂ integration has the same sign and the range of temperature change is at least 1 °C.

found. The strong correlation in certain regions of the globe, however, makes the combination between temperature- and precipitation change an interesting quantity to evaluate in observations.

3.3. DAILY TEMPERATURE CONTRAST CHANGE

Karl *et al.* (1993) as well as Kukla *et al.* (1994) have stressed that there is some indication that the observed global warming is partially caused by a higher increase of the daily minimum temperature than that of the daily maximum temperature. This causes a decrease of the daily amplitude of temperature. Possible reasons for this effect might be changes in cloudiness and/or in the aerosol distribution. A modelling study of the feedback mechanisms involved can be found by Hansen *et al.*, 1994.

The decrease of the daily temperature range (DTR) can also be seen in the model simulations, but not equally strong in all regions: Contrary to the changes in the mean values, the changes in the daily temperature contrast over North America (Figure 8(a), 9(a)) are more pronounced for the $2 * CO_2$ experiment than in the $3 * CO_2$ simulation. In all seasons except for late summer/early fall the daily temperature contrast is diminished. As the change in the daily temperature range is small and does not even have the same sign for the $2 * CO_2$ and the $3 * CO_2$ experiment, it is difficult to find the physical mechanisms behind it. The signal is certainly not as strong in the model as the observations suggest (Karl *et al.*, 1993). This might, however, also be connected by an unfortunate choice of the IPCC region, which leaves out most of the North American continent.

Southern Asia (Figure 8(b), 9(b)) has a strong change of the daily temperature contrast at the $3 * CO_2$ experiment. This change has a pronounced seasonal cycle as well. It is diminished by almost 1.5 K in late fall. The minimum temperature has risen more than the maximum temperature, which is caused by the higher soil moisture due to the increased monsoon precipitation (c.f. section 3.1.2), thereby increasing the watervapor and cloud cover. This simulated effect is endorsed by observational studies, which emphasize the role of the cloud-cover and the watervapor, however for other regions of the globe (Plantico *et al.*, 1990; Kaas and Frich, 1994).

In the Sahel region (Figure 8(c), 9(c)) the change of the daily temperature contrast is not unequivocal. In the $2 * CO_2$ experiment the contrast increases in summer and autumn, while in $3 * CO_2$ it decreases in almost every season. The change in precipitation might be the key to this inconsistent behavior: In the seasons with the strongest difference between $2 * CO_2$ and $3 * CO_2$ (summer and fall) the rain increases with $3 * CO_2$ and decreases with $2 * CO_2$.

In southern Europe (Figure 8(d), 9(d)) the change in the daily temperature contrast increases in the $3 * CO_2$ experiment, particularly in summer, while it stays almost unchanged in the $2 * CO_2$ experiment. In the $3 * CO_2$ experiment, this can be linked to cloudiness: In summer the cloudiness decreases and allows

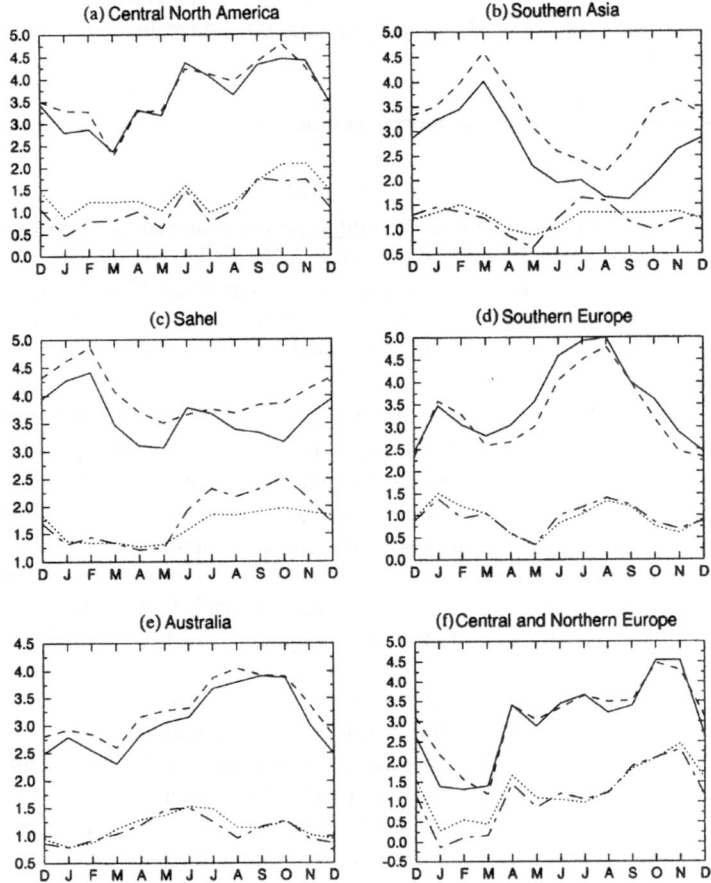

Fig. 8. The change of the annual cycle of the minimum and maximum temperature (°C) for the $2 * CO_2$ and $3 * CO_2$ integration relative to the $1 * CO_2$ experiment for Central North America (a); Southern Asia (b); the Sahel region (c); Southern Europe (d); Australia (e); and Central and Northern Europe (f). Solid: $3 * CO_2$ minus $1 * CO_2$ maximum temperature; dashed: $3 * CO_2$ minus $1 * CO_2$, minimum temperature; dash-dot: $2 * CO_2$ minus $1 * CO_2$, maximum temperature; dotted: $2 * CO_2$ minus $1 * CO_2$, minimum temperature.

during daytime more solar radiation to reach the surface. At the same time the soil-moisture has become so low that the radiation is directly converted into a temperature increase.

The temperature contrast does not increase on average over Central and Northern Europe (Figure 8(f), 9(f)), because the signal is rather unequivocal. This is caused by the peculiarity of the region selected. Taking both, Central and Northern

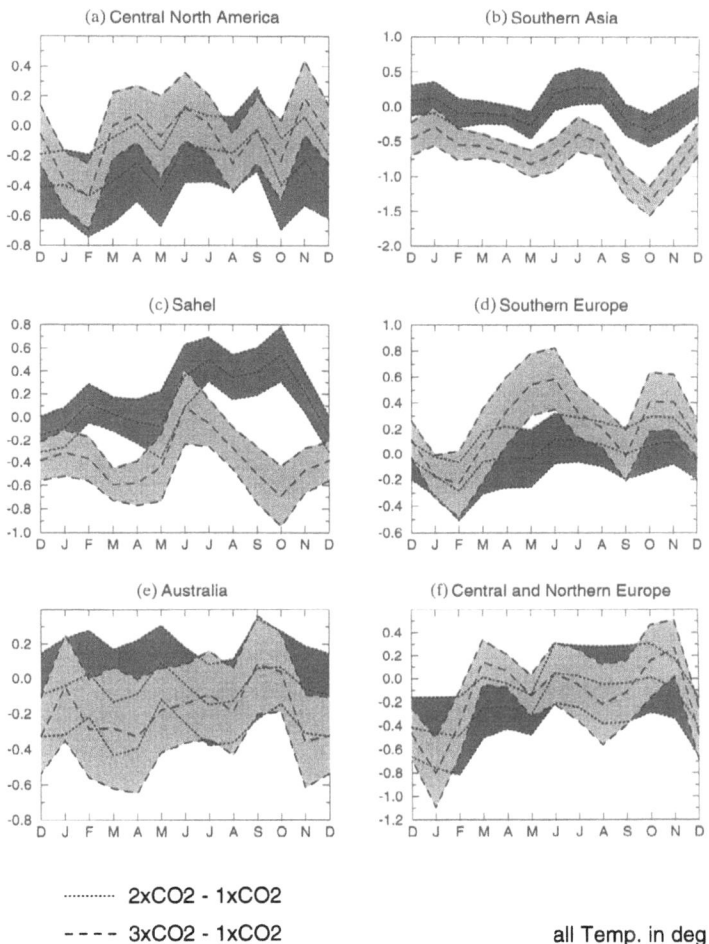

Fig. 9. The change of the annual cycle of the daily temperature contrast (°C) for the 2 * CO₂ (dotted) and 3 * CO₂ (dashed) integration relative to the 1 * CO₂ experiment for Central North America (a); Southern Asia (b); the Sahel region (c); Southern Europe (d); Australia (e); and Central and Northern Europe (f). The shading indicates the confidence limits.

Europe separately, one finds that in Central Europe the contrasts increases, while in Northern Europe it rather decreases (Figure 10).

Cao *et al.* (1992) find in a simulation with a mixed-layer model that for CO_2 doubling over Europe the diurnal temperature range is increased as well, but only in spring.

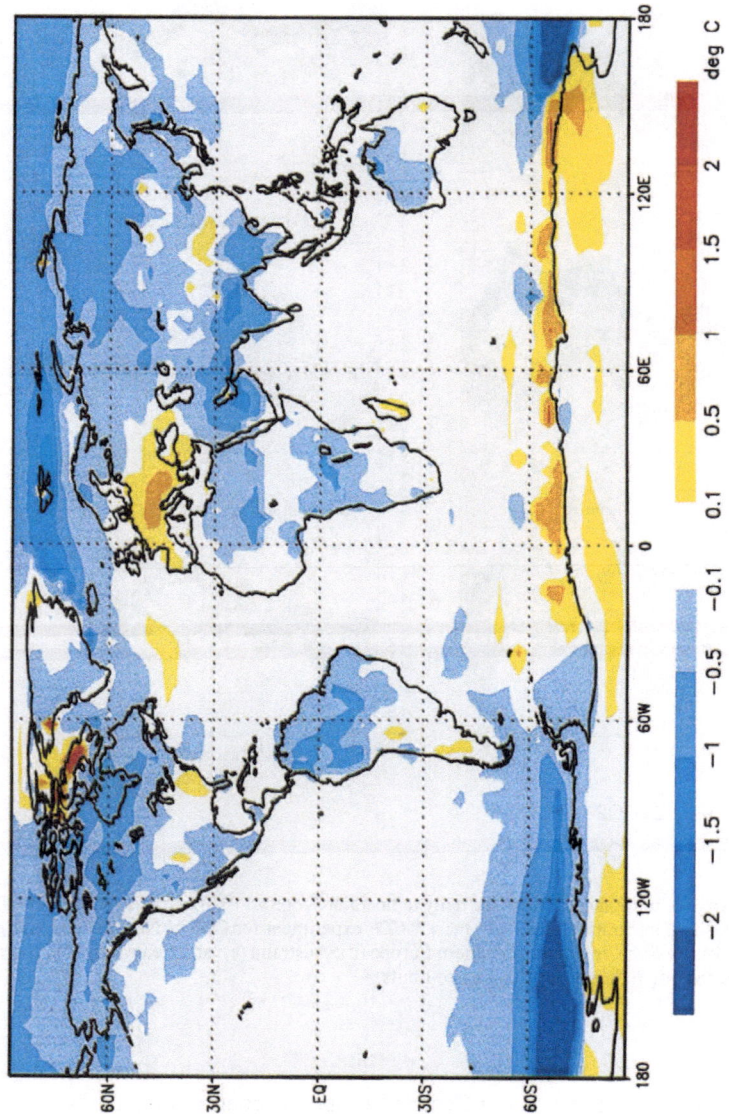

Fig. 10. The regional distribution of the change of the annual mean difference between minimum and maximum temperature for the $3 * CO_2$ integration relative to the $1 * CO_2$ experiment. Regions, where the $2 * CO_2$ and $3 * CO_2$ simulation do not show the same trend or where the changes are not significant, have been blanked out.

This different behavior of the daily temperature contrast in Europe compared to the U.S., can only partially be found in observational studies (Karl *et al.*, 1993; Bücher and Dessens, 1991; Kaas and Frich, 1994: Bradzil *et al.*, 1995). On several locations (f.e. Pyrenees) a lowering of the contrast has been observed, for the Alps only in the valleys (Weber *et al.*, 1994). Generally, the DTR seems to behave in Europe differently than in the U.S. It has to be stressed that in the real world CO_2 has not yet doubled and therefore the response might not yet have been established as a clear signal. There might be other effects like cloud formation due to aerosols masking the real signal.

The change of the daily temperature contrast over Australia (Figure 8(e), 9(e)) is, at least for the $2 * CO_2$ experiment, quite erratic, while a decrease can be found, particularly for the Austral summer, in the $3 * CO_2$ run. The observations suggest a decrease as well (Plummer *et al.*, 1994; Karl *et al.*, 1993).

This Figure 10 shows the regions, where the daily temperature contrast is changed in the $3 * CO_2$ experiment, but only, if the change of the $2 * CO_2$ experiment has the same sign, and only, if the change is statistically significant. Only in these regions an analysis of this quantity is advisable. It can be seen that in the annual mean in almost all land regions of the globe, with the notable exception of middle and southern Europe and Antarctica, there is a clear tendency for a diminished contrast. This is a stable pattern for all seasons. The Arctic undergoes a strong seasonal change with an increase in spring and a strong decrease in fall. This is related to the changed ice melt/ice formation in these transition seasons.

The change of the daily temperature contrast is strongly linked with the change in the total cloud cover, as has been found in observations in the U.S. (Karl *et al.*, 1993) and in the soil moisture (c.f. Table II). More clouds lowers the radiative cooling at the surface at night and more soil moisture damps the warming during daytime, and therefore causes a smaller minimum-maximum difference (Verdecchia *et al.*, 1994). The only exception is region No. 1 (Central North America). The low correlation in this area can be explained by the poorly defined seasonal cycle of the change of the daily temperature contrast. Other effects besides cloud cover change and soil moisture change, which might be responsible for the damping of the DTR like anthropogenic aerosols (Hansen *et al.*, 1994), have not been modelled in this particular study. The error implied by the lack of this forcing term is small according to Hansen *et al.*'s study.

3.4. CHANGES IN THE PRECIPITATION INTENSITY

To estimate the change in the rain intensity, the daily data have been scanned for the rain amount per 24 h interval, which are then separared into classes (> 1, > 2, > 5, > 10, > 20, > 100 mm/day), and then added for every season.

Focussing on the IPCC regions, Figure 11 shows the change of the precipitation classes for all seasons for the control, $2 * CO_2$ and the $3 * CO_2$ simulatlons. The height of each column gives the total precipitation for every season. The

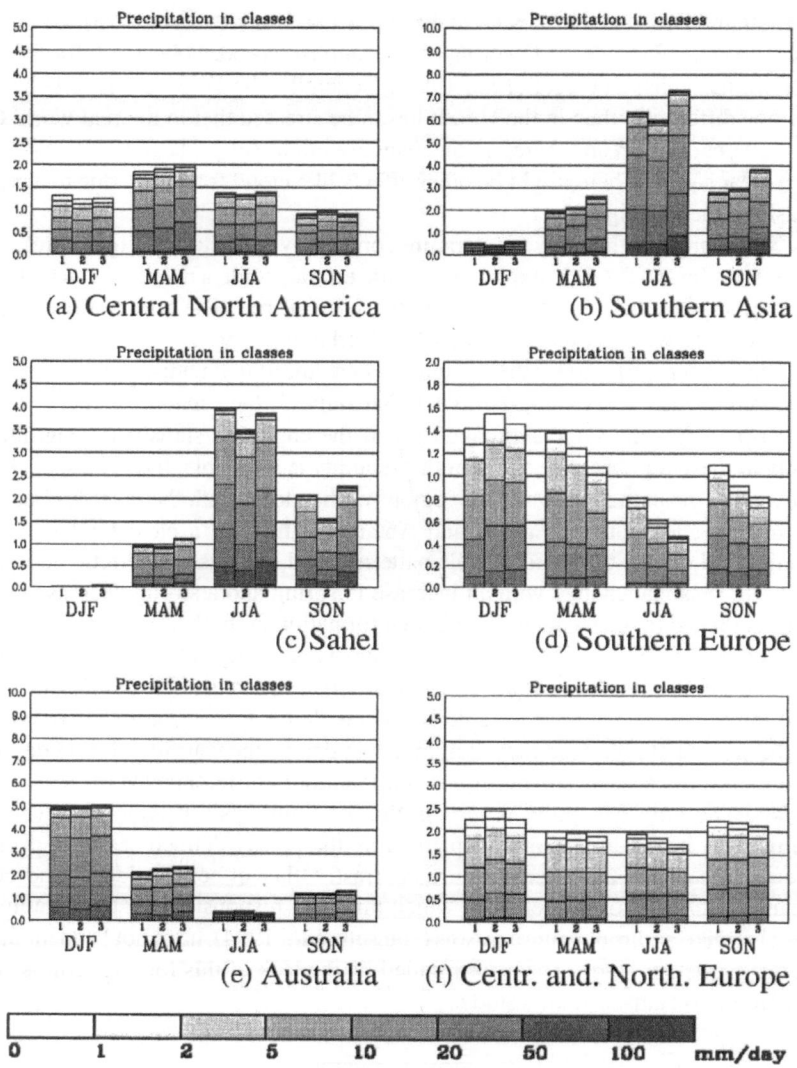

Fig. 11.   The seasonal distribution of precipitation intensity for the 1 * CO₂, 2 * CO₂ and 3 * CO₂ integrations for Central North America (a); Southern Asia (b); the Sahel region (c); Southern Europe (d); Australia (e); and Central and Northern Europe (f). 1 = control; 2 = 2 * CO₂; 3 = 3 * CO₂ experiment. The height of each column indicates the total seasonal precipitation, the subdivisions the share of each precipitation class to the total amount.

TABLE II

Correlation of the annual cycle of the daily temperature contrast change for 3 * CO_2 (in brackets for 2 * CO_2) with the one of the cloud cover and the soil moisture change

No.	Region	Cloud cover change (%)	Soil moisture change
1	Central North America	−28(1)	−47(−50)
2	Southern Asia	−92(−87)	−78(−47)
3	Sahel	−84(−86)	−85(−88)
4	Southern Europe	−86(−51)	−73(−43)
5	Australia	−85(−54)	−75(−92)
6	Central and Northern Europe	−70(−61)	−63(−23)

share of each of the precipitation classes to the total amount is shown via the subdivisions of the columns. The uppermost box represents the class with the smallest precipitation-intensity.

As mentioned before the precipitation amount over Central North America (Figure 11(a)) does not change significantly. It is, however, interesting to note that the rain intensities alter significantly. The class 50–100 mm/day has increased its share on the total precipitation in every season while the share of the rain classes below 10 mm/day stay the same ore even decrease under 3 * CO_2 conditions in every season. Over southern Asia (Figure 11(b)) particularly the rainfall in the summer monsoon season is increased, again by an increase in the high intensity events. This trend is systematic in the 2 * CO_2 and 3 * CO_2 experiments. It is interesting to note that in autumn too the severe rain-events increase. In the 3 * CO_2 experiments even days with precipitation of more than 50 mm/day emerge, which are not present in the control simulation. In the Sahel region (Figure 11(c)) most of the precipitation falls in summer and autumn. The change is not unequivocal, but again, there exists a tendency to increase the strong rain events at the 3 * CO_2 experiment. Southern Europe experiences, with the exception of the winter season, less precipitation under increased CO_2 concentration conditions (Figure 11(d)). The days with intense rain stay about the same or increase marginally. The decrease of overall rainfall can mainly be attributed to a decrease of the weak and medium intensity classes. Australia, as mentioned before, does not show a distinct change of the precipitation (Figure 11(e)). There is a clear indication, that the number of days with high rain intensity increases while the number of days with a low rain intensity decreases. A similar result has been obtained by Whetton et al. (1993), who analyses the change in rainfall intensity over Australia simulated by a number

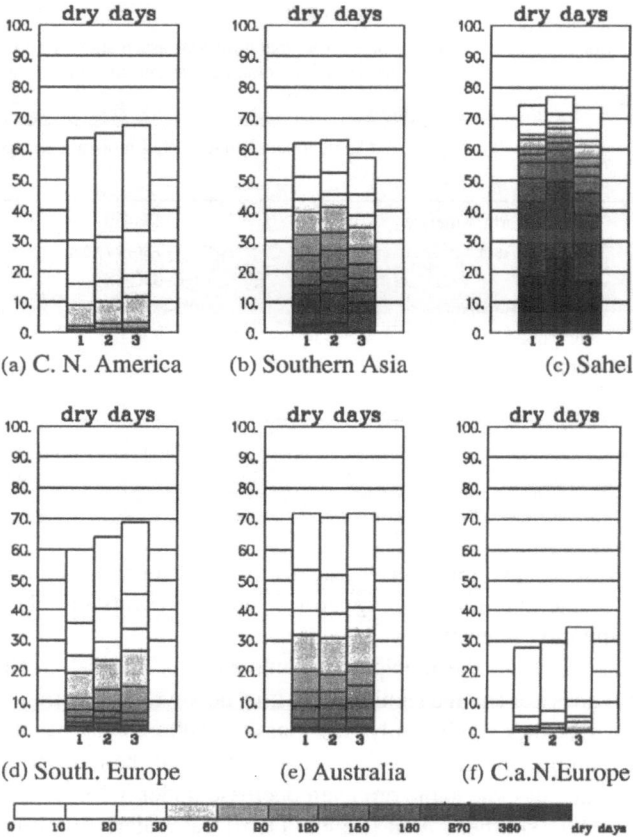

Fig. 12.   The annual mean distribution of consecutive dry days for the 1 * CO₂, 2 * CO₂ and 3 * CO₂ integration for Central North America (a); Southern Asia (b); the Sahel region (c); Southern Europe (d); Australia (e); and Central and Northern Europe (f). 1 = control; 2 = 2 * CO₂, 3 = * CO₂ experiment. The height of each column indicates the total percentage of dry days, the subdivisions the share of each dry spell class to the total percentage.

of equilibrium experiments. In Central and Northern Europe (Figure 11(f)) the total rainfall decreases. The heavy rainfall events increase, particularly in the autumn.

Figure 13 displays those regions of the 3 * CO₂ experiment, where the heavy rain (> 10 mm/day) is changed significantly (t-test) and has the same sign as in the 2 * CO₂ experiment.

Only over oceanic regions the heavy precipitation decreases, while over large areas of the globe, particularly over the regions with tropical rain forest it increases. Since regions like the northern U.S., Canada and Northern Europe have long records not only of the rainfall amount, but also of the rainfall intensity, an analysis of this quantity for these regions might lead to a detection of the climate change.

3.5. CHANGES IN THE FREQUENCY OF DROUGHTS

To get a more detailed view into the frequency of consecutive dry days (precipitation < 0.1 mm/d), 11 classes depending on the number of consecutive dry days have been defined. The classes are: 1 to 10, 11 to 20, 21 to 30, 31 to 60, 61 to 90, 91 to 120, 121 to 150, 151 to 180. 181 to 270, 271 to 360, and > 360 days without precipitation. The simulated years have been scanned for these dry spells and every dry period has been classified. The frequency has been normalized by the total number of days analysed.

In Central North America (Figure 12 (a)) the total number of dry days increases with increased CO_2 concentration. This increase is mainly caused by an increase of very long dry periods. In the 30 year period analysed one would have only a 1% probability of drought of more than three months; while in the 3 * CO_2 experiment this chance is doubled.

In Southern Asia (Figure 12(b)) the number of really dry days is actually decreasing in the 3 * CO_2 experiment, and so is the probability of very long dry spells. In the Sahel region (Figure 12(c)) the number of dry days stays roughly the same, and no marked shift towards longer dryer episodes can be found. Over Southern Europe (Figure 12(d)) the probability of a longer dry spell increases with increased CO_2 concentration, particularly in the 30–60 day class interval. Australia (Figure 12(e)) does not show any marked trend in the change of dry days. This is caused by combining regions with an increase of dry days in the southern part of Australia with regions of an decrease of dry spells in northern Australia. Over Central and Northern Europe (Figure 12(f)) a clear trend can be found towards more dry days and longer dry spells.

To obtain a global overview of the absolute number of dry days combined by their time distribution, we computed the 'average waiting time for the next precipitation' on a randomly chosen day. The significance of this quantity can be explained by an example: Let us consider two gridpoints, both of which have 50% rainy days within one year. In the first grid point it rains exactly every second day, while in the other 180 dry days are periodically followed by 180 rainy days. In the first case the waiting for rain time is $0.5 \times 1 + 0.5 \times O = 0.5$ days, but in the second case it is $\frac{1}{360} \times 180 + \frac{1}{360} \times 179 + \frac{1}{360} \times 178 + ... + \frac{1}{360} \times 1 + \frac{180}{360} \times 0 = 45.25$ days. The 'average waiting time for the next precipitation' is therefore a measure for the length of dry spells. The global distribution and the change of this quantity has been displayed in Figure 14 for the 3 * CO_2 experiment, the change again only for those points, where the sign for the 2 * CO_2 and the 3 * CO_2 experiment coincides. The 'average waiting tlme for the next precipitation' is, as expected, large in the desert regions of the globe, while in the mid-latitudes and the tropics the rainfall almost a daily event. Under climate change conditions, it is increased significantly in the

U. CUBASCH ET AL.

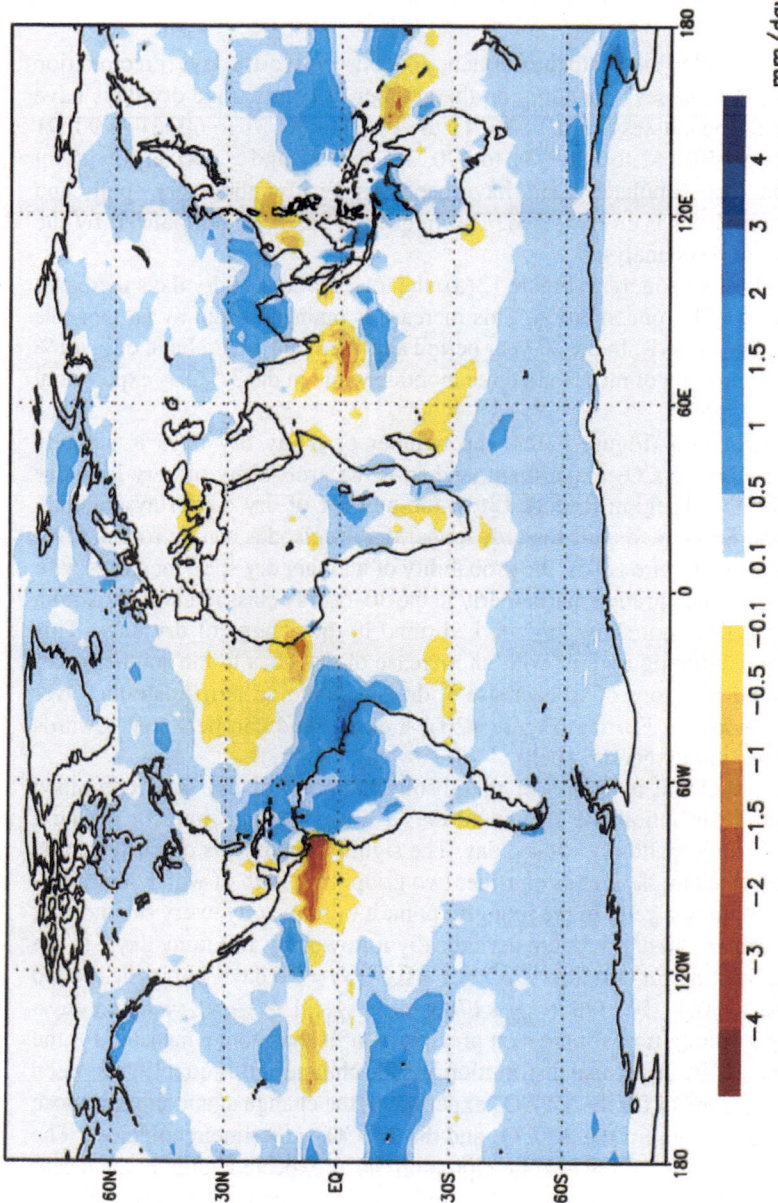

Fig. 13. The regional distribution of the annual mean change of heavy rain (> 10 mm/day) for the 3 * CO_2 integration relative to the 1 * CO_2 experiment. Regions, where the 2 * CO_2 and 3 * CO_2 simulation do not show the same trend or where the changes are not significant, have been blanked out.

mid-latitudes, while it is decreased in the tropics and polar regions. The decrease of the waiting for rain time in some of the desert regions, notably the Sahara, is only of marginal importance and no statistical significance. Here an 'average waiting time for the next precipitation' can be longer than 500 days. The simulations are too short to give these long waiting time much statistical stability. Furthermore, the change is less than a tenth of this value.

4. Discussion

The model simulates in most regions the annual mean temperature climatology realistically with deviations rarely exceeding 3 K and with a realistic phase of the seasonal cycle. In some regions, the model deviations are larger than the simulated change for the climate change scenarios.

The weaker forcing of he $2 * CO_2$ experiment does not allow the changes to emerge as clearly as in the $3 * CO_2$ experiment. Therefore the change for the $2 * CO_2$ experiment is only 30% to 40% of the change of the experiment. This can be attributed to the cold start phenomenon of the simulation, from which the SST data have been derived. This cold start, however, cannot explain that the change in the seasonal cycle is frequently not yet visible in the $2 * CO_2$ experiment.

A comparison with mixed-layer model results shows that the $3 * CO_2$ exeriments resemble rather the results of the equilibrium $2 * CO_2$ experiments than the $2 * CO_2$ time-slice runs. As shown in (IPCC, 1990) the transient models have obtained at the time of CO_2 doubling only about 60% of the equilibrium warming, while by the time they have reached $3 * CO_2$, the transient experiment simulates about the equilibrium temperature change value.

Regional climate changes caused by different SST pattern distribution might have to be taken into consideration as well. As has been shown in Cubasch (1985) a tropical SST anomaly can cause significant changes in the vicinity of the anomaly, and can influence the flow pattern in the mid-latitudes in both hemispheres. However, this effect should not be of much importance, since the tropical SST taken in the tropical belt still correlates almost 80% between the two climate change simulations. The blurred picture of the climate change in Australia is caused by the combination of a tropical and a subtropical region. Here partially the opposite effects in the two different climate zones cancel each other out.

The precipitation simulations are much less reliable than the temperature simulations and bear in many regions not much resemblance with the observations. It has, however, to be noted that not too much confidence can be placed in the observations of precipitation either. The simulated climate change is not very clear and has generally no unambiguous trend. Precipitation as such, as already has been stressed by Santer et al. (1994) is not suited for climate change detection studies and only of limited use for climate impact analysis (von Storch et al., 1993).

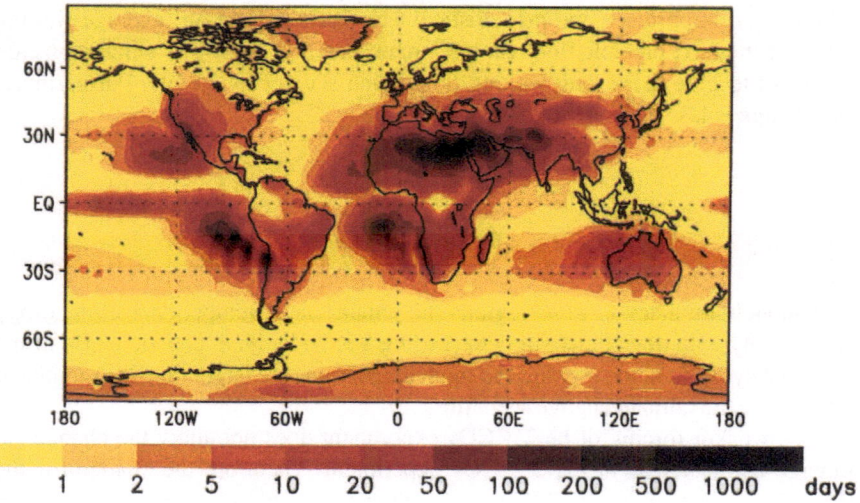

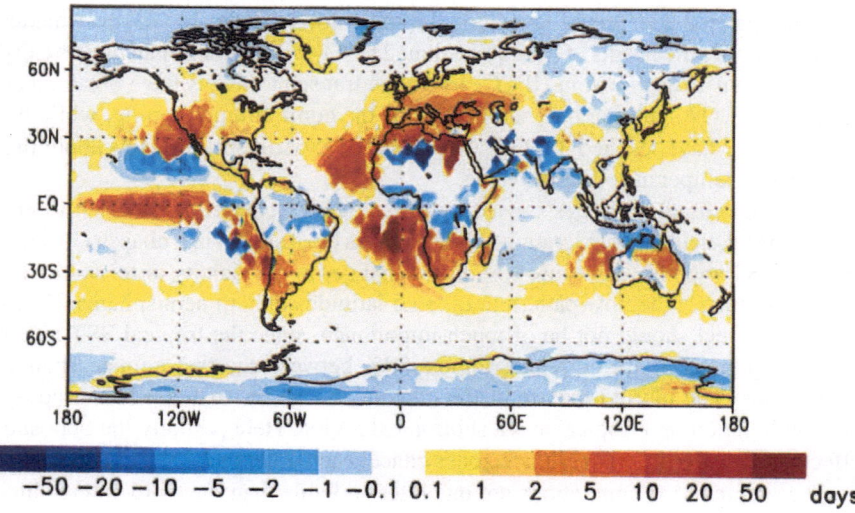

Fig. 14.   The regional distribution of the annual mean of the average waiting time for the next precipitation for the control integration (top) and its change in the 3 * CO$_2$ integration relative to the control experiment (bottom). Regions, where the 2 * CO$_2$ and 3 * CO$_2$ simulation do not show the same trend have been blanked out.

Nearly all continental regions with the exception of the higher latitudes have a negative correlation between the precipitation change and the temperature change, i.e. in those months when the temperature rise is largest, the precipitation change is minimal, and vice versa. This signal is particularly strong in the tropical regions of Africa and South America, and over India, Indonesia and Central and Southern Europe. Over North America this signal is not very distinct. In the polar regions, this correlation is reversed, i.e. in the months with the largest temperature increase also the precipitation increases.

The daily temperature contrast decreases over wide areas of the globe with the exception of Central and Southern Europe. Since this quantity has been measured regularly by meteorological services, it would be worthwhile to analyse it in more detail as has been done up to now. Furthermore, the change of the daily temperature contrast can in most regions be linked to a change in the cloud cover, a quantities archived in the meteorological services and easily observed from satellites.

Similarly the rain intensity and the average waiting time for rain can be used for the detection of climate change. Both are quantities, which have been regularly observed or can easily be derived from regularly observed parameters. Since both are regionally dependent, a comparison of the sign of the change could give an additional confidence into a possibly found climate change.

It has, however, to be stressed that these latter parameters and their changes are strongly dependent on the physical parameterization of the models and the realism of the simulated feedbacks. Most of these parameters have yet to be validated, and, considering the deviations in the simulation of the means of precipitation and temperature, they will certainly display large differences. The next generation of models with a higher resolution (T106 = Gaussian grid of ca. 1.1^0) simulates the climate more realistically (Arpe et al., 1994: Bengtsson et al., 1994), but still has appreciable problems with the hydrological cycle. A time slice integration has been done with this mode, but only for a period of five years and only for the $2 * CO_2$ case. The higher resolution model shows a higher sensitivity to the radiative forcing, i.e. the seasonal cycle is altered already at $2 * CO_2$. However, the small number of simulated years makes it impossible to assign any statistical significance to this result. A new model version (ECHAM4) overcomes some of the problems in the simulation of the temperature over North America (Roeckner, pers. com.) and generally is more realistic. It would be interesting to see if the parameter changes discussed here can be found in simulations of other modeling groups as well.

Acknowledgments

We would like to thank Arno Hellbach and Peter Lenzen for their support in the practical aspects of this work and K. Arpe, E. Roeckner, B. Machenhauer, H. von Storch as well as M. Lal and T. Karl for their helpful discussions. The research has been supported by the German Ministry for Education, Science, Research and Technology (BMBF), the Max-Planck-Gesellschaft, the Freie und Hansestadt Hamburg and the EC Environmental program. The authors are grateful to the staff of the DKRZ for their technical support.

References

Arpe, K., Bengtsson, L., Dümenil, L., and Roeckner, E.: 1994, 'The Hydrologic Cycle in the ECHAM3 Simulations of the Atmospheric Circulation', in Desbois, M., Desalmond, E. (eds.), *Global Precipitation and Climate Change*, Vol. I, **26**, 361–377.

Bardossy, A. and Plate, E. J.: 1992, 'Space-Time Models for Daily Rainfall Using Atmospheric Circulation Pattern', *Water Resource Res.* **28**, 1247–1259.

Begntsson, L., Arpe, K., Roeckner, E., and Schulzweida, U.: 1994, *Climate Predictability Experiments with a General Circulation Model*, Report No. 145, Max-Planck–Institut für Meteorologie, Bundesstr. 55, Hamburg, Germany.

Bengtsson, L., Botzet, M., and Esch, M.: 1994, *Will Greenhouse Gas-Induced Warming over the Next 50 Years Lead to Higher Frequency and Greater Intensity of Hurricanes?*, Report No. 139, Max-Planck-Institut für Meteorologie, Bundesstr. 55, Hamburg, Germany.

Brinkop, S.: 1991, *Inclusion of Cloud Processes in the ECHAM PBL Parameterization*, Large Scale Atmospheric Modelling Rep. No. 9, 5–14, Met. Inst. Univ. Hamburg, Germany.

Bradzil, R., Budikova, M., Auer, I., Boehm, R., Cegnar, T., Fasko, P., Gajic-Capka, M., Koleva, E., Lapin, M., Niedzwiedz, T., Szalai, S., Ustrnul, Z., Weber, R. O., and Zaninovic, K.: 1995, 'Trends of Maximum and Minimum Daily Temperatures in Central and Southeastern Europe – Natural Variability or Greenhouse Gas Signal?', submitted to *Int. J. Climatol.*

Bùcher, A. and Dessens, J.: 1991, 'Secular Trends of Surface Temperature at an Elevated Observatory in the Pyrenees', *J. Clim.* **4**, 859–868.

Cao, H. X., Mitchell, J. F. B., and Lavery, J. R.: 1992, 'Simulated Diurnal Range and Variablity of Surface Temperature in a Global Climate Model for Present and Doubled CO_2 Climates', *J. Clim.* **5**, 920–943.

Claussen, M. and Esch, M.: 1992, *BIOMES Computed from Simulated Climatologies*, Report No. 89, Max-Planck-Institut für Meteorologie, Bundesstr. 55, Hamburg, Germany.

Cubasch, U.: 1985, 'The Mean Response of the ECMWF Global Model to the El Niño Anomaly in Extended Range Prediction Experiments', *Atmosphere-Ocean* **23**, 43–66.

Cubasch, U., Hasselmann, K., Höck, H., Maier-Reimer, E., Mikolajewicz, U., Santer, B. D., and Sausen, R.: 1992, 'Time-Dependent Greenhouse Warming Computations with a Coupled Ocean-Atmosphere Model', *Clim. Dynam.* **8**, 55–69.

Cubasch, U., Santer, B.D., Hellbach, A., Hegerl, G.C., Höck, H., Maier-Reimer, E., Mikolajewicz, U., Stössel, A., and Voss, R.: 1994, 'Monte Carlo Climate Change Forecasts with a Global Coupled Ocean-Atmosphere Model', *Clim. Dynam.* **10**, 1–19.

Cubasch, U., Hegerl, G., Hellbach, A., Höck, H., Mikolajewicz, U., Santer, B. D., and Voss, R.: 1995, 'A Climate Change Simulation Starting 1935', *Clim. Dynam.* **11**, 71–84.

DKRZ: 1993, *The ECHAM3 Atmospheric General Circulation Model*, Techn. Rep. No. 6, DKRZ, Budensstr. 55, Hamburg, Germany.

Dümenil, L. and Todini, E.: 1992, 'A Rainfall-Runoff Scheme for Use in the Hamburg GCM', *EGS Series on Hydrol. Sc.* **1**, 129–157.

Fichefet, T. and Tricot, C.: 1992, 'Influence of the Starting Date of Model Integration on Projections of Greenhouse-Gas Induced Cliamte Change', *Geophys. Res. Lett.* **19**, 1771–1774.

Giorgi, F. and Mearns, L.: 1991, 'Approaches to the Simuylation of Regional Climate Change: A Review', *J. Geophys. Res.* **29**, 191–216.

Gleckler, P. J., Randall, D. A., Boer, G., Colmann, R., Dix, M., Galin, V., Helfand, M., Kiehl, J., Kitho, A., Lau, W., Liang, XZ., Lykossov. B., McAvaney, B., Miyakoda, L., and Planton, S.: 1994, 'Cloud-Radiative Effects on Implied Oceanic Energy Transports as Simulated by Atmospheric General Circulation Models', PCMDI Report No. 15, PCMDI/LLNL, Livermore, Ca., U.S.A.

Grotch, S. L. and McCracken, M. C.: 1991, 'The Use of General Circulation Models to Predict Regional Climate Change', *J. Clim.* **4**, 286–303.

Hansen, J., Sato, M., and Ruedy, R.: 1994, 'Long-Term Changes of Diurnal Temperature Cycle: Implications about Mechanisms of Global Climate Change', in: Kukla, G. *et al.* (eds.), *Asymmetric Change of Daily Temperature Range*, DOE Conf-9309350.

Hasselmann, K., Sausen, R., Maier-Reimer, E., and Voss, R.: 1992, 'On the Cold Start Problem in Transient Simulations with Coupled Ocean-Atmosphere Models', *Clim. Dynam.* **9**, 53–61.

Hense, A., Kerschgens, M., and Raschke, E.: 1982, 'An Economical Method for Computing Radiative Transfer in Circulation Models', *Quart. J. Roy. Met. Soc.* **108**, 231–252.

IPCC: 1990, *Climate Change: The IPCC Scientific Assessment*, Houghton, J., Jenkins, G. J., and Ephraums, J. J. (eds.), Cambridge University Press, 364 pp.

IPCC: 1992, *Climate Change: The Supplementary Report to the IPCC Scientific Assessment*, Houghton, J., Callendar, B. A., and Varney, S. K. (eds.), Cambridge University Press, 198 pp.

Kaas, E. and Frich, P.: 1994, 'DTR and Cloud Cover over the Nordic Countries: Observed Trends and Estimates for the Future', in Kukla, G. *et al.* (eds.), *Asymmetric Change of Daily Temperature Range*, DOE Conf-9309350.

Karl, T., Jones, P. D., Knight, R. W., Kukla, G., Plummer, N., Rzuvaev, V. N., Gallo, K. P., Lindesay, J., and Charlson, R. J.: 1993, 'Asymmetric Trends of Daily Maximum and Minimum Temperature', *Bull. Am. Meteor. Soc.* **74**, 1007–1023.

Karl, T. R., Knight, R. W., and Plummer, N.: 1995, 'Are Temperatures Becoming More Variable and Precipitation More Extreme', *Review Nature*. Karl, T. R., Wang, W.-C., Schlesinger, M. E., and Knight, R. W.: 1990, 'A Method of Relating General Circulation Model Simulated Climate to the Observed Local Climate, Part I: Seasonal Statistics', *J. Clim.* **3**, 1053–1079.

Kukla, G., Karl, T. R., and Riches, M. R. (eds.): 1994, *Asymmetric Change of Daily Temperature*, Proceedings of the MINIMAX workshop, College Park, Maryland, DOE report GCR, CONF-9309350.

Lal, M., Cubasch, U., and Santer, B. D.: 1994, 'Effect of Global Warming on Indian Monsoon Simulated with a Coupled Ocean-Atmosphere General Circulation Model', *Current Sc.* **66**, 430–48.

Legates, D. R. and Willmott, C. J.: 1990a, 'Mean Seasonal and Spatial Variability in Gauge Corrected Global Surface Air Temperature', *J. Climatol.* **41**, 11–21.

Legates, D. R. and Willmott, C. J.: 1990b, 'Mean Seasonal and Spatial Variability in Gauge Corrected Global Precipitation', *J. Climatol.* **10**, 111–127.

Lohmann, U., Sausen, R., Bengtsson, L., Cubasch, U., Perlwitz, J., and Roeckner, E.: 1993, 'The Köppen Climate Classification as a Diagnostic Tool for General Circulation Models', *Clim. Res.* **3**, 177–193.

Louis, J. F.: 1979, 'A Parametric Model of Vertical Eddy Fluxes in the Atmosphere', *Boundary Layer Meteorol.* **17**, 187–202.

Mahfouf, J. F., Cariolle, D., Royer, J.-F., Geleyn, J.-F., and Timbal, B.: 1994, 'Responses of the Meteo-France Climate Model to Changes in CO_2 and Sea Surface Temperature', *Clim. Dynam.* **9**, 345–362.

Miller, M. J., Palmer, T. N., and Swinbank, R.: 1989, 'Parameterization and Influence of Sub-Grid Scale Orography in General Circulation and Numerical Weather Prediction Models', *Met. Atm. Phys.* **40**, 84–109.

Parey, S.: 1994, *2 * CO_2 and 3 * CO_2 Time Slice Experiments. Change in Mean and Variability – 19th General Assembly of the European Geophysical Society – Grenoblè*, April 1994.

Perlwitz, J., Cubasch, U., and Roeckner, E.: 1994, *Simulation of Greenhouse Warming with the ECHAM3 Model Using the Time-Slice Method*, Report, Max-Planck-Institut für Meteorologie, Bundesstr. 55, Hamburg, Germany, *(in preparation)*.

Plummer, N., Lin, Z., and Torok, S.: 1994, 'Recent Changes in the Diurnal Temperature Range over Australia', in: Kukla, G. *et al.* (eds), *Asymmetric Change of Daily Temperature Range*, DOE Conf-9309350.

Roeckner, E., Arpe, K., Bengtsson, L., Brinkop, S., Dümenil, L., Esch, M., Kirk, E., Lunkeit, F., Ponater, M., Rockel, B., Sausen, R., Schlese, U., Schubert, S., and Windelband, M.: 1992,

Simulation of the Present-Day Climate with the ECHAM Model: Impact of Model Physics and Resolution, Report No. 93, Max-Planck-Institut für Meteorologie, Bundesstr. 55, Hamburg, Germany.

Rockel, B., Raschke, E., and Weynes, B.: 1991, 'A Parameterization of Broad Band Radiative Transfer Properties of Water, Ice and Mixed Clouds', *Beitr. Phys. Atmos.* **64**, 1–12.

Santer, B. D., Cubasch, U., Mikolajewicz, U., and Hegerl, G.: 1993, *The Use of General Circulation Models in Detection Climate Change Induced by Greenhouse Gases*, PCMDI Report No. 10, PCMDI/LLNL, Livermore, Ca., U.S.A.

Santer, B. D., Brüggemann, W., Cubasch, U., Hasselmann, K., Maier-Reimer, E., and Mikolajewicz, U.: 1994, 'Signal-to-Noise Analysis of Time-Dependent Greenhouse Warming Experiments. Part 1: Pattern Analysis', *Clim. Dynam.* **9**, 267–285.

Schlesinger, M. E. and Mitchell, J. F. B.: 1987, 'Climate Model Simulations of the Equilibrium Response to Increased Carbon Dioxide', *Rev. Geophys.* **25**, 760–798.

Tiedtke, M.: 1989, 'A Comprehensive Mass Flux Scheme for Cumulus Parameterization in Large-Scale Models', *Mon. Wea. Rev.* **117**, 1779–1800.

Verdecchia, M., Visconti, G., Giorgi, F., and Marinucci, M. R.: 1994, 'Diurnal Temperature Range for a Doubled Carbon Dioxide Concentration Experiment: Analysis of Possible Physical Mechanisms', *Geophys. Res. Let.* **21**, 1527–1530.

von Storch, H., Zorita, E., and Cubasch, U.: 1993, 'Downscaling of Global Climate Change Estimates to Regional Scales: An Application to Iberian Rainfall in Wintertime', *J. Clim.* **6**, 1161–1171.

Wever, R. O., Talkner, P., and Stefaniki, G.: 1994, 'Asymmetric Diurnal Temperature Change in the Alpine Region', *Geophys. Res. Let.* **21**, 673–676.

Whetton, P. H., Fowler, A. M., Haylock, M. R., and Pittock, A. B.: 1993, 'Implications of Climate Change Due to the Enhanced Greenhouse Effect on Floods and Droughts in Australia', *Clim. Change* **25**, 289–317.

Wigley, T. M. L., Jones, P. D., Briffa, K. R., and Smith, G.: 1990, 'Obtaining Subgrid Scale Information from Corase-Resolution General Circulation Model Output', *J. Geophys. Res.* **95**, 1943–1953.

(Received 25 January, 1995; in revised form 12 July, 1995)

MONITORING CHANGES OF CLOUDS

WILLIAM B. ROSSOW

NASA Goddard Institute for Space Studies, New York, NY 10025, U.S.A.

and

BRIAN CAIRNS

Columbia University, New York, NY 10027, U.S.A.

Abstract. An analysis of the spatial and temporal scales of cloud variability and their coupling provided by the results from existing cloud observing systems allows us to reach the following conclusions about the necessary attributes of a cloud monitoring system. (1) Complete global coverage with uniform density is necessary to obtain an unbiased estimate of cloud change and an estimate of the reliability with which that change can be determined. (2) A spatial sampling interval of less than 50 km is required so that cloud cover distributions will generally be homogeneous, or statistically homogeneous, within a sample. (3) A sampling frequency of at least six times a day ensures not only that the diurnal and semi-diurnal cycles are not aliased into long term mean values, but also that changes in them can be monitored. (4) Since estimated climate changes are only evident on a decadal time-scale, unless cloud monitoring is continuous with a record length greater than 10 years and has very high precision ($\approx 1\%$) instrument calibration with overlapping observations between each pair of instruments, it will not be possible either to detect or to diagnose the effects of cloud changes on the climate.

1. Introduction

Proper monitoring of changes in clouds that can feedback on climate changes must account for the large variety of cloud characteristics and their complex space-time variations. Because clouds on Earth are formed by condensation of water vapor, their variations are under control of and exhibit similar variations as the atmospheric motions (e.g., Rossow, 1978). The key characteristics of atmospheric motions are that they vary over a large range of space-time scales and that variations at particular space-time scales are coupled. Thus, clouds vary on time scales from ~ 10 min to (at least) ~ 10 yr and on space scales from ~ 30 m to $\approx 40,000$ km; but the smallest/largest time scales are associated with the smallest/largest space scales (Figure 1). Once we know something about the spectrum of cloud variations with scale, we can design an observing system to monitor cloud changes that is not a 'brute force' approach that simply resolves the smallest scales and covers the largest scales. In fact, the coupling of space-time scales makes this approach very impractical, requiring global images at 5 min intervals with a spatial resolution of 15 m. However, we need to do more than merely detect change; rather we need to be able to evaluate the effects of such changes on the climate, which requires more detailed observations than needed for change detection.

The shapes of the distributions of cloud properties are not even approximately Gaussian (Rossow and Schiffer, 1991). Surface and satellite determinations of cloud areal coverage both show that the distribution of this quantity is bi-modal

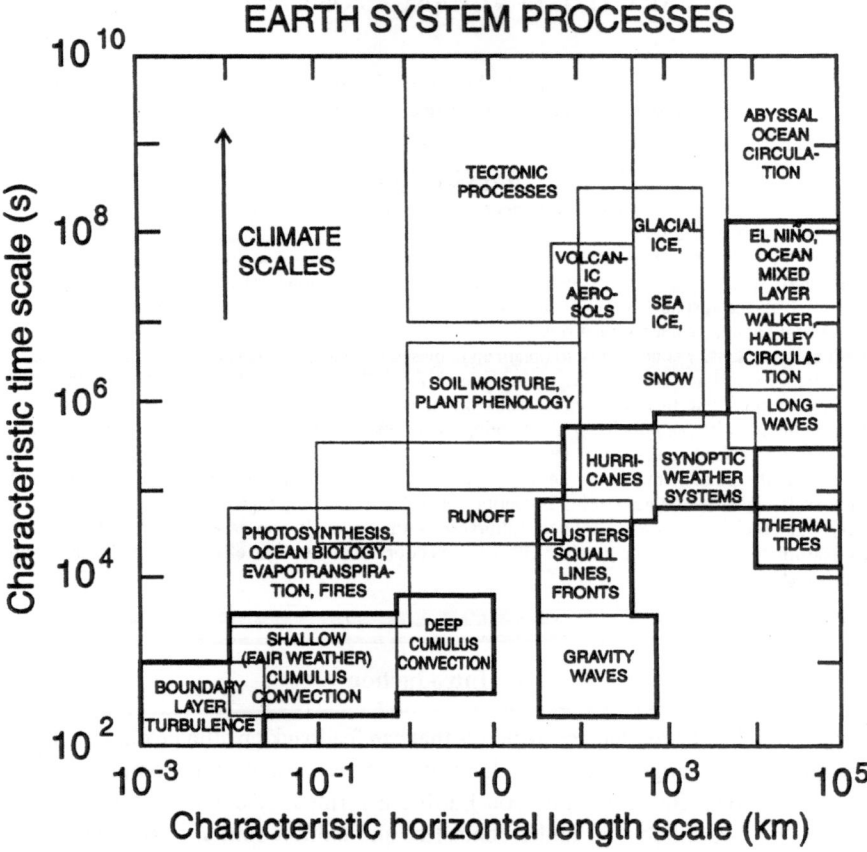

Fig. 1. Schematic of spatial and temporal scales associated with Earth system processes.

at smaller ($\lesssim$ 200 km) spatial scales (Figure 2). The optical thickness (or water content) distribution is monomodal, but very asymmetric with the largest values occurring very rarely and being more than 40 times larger than the most frequently occurring values (Figure 3a). When plotted on a scale that is linear in the radiative effect of cloud optical thickness (Figure 3b), the distribution is very broad but still shows a preponderance of optically thin clouds. The distribution of cloud top pressures viewed from satellites (Figure 4) is almost uniform over the range from 300 to 800 mb with a weak concentration at higher values (more low-level clouds). Obscuration of lower level clouds by higher clouds reduces the concentration at lower levels somewhat. There is also a small secondary concentration near the tropopause (100 mb) but the reality of this feature is uncertain (Liao et al, 1995a, b).

COMPARISON OF ISCCP AND SURFACE CLOUD AMOUNTS

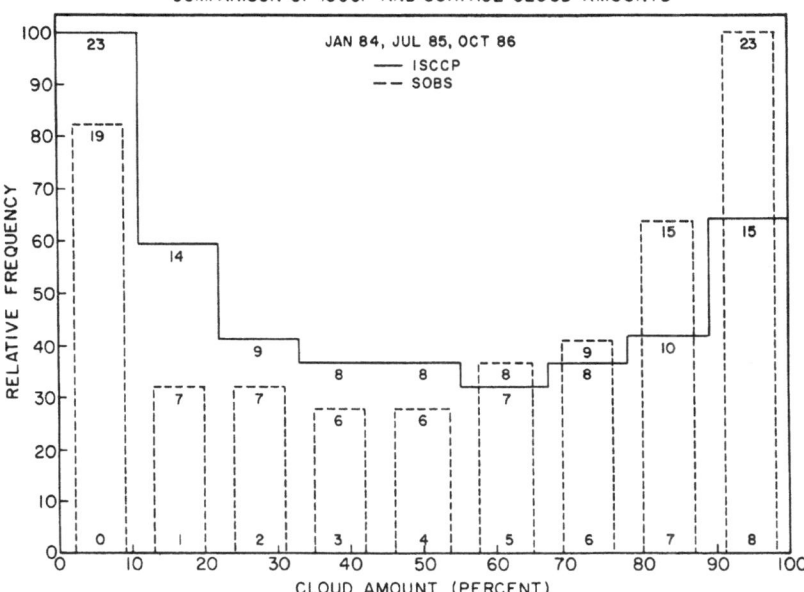

Fig. 2. Frequency distribution of cloud amounts in each octa (12.5%) over land from individual surface observations (dashed line) and the ISCCP analysis (solid line) for January 1984, July 1985 and October 1986. The ISCCP analysis is for areas about 280 km across and the surface observations represent areas about 50 km across. Fractions of total population in percent in each cloud amount interval are indicated by numbers near the top of each bar.

These one dimensional distributions (Figures 3 and 4) are misleading because the variations of these two basic cloud properties are correlated, producing more complicated structures that can be described in terms of the occurrence of different cloud types (Rossow and Schiffer, 1991). Figure 5 contrasts the distributions of cloud properties in the tropics, subtropical oceans and midlatitudes. Since the effects of clouds on the planetary radiation budget and hydrological cycle are strongly non-linear functions of these cloud properties, the dominant cloud types in the radiation budget are not the dominant clouds in the hydrological cycle. The most frequently occurring clouds are lower-level and lower optical thickness, so that they dominate the radiation budget on average (Rossow and Zhang, 1995); whereas the key cloud types in the hydrological cycle are the relatively rare precipitating systems characterized by high optical thicknesses and high top heights (Lin and Rossow, 1994). Moreover, the relation of cloud properties to atmospheric motions is more likely to involve variations of cloud types expressed as correlated changes in several cloud properties (cf. Lau and Crane, 1995). Thus, cloud changes cannot be adequately described by an average and standard deviation of the cloud properties, determined separately; rather cloud changes must be described by **correlated**

ISCCP-C1 April 1988

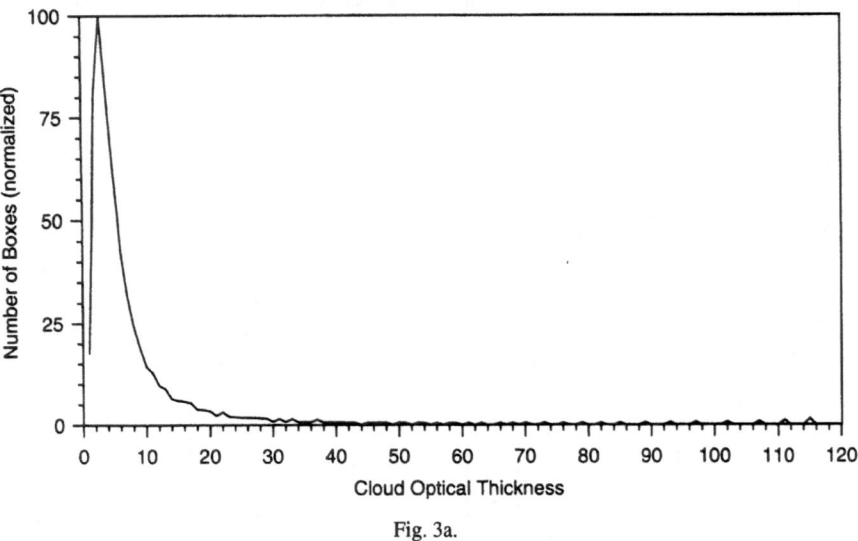

Fig. 3a.

ISCCP-DX NOAA-9 Ascending April 1987

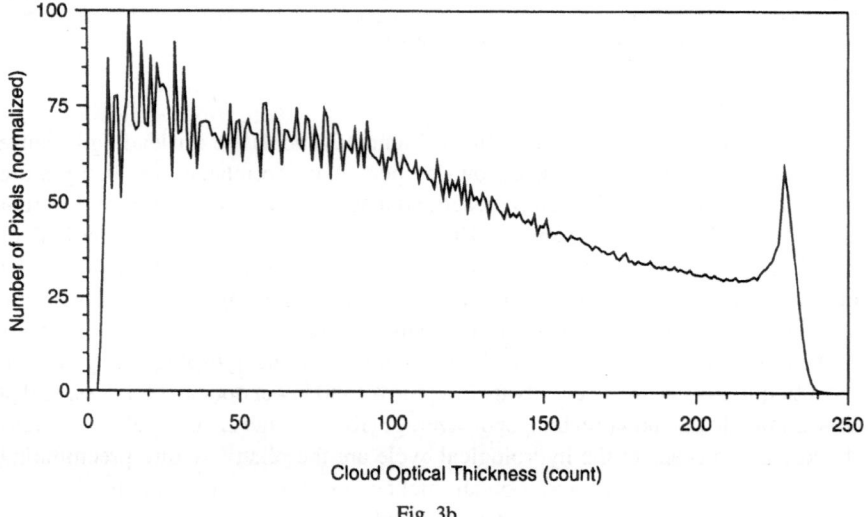

Fig. 3b.

Figs. 3(a)–(b). Frequency distribution over the whole globe of: (a) cloud optical thickness for April 1988; and (b) cloud spherical albedos (proportional to optical thickness count values in the ISCCP dataset) for April 1987 from ISCCP. In (a) individual values have been averaged over regions about 280 km in size, but in (b) the distribution of individual values is shown directly.

ISCCP-C1 April 1988

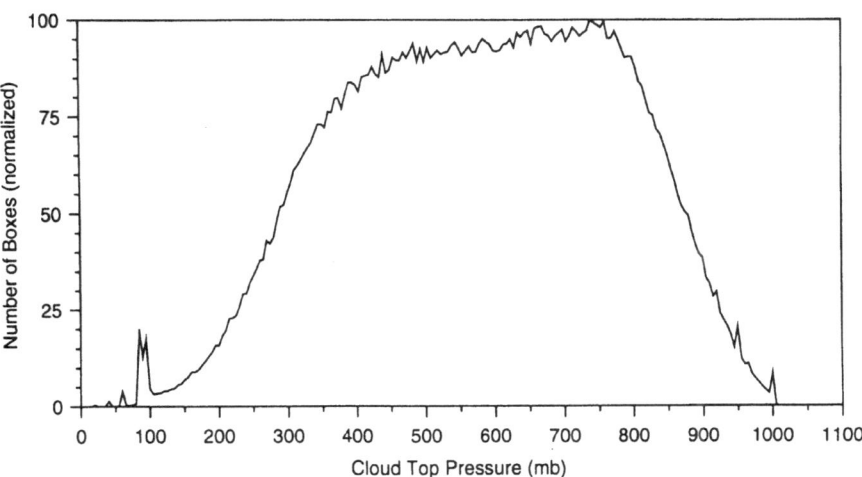

Fig. 4. Frequency distribution over the whole globe of cloud top pressure (in millibars) inferred by ISCCP every three hours in April 1988. Individual values have been averaged over regions about 280 km in size.

changes in the shapes of the distributions of cloud properties, e.g., changes in amounts of cloud types.

Since the effects of cloud changes on the radiation and hydrological budgets also vary with the environment in which the cloud occurs, monitoring must also describe changes in the geographic distribution of clouds. For example, a shift of cloud cover from land to ocean without changing the total would alter the radiation budget. Moreover, systematic diurnal and seasonal changes in solar radiation and atmospheric temperature also alter cloud effects on the radiation and hydrological budgets, so that monitoring must also describe the distribution of clouds relative to these two time scales. For example, a shift of cloud cover from day to night or from winter to summer without changing the total would alter the radiation budget. In fact, any significant correlation of cloud variations with systematic variations of the atmosphere and surface can be crucial to determining their effects on the climate because such variations would change the radiative and latent heating/cooling that drives the atmospheric circulation. That cloud variations are, themselves, controlled by the atmospheric motions guarantees that many such correlations are possible. These complex correlations give rise to the possibility of many feedbacks of clouds on the climate, making their determination crucial to accurate projections of climate change.

The advent of a new global, long-term satellite cloud climatology, produced by the International Satellite Cloud Climatology Project (ISCCP), makes possible the systematic investigation of the cloud variations, including changes in cloud cover,

DISTRIBUTION OF CLOUD PROPERTIES

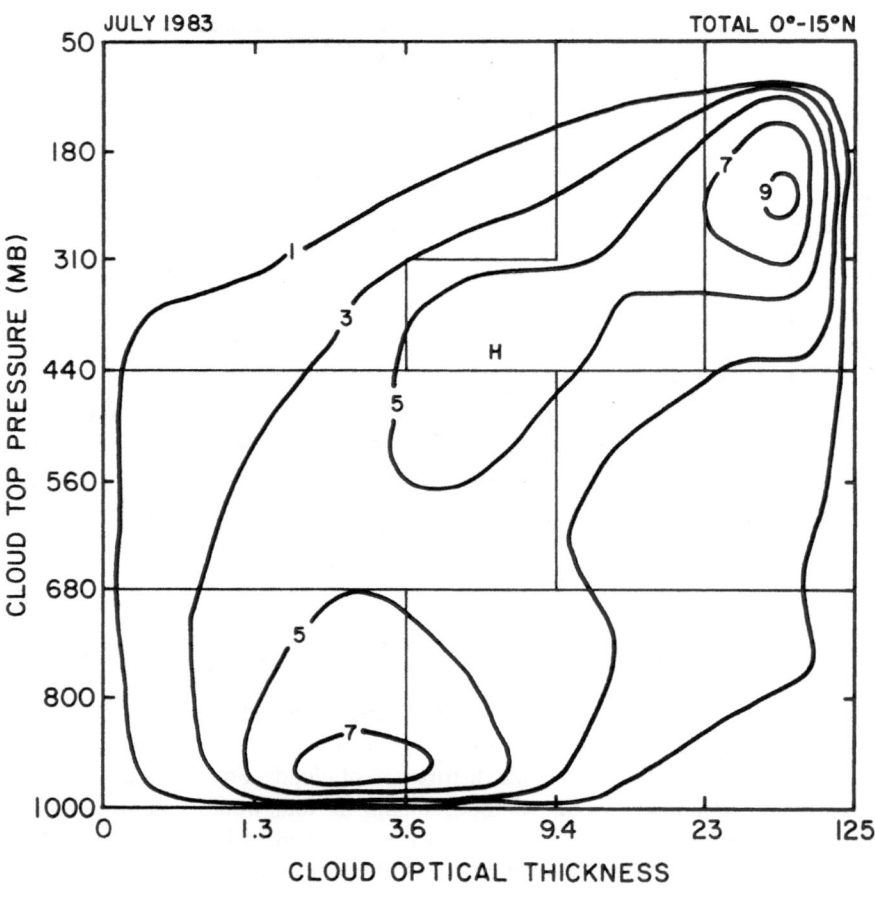

Fig. 5a.

top height (or pressure) and optical thickness (Rossow and Schiffer, 1991), over a very large range of space and time scales. These parameters are the main quantities governing the effects of clouds on Earth's radiation budget and are related to the main quantities that determine the role of clouds in the hydrological cycle. In particular, the cloud optical thickness, which is being measured systematically for the first time by ISCCP (cf. Rossow and Lacis, 1990), is a function of the cloud water content and vertical extent. This comprehensive dataset can also be used to draw together many previously isolated results to examine the full range of cloud variability. We review what is known about smaller scale cloud variations and illustrate some of the larger scale changes using the ISCCP dataset to determine the necessary characteristics of a cloud monitoring system.

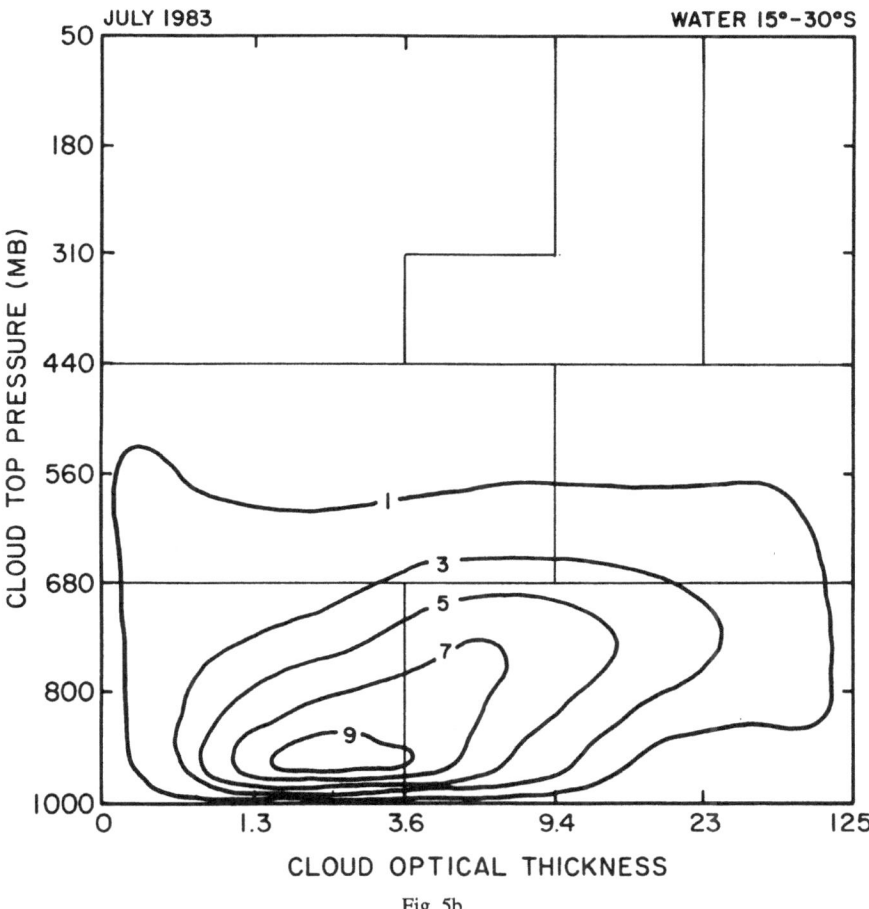

Fig. 5b.

2. Characteristics of Cloud Observing Systems

Clouds have been observed from the surface, from instrumented aircraft, and, most recently, from satellites. Each of these systems has characteristic space-time sampling patterns. Rawinsondes (Elliot and Gaffen, 1991; Gaffen *et al.*, 1991) and most lidars (cf. Sassen, 1991) provide vertical profiles at one point. The former are typically launched only once every 12 hr while the later make measurements with time resolution better than 10 s but only for short periods of time. Surface weather observers (Warren *et al.*, 1986, 1988) and typical radiometers and cameras describe the horizontal variations of clouds over a domain that is approximately 30–50 km across (Barrett and Grant, 1979). Most such data are available with time

[181]

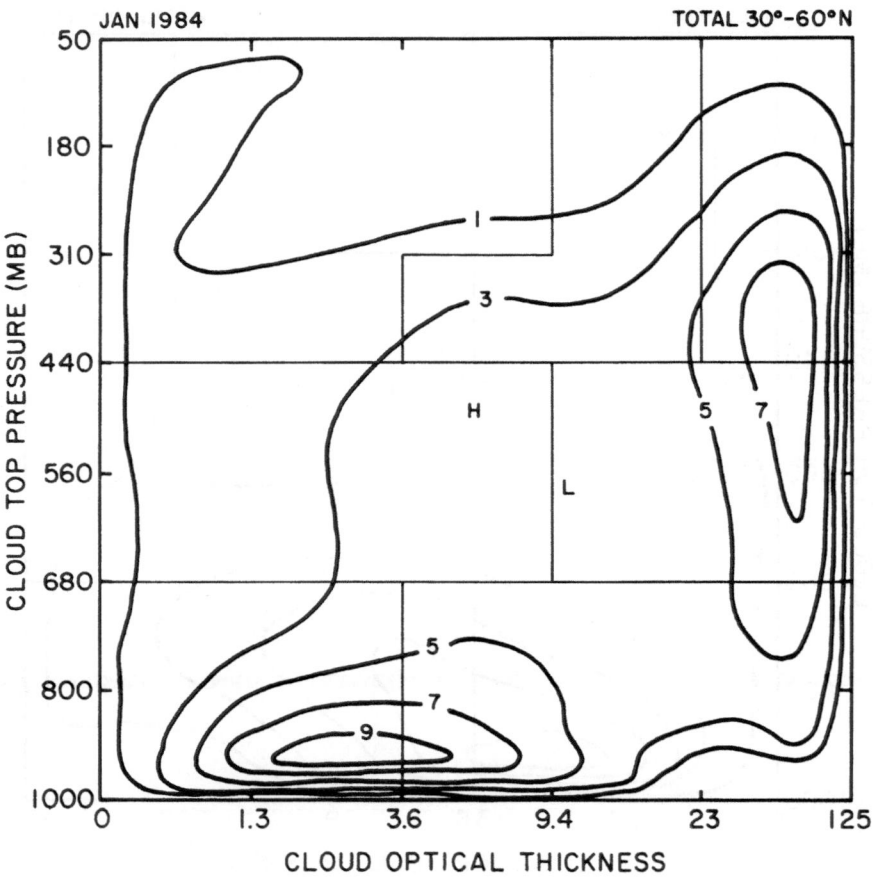

Fig. 5c.

Figs. 5(a)–(c). Two dimensional frequency distribution of cloud top pressure and optical thickness for three latitude zones: (a) the tropics between 0–15° N during July 1983; (b) subtropical oceans between 15–30° S during July 1983; and (c) northern midlatitudes between 30–60° N during January 1984. Contours indicate relative frequency with a peak value of 10.

resolutions of at least 3 hr. Radars (e.g., Houze *et al.*, 1989; Kropfli *et al.*, 1995) often scan a volume about 50–200 km across by 10–20 km deep with very high time resolution, though these data are not systematically analyzed. The major limitation of surface-based cloud datasets is that they are predominantly from populated land areas, leaving large portions of some continents and the oceans unobserved (Warren *et al.*, 1986, 1988). Collection of ship observations over many decades have probably provided accurate information for most of the northern hemisphere ocean, however. Thus, surface cloud observations can be approximately characterized as providing long time records of point-like measurements with relatively high time-space resolution and relatively complete coverage of most land areas but incomplete global coverage. The only surface-based observations that have been analyzed systematically are surface weather observer reports (Warren *et al.*, 1986, 1988; Hahn *et al.*, 1994).

Aircraft can provide *in situ* cloud measurements with fairly high spatial resolution and can cover somewhat larger areas ($\sim$ 500 km) than observed from the surface; however, their spatial sampling is much poorer unless many aircraft work together and their time sampling is usually equivalent to a single sample. Aircraft can be used to probe either vertical or horizontal structures, but multiple aircraft are required to do both at the same time. Few systematic analyses of observations from a large number of aircraft cloud observations have been undertaken (cf., Cox *et al.*, 1987).

Satellite instruments more easily cover the largest space and time scales and provide a variety of cloud measurements (Rossow, 1989). Satellites in polar orbits can provide complete global coverage, albeit with relatively low time resolution ($\approx$ 12 hr for a single orbiter). On the other hand, satellites in geostationary orbit can provide very high time resolution ($\approx$ 15–30 min) over large areas; four to five such satellites can cover most of the globe up to latitudes of about 55° in each hemisphere. Spatial resolution is highest for the imaging radiometers: image pixel sizes for weather satellites range from 1–5 km. Although satellites intended to study land surfaces have resolutions of 10–30 m, such data are rarely collected and analyzed for the whole globe. The coupling of the space-time scales of cloud variations constrains the meaningful resolution of satellite observations (Salby, 1989). A polar orbiting instrument with a sampling of about 12 hrs can only meaningfully describe cloud variations on spatial scales > 1000 km. An instrument in geostationary orbit with a time resolution of 30 min can describe cloud variations down to spatial scales $\sim$ 5 km but cannot view scales $\gtrsim$ 5,000 km.

One interesting point regarding the issue of coupling of space-time scales is whether randomized sampling might provide a practical compromise between high resolution and aliasing. Theoretically, if truly random sampling in space and time were possible with a satellite observing system, unaliased estimates of the measured quantities could be obtained, but at the expense of substantially increased noise in the estimates of monthly mean quantities (Gaster and Roberts, 1977). This statement should be qualified by noting that most analyses of such issues presume that sampling is truly random, a Poisson point process, or a renewal process (Shapiro and Silverman, 1960), which would not be strictly true for a satellite observing system. Random sampling does not, therefore represent a universal panacea, particularly if noise is a concern.

Several comprehensive satellite cloud datasets now exist (Stowe *et al.*, 1988, 1989; Rossow and Lacis, 1990; Mokhov and Schlesinger, 1993, 1994), but the ISCCP datasets are higher resolution and report more cloud properties (Rossow

CLOUD AMOUNT FROM ISCCP AND SURFACE OBSERVATIONS

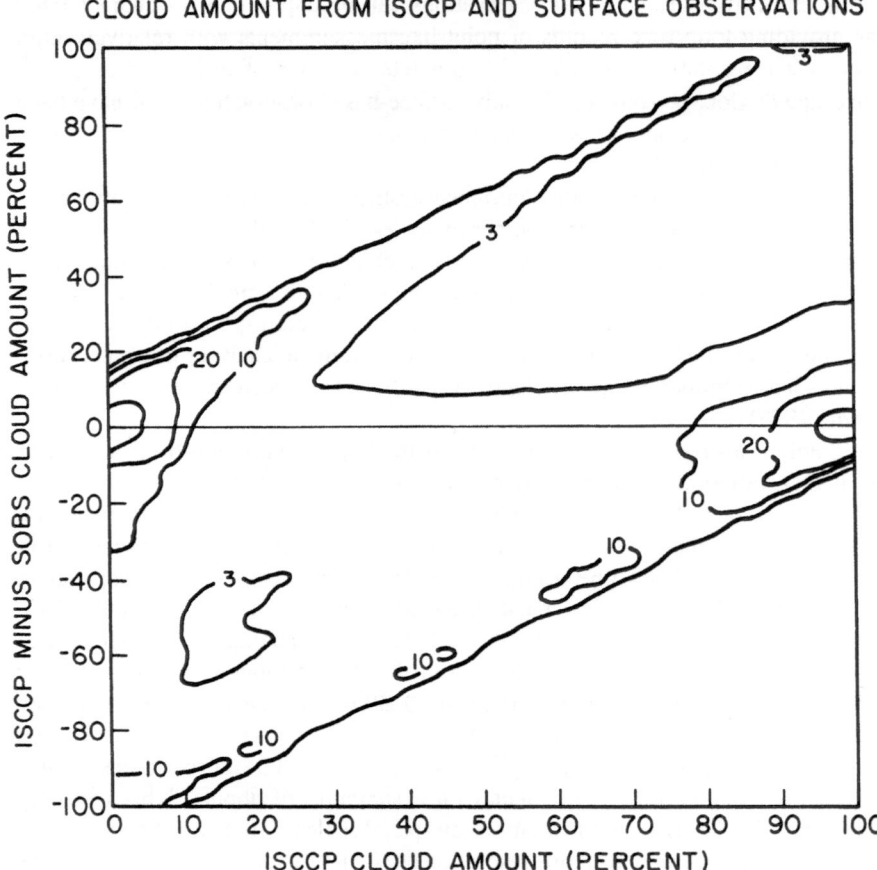

Fig. 6.   Two dimensional frequency distribution of the difference between ISCCP and surface-observed estimates of cloud amount and the corresponding ISCCP cloud amount. Contours indicate frequencies in percent relative to the maximum value; unlabeled contours near (0,0) and (0,100) are for 50%.

and Schiffer, 1991). The ISCCP climatology is global, covers eight years at present, and has a space-time resolution of 30 km – 3 hr.

3. Characteristic Variations of Clouds

Comparison of cloud cover fractions determined at two different spatial scales, one about 50 km and one about 280 km, shows a trifurcation of the difference distribution (Figure 6)) when plotted against the cloud fraction from either source (Rossow *et al.*, 1993). The cases forming the two slanting branches are caused by stratiform cloud types that either completely or partially cover the larger area observed from satellites and appear as either completely clear or completely overcast skies in the more 'point-like' surface observations. In other words, the cloud cover is partial only at larger scales. Since the location of the surface observer is essentially random within the larger domain, the disagreement between the two measurements is almost always the maximum possible when the larger area is only partially covered. The cases forming the horizontal branch with near-zero differences are caused by 'broken' cloud types that produce partial coverage at smaller scales. In this case, the disagreements between the two measurements appears random with little bias. The relative proportion of these groups is consistent with the populations of stratiform and cumuloform cloud types reported in surface observations over land (Warren *et al.*, 1986). This result implies significant numbers of clouds with sizes < 50 km and > 280 km. The practical limitations on surface and aircraft observations, as well as the cost of analysis of very high resolution satellite imgaes, has tended to divide cloud studies into those considering scales smaller than about 200 km and those considering larger scales.

3.1. SMALLER SCALE VARIATIONS

Many studies of the smaller scale structure of clouds have focused on the possible biases in estimating fractional cover using 'low' resolution satellite instruments. A recent study by Wielicki and Parker (1992) using LANDSAT images with a spatial resolution of 28 m shows that highly broken marine boundary layer clouds, characterized by size distributions with broad peaks in the range between 0.5 km and 5 km, are also characterized by a continuum of optical thicknesses (see also Cahalan *et al.*, 1994). There is also a tendency for the average optical thicknesses to increase as the average cloud cell size increases. In another study, lower average values of cloud cover (<0.3) are found to be associated with more frequent occurrence of lower optical thicknesses (Harshvardhan *et al.*, 1994). Consequently, the overestimate of cloud cover by counting cloudy pixels in lower resolution satellite images is partially offset by the failure to detect some of the optically thinner clouds (Wielicki and Parker, 1992). This means that the cloud detection threshold can be tuned *for a particular type of cloudiness* to obtain accurate cloud cover. The key implication of this result is that smaller scale cloud variations may be more appropriately considered in terms of variations of optical thickness, where clear sky is associated with the lowest value, rather than in terms of discrete objects with distinct and well-defined boundaries.

Cloud size distribution studies at spatial scales < 200 km have focused predominantly on marine boundary layer cloudiness (however, see Parker *et al.* (1986) for a study of land boundary layer clouds and Kuo *et al.* (1988) for a study of cirrus clouds). The size distributions are generally continuous power laws with negative exponents < −1 and often exhibit changes in slope in the size range from 0.5 km to 5 km that have been described as 'peaks' (Cahalan and Joseph, 1982; Parker *et al.*, 1986; Wielicki and Welch, 1986; Welch *et al.*, 1988: Joseph and Cahalan, 1990, Sengupta *et al.*, 1990; Weger *et al.*, 1992; Zhu *et al.*, 1992; Weger *et al.*, 1993; Lee *et al.*, 1994). These distributions are generally such that most of the area is occupied by the larger clouds (exponent > –2), except at scales smaller than the 'peak' scale, where the exponent approaches –3. Moreover, clustering also appears to be common. Organization of the smaller cloud elements into clusters ~ 10–50 km in size suggests influence by the boundary layer dynamics. The study by Lee *et al.* (1994) also shows clustering at scales < 3 km, which they interpret as reflecting the dynamical forcing for convection. At intermediate scales (3–10 km), the spatial distribution appears random. In these particular results, the radiance threshold is selected at the median reflectivity, so that it is optical thickness variations that are being examined. The fact that the spatial distribution and size of the clusters resembles that of the larger cloud elements suggests instead that the clusters containing very small (< 1 km) elements are dissipating forms of the larger clouds. Since all of these studies are based on analyses of single images, the size spectra mix newly formed, mature and decaying clouds together in a way that may be misleading. There have been no studies that examine the correlation of the space and time variations of clouds at these smaller scales.

A key characteristic of these highly broken boundary layer clouds with very small elements may have been missed because all of the studies cited above use single views of domains that are only 100–200 km in size (the LANDSAT scene size). This characteristic was suggested by the sampling study of Seze and Rossow (1991a, b), where the radiance variation statistics of such clouds from 5 km resolution satellite images were compared with statistics from the same data sampled to spatial intervals of 30 km (like the ISCCP dataset). The quantitative similarity of these statistics was explained by assuming that such small-scale broken clouds tends to occur in 'fields' that are large enough (many hundreds of kilometers) that the sampled dataset still contains a sufficient sample to portray the variation statistics accurately. This conclusion requires that the small scale statistics be homogeneous over the larger spatial scale of the whole cloud field. That this might be true is suggested in the study of Lee *et al.* (1994), where the positions of the clouds and cloud clusters become essentially random at scales > 5 km. This conclusion needs further confirmation, but it explains the ability of the sampled ISCCP dataset to describe cloud amounts even for broken cloud types (Rossow *et al.*, 1993).

In general, the depth of the troposphere is only 10–15 km, so that typical cloud horizontal dimensions are far larger than their vertical dimensions, i.e., clouds form layers. Even the smaller boundary layer clouds tend to have vertical extents that are smaller than their horizontal dimensions. The dynamic coupling of atmospheric vertical structure to the horizontal distribution of clouds is often neglected in small scale studies, even though the boundary layer depth may well determine the horizontal size spectrum of the clouds over land and ocean (Kaimal *et al.*, 1976; Nicholls, 1989). An additional important feature of cloud vertical structure is the frequent occurrence of multiple layer clouds. Surface observations suggest that at least half of cloud occurrences involve two layers (Warren et al., 1985). A preliminary analysis of a limited rawinsonde dataset reaches a similar conclusion and notes that the typical separation distance of these two layers is of the same order as the average layer thicknesses (Wang and Rossow, 1995).

In the time domain the most important, coherent, high frequency variation of clouds is the diurnal cycle (Hendon and Woodbury, 1993; Salby et al., 1991; Cairns, 1995). This is demonstrated in Figure 7 where we show the power spectra of cloud top pressures for the zonal average at a latitude of 45° N and for a single small region ($\approx$80,000 km^2) in southwest Europe at the same latitude. It is apparent from Figure 7b that mesoscale variability provides much more localized variability than the diurnal cycle. However mesoscale variability does not have the same global temporal coherence that the diurnal variations do. This is shown by Figure 7a where mesoscale variability is suppressed in this large scale average compared with the diurnal cycle. Thus, although the local variability explained by the diurnal cycle may not be as great as mesoscale variability, the spatially coherent phase of the diurnal cycle (Cairns, 1995) means that it is of particular importance when considering the sampling issues relevant to monitoring. The spectra shown in Figure 7 are estimated using a multitaper method with five windows and their variance is therefore well estimated by a chi-square distribution with ten degrees of freedom (Thomson, 1990). Since the spectra are logarithmically plotted the reader can readily estimate confidence intervals that apply to any frequent point (Priestly, 1981). The only significant line components in these spectra are the diurnal and semi-diurnal components.

A better understanding of the smaller scale temporal variability of clouds and their regional variations can be obtained from an EOF analysis of global, daily mean maps of cloud amount, optical thickness and cloud top pressure. Using a 90-day record from boreal spring, we find that local day-to-day variations account for about 60% of the variability, more than the time-mean regional variations; however, these rapid time variations mostly form a 'noise-like' distribution of principal components, each of which only explains 1–2% of the variance. This indicates little global coordination of the higher frequency (daily) cloud variations.

As already suggested in Figure 7, the diurnal cycle is the strongest coherent variation of clouds on time scales less than one season. The relative importance of the diurnal variations compared with lower frequency fluctuations can be evaluated by fitting a model, consisting of a mean, a diurnal and a semi-diurnal component, to the seasonally averaged diurnal variation of clouds from the ISCCP dataset. Although there is some variation in the relative strength of the three terms for different seasons and cloud types, approximately 75% of the global cloud variation is explained by the seasonal mean, 15% is explained by the diurnal component, and 5% is explained by the semi-diurnal component (Cairns, 1995). The magnitude and phase of the diurnal cycle are highly variable with location and the diurnal cycle is not a simple harmonic oscillation (Kondragunta and Gruber, 1994; Cairns, 1995).

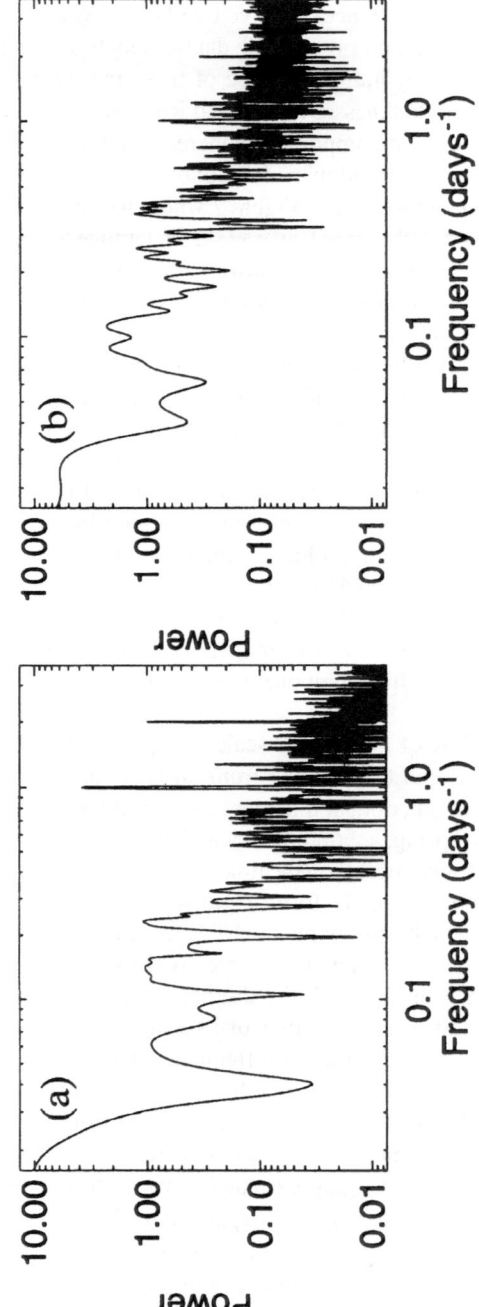

Fig. 7 Temporal spectra of cloud top pressure variations for December 1987 to March 1988 from measurements at three hour intervals (ISCCP C1 data). (a) Spectrum of a time series that was interpolated and then shifted to local time at each longitude and spatially averaged around the entire 45° N latitude zone; (b) Spectrum of a time series that was averaged over a 400 × 400 km² area in southwest Europe at the same latitude as (a).

Low cloud amount over land has a well-defined diurnal maximum near 1330 Local Standard Time (LST), while over oceans there is broad maximum near 0800 LST. The amplitude of low cloud variation is approximately 10–15% over land areas, but only a few percent over oceans (Carlson *et al.*, 1995). Middle-level cloud amount has a maximum at around 0500 LST over tropical land areas while high cloud amount peaks between 1800 LST and midnight. Over oceans in the convergence zones (Intertropical – ITCZ, South Pacific – SPCZ, South Atlantic – SACZ), high cloud amount is maximum in early evening (1600 to 2000 LST), but there is also an early morning maximum in deep convective clouds (Fu *et al.*, 1990). The amplitude of diurnal variations of middle- and high-level clouds are generally <5% on average (Carlson *et al.*, 1995). The principal seasonal variation in the geographic distribution of larger diurnal amplitudes is for the summer hemisphere to have the stronger diurnal cycles. The asymmetry of the diurnal cycle, its varying phase with cloud type, and its amplitude and phase variations with location all emphasize the importance of proper sampling of this time scale in any cloud observations to prevent aliasing of these scales into low frequency variability.

3.2. LARGER SCALE VARIATIONS

Figure 2 shows that the cloud amount frequency distribution is bimodal, even at a spatial scale of 280 km (equivalent to 2.5° latitude-longitude at the equator), and that 15% of cloud systems are large enough to completely cover an area of this size (Figure 2 shows observations over land; over ocean more than 30% of clouds are large enough to completely cover 280 km areas – Rossow and Schiffer, 1991). Figure 8 shows the evolution of the ISCCP cloud amount frequency distribution in three latitude zones as the observations are averaged over progressively larger areas. In the tropics (Figure 8a), a nearly monomodal distribution with a peak near 50% is apparent when the observations are averaged over 10° latitude-longitude; however, there are still occurrences of complete overcast even at this scale (≈ 1100 km). In northern midlatitudes a mode at about 75% only appears in the distribution averaged over 20° , but completely overcast cases are still frequent (Figure 8b). In southern midlatitudes where average cloud amount is very high, there is only a suggestion of a mode at 90% in the distribution averaged over 20° (Figure 8c). The monomodal distribution shape appears at a smaller scale in the tropics than in midlatitudes, but all distributions begin to exhibit this shape at a scale of ≈ 2200 km.

A complementary way to examine the large scale variability of clouds is to look at the spatial spectrum around a latitude circle (Zangvil, 1975): Figure 9 shows representative spectra at the equator, 30° N and 45° N. Each spectrum is formed from 100 realizations, which are the 3 hr ISCCP maps of cloud amount for the winter season of 1987–1988, and estimated using a multi-taper method (Thomson, 1990). The gross behavior of the spectrum is a power law with an exponent near –2; the –5/3 line representing a Kolmogorov turbulence spectrum

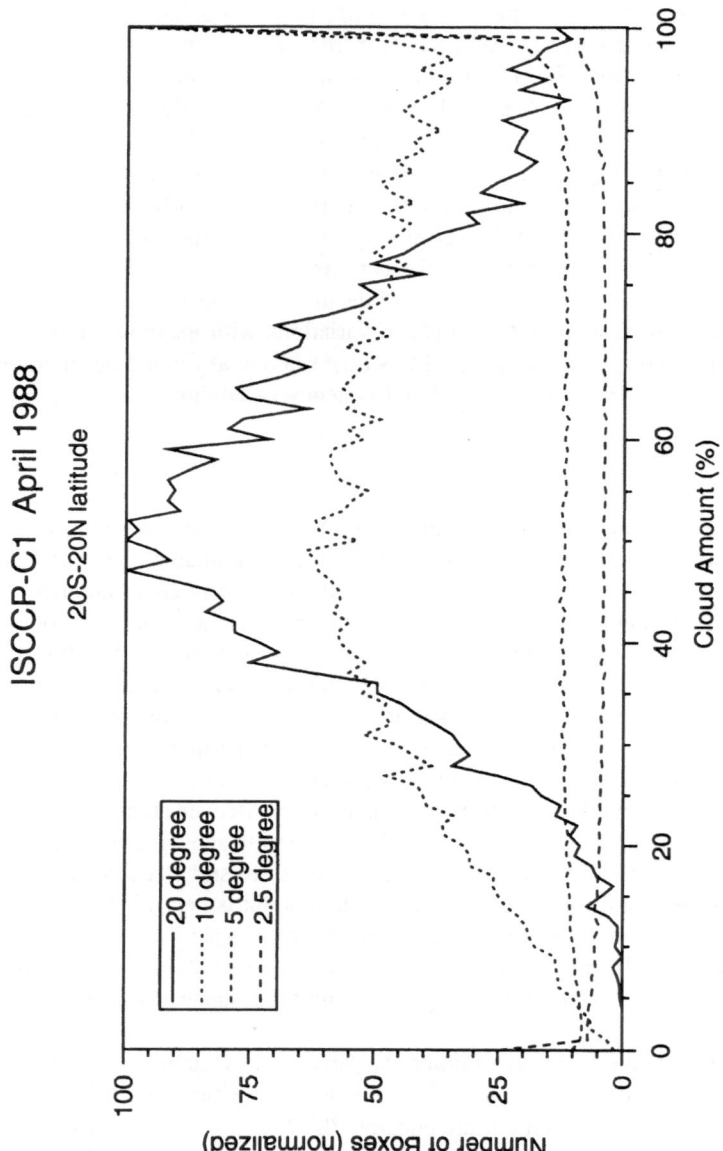

Fig. 8a.

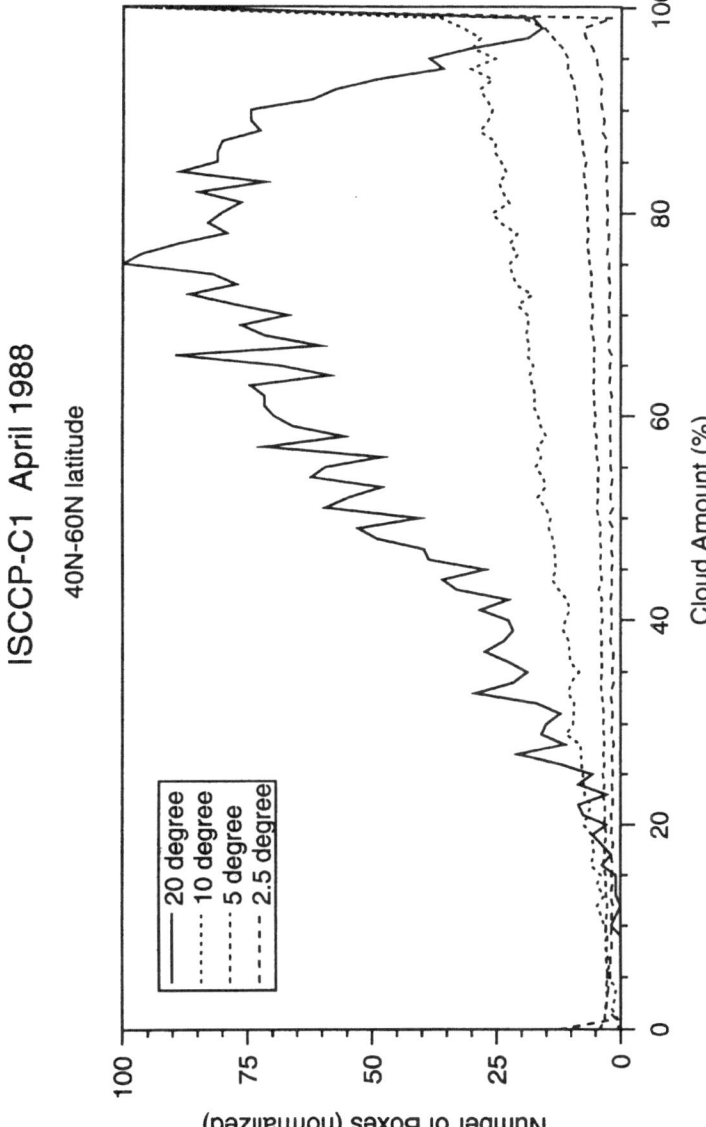

ISCCP-C1 April 1988

40N-60N latitude

Cloud Amount (%)

Number of Boxes (normalized)

20 degree
10 degree
5 degree
2.5 degree

Fig. 8b.

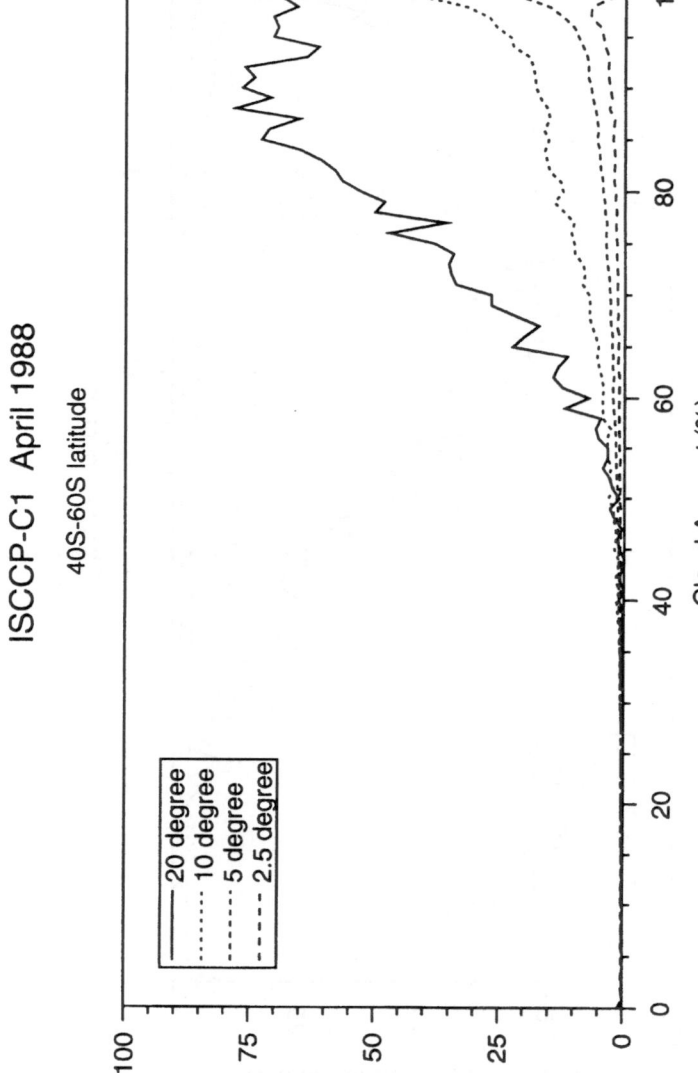

Fig. 8(a)–(c). Frequency distributions of ISCCP cloud amounts at three hour intervals for April 1988 averaged over four different map resolutions: 2.5°, 5.0°, 10° and 20° in the latitude zone (a) 20° –20° N; (b) 40° N–60° N; and (c) 40° S–60° S.

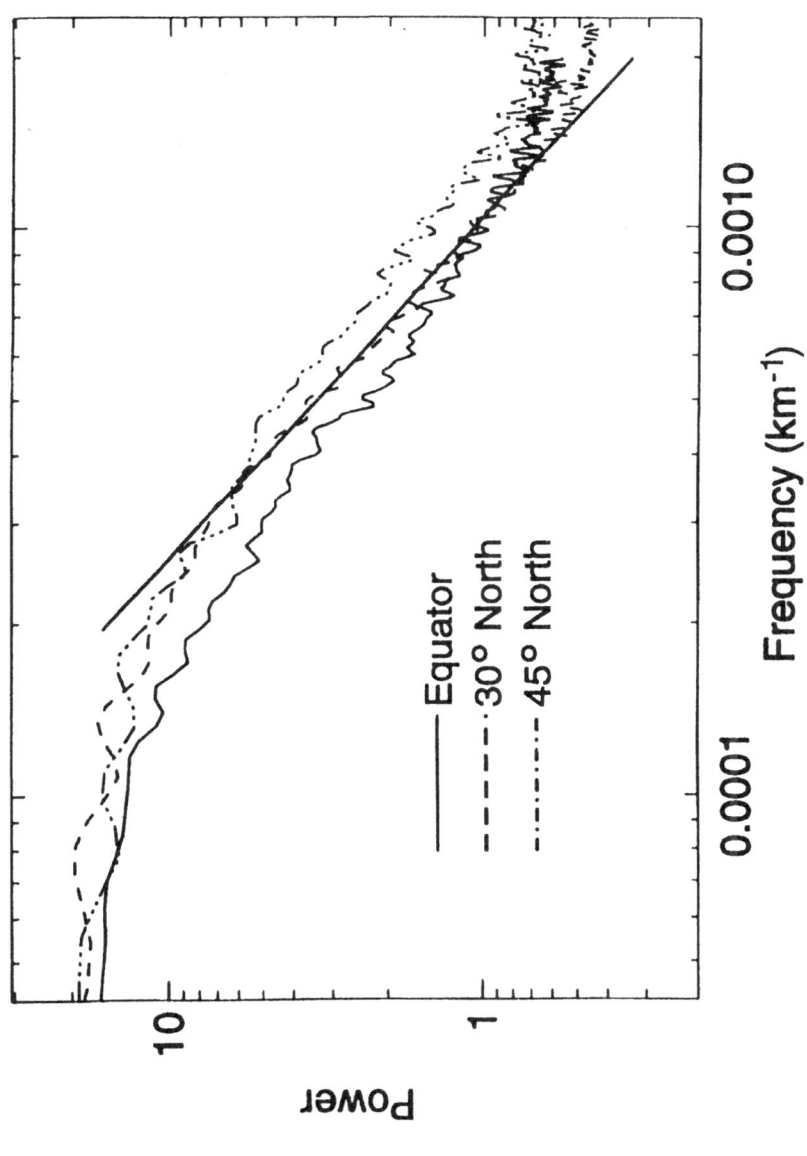

Fig. 9 Spatial power spectra of ISCCP cloud amount variations averaged over 100 'snapshots' at three hour intervals for the equatorial zone (solid line). 30° N (dashed line) and 45° N (dash-dot line). A straight solid line with a slope of $-5/3$ is shown for reference.

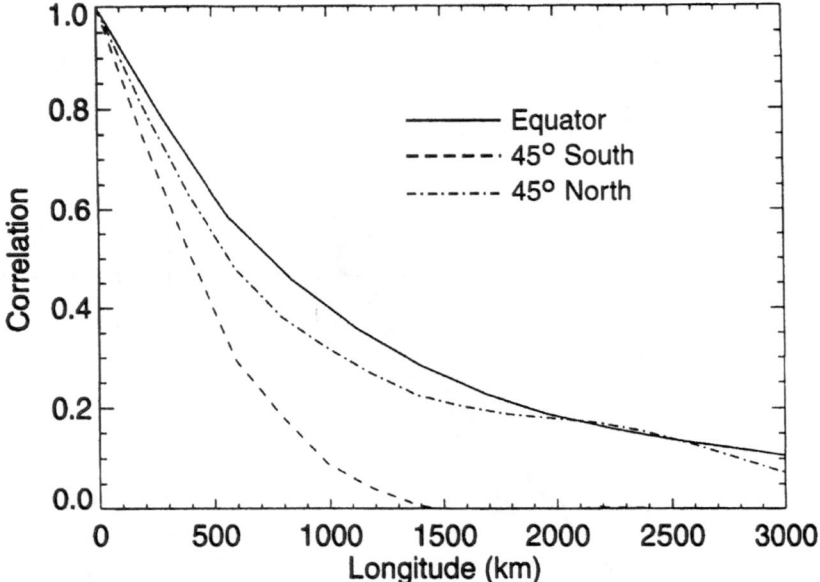

Fig. 10. Circular autocorrelation functions of ISCCP cloud amount variations averaged in a Z-statistic, where $Z = 0.5\log_e[(1+r)/(1-r)]$, over 100 'snapshots' at three hour intervals for the equatorial zone (solid line), at 45° S (dashed line) and at 45° N (dash-dot line).

is shown for reference. The power law appears to break down at an outer scale of ≈ 5000 km. Another way to examine the higher frequency behavior of cloud amount is through the circular function, which is shown for the equator and 45° N and S in Figure 10. These correlation functions are also the averages of 100 realizations at 3 hr time resolution. At the equator the correlation length scale is ≈ 1000 km, but at midlatitudes it is only ≈ 750 km and ≈ 500 km in the north and south, respectively. Despite the somewhat smaller size of cloud systems in the tropics suggested by Figure 8, these tropical systems appear to be coordinated by the longer waves so as to produce a somewhat larger correlation length scale. The shorter correlation length scales for **variations** in southern midlatitudes as compared with northern midlatitudes reflects the weaker stationary wave activity in that hemisphere: the stationary waves are generally longer wavelength than the transient waves (Pandolfo, 1993). Even with a smaller correlation length scale, the cloudiness is generally more nearly complete in southern midlatitudes (cf. Figure 8c).

The tendency of clouds to form 'thin' layers (i.e., to have vertical extents smaller than their horizontal extents) is illustrated by comparing the variability of ISCCP cloud top pressures within small local regions of about 280 km size with the geographic variation of cloud top pressures averaged over these small domains (Figure 11). Generally, the local variations of cloud top pressure are less than half

ISCCP-C1 April 1988

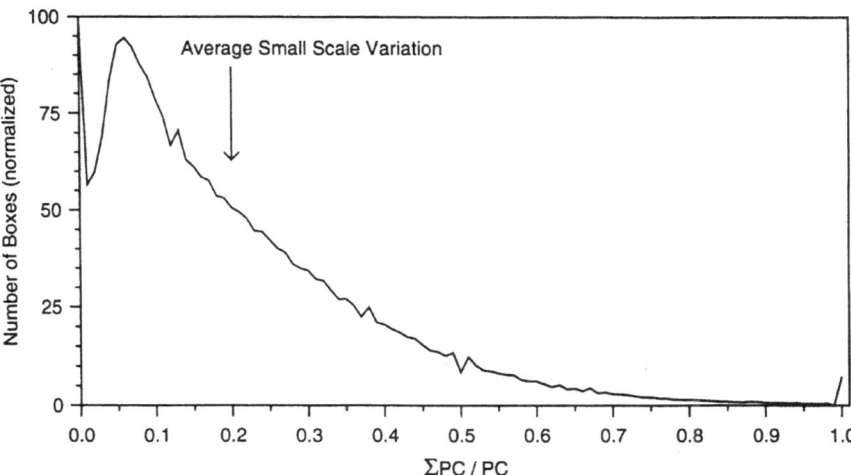

Fig. 11. Frequency distribution of the subgrid scale (< 280 km) standard deviation (ΣPC) of ISCCP cloud top pressures normalized by the corresponding mean cloud top pressure (PC). The standard deviation of the mean values over the globe is indicated by an arrow.

as large as they are from region to region; however, there is a significant number of systems with much larger local varibilities. In contrast, the magnitude of local variability of cloud optical thickness (not shown) is only about one quarter of the regional variations.

The importance of cloud top pressure variations is also apparent in its temporal power spectrum, averaged around the equator, when compared with the power spectrum of cloud amounts (Figure 12). The suppressed diurnal and semi-diurnal cycles in ISCCP total cloud amount result from the different phases of different cloud types (Cairns, 1995); however, these changes of cloud types appear as stronger, more coherent changes in the cloud top pressure over the day in the tropics (cf. Fu *et al.*, 1990). The suppression of the diurnal cycle in cloud amount is also related to the averaging with longitude: Figure 13 shows the autocorrelation of the zonally averaged cloud amount time series (solid line) and of a small region in equatorial Africa. The zonal average ISCCP cloud amount is correlated over ≈ 2 days while the correlation over Africa exhibits a diurnal fluctuation. Unlike the midlatitude example (Figure 7), the diurnal cycle is more apparent locally in the tropics because phase differences among different locations nearly eliminate any variation in the zonal average. This demonstrates, again, the association of short time scales with small spatial scales.

The larger scale (> 30 days) temporal variability of clouds can be compared to their regional variations by employing an EOF analysis of global, monthly mean

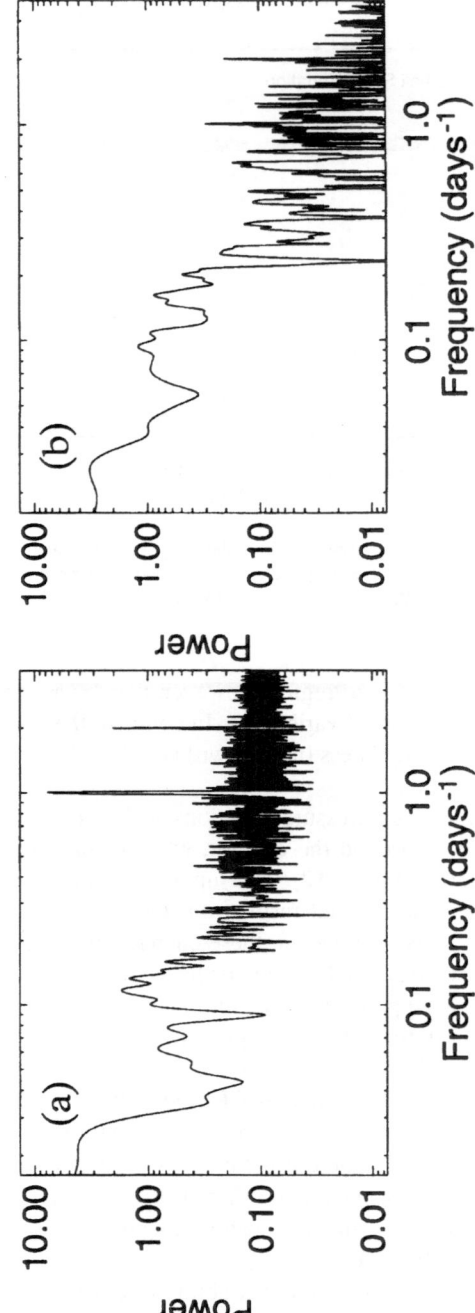

Fig. 12 Temporal power spectra of ISCCP cloud top pressures (a); and cloud amounts (b) for the period December 1987 to March 1988, using three hour samples. The data were interpolated and then shifted to local time at each longitude and spatially averaged around an equatorial zone 5° wide.

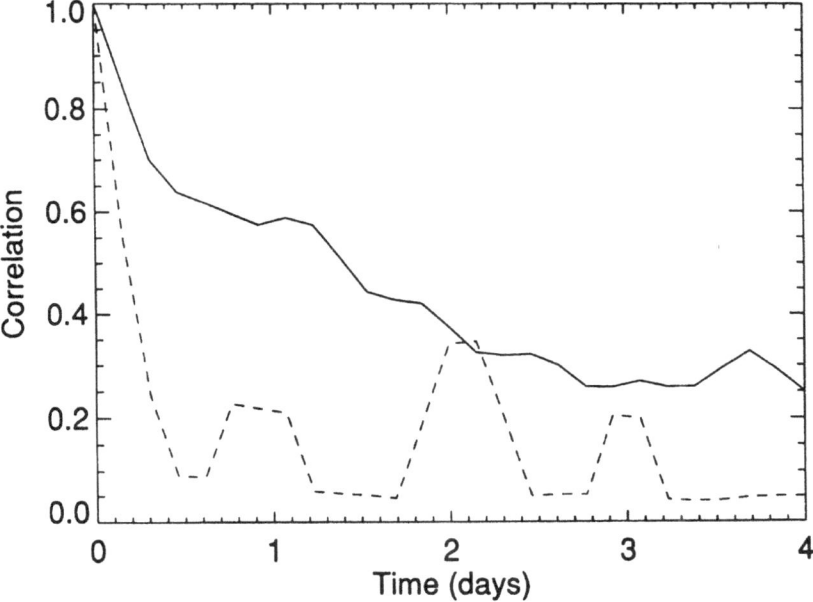

Fig. 13. Temporal autocorrelation functions of cloud amount variations for December 1987 to March 1988, using three hour samples from ISCCP for a time series that was interpolated and then shifted to local time at each longitude and spatially averaged around the equatorial zone (solid line) and a time series that was averaged over a 500×500 km^2 area in equatorial Africa at the same latitude (dashed line).

maps of cloud amount, optical thickness and cloud top pressure from ISCCP. We find that the first principal component of all three cloud properties is the time mean geographic variation (Figure 14). In the case of cloud amount and cloud top pressure, the regional variations account for about 65–70% of the total variance; but regional variations of cloud optical thickness account for only about 35% of the total. The next two principal components represent the annual cycle of cloud properties (Figure 14), accounting for another 15% of the variance of cloud amount and top pressure. The annual cycle of cloud optical thickness, on the other hand, accounts for almost 25% of the total variance. A few percent of the variation of cloud top pressure and optical thickness appears in a semi-annual cycle. Over longer periods of time, the dominant fluctuation in cloud properties is associated with the El Niño/Southern Oscillation (ENSO) phenomenon that accounts for 2–4% of the total variation in high cloud amount. The temporal variation of the 'ENSO' eigenvector of high cloud amount is shown for the NIMBUS-7 and ISCCP data records in Figure 15. That these variations are associated with the ENSO variation can be deduced only from the combined data record – each dataset alone has only one example making the interpretation ambiguous – showing the value of long continuous data records for diagnosing climate change.

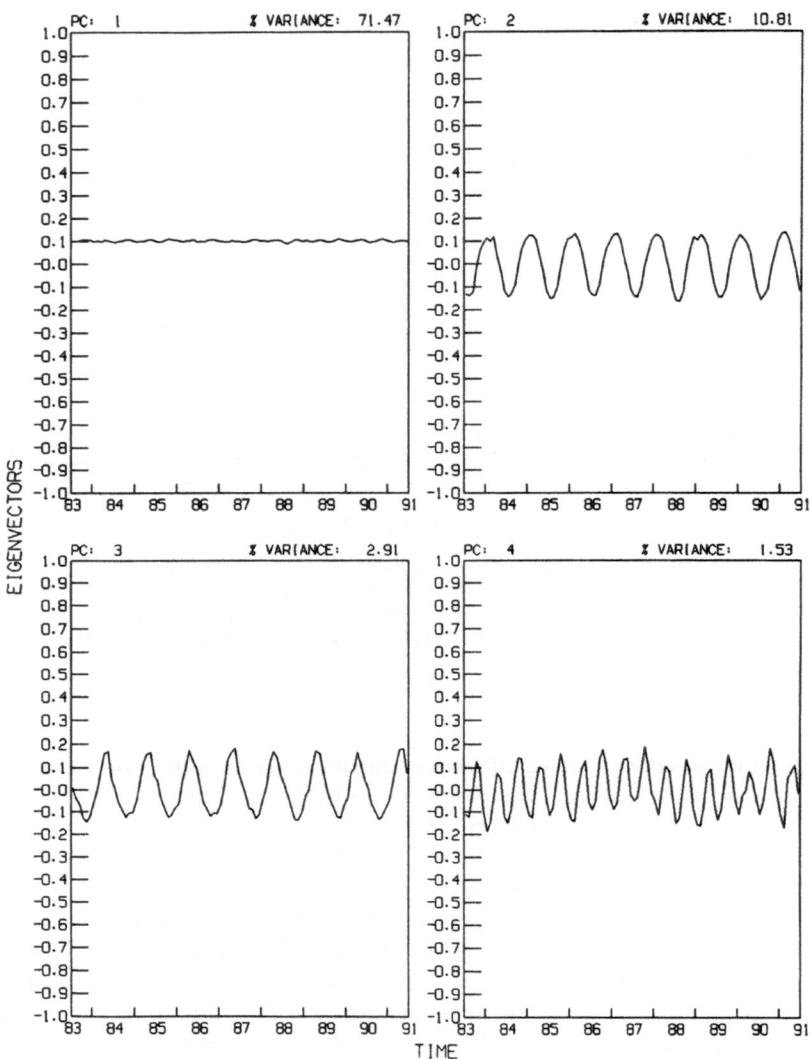

Fig. 14a.

Figure 16a shows an infrared image of the western equatorial Pacific region from GMS-4 satellite with a spatial resolution of 5 km (these data were prepared for the TOGA-COARE, Flament and Bernstein, 1993). This area was selected to illustrate the coupling of space and time scales because it is dominated by smaller

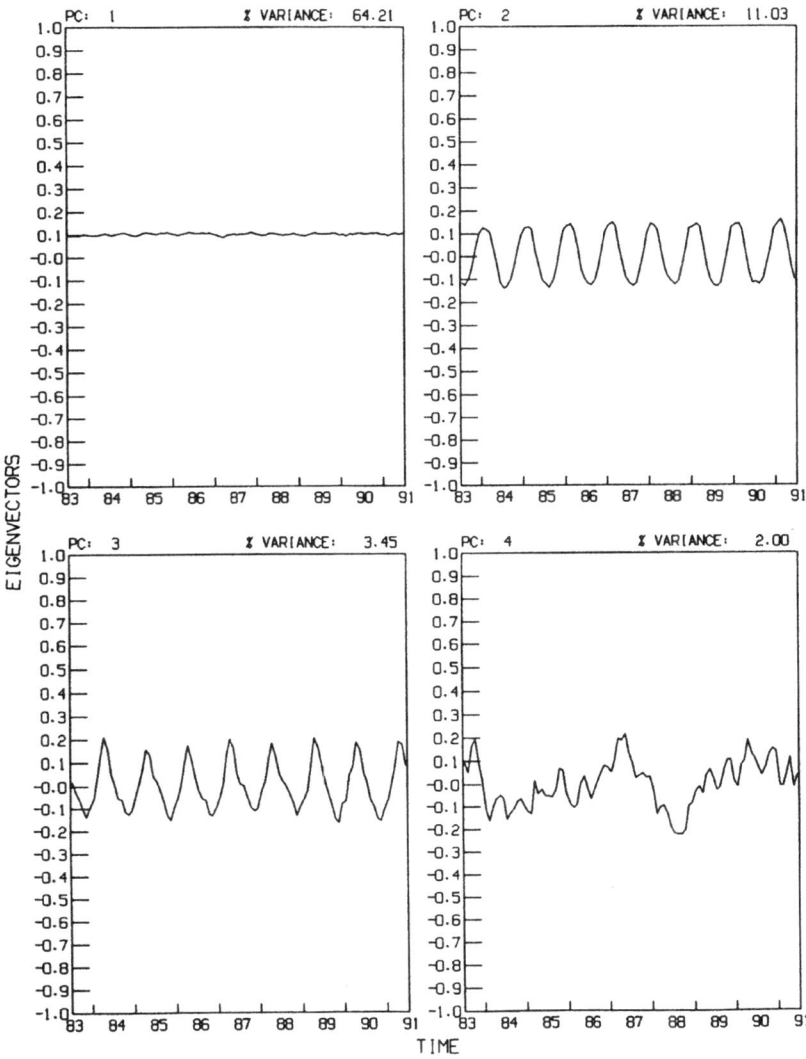

EIGENVECTORS FROM PRINCIPAL COMPONENT ANALYSIS
MONTHLY MEAN CLOUD TOP PRESSURE

Fig. 14b.

scale motions and convective cloud structures. Hourly images like the one in Figure 16a are averaged over different time intervals: Figures 16b and 16c show the 24-hr and 240-hr averages. The visual impression of 'smoothing' or removal of smaller spatial scales as the averaging time period is increased is quantified in Figure 17,

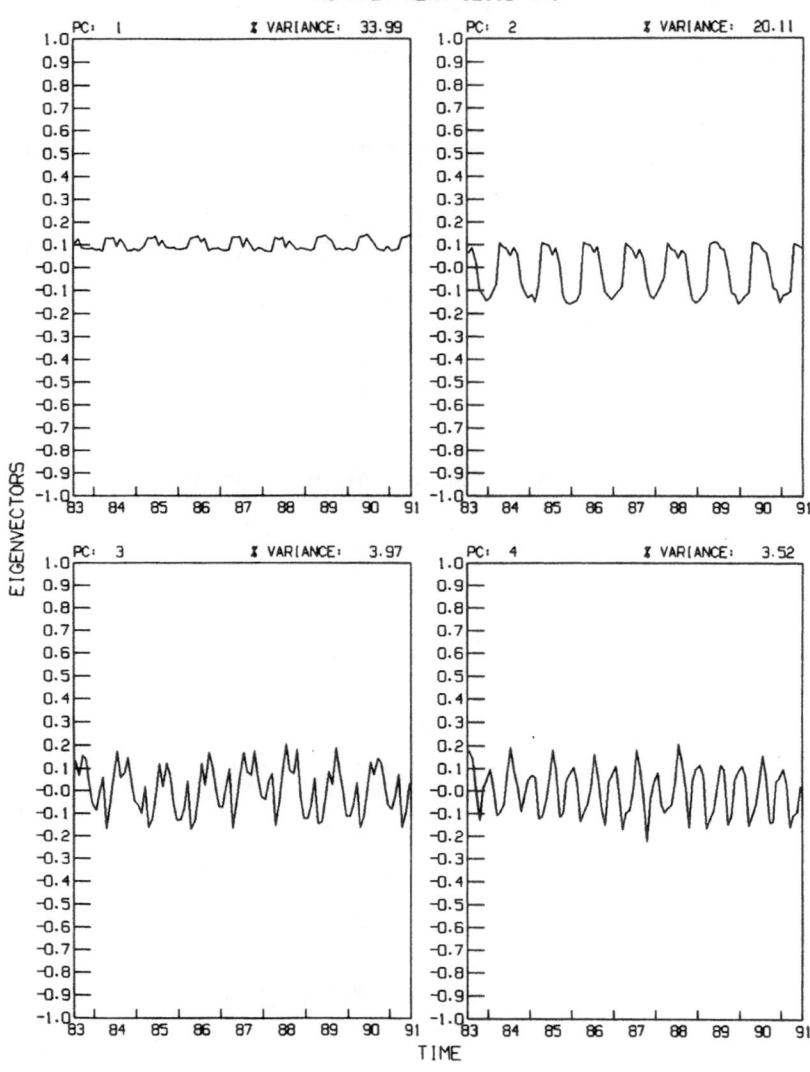

Fig. 14c.

Figs. 14(a)–(c). Time series derived from a principal components analysis of ISCCP global, monthly mean data for the period July 1983 to June 1991 for: (a) cloud amount; (b) cloud top pressure; and (c) cloud optical thickness. The first four significant eigenvectors are shown and the percent of the total variance that they explain is indicated.

which shows the evolution of the one dimensional Fourier power spectra. Most of the power is actually at spatial scales > 500 km, despite the visual predominance of small scale 'texture'. This means that the dominant variations of clouds that cause significant changes in radiative fluxes occur at these larger space scales. The slope of the spectrum is about –5/3 at scales larger than 300 km, but about –2.5 at smaller scales. Time averaging reduces the power at all scales < 500 km by at least one order of magnitude. Notably, there is not much difference between the spectrum for 24-hr and 72-hr averaging periods, suggesting little spatial variability associated with times in this range. Most of the spatial variability at scales < 500 km is eliminated when averaging over one complete diurnal cycle, but about half the decrease is produced by averaging over 3 hr for scales < 100 km. The largest scales (> 1000 km) are not affected until averaging over 240 hr. The spectral slope of the time-averaged data is about –2.

4. Sampling Uncertainties

Monitoring of long-term changes in global cloudiness that are **significant** to climate change requires observations that provide adequate sampling of the variability of clouds described above. The requisite statistical accuracy obtained from a cloud observing system can be estimated by determining the magnitude of changes in global mean cloud properties that would produce changes in the mean radiation balance of Earth that are >0.5 Wm^{-2}, about 25% of the forcing already produced by increased greenhouse gas abundances (cf. Hansen *et al.*, 1993). Based on calculations of the radiation budget using the observed atmospheric, surface and cloud properties (see also Zhang *et al.*, 1995), such a change in the radiation budget would be produced by changes of cloud amount, top pressure, optical thickness and particle radius of $\approx 1\%$, ≈ 10 mb, ≈ 0.15 and ≈ 0.3 μm, respectively. Figure 18 shows a possible 'signal' that should be detected by a cloud monitoring system: the eight year record of global cloud cover (deviation from the mean) from ISCCP shows a slow variation of $\pm 2\%$ that appears to be associated with the 'cycle' of El Niño events during this period (see Figure 15). The $\sim 1\%$ variations on shorter time scales (roughly month-to-month) may represent the 'noise-level' in these results.

If the purpose of monitoring cloud changes goes beyond merely detecting change to diagnosing the effects of the change, then the observing system must not only provide sufficient sampling density but also provide **complete** global coverage. The reason for this additional requirement is that, even when the global mean cloud cover does not change, the location of the clouds can: incomplete observation of the globe would not distinguish between these two cases. To illustrate the effect of partial coverage, we re-calculated the 'global' monthly mean cloud amounts from the ISCCP dataset shown in Figure 18 using the spatial coverage obtained in the analysis of surface temperatures by Hansen and Lebedeff (1988),

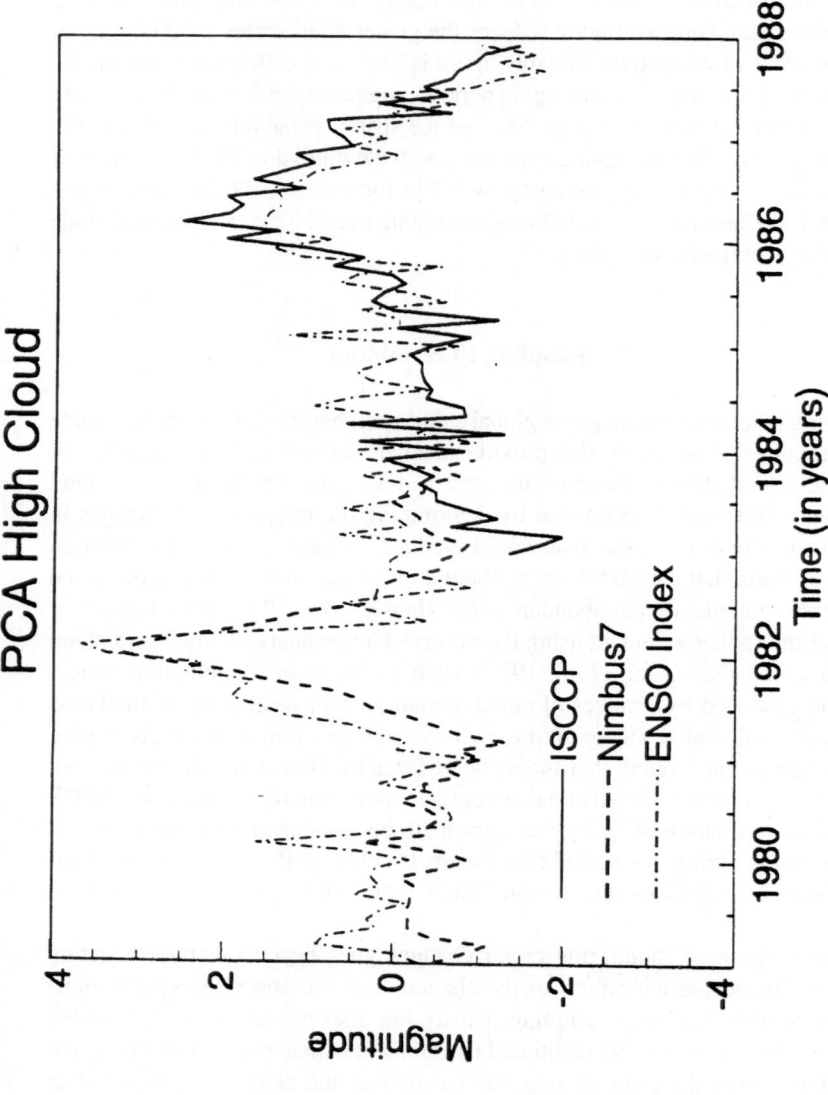

Fig. 15. Time series associated with the first principal components from the ISCCP (solid line) and Nimbus 7 (dashed line) of monthly mean high cloud amount anomalies. The dash-dot line shows the Southern Oscillation index (difference between sea level pressure at Darwin and Tahiti). The principal component analysis was applied to the global gridpoint normalized data sets.

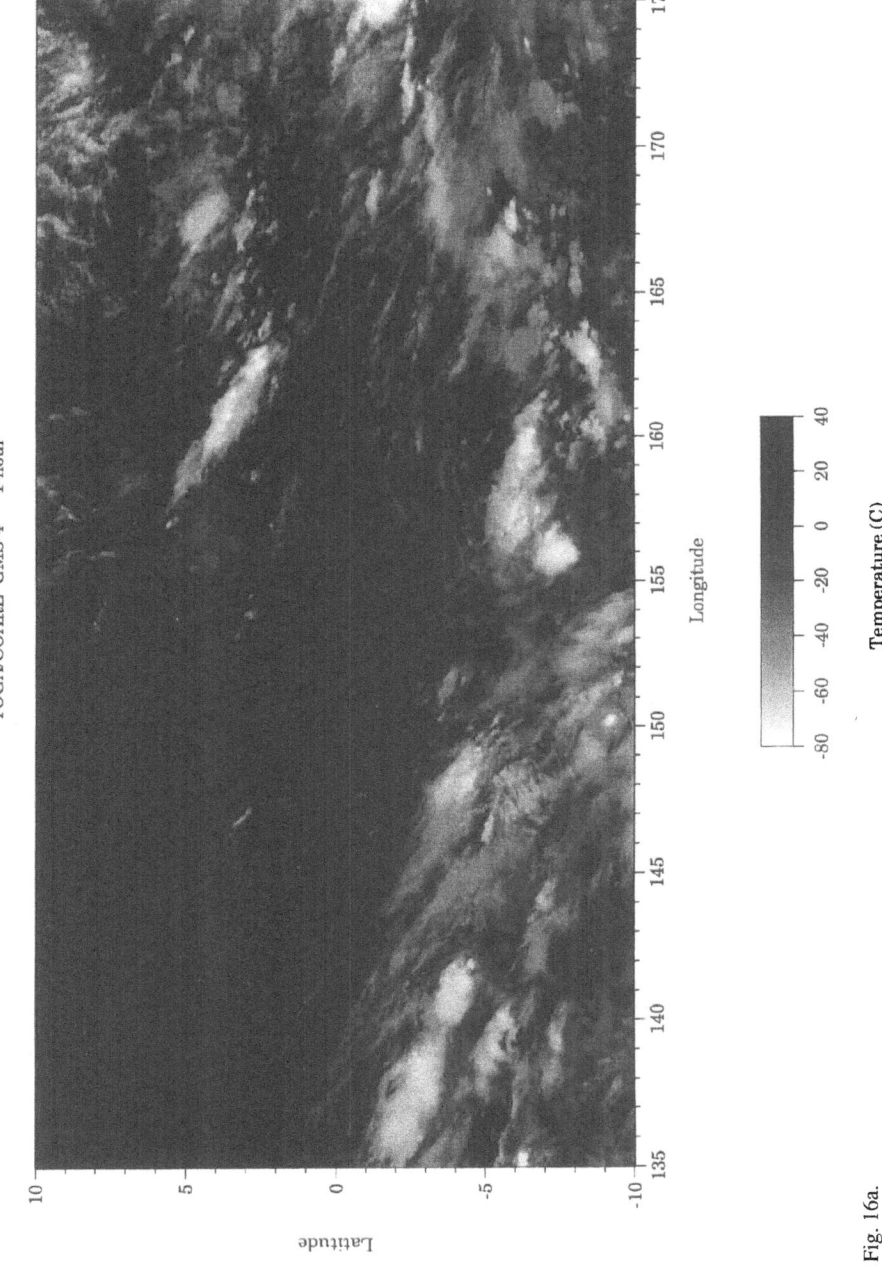

Fig. 16a.

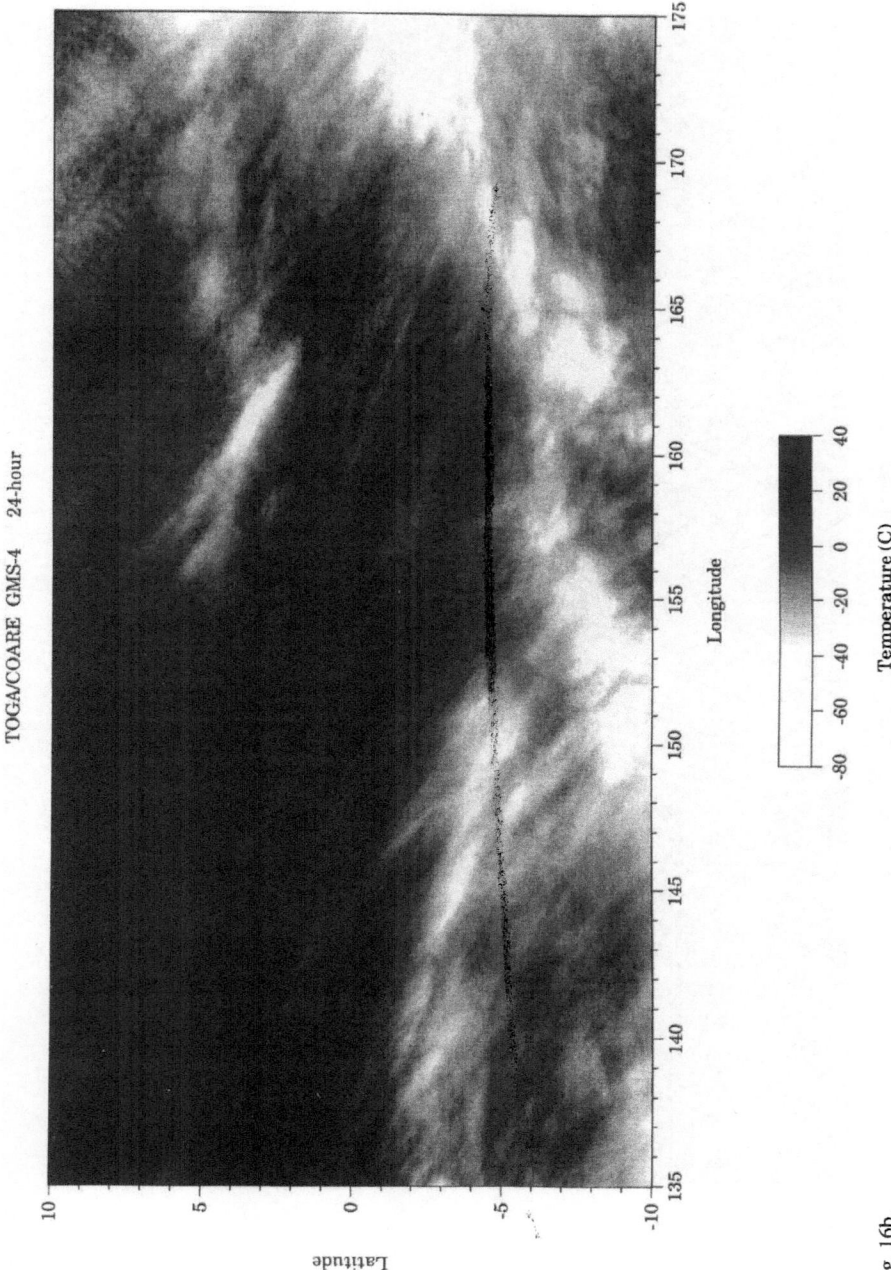

Fig. 16b.

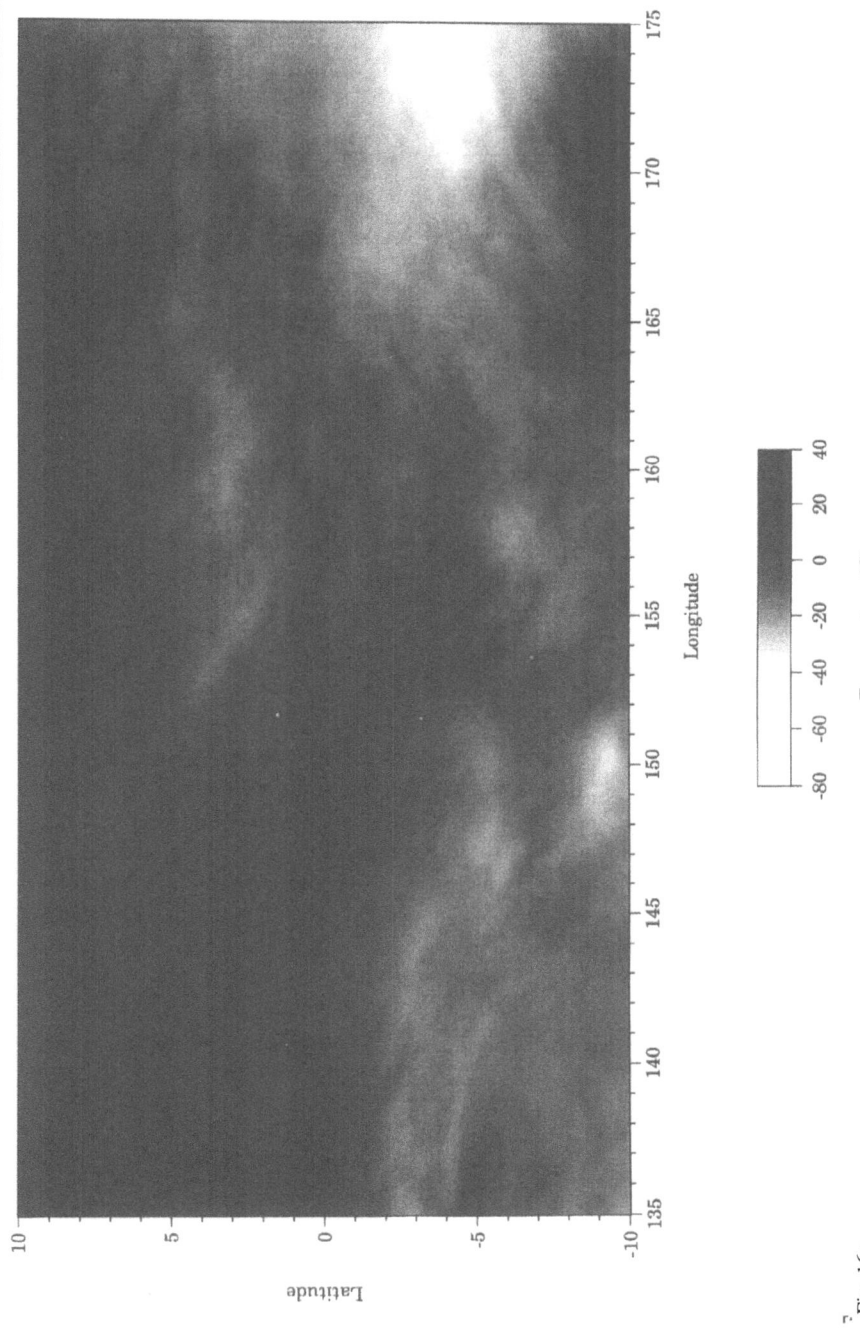

Temperature (C)

Fig. 16c.

Fig. 16(a)–(c). Grey scale representation of infrared brightness temperature images from the Japanese GMS-4 satellite for the western equatorial Pacific (TOGA-COARE) region at a spatial resolution of 5 km (Flament and Bernstein, 1993): (a) single image; (b) average of 24 images; and (c) average of 240 images.

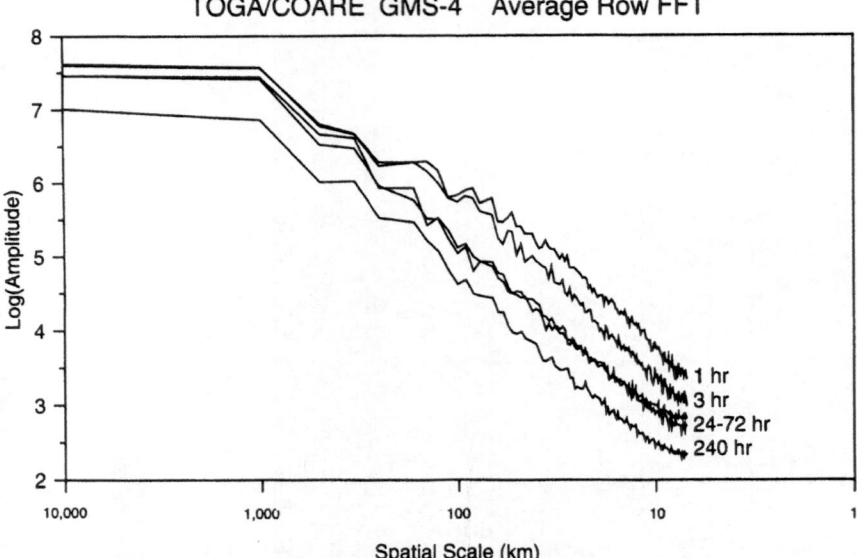

Fig. 17. Spatial power spectra of the infrared brightness temperatures for longitudinal strips through the western equatorial Pacific region shown in Figure 16 (averaged over all latitudinal bands to reduce variance of the spectral estimate). The spatial spectra are evaluated after the brightness temperatures have been averaged over the time intervals indicated: 1 hour (single image), 3 hours, 24 hours, 72 hours and 240 hours.

which covers about 80% of the globe by assuming that each observation represents rather large areas (∼ 1000 km across). If we compare months that have nearly identical original average cloud amounts (within 0.5%), the range of average cloud amounts produced from the partial dataset is almost three times larger. In other words, the apparent cloud amount variation is almost three times 'noisier' than it actually is. Hence, diagnostic monitoring of clouds requires globally complete observations that are only feasible from satellites.

The sampling of clouds required to detect changes must also be 'dense' enough to acquire proper statistics for the significant time and space scales of cloud variation. In the previous section we have shown that the smallest variation time scale that persists in the long-term statistics is the diurnal scale (actually the semi-diurnal scale – Figure 12), whereas most of the variation at spatial scales < 500 km is eliminated in long-term averages (Figure 17). Thus, cloud observations must provide an unbiased sample of the diurnal cycle, but only need a sufficiently small spatial sampling interval to get enough samples of the smaller spatial scale variations.

Global coverage and diurnal sampling cannot be accomplished with observations from one satellite (Salby, 1982); hence, the minimum satellite observing system for cloud monitoring is two satellites (Figure 19) (Brooks *et al.*, 1986; McConnell and North, 1987; Shin and North, 1988; Bell *et al.*, 1990). One satellite

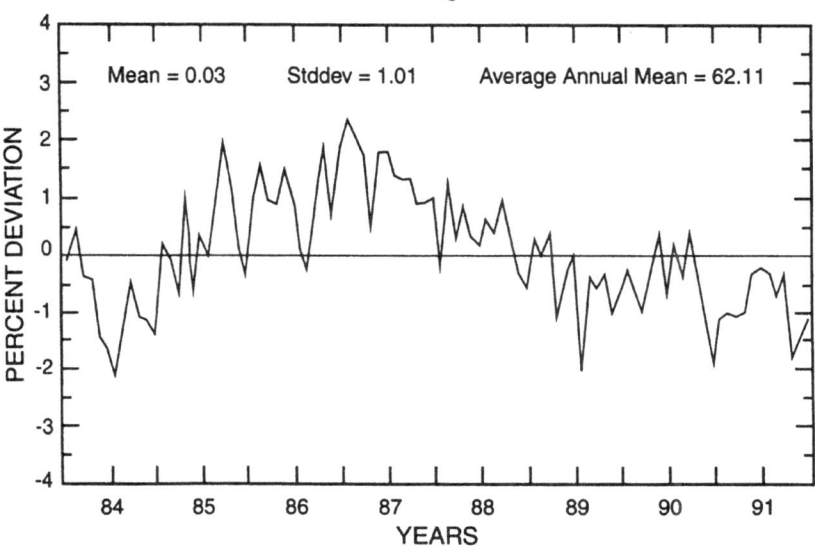

MONTHLY CLOUD AMOUNT DEVIATIONS
JUL 83 through JUN 91

Fig. 18. Variation of the globally averaged, monthly mean cloud amount deviation from the average over the period July 1983 through June 1991 from the ISCCP dataset.

provides global coverage at two times of day and one satellite provides statistical coverage of the diurnal variations at lower latitudes. This minimum system is sufficient to produce accurate averages, including an accurate average diurnal cycle; however, this system is not sufficient for diagnosis of the effects of cloud changes because it cannot describe possible changes in the diurnal variation of clouds. Although an **average** diurnal cycle can be obtained statistically, its accurcay depends on an assumption that the diurnal cycle has not changed over some time period. Hence, the diagnostic requirement demands a satellite system that actually resolves the diurnal time scale, which can be done with three, properly separated, sun-synchronous polar orbiters (Figure 19). Similarly although sampling along the satellite nadir track is sufficient for monitoring changes in global mean cloud properties (Hansen *et al.*, 1993), in order to detect possible shifts in storm tracks or near-coastal cloudiness, observations from scanning instruments are preferable.

To test various cloud observing strategies, we simulated the observations of two satellites with the orbits like those illustrated in Figure 19 (upper panel) by sampling the full ISCCP cloud dataset that has an effective resolution of 30 km and 3 hr. The ISCCP dataset is, itself, a sample of satellite observations with an original spatial resolution of about 5 km; but this sample has been shown to capture the statistics of the original 5 km data (Seze and Rossow, 1991a, b). For this test, we collect 6–9

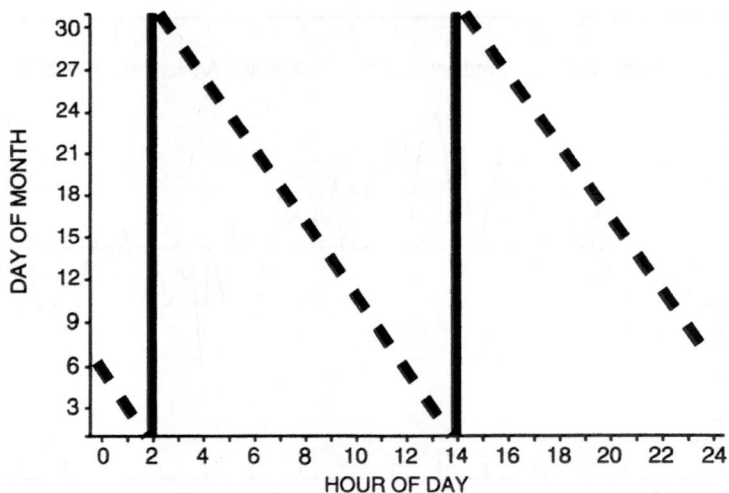

Fig. 19. Two alternative satellite sampling strategies for adequate diurnal sampling illustrated by showing the coverage of time-of-day for each day of the month. The upper panel shows the sampling from one sun-synchronous polar orbiter and an inclined orbiter that drifts in diurnal phase during the month, while the lower panel shows the sampling from three sun-synchronous polar orbiters.

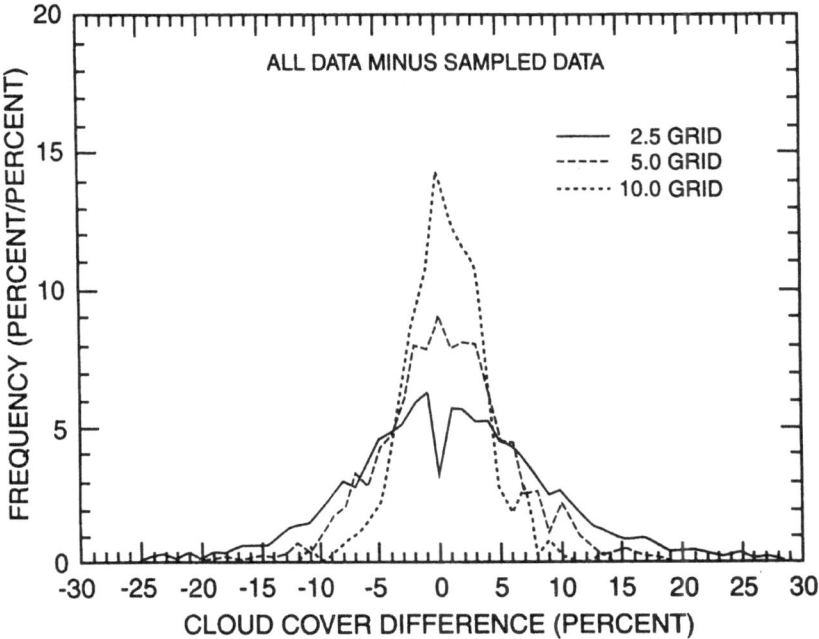

Fig. 20.   Frequency distribution over the whole globe of the differences in the regional monthly mean cloud amounts produced from the combined nadir samples of a polar orbiter and an inclined orbiter compared with the full ISCCP sampling. Both datasets are averaged at three different map grid resolutions: 2.5° (solid line), 5.0° (dashed line) and 10° (dotted line).

pixels at each location along the satellite ground track to represent nadir sampling from two satellite orbits. Since we only select a subset of the ISCCP pixels, there is no measurement error, only sampling error in the comparison of the sampled and original data. We focus on cloud cover because its bi-modal frequency distribution (Figure 2) implies a very large natural variability ($\approx$ 30–40%) that produces large sampling effects. The frequency distribution can be thought of as a probability distribution for a single sample (Warren *et al.*, 1986, 1988), so that more than 1000 samples are required to reduce sampling uncertainty below 1%.

To evaluate the accuracy of the sampled dataset, we first calculate the monthly mean cloud cover for each region on Earth on three map grids: 2.5°, 5° and 10°. This is done for the full ISCCP and sampled datasets. We also calculate these averages over three months (one season). The mapped values from the sampled dataset are compared with the full ISCCP values (considered to be the truth) and the frequency distribution of the differences collected. Figure 20 shows these difference distributions for one month averages. Enlarging the averaging domain from 2.5° to 10° decreases the standard deviation of the differences from $\approx$ 8% to $\approx$ 3% and averaging over three months (not shown) decreases the standard deviations from $\approx$ 8% to $\approx$ 5% for the 2.5° map grid and from $\approx$ 3% to $\approx$ 2% for the 10° map grid.

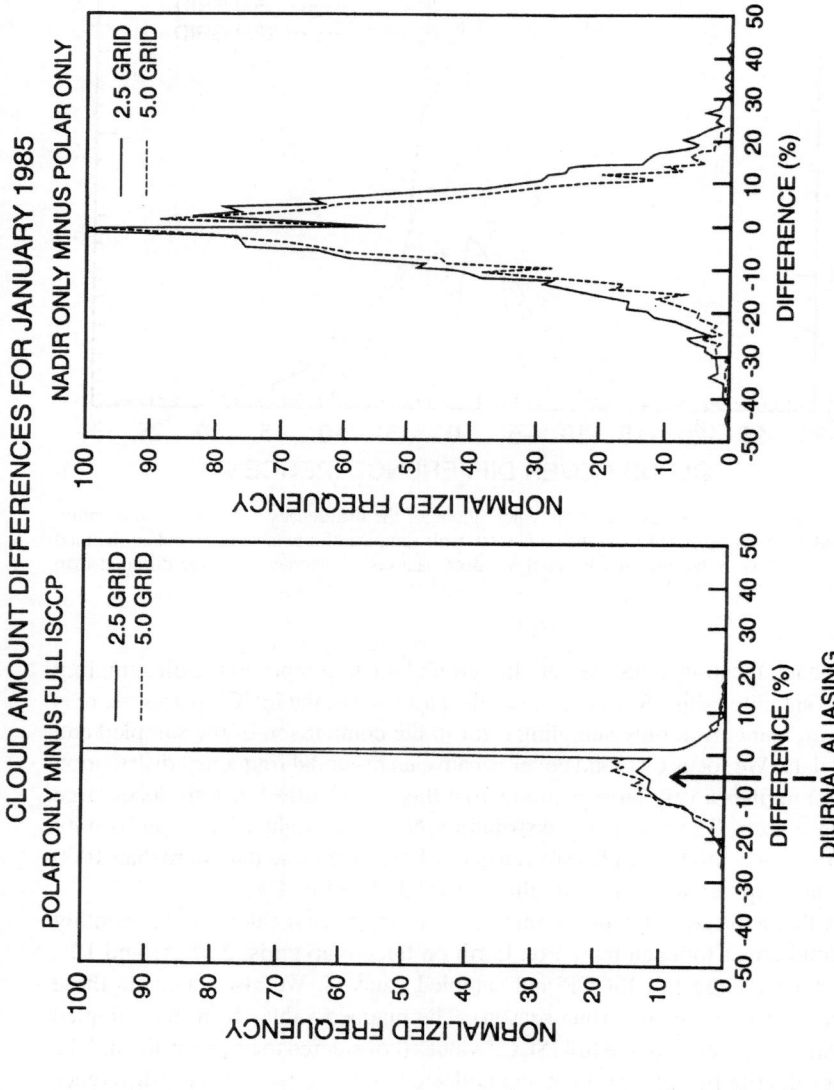

Fig. 21. Frequency distribution over the whole globe of the differences in the regional monthly mean cloud amounts produced from a single polar orbiter compared with the full ISCCP sampling (left panel). The right panel shows the differences between the full and sampled polar orbiter cloud amounts. Results are shown for two map grid resolutions, 2.5° (solid lines) and 5.0° (dashed lines).

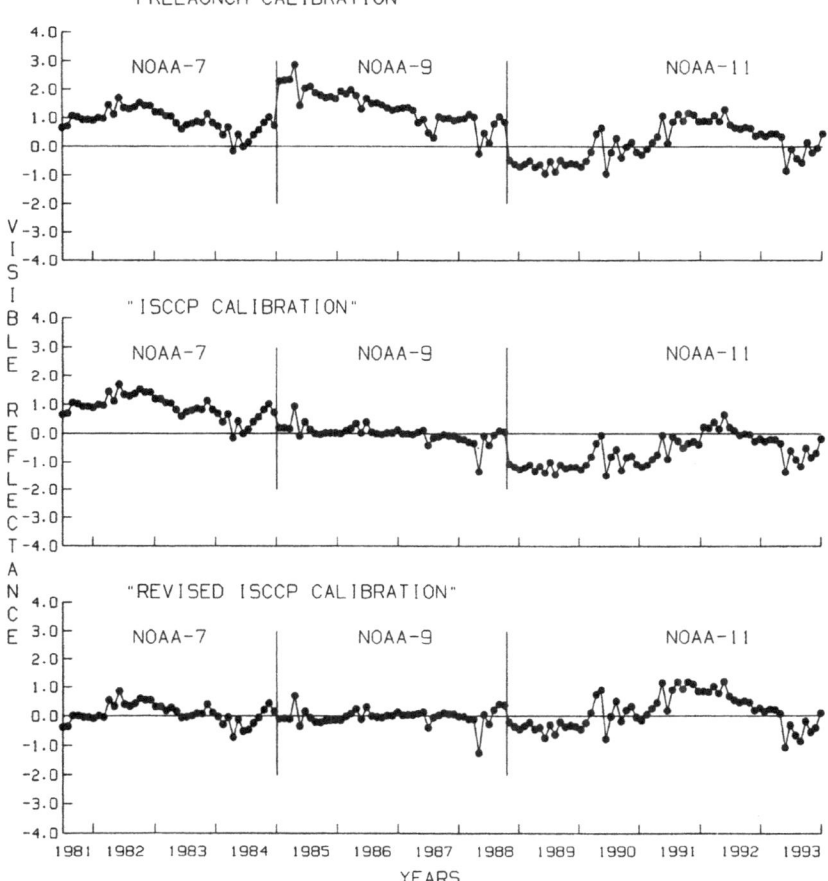

Fig. 22. Variation of the globally averaged, surface visible reflectances over the period July 1983 through December 1993 as deviations from the average values over the NOAA-9 results in the middle panel. The upper panel shows the prelaunch calibration for NOAA-7, NOAA-9 and NOAA-11 and next two panels show the results of two analyses to remove the calibration discontinuities. The remaining anomaly in 1991–1992 is caused by the Pinatubo volcanic aerosol.

The bias error in the global mean cloud amount is < 0.5% for all cases. Thus, the sampling error by the minimum (two-satellite) observing system would just allow measurement of the cloud amount changes shown in Figure 18.

The bias error introduced by incomplete diurnal sampling is illustrated by comparing the results obtained only from the sun-synchronous polar orbiter (samples roughly at 0230 and 1430 local time) with the full ISCCP results (cf. Salby, 1988b; Bell *et al.*, 1990); Figure 21 shows the distribution of differences in monthly mean cloud amounts. Although the global mean bias is not too large, the shape of the

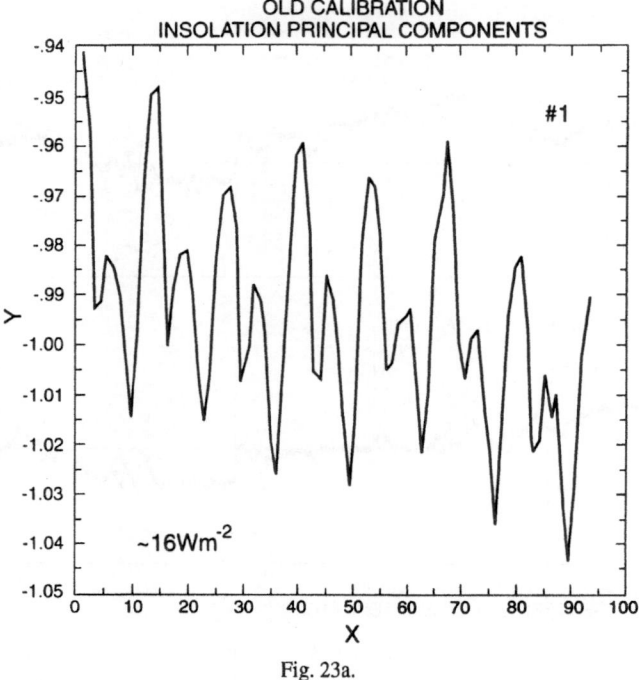

Fig. 23a.

difference histogram clearly shows that biases are much larger in some regions
where the amplitude of the diurnal cycle is larger.

5. Calibration

Since small changes in the physical properties of clouds can produce important
changes in the radiation balance, a cloud monitoring system must also maintain
a uniform relative calibration over the whole data record. For example, to detect
changes of cloud top pressure of 10 mb requires a relative calibration accuracy of
satellite-measured IR radiances of about 0.2% (≈ 0.5 K in brightness temperature);
to detect a change of cloud optical thickness of 0.15 requires a relative calibration
accuracy of visible radiances of about 1%. To monitor changes in cloud types
requires that the calibration be maintained consistently over all wavelengths used.
With today's satellite instruments, these relative accuracies are not attainable from
independent information. Figure 22 shows three versions of the visible radiance
calibration for the ISCCP dataset; only the 'pre-launch' calibration is independent
information. The remaining two calibrations use the time record of observations
of Earth's surface reflectance to remove differences between different radiometers
in the series and to remove sensor sensitivity changes. Note, however, that this

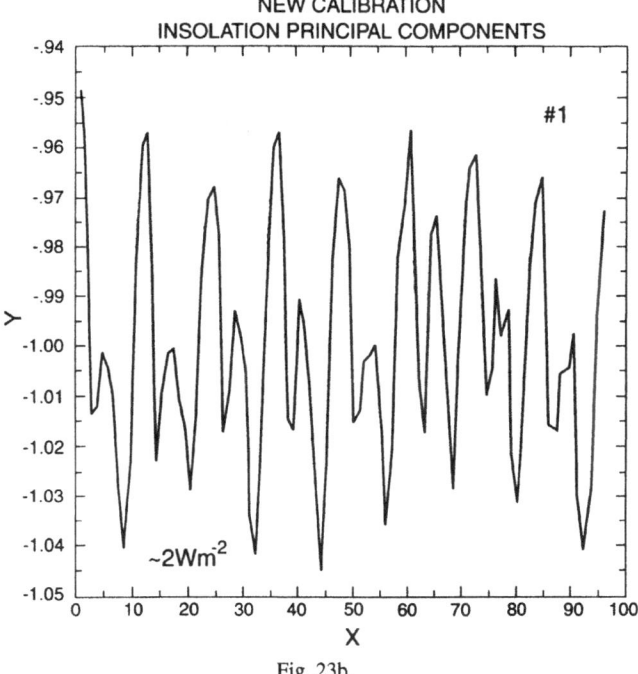

Fig. 23b.

Figs. 23(a)–(b). Time series associated with the first principal components of monthly mean surface solar insolation calculated from ISCCP cloud properties with: (a) the initial ISCCP calibration (second panel in Figure 22); and (b) the revised ISCCP calibration (third panel in Figure 22). The magnitude of the trend in (a) is about 16 Wm^{-2} and in (b) no more than 2 Wm^{-2}. The x-axis represents a time period of eight years, July 1983–June 1991.

procedure must assume that *Earth is, on average, a constant radiometric target*, hence, there is no independent determination as to whether this is actually the case. Figure 23a shows the first principal component of the monthly mean surface solar insolation determined from the ISCCP cloud properties derived with the second calibration. The presence of two small residual (< 10%) discontinuities in the second calibration produces a spurious trend in the surface insolation record with an overall variation of about 16 Wm^{-2} ($\approx$ 8% of the global annual mean value). Removal of these two discontinuities in the third calibration also removes the trend (to within 2 Wm^{-2}) in the surface insolation record (Figure 23b). The current ISCCP results exhibit an apparent decreasing trend in cloud optical thickness and cloud top temperature, most of which appears to be associated with the calibration discontinuities (Klein and Hartmann, 1993).

6. Necessary Attributes of Cloud Monitoring System

Based on this review and assessment of the characteristics of cloud variations, we can briefly outline some of the **necessary** attributes of a cloud monitoring system.

1. complete global coverage with uniform density;
2. spatial sampling interval $\leq$ 50 km;
3. sampling frequency $\geq$ 6 times per day;
4. record length > 10 years with uniform density.

To maintain uniform instrument calibration to the needed very high precision ($\approx$ 1%), overlapping observations between each pair of instruments in the series are essential. Moreover, if actual climate changes are to be detected, calibration must be obtained from information independent of the climate.

References

Barrett, E. C. and Grant, C. L.: 1979, 'Relations between Frequency Distributions of Cloud over the United Kingdom Based on Conventional Observations and Imagery from LANDSAT 2', *Weather* **34**, 416–424.

Bell, T. L.: 1987, 'A Space-Time Stochastic Model of Rainfall for Satellite Remote-Sensing Studies', *J. Geophys. Res.* **92**, 9631–9643.

Bell, T. L., Abdullah, A., Martin, R. L., and North, G. R.: 1990, 'Sampling Errors for Satellite-Derived Tropical Rainfall: Monte Carlo Study Using a Space-Time Stochastic Model', *J. Geophys. Res.* **95**, 2195–2206.

Brest, C. L. and Rossow, W. B.: 1992, 'Radiometric Calibration and Monitoring of NOAA AVHRR Data for ISCCP', *Int. J. Remote Sensing* **13**, 235–273.

Brooks, D. R., Harrison, E. F., Minnis, P., Suttles, J. T., and Kandel, R. S.: 1986, 'Development of Algorithms for Understanding the Temporal and Spatial Variability of the Earth's Radiation Balance', *Rev. Geophys.* **24**, 422–438.

Cahalan, R. F. and Joseph, J. H.: 1982, 'Fractal Statistics of Cloud Fields', *Mon. Wea. Rev.* **117**, 261–272.

Cahalan, R. F., Ridgeway, W., Wiscombe, W. J., Bell, T. L., and Snider, J. B.: 1994, 'The Albedo of Fractal Stratocumulus Clouds', *J. Atmos. Sci.* **51**, 2434–2455.

Cairns, B.: 1995, 'Diurnal Variations of Cloud from ISCCP Data', *Atmos. Res.* **37**, 133–146.

Carlson, B. E., Cairns, B., and Rossow, W. B.: 1995, 'Spatial and Temporal Characterization of Diurnal Cloud Variability Using ISCCP', *J. Clim.*, (submitted).

Cox, S. K., McDougal, D. S., Randall, D. A., and Schiffer, R. A.: 1987, 'FIRE – the First ISCCP Regional Experiment', *Bull. Amer. Meteor. Soc.* **68**, 114–118.

Elliott, W. P. and Gaffen, D. J.: 1991, 'On the Utility of Radiosonde Humidity Archives for Climate Studies', *Bull. Amer. Meteor. Soc.* **72**, 1507–1520.

Flament, P. and Bernstein, R.: 1993, *Images from the GMS-4 Satellite During TOGA-COARE (November 1992 to February 1993)*, Tech. Rep. 93–06, School of Ocean and Earth Sciences and Technoloy, Univ. Hawaii, Honolulu, 20 pp. with two CDROMs.

Fu, R., Del Genio, A. D., and Rossow, W. B.: 1990, 'Behavior of Deep Convective Clouds in the Tropical Pacific Deduced from ISCCP Radiance Data', *J. Clim.* **3**, 1129–1152.

Gaffen, D. J., Barnett, T. P., and Elliott, W. P.: 1991, 'Space and Time Scales of Global Tropospheric Moisture', *J. Clim.* **4**, 989–1008.

Gaster, M. and Roberts, J. B.: 1977, 'The Spectral Analysis of Randomly Sampled Record by a Direct Transform', *Proc. R. Soc. Lond. A.* **354**, 27–58.

Hahn, C. J., Warren, S. G., and London, J.: 1994, *Edited Synoptic Cloud Reports from Ships and Land Stations over the Globe, 1982–1991*, Carbon Dioxide Information Analysis Center, Oak Ridge National Laboratory, Oak Ridge, Tennessee, 47 pp.

Hansen, J. E. and Lebedeff, S.: 1988, 'Global Trends of Measured Surface Air Temperature', *J. Geophys. Res.* **92**, 13, 345–372.

Hansen, J., Rossow, W., and Fung, I.: 1993, *Long-Term Monitoring of Global Climate Forcings and Feedbacks*, NASA Conference Publication 3234, 89 pp.

Harshvardhan, Wielicki, B. A., and Ginger, K. M.: 1994, 'The Interpretation of Remotely Sensed Cloud Properties from a Model Parameterization Perspective', *J. Clim.* **7**, 1987–1998.

Hendon, H. H. and Woodbury, K.: 1993, 'The Diurnal Cycle of Tropical Convection', *J. Geophys. Res.* **98**, 16, 623–16, 637.

Houze, R. A., Rutledge, S. A., Bifferstaff, M. I., and Smull, B. F.: 1989, 'Interpretation of Doppler Weather Radar Displays of Midlatitude Mesoscale Convective Systems', *Bull. Amer. Meteor. Soc.* **70**, 608–619.

Joseph, J. H. and Cahalan, R. H.: 1990, 'Nearest Neighbor Spacing of Fair Weather Cumulus Clouds', *J. Appl. Meteor.* **29**, 793–805.

Kaimal, J., Wyngaard, J. C., Haugen, D. A., Cote, O. R., Izumi, Y., Caughey, S. J., and Readings, C. J.: 1976, 'Turbulence Structure in the Convective Boundary Layer', *J. Atmos. Sci.* **33**, 2152–2169.

Klein, S. A. and Hartmann, D. L.: 1993, 'Spurious Trends in the ISCCP C2 Dataset', *Geophys. Res. Lett.* **20**, 455–458.

Kondragunta, C. R. and Gruber, A.: 1994, 'Diurnal Variations of the ISCCP Cloudiness', *Geophys. Res. Lett.* **21**, 2015–2018.

Kropfli, R. A., Matrosov, S. Y., Uttal, T., Orr, B. W., Frisch, A. S., Clark, K. A., Bartram, B. W., Reinking, R. F., Snider, J. B., and Martner, B. E.: 1995, 'Studies of Cloud Microphysics with Millimeter Wave Radar', *Atmos. Res.* **35**, 299–313.

Kuo, K. S., Welch, R. M., and Sengupta, S. K.: 1988, 'Structural and Textural Characteristics of Cirrus Clouds Observed Using High Spatial Resolution LANDSAT Imagery', *J. Appl. Meteor.* **27**, 1242–1260.

Lau, N.-C. and Crane, M. W.: 1995, 'A Satellite View of the Synoptic-Scale Organization of Cloud Properties in Midlatitude and Tropical Circulation Systems', *Mon. Wea. Rev.* **123**, 1984–2006.

Lee, J., Chou, J., Weger, R. C., and Welch, R. M.: 1994, 'Clustering, Randomness, and Regularity in Cloud Fields, 4, Stratocumulus Cloud Fields', *J. Geophys. Res.* **99**, 14, 461–480.

Liao, X., Rossow, W. B., and Rind, D.: 1995a, 'Comparison between SAGE II and ISCCP High-Level Clouds, Part I: Global and Zonal Mean Cloud Amounts', *J. Geophys. Res.* **100**, 1121–1135.

Liao, X., Rossow, W. B., and Rind, D.: 1995b, 'Comparison between SAGE II and ISCCP High-Level Clouds, Part II: Locating Cloud Tops', *J. Geophys. Res.* **100**, 1137–1147.

Lin, B. and Rossow, W. B.: 1994, 'Observation of Cloud Liquid Water Path over Oceans: Optical and Microwave Remote Sensing Methods', *J. Geophys. Res.* **99**, 20, 907–20, 927.

Machado, L. A. T., Desbois, M., and Duvel, J-P.: 1992, 'Structural Characteristics of Deep Convective Systems over Tropical Africa and Atlantic Ocean', *Mon. Wea. Rev.* **120**, 392–406.

Machado, L. A. T., Duvel, J-P., and Desbois, M.: 1993, 'Diurnal Variations and Modulation by Easterly Waves of the Size Distribution of Convective Cloud Clusters over West Africa and the Atlantic Ocean', *Mon. Wea. Rev.* **121**, 37–49.

Machado, L. A. T. and Rossow, W. B.: 1993, 'Structural Characteristics and Radiative Properties of Tropical Cloud Clusters', *Mon. Wea. Rev.* **121**, 3234–3260.

McConnell, A. and North, G. R.: 1987, 'Sampling Errors in Satellite Estimates of Tropical Rain', *J. Geophys. Res.* **92**, 9567–9570.

Mokhov, I. I. and Schlesinger, M. E.: 1993, 'Analysis of Global Cloudiness, 1, Comparison of ISCCP, Meteor and Nimbus 7 Satellite Data', *J. Geophys. Res.* **98**, 12, 849–12, 868.

Mokhov, I. I. and Schlesinger, M. E.: 1994, 'Analysis of Global Cloudiness, 2, Comparison of Ground-Based and Satellite-Based Cloud Climatologies', *J. Geophys. Res.* **99**, 17, 045–17, 065.

Nicholls, S.: 1989, 'The Structure of Radiatively Driven Convection in Stratocumulus', *Quart. J. Roy. Meteorol. Soc.* **115**, 487–511.

Pandolfo, L.: 1993, 'Observational Aspects of the Low-Frequency Intraseasonal Variability of the Atmosphere in Middle Latitudes', *Adv. Geophys.* **34**, 93–174.

Parker, L., Welch, R. M., and Musil, D. J.: 1986, 'Analysis of Spatial Inhomogeneities in Cumulus Clouds Using High Spatial Resolution LANDSAT Data', *J. Clim. Appl. Meteorol.* **25**, 1301–1314.

Priestly, M. B.: 1981, *Spectral Analysis and Time Series*, Academic, New York.

Rossow, W. B.: 1978, 'Cloud Microphysics: Analysis of the Clouds of Earth, Venus, Mars, and Jupiter', *Icarus* **36**, 1–50.

Rossow, W. B.: 1989, 'Measuring Cloud Properties from Space: A Review', *J. Clim.* **2**, 201–213.

Rossow, W. B. and Lacis, A. A.: 1990, 'Global, Seasonal Cloud Variations from Satellite Radiance Measurements. Part II: Cloud Properties and Radiative Effects', *J. Clim.* **3**, 1204–1253.

Rossow, W. B. and Garder, L. C.: 1993a, 'Cloud Detection Using Satellite Measurements of Infrared and Visible Radiances for ISCCP', *J. Clim.* **6**, 2341–2369.

Rossow, W. B. and Garder, L. C.: 1993b, 'Validation of ISCCP Cloud Detections', *J. Clim.* **6**, 2370–2393.

Rossow, W. B. and Schiffer, R. A.: 1991, 'ISCCP Cloud Data Products', *Bull. Amer. Meteor. Soc.* **72**, 2–20.

Rossow, W. B., Walker, A. W., and Garder, L. C.: 1993, 'Comparison of ISCCP and Other Cloud Amounts', *J. Clim.* **6**, 2394–2418.

Rossow, W. B. and Zhang, Y-C.: 1995, 'Calculation of Surface and Top-of Atmosphere Radiative Fluxes from Physical Quantities Based on ISCCP Datasets, Part II: Validation and First Results, *J. Geophys. Res.* **100**, 1167–1197.

Salby, M. L.: 1982, 'Sampling Theory for Asynoptic Satellite Observations. Part I: Space-Time Spectra, Resolution, and Aliasing', *J. Atmos. Sci.* **39**, 2577–2600.

Salby, M. L.: 1988a, 'Asynoptic Sampling Considerations for Wide-Field-of-View Measurements of Outgoing Radiation. Part I: Spatial and Temporal Resolution', *J. Atmos. Sci.* **45**, 1176–1183.

Salby, M. L.: 1988b, 'Asynoptic Sampling Considerations for Wide-Field-of-View Measurements of Outgoing Radiation. Part II: Diurnal and Random Space-Time Variability', *J. Atmos. Sci.* **45**, 1184–1204.

Salby, M. L.: 1989, 'Climate Monitoring from Space: Asynoptic Sampling Considerations', *J. Clim.* **2**, 1091–1105.

Salby, M. L., Hendon, H. H., Woodbury, K., and Tanaka, K.: 1991, 'Analysis of Global Cloud Imagery from Multiple Satellites', *Bull. Amer. Meteor. Soc.* **72**, 467–480.

Sassen, K.: 1991, 'The Polarization Lidar Technique for Cloud Research: A Review and Current Assessment', *Bull. Amer. Meteor. Soc.* **72**, 1848–1866.

Sengupta, S. K., Welch, R. M., Navar, M. S., Berendes, T. A., and Chen, D. W.: 1990, 'Cumulus Cloud Field Morphology and Spatial Patterns Derived from High Spatial Resolution LANDSAT Imagery', *J. Appl. Meteor.* **29**, 1245–1267.

Seze, G. and Rossow, W. B.: 1991a, 'Time-Cumulated Visible and Infrared Radiance Histograms Used as Descriptors of Surface and Cloud Variations', *Int. J. Remote Sensing* **12**, 877–920.

Seze, G. and Rossow, W. B.: 1991b, 'Effects of Satellite Data Resolution on Measuring the Space-Time Variations of Surfaces and Clouds', *Int. J. Remote Sensing* **12**, 921–952.

Shapiro, H. S. and Silverman, R. A.: 1960, 'Alias-Free Sampling of Random Noise', *J. SIAM* **8**, 225–248.

Shin, K.-S. and North, G. R.: 1988, 'Sampling Error Study for Rainfall Estimate by Satellite Using Stochastic Model', *J. Appl. Meteor.* **27**, 1218–1231.

Stowe, L. L., Wellemeyer, C. G., Eck, T. F., Yeh, H. Y. M. and the NIMBUS-7 Cloud Data Processing Team: 1988, 'NIMBUS-7 Global Cloud Climatology. Part I: Algorithms and Validation', *J. Clim.* **1**, 445–470.

Stowe, L. L., Yeh, H. Y. M., Eck, F. T., Wellemeyer, C. G., Kyle, H. L. and the NIMBUS-7 Cloud Data Processing Team: 1989, 'NIMBUS-7 Global Cloud Climatology. Part II: First Year Results', *J. Clim.* **2**, 671–709.

Thomson, D. J.: 1990, 'Quadratic-Inverse Spectrum Estimates: Application to Paleoclimatology', *Phil. Trans. Roy. Soc. Lond.* **A332**, 539–597.

Wang, J. and Rossow, W. B.: 1995, 'Determination of Cloud Vertical Structure from Upper Air Observations', *J. Appl. Meteor.*, (in press).

Warren, S. G., Hahn, C. J., and London, J.: 1985, 'Simultaneous Occurrence of Different Cloud Types', *J. Clim. Appl. Meteor.* **24**, 658–667.

Warren, S. G., Hahn, C. J., London, J., Chervin, R. M., and Jenne, R. L.: 1986, *Global Distribution of Total Cloud and Cloud Type Amounts over Land*, 29 pp. + 200 maps, (NTIS number DE87–00–6903).

Warren, S. G., Hahn, C. J., London, J., Chervin, R. M., and Jenne, R. L.: 1988, *Global Distribution of Total Cloud and Cloud Type Amounts over the Ocean*, 42 pp. + 170 maps, (NTIS number DE90–00–3187).

Weger, R. C., Lee, J., Zhu, T., and Welch, R. M.: 1992, 'Clustering, Randomness, and Regularity in Cloud Fields, 1, Theoretical Considerations', *J. Geophys. Res.* **97**, 20, 519–20, 536.

Weger, R. C., Lee, J., and Welch, R. M.: 1993, 'Clustering, Randomness, and Regularity in Cloud Fields, 3, The Nature and Distribution of Clusters', *J. Geophys. Res.* **98**, 18, 449–18, 463.

Welch, R. M., Kuo, K. S., Wielicki, B. A., Sengupta, S. K., and Parker, L.: 1988, 'Marine Stratocumulus Cloud Fields off the Coast of Southern California Observed Using LANDSAT Imagery. Part I: Structural Characteristics', *J. Appl. Meteor.* **27**, 341–362.

Wielicki, B. A. and Parker, L.: 1992, 'On the Determination of Cloud Cover from Satellite Sensors: The Effects of Sensor Spatial Resolution', *J. Geophys. Res.* **97**, 12, 799–12, 823.

Wielicki, B. A. and Welch, R. M.: 1986, 'Cumulus Cloud Field Properties Derived Using LANDSAT Digital Data', *J. Clim. Appl. Meteor.* **25**, 261–276.

Zangvil, A.: 1975, 'Temporal and Spatial Behavior of Large-Scale Disturbances in Tropical Cloudiness Deduced from Satellite Brightness Data', *Mon. Wea. Rev.* **103**, 904–920.

Zhang, Y.-C., Rossow, W. B., Lacis, A. A.: 1995, 'Calculation of Surface and Top-of-Atmosphere Radiative Fluxes from Physical Quantities Based on ISCCP Datasets, Part I: Method and Sensitivity to Input Data Uncertainties', *J. Geophys. Res.* **100**, 1149–1165.

Zhu, T., Lee, J., Weger, R. C., and Welch, R. M.: 1992, 'Clustering, Randomness, and Regularity in Cloud Fields, 2, Cumulus Cloud Fields', *J. Geophys. Res.* **97**, 20, 537–20, 558.

(Received 23 January, 1995; in revised form 17 July, 1995)

ON DETECTING LONG-TERM CHANGES IN ATMOSPHERIC MOISTURE

WILLIAM P. ELLIOTT

Air Resources Laboratory, NOAA, 1315 East West Hwy., Silver Spring MD 20910, U.S.A.

Abstract. Long-term temperature changes are expected to give rise to changes in the water vapor content of the atmosphere, which in turn would accentuate the temperature change. It is thus important to monitor water vapor in the troposphere and lower stratosphere. This paper reviews existing data for such an endeavor and the prospects for improvement in monitoring.

In general, radiosondes provide the longest record but the data are fraught with problems, some arising from the distribution of stations and some from data continuity questions arising from the use of different measuring devices over both time at one place and over space at any one time. Satellite records are now of limited duration but they will soon be useful in detecting changes. Satellite water vapor observations have their own limitations; there is no one system capable of measuring water vapor over all surfaces in all varieties of weather. Among the needs are careful analysis of existing records, the collection of metadata about the measuring systems, the development of a transfer standard radiosonde system, and the commitment to maintaining an observing system dedicated to describing any climate changes worldwide.

1. Introduction

Water vapor plays a major role in the dynamics of the atmosphere's circulation as well as in radiation exchange within the atmosphere. A large portion of the energy transferred between the surface and the free atmosphere is in the form of latent heat. The redistribution of this latent heat and its realization through condensation and precipitation is a main energy source for the general circulation. Water vapor is also the most important of the greenhouse gases in the overall control of climate, and its condensate, clouds, modulate the radiative energy transfer. Thus changes in water vapor concentration have major effects on climate.

Models estimating effects of greenhouse gas increases portray an increasing water vapor concentration as the atmosphere warms, 20% to 30% as CO_2 doubles in various models (Meehl and Washington, 1990; Hansen *et al.*, 1984). Some such result would be expected if only because the saturation vapor pressure increases with temperature. The increased moisture content in turn would increase the warming. This positive feedback of water vapor is one of the largest factors calculated to amplify the effects of increased greenhouse gas concentrations.

It is not only the total quantity of water or its geographical distribution that is of interest; the vertical distribution is also important. Despite the small absolute amount of water possible in the cold of the upper troposphere and stratosphere, water vapor changes there have a disproportionate effect on radiation exchange and surface temperature. It is, approximately, the fractional change in water vapor

Climatic Change **31**: 349–367, 1995.

rather than the absolute change that is important (Shine and Sinha, 1991). Thus not only total water vapor should be monitored but its vertical distribution as well, both for long-term changes and for understanding the dynamics of the atmosphere. How well we can monitor water vapor now and the prospects for the future are the topics of this paper.

This note stresses long-term monitoring. Only brief mention will be made of measuring systems that are not now, or apt to be, widely deployed. Stress will also be placed on sources of bias in observations and possible changes in bias. For detecting long-term climate changes, no matter what their cause, homogeneous records are necessary. Indeed, homogeneity is more important than absolute accuracy, given a reasonable level of the latter. Thus observations that have large random errors may still be useful for detecting change and introducing more accurate instruments, if they produce an offset in the time series, introduce analysis problems which have to receive attention.

The next part will review some of the water vapor variables, both measured and calculated, and what we know of their distribution. Then will follow a review of the available observations, and their current limitations, focusing on routine surface and upper-air radiosonde measurements made by the world's weather services as well as data routinely collected from space-borne instruments. Next, the outlook for the next few years will be discussed, and finally some thoughts on what might be done to improve monitoring.

2. Background

2.1. VARIABLES AND ALGORITHMS

There are a number of quantities used to express the amount of water vapor in the air. As most depend in one fashion or another on the temperature of the parcel being considered, almost any measurement of a water vapor variable is accompanied by a measurement of temperature. The usual reported variable is the dewpoint temperature, T_d, but this is not always measured directly. Rather, it is often calculated from measurements of temperature and either wet-bulb temperature, T_w, or relative humidity, RH. The latter is taken as the ratio of the actual vapor pressure to the vapor pressure that would saturate the air at its present temperautre, the saturation vapor pressure or s.v.p.

The specific humidity, q, and its close relative, the mixing ratio, are often calculated. (These quantities are, respectively, the ratio of the mass of water vapor in a given volume to the total mass in the volume, the specific humidity, or to the mass of dry air, the mixing ratio.) The integral of q between pressure surfaces gives the precipitable water, PW, (or column integrated water vapor) between the pressure levels or in the entire column. Values of PW can also be estimated from remotely sensed radiances either from satellites, aircraft or from the ground. There are additional remote measuring techniques, including active ones such as the various lidars, that are quite useful as research tools. Their cost will likely restrict their use to research for some time and they are not discussed further here.

The relationship that is basic to all calculations of water vapor quantities is that between temperature and the saturated vapor pressure, a function only of temperature. This relationship, the Clausius-Claperyon equation, is non-linear and cannot be integrated analytically. Furthermore, laboratory measurements of saturated water vapor (over a plain surface of pure water) are not made at temperatures below freezing (although relative humidity is usually expressed as the ratio of vapor pressure to s.v.p. over water at all temperatures). There are a number of approximations to the s.v.p.-temperature relation (see, e.g., Gueymard, 1993), ultimately based on laboratory measurements and extrapolated to below-freezing temperatures.

2.2. WATER VAPOR DISTRIBUTION

The source and the sink for almost all water in the atmosphere is the earth's surface, through evaporation and precipitation. Oxidation of methane appears to be an important additional source of stratospheric water, and this probably contributes to the slight increase with height of water vapor concentration in the stratosphere. As methane emissions appear to be increasing at the ground (Steele *et al.*, 1992; Khalil and Rasmussen, 1993), they could contribute to an increase in water in the stratosphere. Additions from volcanoes and other geologic activity, while probably important over vast time scales, have not been shown to affect significantly the present distribution or its changes.

If the water vapor in the air were all condensed, the average depth of the condensate would be about 2.5 cm. Above polar regions the mean PW is about 0.5 cm and near the equator it averages about 5 cm. Globally, about half of all the moisture in the atmosphere is between sea level and 850 mb and only about 5–7% is above 500 mb. The amount in the stratosphere is probably less than 1% of the total.

In the stratosphere the volume mixing ratio is only a few parts per million (ppmv) with frost-point tempertures as low as –90 to –100 °C. Near the surface in the tropics, on the other hand, the mixing ratio can be over 20,000 ppmv. This range of at least 4 orders-of-magnitude makes extreme demands on sensors, demands which few can meet, and so different techniques are called for at different heights. Furthermore, because precipitation rapidly removes water from the atmosphere, the residence time of water vapor in the atmosphere is only about 10 days, and its horizontal distribution is quite variable. Thus observations at many locations over a substantial period of time are required to establish its distribution.

3. Long-Term Records

A few records of surface temperature extend for several centuries, and some estimates of global surface temperatures extend over a century. The same is not true of humidity records, either because the records were not maintained or the quantity was not measured. In addition, the technology of routine humidity measuring has changed a good bit over the years, more so than of temperature. This is partly because there are several humidity quantities that can be measured directly: RH, T_d, and T_w, for instance. Changes in instrumentation and recording practices require care to identify because changes they can inject into the records can be confused with changes in climate.

Another prominent issue in the evaluation of humidity records is the algorithms used both to convert electro-mechanical signals to meteorological variables and to convert these variables to one another can and have changed. The introduction of computers into data handling and processing allowed more accurate computational methods to replace less accurate, but less cumbersome, ones. However, changes in processing algorithms can lead to subtle differences in calculated values which could then appear as apparent climate changes (Elliott and Gaffen, 1993; Wade, 1994).

3.1. SURFACE OBSERVATIONS

There are some records of surface humidity extending back into the last century in the U.S. and likely in some other countries, at least in the Northern Hemisphere. In the U.S. the early observations were of wet- and dry-bulb temperatures. The thermodynamics of wet-bulb thermometers is complicated; conversion to relative humidity and dewpoint was accomplished with tables and slide-rules. After about 1960, the Weather Service installed hygrothermometers which give dewpoint directly, using absorption of water by lithium chloride crystals which changes their electrical properties. In 1984 the Weather Service adopted dewpoint hygrometers. Currently the Weather Service is replacing these with newer versions in the ASOS (Automatic Surface Observing System) program. Other nations have likely undergone similar changes over the years. The Historical Climate Network of the U.S. does not make humidity measurements so this network does not offer an alternative to the normal weather stations.

As with estimating surface temperature trends, station histories need to be considered in evaluating water vapor records. Moves of the station, instrument replacement and changes in the surroundings would affect the records. The effect of urbanization, the urban heat island, leads to rural-urban differences in humidity as well as temperature (Lee, 1991). Increases in humidity could also come about from increased irrigation or the construction of a nearby reservoir. Such local changes could obscure a regional or global signal, as is true with temperature.

A source of long-term humidity data over the oceans is the Comprehensive Ocean-Atmosphere Data Set (COADS; Woodruff *et al.*, 1987). There are humidity data taken from research vessels and merchant ships but how widespread these are needs investigation. There has been no attempt known to the author to estimate global or even regional long-term surface moisture changes. Brazel and Balling (1986) examined the 1896–1984 humidity record for Phoenix, AZ, to seek local influences and they did find a decrease in RH accompanying the urban warming in Phoenix, but little change in dewpoint.

The whole topic of examining moisture changes at the surface over the globe, or a substantial part of it, would first require a search of the meteorological archives to determine what would be feasible.

3.2. UPPER AIR (RADIOSONDES)

Most of our knowledge of tropospheric water vapor (see e.g. Peixoto and Oort, 1992) comes from routine radiosonde observations taken for weather forecasts. Observations have been made globally since World War II but the bulk of useful humidity data begin in 1958 when the present observing times of 00 and 12 UTC were adopted. The main purpose of these observations is weather forecasts so maintaining long-term homogeneity of the observational record has not been a primary concern. Because frequent improvements in technology have been introduced, separating true long-term trends or variations from the effects of differing observation techniques is difficult.

There are about 700–800 regularly reporting radiosonde stations reporting once or twice a day, although there are additional stations that send reports irregularly. The distribution of these stations leaves large gaps in coverage, Figure 1. The stations shown are those from which useful data can be obtained over part of the period since 1973. Some are no longer operating, particularly the Ocean Weather Stations in the North Pacific and North Atlantic. Since the almost complete demise of these OWS we are at the mercy of the distribution of islands for observations in remote parts of the oceans, so that the Southern Ocean, the Eastern Pacific and the South Atlantic are particularly under-sampled. (Several nations are making radiosonde observations from moving merchant vessels following particular tracks, which provide augmentation of observations along these routes.) Even some continental areas are not well covered; Africa and South America in particular.

The expendable nature of the instruments and the number used daily means the instruments cannot be expensive or need highly trained scientists to operate them. These constraints limit the quality of the instrumentation. The current humidity sensors respond more slowly at the low temperatures of the upper troposphere and stratosphere and are generally less reliable under these conditions. (This is why, until recently, the U.S. did not report humidity values when the temperature was below –40 °C.) Above about 500 mb the reliability of radiosonde moisture

[223]

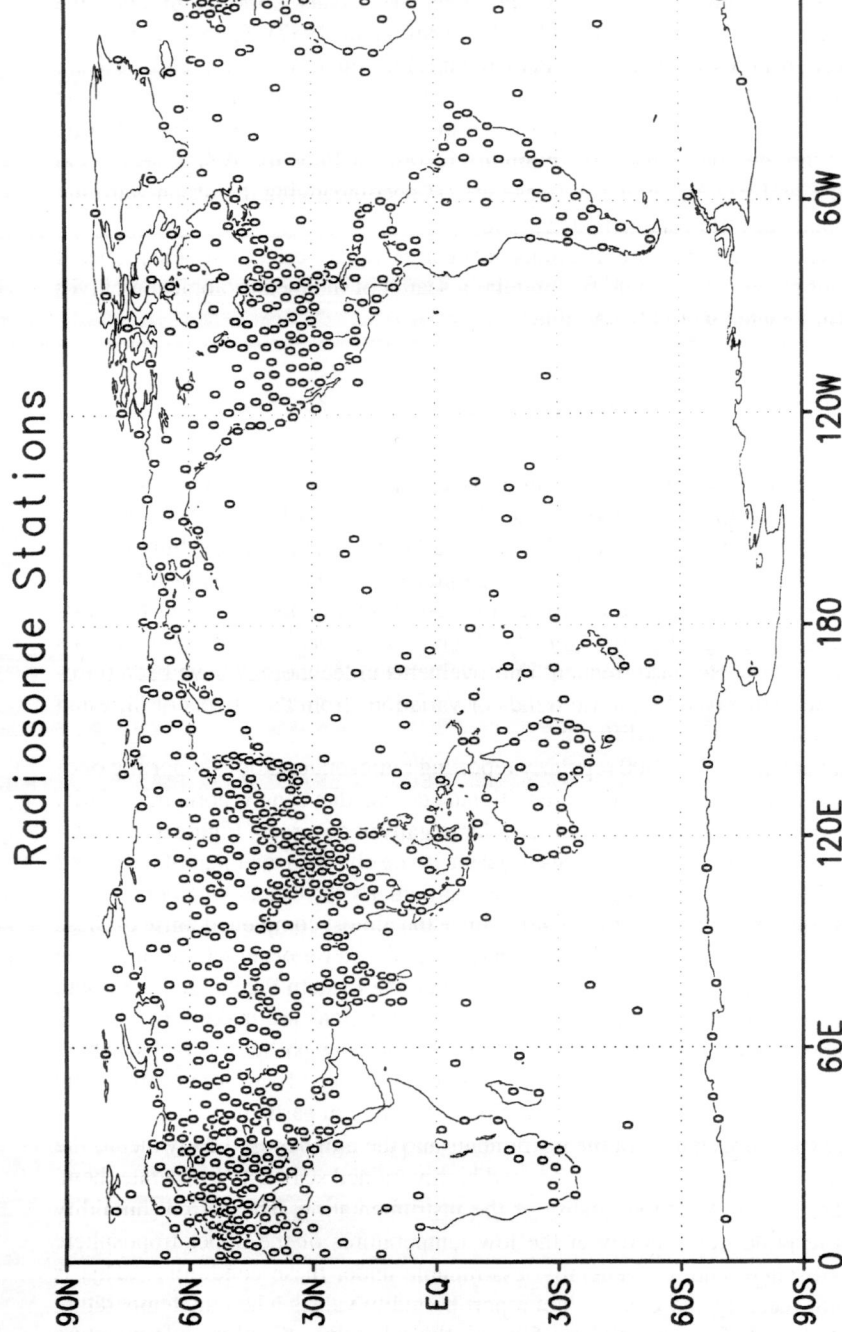

Fig. 1. Location of land-based radiosonde stations that have relatively long records during the 1973–94 period. Also included are some Ocean Weather Stations, most of which are no longer active.

data diminishes although in the tropics the temperatures remain warm enough for reasonable estimates up to about 300 mb.

A more serious problem with radiosonde data stems from the variety of instruments and reporting practices that are now or have been in use. About ten different manufacturers supply radiosondes to the world's weather services. Some of these use quite different sensors to measure relative humidity. The most widely used sonde, that manufactured by Väisälä Oy, uses a device that changes capacitance with changes in humidity. The sonde most widely used in the U.S., as well as throughout much of the north Pacific and Latin America, is made by VIZ Corp. and uses a carbon-based element whose resistance changes with humidity. Others use other elements whose resistnace or physical dimensions change with humidity, such as goldbeater's skin.

There have been and continue to be intercomparisons of some of these instruments (Nash and Schmidlin, 1987) but it is difficult to adjust the data to a common standard. Schmidlin (personal communication) states that most tested instruments give comparable values at humidities between 20% and 80% but outside this range differences can become significant. Some of the differences result from different reactions to wetting in clouds. Also some come about from the algorithms used in data reduction, as discussed below.

Furthermore, most nations have changed humidity sensors and other radiosonde components over the decades, in addition to changing suppliers. A particular example (Gaffen, 1992) comes from Adelaide, Australia, Figure 2, where there have been frequent changes of sonde supplier. Gaffen (1993) has assembled a useful summary of some changes in world-wide radiosonde in use since World War II.

Changes in the design of the enclosure and even changes in the length of the attachment to the balloon can have noticeable effects. In 1965, when the U.S. changed from a lithium chloride sensor to the carbon hygristor, a new design of the sonde's case was introduced. This led to the notorious situation shown by Elliott and Gaffen (1991), where the data at Hilo HI demonstrate the dramatic effect of allowing sunlight into the humidity sensor housing. Despite the fact that the problem was recognized early (Morrissey and Brousaides, 1970), a redesigned housing was not introduced until 1973.

Changes are not confined solely to measuring instruments. There have been changes in data processing algorithms and algorithms for converting from one moisture variable to another. Two sets of algorithms are needed. The first converts the electrical signals from the sonde to meteorological quantities and the second converts these quantities to others that are desired. Wade (1994) gives a description of the changes in the U.S. calcultion of low humidities from the sonde's electrical signals, where an apparent error has been carried through for years. He further notes that currently the calculations still require improvement (which is now under consideration). What is sobering about Wade's discussion is the realization that many of the problems he describes have been known for some time but making the corrections seems to present inordinate bureaucratic difficulties.

[225]

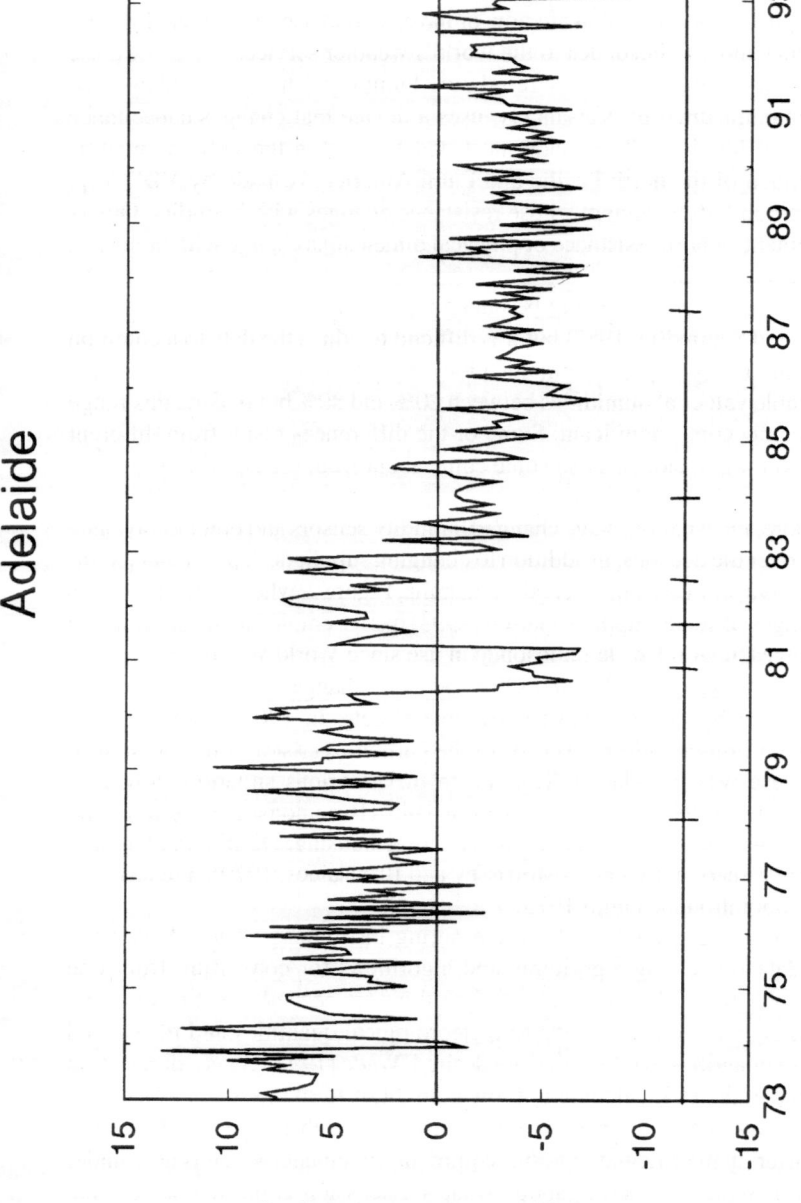

Fig. 2.   Time series of 500 mb dewpoint anomalies at Adelaide, Australia. The vertical bars along the x-axis indicate dates when changes in radiosonde types occurred.

Another source of potential problems is related to the use of different algorithms to convert relative humidity, to which the sonde's sensor responds, to dewpoint depression which is the quantity reported. Elliott and Gaffen (1993) show that this can be a source of small inhomogeneities in both horizontal moisture fields, as different nations use different conversions, and in time when new algorithms are adopted.

Not all nations report similar observed values in the same way. Until recently, the U.S. practice was to report humidity data at temperatures below $-40\,°C$ as 'missing' and all relative humidities less than 20% as 19%, (or dewpoint depression as '30', which is not the same RH value). These practices were dropped in Oct. 1993 by the U.S. and this will create a small discontinuity in records of monthly mean values. Some other nations have used similar rules and they could change also.

Most (but not all) of these changes represent improvements in sensors or other practices and so are to be welcomed. Nevertheless they make it difficult to sep- arate climate changes from changes in the measurement programs. In the 1940s untreated human hair was the humidity sensor on most sondes but values were virtually useless at temperatures below freezing because the lag time at these tem- peratures was so large. Since then, there have been several generations of sensors and now sensors have much faster response times. Whatever the improvements for weather forecasting, they do leave the climatologist with problems. Because relative humidity generally decreases with height slower sensors would indicate a higher humidity at a given height than today's versions (Elliott et al., 1994). This effect would be particularly noticeable at low temperatures where the differences in lag are greatest. A study by Soden and Lanzante (submitted) finds a moist bias in upper troposphere radiosondes using slower responding humidity sensors relative to more rapid sensors, which supports this conjecture. Such improvements would lead the unwary to conclude that somepart of the atmosphere had dried over the years.

Despite these problems there have been attempts to estimate changes in tropo- spheric water vapor from radiosonde observations. Hense et al. (1988) report an upward trend of moisture in the 700–500 mb layer in the western Pacific from 1965 to 1986 but there are questions from the '65–'72 period because the U.S. changed instrumentation in 1973. Elliott et al. (1991) document a moisture increase in the equatorial Pacific from 1973–86 and Gaffen et al. (1991) using more stations and an empirical orthogonal function analysis also found an increase in moisture in the tropics, a finding supported by Gutzler (1992) in a study of 4 tropical island stations. He found an increase in precipitable water of about 6% per decade for the period 1973–1988. Gaffen et al. (1992) computed trends for some 35 stations with reasonably homogeneous records around the globe. Most of these stations showed an increase in precipitable water during the period 1973–1990, with the largest and most statistically significant trends again at tropical stations, where increases as large as 13% per decade were found. In a forthcoming study, Ross and Elliott (in preparation) find statistically significant increases in PW, of 3 to 7% per decade, over most of North America, except for north and east Canada.

There has been one series of observations of stratospheric moisture of sufficient length to begin examining trends. Oltmans and Hofmann (1995) report data from balloon-borne frost-point hygrometers from Boulder, Colorado for 1981–1994 which show increases in stratospheric water vapor concentration at heights between 9 and 27 km with the most statistically significant values, which are between 0.5% and 1% per year, between 16 and 22 km. The increases are large enough that they may not be entirely caused by increased oxidation of methane.

Thus there is some observational evidence for increases in moisture content in the troposphere and perhaps in the stratosphere over the last 2 decades. Because of limitations of the data sources and the relatively short record length, further observations and careful treatment of existing data will be needed to confirm a global increase.

3.3. UPPER AIR (SATELLITES)

Most water vapor observations from satellites are too recent to give much help in estimating past changes in moisture. Techniques for extracting moisture information from satellite-observed radiances are rapidly improving, however, and can be applied to observations collected in the past. Unfortunately, there is no system in present operation which is able to measure total column water vapor in all conditions; that is, over both ocean and land in both clear and cloudy regions. Furthermore, only rough estimates of the vertical distribution of water vapor are now possible. A useful summary of remote sensing techniques, including satellite and ground-based observations, can be found in Starr and Melfi, 1991.

The longest record, since 1979, is from the Tiros Operational Vertical Sounder (TOVS) system. The system employs several of the High-resolution Infrared Radiation Sounder (HIRS) channels for water vapor (Smith et al., 1979; Wu et al., 1993). As the observed radiances are affected by temperature, temperature must also be retrieved. Moisture retrievals are not produced in overcast areas or with cloud cover greater than about 75%. In partly cloudy scenes (up to about 75% cloud cover) the measurements are valid only in the cloud-free areas. The channels of the HIRS-2 sounding system can resolve the precipitable water into three layers, approximately the 1000–700, 700–500, and 500–300 mb layers.

At first, radiosonde data were used to produce the temperature and moisture profiles from the HIRS observations through statistical regressions. Physically based techniques are now being used that reduce the reliance on nearby radiosondes. Some of these use Numerical Weather Prediction runs for the first-guess. Accuracy is believed to be within about ±20% of the radiosonde values for total PW; less is known of the accuracy in the upper layers, partly because of the uncertainties of the radiosonde data there.

The Stratospheric Aerosol and Gas Experiment (SAGE II) instrument on board the Earth Radiation Budget Satellite has provided measurements of water vapor in the stratosphere since late 1984. The technique depends on solar occultation and allows water vapor to be calculated with 1-km vertical resolution from the mid-troposphere to 45 km. It has the best precision of any of the current instruments aboard satellites. The spatial and temporal sampling is somewhat limited and the method does not permit measurements below cloud tops and reaches down to about 5 km only one-half the time. It has provided the best estimates of the global climatology of water vapor in the stratosphere (McCormick *et al.*, 1993).

Other observations of total column water have been provided by microwave instruments (SSMR, SSM/I) aboard several satellites (Liu *et al.*, 1992). While these instruments can 'see' through most clouds, except in heavy precipitation, they do not provide vertical resolution and their use for water vapor is limited to over-water regions. (The microwave emissions from land surfaces are too irregular to separate them from the atmospheric emissions.) SSMR observations were made from 1978–1984 while SSM/I began in 1987. SSM/I will continue to be used on future satellites.

The water vapor IR channels on the GOES/VAS and METEOSAT give good spatial descriptions of moisture features. The vertical resolution is relatively coarse, being comparable to that of TOVS and it requires a large effort to quantify the accuracy of the moisture retrievals. Nevertheless, strides are being made in this regard and information on water vapor over the regions viewed by these satellites may be forthcoming.

Much of the original data from many of the satellites have been archieved. This makes it quite possible that the ongoing work on improving the moisture-retrieval algorithms will allow these 'old' data to be reworked to obtain better estimates of water vapor from the periods of operation. However, when new satellites replace older ones, even those launched with the same types of instruments, care is necessary to ensure data continuity. Changes in Equator-crossing times and deterioration of calibration in space are other potential problems. Satellite observations do not escape data homogeneity questions.

3.4. ANALYZED DATA FIELDS

It has been suggested that the four-dimensional data assimilation systems used to initialize models for forecasts could provide a physically consistent data set for analyzing trends of several quantities, including moisture. However, this attractive idea has problems similar to the measurements, namely, that there have been changes over the years in the models used to assimilate the data (Trenberth and Olson, 1988) which produces changes in the fields that can mimic true changes in climate. Some hope lies in the so-called Reanalysis Project (Kalnay and Jenne, 1991) whose goal is to produce consistent analyses using the same data assimilation procedures on the archived data sets. This project will still have to cope with

[229]

changes in the instruments that produced the original values if climate change is to be monitored by this procedure.

3.5. METADATA

If all observations were taken with the same unchanging instruments and recorded with the same procedures there would be little use for details about them for climate change purposes. Such is not the case, of course, and so knowledge of the changes and how they might affect the record are necessary. Unfortunately, such information is often buried in the archives of the respective weather services or not available at all. Elliott and Gaffen (1991) give some history of the changes in U.S. radiosondes as they affect the moisture record and Gaffen (1993) summarizes some information on radiosonde changes world-wide. Neither of these compilations is complete; a number of nations did not respond to Gaffen's requests for information and there have been recent changes in the U.S. practices, see e.g. Wade (1994).

Metadata about satellite observations and also changes in data assimilation procedures have presumably been recorded but the information is scattered. Individual data sets are usually accompanied by some information about the observing techniques. Whether a systematic record is being kept, including information about cross calibrations among the satellites, and whether research workers in the future will know how to obtain it is not known to this author.

3.6. SUMMARY

All reasonably lengthy records of water vapor that could be examined for trends present problems. The surface data are subject to the same problems as the surface temperature data; station moves and changes in surroundings. In addition, changes in humidity sensors have been even more frequent than those of temperature sensors. Surface humidity records have not been examined save at a few locations, so their utility for trend detection is not known.

Radiosondes are the source of the longest records of upper-air moisture but their global distribution has large gaps and their ability to provide reliable data above 500 mb (about 6 km) is suspect. In addition the frequent changes in nations's observing programs makes identifying climate changes difficult.

Besides the relative brevity of their records, satellite observations of moisture cannot be made over all surfaces in all conditions. At present the vertical resolution of the observations is not as good as desired. They, too, suffer from changes in sensors as newer satellites replace those whose useful life ends. Nevertheless satellites provide data at heights where radiosonde moisture measurements are inadequate and in locations where radiosondes are not available.

In addition to being subject to all the problems with continuity of the underlying observation, analyzed NWP fields experience changes in analysis and data assimilation procedures.

Information about changes in instrumentation and procedures is useful for identifying potential discontinuities in climate records. However this information is lacking in many places and incomplete or even erroneous in some instances.

4. Near-Term Trends in Observations

There have been substantial changes in all the observing systems in the past decade and this is likely to continue for the next decade and beyond. The brief survey of anticipated changes emphasizes U.S. plans as these are best known to the author. The U.S. is unlikely to be the only nation making changes and those anlayzing data records must be alert for future changes.

4.1. SURFACE OBSERVATIONS

In the U.S. replacement of manual surface observations by ASOS is going on now and scheduled for completion in 1996. This activity is creating turmoil with climate records of surface quantities. The wholesale substitution of instruments, designed to operate unattended and with minimal maintenance, has already caused some recorded surface temperatures to be 1–2 °C too high at some locations (Kessler *et al.*, 1993). One can expect similar problems exist with the humidity records there, not only because the temperature records are compromised and so conversions to other humidity variables questionable, but the exposure problems of the temperature sensors may also be found with the humidity sensors. Additionally, the possibility exists that some of the new sensors will be at different heights than the older ones.

4.2. RADIOSONDES

The U.S. is also changing its radiosonde network as part of its Modernization Program. When it is finished somewhere between 1/3 and 1/2 of the radiosonde stations will have been moved from their locations of a few years ago (or disappeared). In addition there have been, and likely will be more, changes in the radiosonde supplier. For many years the sondes were manufactured by the VIZ Corp. which used a carbon-based hygristor. Between 1989 and 1995 sondes of another vendor were used at some locations, mainly in the western U.S. Although these used the same humidity sensor, there were differences in their reported humidity values. These have now been replaced with VIZ sondes. It is now planned to replace VIZ sondes with Väisälä sondes at some locations later in 1995. Also the VIZ sondes themselves are in some cases newer versions. Some of these changes will be accompanied by algorithm changes, as mentioned earlier. The U.S. is also developing a next generation of radiosondes, whose introduction into the network could occur in this decade.

Overseas the situation is a bit less tumultuous but changes are also taking place. Sondes from Väisälä are slowly replacing those manufactured in other countries, particularly in Europe. Both Japan and China, which make their own sondes, are experimenting with different humidity sensors (Schmidlin, personal communication). Some nations newly formed from the former Soviet Union are changing supplier; others may follow.

At the same time, the major suppliers of radiosondes continue to improve their products. Humidity sensors are receiving attention and we can hope for better performance at high altitudes and at low humidities. Some attention is being paid to improving data handling algorithms, also.

A problem that could result in fewer observations in the future arises from potential costs, particularly to the less-developed nations. There may be requirements for narrower band-widths for radiosonde transmitters because the part of the frequency spectrum available to radiosondes could be curtailed to make room for commercial broadcasts. This could increase the cost of sondes and ground stations which might reduce the number of observations. Furthermore, the use of GPS equipment for wind calculations would be another additional expense.

4.3. SATELLITES

Improved sensors have recently been launched or are scheduled in the near future. Advanced Microwave Sounding Units (AMSU A and B) should produce better water vapor profiles as will the SSM/T-2 aboard the DMSP satellites. There are also improvements in the moisture sensing of the geostationary satellites being launched for NOAA. Beyond these, there are plans for improvements in the EOS series. It is difficult to say when, with the budget restrictions now prevailing, the EOS satellites will be in orbit or what instruments they will carry.

We can also expect improvements in satellite retrieval of water vapor data. This is an on-going effort and, while not without cost, much more can be reasonably hoped for, both in processing the newer observations and in gleaning more from the older ones. In the latter effort the NOAA-NASA Pathfinder program (Ohring and Booth, 1995) should be particularly useful. Archived operational satellite observations, as far back as 1978, are being reprocessed with the best available calibrations and community consensus algorithms to produce a research quality data set.

5. What More Needs To Be Done?

The focus of this discussion is long-term changes in moisture, i.e. changes over several decades. This means observations and their analysis will have to cope with shorter term changes, such as those associated with ENSO phenomena and occasional volcanic eruptions, as well as instrumental noise and the vagaries of weather. The magnitude of these signals may well be greater than the signal from greenhouse gas increases or other long-term climate adjustments. It will require sustained observations over several decades to detect long-term moisture increases so there is real importance to extracting what one can from existing records. To discard the past data, with all their imperfections, would condemn us to begin anew the recording of climate changes. A perceived lack of data might well be used to delay any policy decisions beyond the time when they could be effective, or, if fears about climate warming ultimately prove to be exaggerated, lead prematurely to unnecessary restrictions.

There will be improvements in measuring techniques in the coming decades and there may be now-unthought-of developments. Continuity of data records as the old gives way to the new should be a major concern. Climatologists should not reject improvements on the grounds that the record will be compromised, but they should insist that the consequences of the introduction of new devices and procedures be well understood and ways be established to blend new data into the time series before the 'old' methods are abandoned.

The discussion in the preceding two paragraphs, and some of what follows, applies equally well to all climate data, not just moisture. Changes in surface temperature and precipitation receive the most attention as these affect the public and are the most widely measured. Water vapor content is not of direct public concern except as it occasionally affects comfort. From the standpoint of monitoring climate change, however, moisture may be better monitored above the surface rather than at the surface. Changes aloft should more readily reflect broad scale changes rather than reflect changes in local conditions.

Radiosondes and recently satellites provide almost all our knowledge of moisture changes. They now complement each other and will continue to be the main sources of data. To extract the most information from past radiosonde and satellite observations we need more information on the histories of changes in equipment and procedures. As noted above, Gaffen's (1993) summary of what we know about radiosondes is not complete and she is continuing to collect information. Some nations did not heed the original WMO request for information; their contributions are still needed. Furthermore, changes in radiosondes and satellite observations and data handling procedures will continue. There should be one repository of such metadata for scientists to consult and it should be kept up-to-date. This requires the active participation of all concerned. Perhaps the WMO should consider undertaking this function.

Wade (1994) shows how examining the algorithms used in transforming the electrical signals from radiosondes into humidity data can result in substantial improvement in the data without touching the instruments themselves. This shows that better data can be acquired in the upper troposphere and lower stratosphere, a region of particular interest and one where the need for better data is particularly acute. In addition, work on improving satellite water vapor retrieval algorithms also should continue. These will not only provide improved data in the future, they will add value to past observations. Improving all data handling algorithms is a relatively low-cost way of gaining additional observational power.

Because we are not likely to see nations adopt one radiosonde type any time soon, there is great need for some means of comparing them, a transfer standard against which all could be compared. This would allow one to adjust data from differing sources to a common reference, although that reference may not be absolutely accurate. It would also assist in ensuring continuity of data over time. Development of such a system is going on at the National Center for Atmospheric Research (Dabberdt, personal communication) and should be encouraged. When this or some other system is available, careful thought must be given to how it is operated, in what conditions and locations, etc.

A promising method of acquiring information on global distribution of total column water vapor (and possibly some vertical resolution) can be found in analyses of signals from satellites in the Global Positioning System (GPS; Rocken et al., 1993). The effect of moisture on the refractive index of the air can be extracted from signal delays. This technique should be explored further as it could provide a relatively inexpensive network of observations because the satellites will be in place for navigational purposes.

A useful effort would be a comparison of a number of water vapor measuring devices in the field. These would include ground-based and aircraft lidars (e.g. Differential absorption and Raman scattering lidars) and other sensors not necessarily appropriate for deployment in a monitoring mode but valuable for calibrating those that are. Accompanied by many ancillary measurements and in such a way that satellites could be overhead at times, such a program would contribute to the development of improved water vapor retrieval algorithms and help evaluate a variety of measuring techniques. Such a program is a goal of GVaP (GEWEX Water Vapor Project). GVaP is, at present, a loose confederation of scientists working on monitoring, as well as understanding, water vapor and climate (Starr and Melfi, 1992).

It would be desirable if one space-borne instrument or technique could be developed that would allow water vapor to be calculated in all weather conditions and above all surfaces. Barring that, combinations of data from microwave and infrared instruments, blended with radiosonde observations, will be needed. These must consider the shortcomings of each data source in the analysis. Such an undertaking is discussed by Vonder Haar et al. in the August, 1994 issue of the GEWEX News (and is available from Vonder Haar at Colorado State Univ. or the GEWEX Project Office (IGPO) Suite 203, 409 Third St. SW, Washington DC 20024).

One approach to monitoring low frequency, global changes in the upper air is the establishment of selected climatological sites around the world using high quality radiosonde instruments and other sensing systems. Their function would be as climate monitors and not primarily to serve as additional sites for the global weather observing network. They could, of course, be co-located with such stations. Because climate monitoring is their function they would not be tied to the 00 and

12 UTC schedule; observations at local noon and midnight might be considered. Observations might also be coordinated with satellite observations.

In any event, considerable planning would need to go into such an enterprise and this is not the place to try to lay out such a network. Deciding how many stations and where they should go could be helped by modeling studies designed to address the question. An order-of-magnitude estimate of the number can be found in Angell's (1988) 63-station network. The estimates of temperature trends from this group of stations compares well with results using all the 700–800 radiosonde stations (Oort and Liu, 1993). Trenberth and Olson (1991) note that this network did a reasonable job of picking up the low frequency fluctuations in temperature over a 9-year period.

In summary then, the recommendations are: (1) collect in one location as much metadata as possible about both radiosonde and satellite observations (surface metadata could be included, as well); (2) continue to improve humidity sensors for both radiosondes and satellites, and data reduction algorithms, but give equally serious attention to continuity of the long-term record; (3) develop a reference radiosonde system to be a transfer standard for both present sondes and new ones as they are developed; (4) plan for a small network of upper-air climate stations, committed to be maintained for decades, perhaps with no determined end.

Probably the strongest recommendations is for the world's weather services to recognize the importance of climate, as distinct from weather forecasting, and to assume responsibility for detecting any changes in it.

Acknowledgements

The author was greatly helped by reviews of drafts of this manuscript from James Angell, John Bates, Dian Gaffen, Arnold Gruber and M. Patrick McCormick. Francis Schmidlin and Charles Wade gave freely of their time in conversations about specific problems with radiosondes. Presentations at the recent Chapman Conference on Water Vapor in the Climate System, held at Jekyll Island GA, 25–28 Oct., 1994, provided many insights into all the problems discussed above.

References

Angell, J. K.: 1988, 'Variations and Trends in Tropospheric and Stratospheric Global Temperatures, 1958–87', *J. Clim.* **1**, 1296–1313.
Brazel, S. W. and Balling, R. C., 1986, 'Temporal Analysis of Long-Term Atmospheric Moisture Levels in Phoenix, Arizona', *J. Clim. Appl. Meteor.* **25**, 112–117.
Elliott, W. P. and Gaffen, D. J.: 1991, 'On the Utility of Radiosonde Humidity Archives for Climate Studies', *Bull. Amer. Meteor. Soc.* **72**, 1507–1520.
Elliott, W. P. and Gaffen, D. J.: 1993, 'Effects of Conversion Algorithms on Reported Upper-Air Dewpoint Depressions', *Bull. Amer. Meteor. Soc.* **74**, 1323–1325.
Elliott, W. P., Gaffen, D. J., Kahl, J. D., and Angell, J. K.: 1994, 'The Effects of Moisture on Layer Thicknesses Used to Monitor Global Temperatures', *J. Clim.* **7**, 304–308.

Elliott, W. P., Smith, M. S., and Angell, J. K.: 1991, 'Monitoring Tropospheric Water Vapor Changes Using Radiosonde Data', in Schlesinger, M. E. (ed.), *Greenhouse-Gas-Induced Climate Change: A Critical Appraisal of Simulations and Observations*, 311–328, Elsevier, Amsterdam, 615 pp.

Gaffen, D. J.: 1992, 'Observed Annual and Interannual Variations in Tropospheric Water Vapor', *NOAA Tech. Memo. ERL ARL-198*, NOAA Air Resources Lab., Silver Spring, MD, 162 pp.

Gaffen, D. J.: 1993, 'Historical Changes in Radiosonde Instruments and Practices', *Instruments and Observing Methods, Report #50*, WMO/TD-541, World Meteorological Organization, Geneva, 123 pp.

Gaffen, D. J.: 1994, 'Temporal Inhomogeneities in Radiosonde Temperature Records', *J. Geophys. Res.* **99**, 3667–3676-

Gaffen, D. J., Barnett, T. P., and Elliott, W. P.: 1991, 'Space and Time Scales of Global Tropospheric Moisture', *J. Clim.* **4**, 989–1008.

Gaffen, D. J., Elliott, W. P., and Robock, A.: 1992, 'Relationships between Tropospheric Water Vapor and Surface Temperature as Observed by Radiosondes', *Geophys. Res. Lett.* **19**, 1839–1842.

Gueymard, C.: 1993, 'Assessment of the Accuracy and Computing Speed of Simplified Saturation Vapor Equations Using a New Reference Dataset', *J. Appl. Meteor.* **32**, 1294–1300.

Gutzler, D.: 1992, 'Climatic Variability of Temperature and Humidity over the Tropical Western Pacific', *Geophys. Res. Lett.* **19**, 1595–1598.

Hansen, J., Lacis, A., Rind, D., Russell, G., Stone, P., Fung, I., Reudey, R., and Lerner, J.: 1984, 'Climate Sensitivity: Analysis of Feedback Mechanisms', in Hansen, J. E. and Takahashi, T. (eds.), *Climate Processes and Climate Sensitivity, Maurice Ewing Series* **5**, American Geophysical Union, Washington, D.C., 368 pp.

Hense, A., Krahe, P., and Flohn, H.: 1988, 'Recent Fluctuatiolns of Tropospheric Temperature and Water Vapor Content in the Tropics', *Meteorol. Atmos. Phys.* **38**, 215–227.

Kalnay, E. and Jenne, R.: 1991, 'Summary of the NMC/NCAR Reanalysis Workshop of April 1991', *Bull. Amer. Meteor. Soc.* **72**, 1897–1904.

Kessler, R. W., Bosart, L. F., and Gaza, R. S.: 1993, 'Recent Maximum Temperature Anomalies at Albany, New York: Fact or Fiction?', *Bull. Amer. Meteor. Soc.* **74**, 215–227.

Khalil, M. A. K. and Rasmussen, R. A.: 1993, 'Decreasing Trend of Methane: Unpredictability of Future Concentrations', *Chemosphere* **26**, 803–814.

Lee, D. O.: 1991, 'Urban-Rural Humidity Differences in London', *Int. J. Climat.* **11**, 577–582.

Liu, W. T., Tang, W., and Wentz, F. J.: 1992, 'Precipitable Water and Surface Humidity over Global Oceans from Special Microwave Imager and European Center for Medium Range Weather Forecasts', *J. Geophys. Res.* **97**, 2251–2264.

McCormick, M. P., Chiou, E. W., McMaster, L. R., Chu, W. P., Larsen, J. C., Rind, D., and Oltmans, S.: 1993, 'Annual Variation of Water Vapor in the Stratosphere and Upper Troposphere Observed by the Stratospheric Aerosol and Gas Experiment II', *J. Geophys. Res.* **98**, 4867–4874.

Meehl, G. A. and Washington, W. M.: 1990, 'CO_2 Climate Sensitivity and Snow-Sea-Ice Albedo Parameterization in an Atmospheric GCM Coupled to Mixed-Layer Ocean Model', *Clim. Change* **16**, 283–306.

Morrissey, J. F. and Brousaides, F. J.: 1970, 'Temperature-Induced Errors in the ML-476 Humidity Data', *J. Appl. Meteor.* **8**, 805–808.

Nash, J. and Schmidlin, F. J.: 1987, *WMO International Radiosonde Intercomparison (U.K., 1984, U.S.A., 1985) Final Report*, World Meteorological Organization, Instruments and Observing Methods Report No. 30, WMO/TD-No. 195.

Ohring, G. and Booth, A. L.: 1995, 'The NOAA Pathfinder Program', *Adv. Space Res.* **16**, (10)15–(10)20.

Oltmans, S. J. and Hofmann, D. J.: 1995, 'Increase in Lower Stratospheric Water Vapor at a Mid-Latitude Northern Hemisphere Site from 1981 to 1994', *Nature* **374**, 146–149.

Oort, A. H. and Liu, H.: 1993, 'Upper-Air Temperature Trends over the Globe, 1958–1989', *J. Clim.* **6**, 292–307.

Peixoto, J. P. and Oort, A. H.: 1992, *Physics of Climate*, American Institute of Physics, New York, 520 pp.

Rocken, C., Ware, R., Van Hove, T., Solheim, F., Albers, C., Johnson, J., Bevis, M., and Businger, S.: 1993, 'Sensing Atmospheric Water Vapor with the Global Positioning System', *Geophys. Res. Lett.* **20**, 2631–2634.

Shine, K. P. and Sinha, A.: 1991, 'Sensitivity of the Earth's Climate to Height-Dependent Changes in Water Vapor Mixing Ratio', *Nature* **354**, 382–384.

Smith, W. L., Woolf, H. M., Hayden, C. M., Wark, D. W., and McMillin, L. M.: 1979, 'The TIROS-N Operational Vertical Sounder', *Bull. Amer. Meteor. Soc.* **60**, 1177–1187.

Starr, D. O'C. and Melfi, S. H. (eds.): 1991, *The Role of Water Vapor in Climate*, NASA Conference Publication 3120, NASA Code NTT-4 Washington, DC, 50 pp.

Starr, D. O'C. and Melfi, S. H. (eds.): 1992, *Implementation Plan for the Pilot Phase the GEWEX Water Vapor Project (GVaP)*, IGPO Pub. Series No. 2, International GEWEX Project Office, Washington, DC, 29 pp.

Steele, L. P., Dlugokencky, E. J., Lang, P. M., Tans, P. P., Martin, R. C., and Masarie, K. A.: 1992, 'Slowing Down of the Global Accumulation of Atmospheric Methane During the 1980's', *Nature* **358**, 313–316.

Trenberth, K. E. and Olson, J. G.: 1988, 'An Evaluation and Intercomparison of Global Analyses from the National Meteorological Center and the European Centre for Medium Range Forecasts', *Bull. Amer. Meteor. Soc.* **69**, 1047–1057.

Trenberth, K. E. and Olson, J. G.: 1991, 'Representativeness of a 63-Station Network for Depicting Climate Changes', in Schlessinger, M. E. (ed.), *Greenhouse-Gas-Induced Climate Change: A Critical Appraisal of Simulations and Observations*, 249–260, Elsevier, Amsterdam, 615 pp.

Wade, C. G.: 1994, 'An Evaluation of Problems Affecting the Measurement of Low Relative Humidity on the United States Radiosonde', *J. Atmos. Oceanic Technol.* **11**, 687–700.

Woodruff, S. D., Slutz, R. J., Jenne, R. L., and Steurer, P. M.: 1987, 'A Comprehensive Ocean-Atmosphere Data Set', *Bull. Amer. Meteor. Soc.* **68**, 1239–1250.

Wu, X., Bates, J. J.: and Khalsa, S. J. S.: 1993, 'A Climatology of the Water Vapor Band Brightness Temperature from NOAA Operational Satellites', *J. Clim.* **6**, 1282–1300.

(Received 23 January, 1995; in revised form 22 June, 1995)

LONG-TERM OBSERVATIONS FOR MONITORING OF THE CRYOSPHERE

JOHN E. WALSH

Department of Atmospheric Sciences, University of Illinois, Urbana, IL 61801, U.S.A.

Abstract. Variations of the cryosphere over decadal-to-century timescales are assessed by a survey of data on sea ice, snow cover, glaciers and ice sheets, permafrost and lake ice. The recent variations are generally consistent across the different cryospheric variables, especially when placed into the context of variations of temperature and precipitation. The recent warming over northern land areas has been accompanied by a decrease of snow cover, particularly during spring; the retreat of mountain glaciers is, in an aggregate sense, compatible with the observed warming; permafrost extent and lake ice duration show similar variations in areas for which data are available. Corresponding trends are not apparent, however, in data for some regions such as eastern Canada, nor in hemispheric sea ice data, especially for winter. The data also suggest an increase of snowfall over high latitudes, including the Antarctic ice sheet.

Estimates of both the climatic and the statistical significance of the recent variations are hampered by data inhomogeneities, the shortness of the records of many variables and the absence of central archives for data on several variables. The potential of monitoring by satellite remote sensing has been realized with several variables (extent of sea ice, snow cover). Other cryospheric variables (snow depth, ice sheet elevation, lake ice, mountain glaciers) may be amenable to routine monitoring by satellites pending advances in instrumentation, modifications of satellite orbit, and further developments in signal detection algorithms. The survey of recent variations leads to recommendations concerning the use of historical data, *in situ* measurements, and remote sensing applications in the monitoring of the cryosphere.

1. Introduction

The greenhouse-induced temperature changes projected by global climate models are largest in the polar regions, where the ice phase is a prominent feature of the ocean-land-atmosphere system. For a doubling of CO_2, the projected near-surface warming of the polar regions is typically 2–3 times the global mean warming of 1.5°–4.5 °C and, in some of the more recent model experiments, tends to be considerably stronger in the Arctic than in the Antarctic (IPCC, 1992, p. 106). The projected warming of the polar regions is strongest in autumn and winter, and is much smaller in summer (IPCC, 1990, p. 140). Among the factors contributing to the models' warming is the temperature-ice-albedo feedback, which almost certainly involves changes in cloudiness. Other contributing factors may be the water vapor/temperature feedback and the breakdown of the surface temperature inversions that are frequently present over snow and ice surfaces. The fact that the model-projected warming is strongest in autumn and winter is largely a consequence of the retreat of sea ice and the delayed freeze-up. The latter occurs because the earlier and more extensive retreat enhances the ocean heat storage

Climatic Change **31**: 369–394, 1995.
© 1995 *Kluwer Academic Publishers.*

during summer sufficiently to delay the autumnal freeze-up and limit the winter ice thickness.

Global climate models also project an increase of precipitation in high latitudes as greenhouse forcing increases (IPCC, 1990, p. 143). Given the relatively large projected increases of temperature, an increase of humidity and an enhancement of the hydrologic cycle are not surprising. In general, the models that predict the largest global mean warming tend to predict the largest increases of precipitation. Over high-latitude land areas, the projected increases of precipitation are 10–20% when CO_2 is doubled. In at least one model (Manabe *et al.*, 1992), the increase of precipitation over the Southern Ocean is sufficient to enhance the stratification of the upper ocean, which in turn reduces the oceanic heat flux to the surface and thus offsets the greenhouse warming locally.

Given the large magnitudes and the potential cryospheric implications of the model-projected increases of temperature and precipitation in high latitudes, the monitoring of the cryosphere is an important part of the strategy for the detection of global change. In this paper we assess the current status of observational data on several components of the cryosphere: sea ice, continental snow cover, ice sheets and glaciers, permafrost and lake ice. In each case, we summarize the key observational datasets, recent variations and/or trends (if any), the major observational uncertainties in the trend assessments, and the extent to which the recent variations and/or trends are compatible with those of other variables and with greenhouse scenarios. Because variations of the cryospheric components are generally driven by variations of atmospheric temperature and precipitation, the assessment of recent variations is presented in the context of recent atmospheric variations. We also suggest priorities for enhancing the observing system in order to provide long-term datasets of adequate quality and homogeneity for the detection of future changes.

While the cryosphere consists of a diverse set of variables and has correspondingly diverse observational requirements, there are several noteworthy commonalities. First, *in situ* observations are generally difficult to obtain because the regions in which cryospheric variables dominate the near-surface regime are often in the world's harshest environments. Second, and largely because of the difficulty of *in situ* measurements, remote sensing techniques have tremendous potential for the monitoring of cryospheric variables. Remote sensing is indeed used routinely in the mapping of sea ice, snow cover and lake ice. However, algorithmic uncertainties have limited many of the potential applications of remote sensing, and some of the climatically most important variables are poorly suited to applications of current remote sensing technology. Thus *in situ* data has an essential role in the continued development of remote sensing techniques. Finally, the existing databases of *in situ* data tend to be quite heterogeneous in spatial and temporal coverage as well as in format and accessibility. An assessment of the availability of existing data on most of the cryospheric variables will point to a need for syntheses or integrations of the diverse datasets that now exist.

2. Observations of Recent Variations

2.1. SEA ICE

In the context of climate monitoring, one can distinguish several potentially important characteristics of sea ice:
- areal coverage of sea ice;
- ice extent;
- thickness;
- albedo;

(The areal extent of sea ice is generally defined as the ocean area poleward of either the ice edge or a prescribed contour of sea ice concentration. It differs from the areal coverage by the amount of ice-free ocean area poleward of the ice edge.) In view of the model projections of decreased duration of sea ice coverage in many high-latitude areas, a fifth characteristic assumes potential importance: the seasonal duration of the ice cover at a point or, conversely, the length of the ice-free season at a point that experiences a seasonal sea ice cover.

The period of relatively homogeneous data on the areal coverage and extent of sea ice begins in the early 1970s, when: (a) passive microwave satellite imagery became available, permitting all-weather and day-night observations of sea ice in both polar regions; and (b) the ongoing series of weekly sea ice analyses, based largely on passive microwave imagery, were begun by the Navy-NOAA Joint Ice Center. More than twenty-two years of such analyses now exist for both the Arctic and the Antarctic. Figures 1 and 2 show the time series of February and August ice extent for the Arctic and Antarctic as derived by the Climate Analysis Center from the Joint Ice Center's weekly ice charts. There is considerable interannual variability in each time series, and there are no systematic trends in either hemisphere or season. The means for the most recent ten years (1984–1993) show the following changes relative to the means for the first 10 years (1973–1982) of the time series for winter (February in the Arctic, August in the Antarctic) and summer (August in the Arctic, February in the Antarctic):

	Winter	Summer
Arctic	–0.5%	–3.2%
Antarctic	–0.4%	–9.2%

While all four changes are negative, implying a decrease of ice extent from 1973–82 to 1984–93, the winter values are essentially zero. Moreover, the relatively large decrease (–9.2%) during the Antarctic winter is reduced to –3.6% if 1973 is replaced by 1983 in the first 10 year period. A very similar seasonality of recent decadal-scale variations was found by Chapman and Walsh (1993) in an extended Arctic sea ice dataset spanning the 1961–1990 period.

More detailed analyses have been performed for the years 1978–1987, during which the Nimbus-7 SMMR (Scanning Multichannel Microwave Radiometer)

Arctic sea ice extent

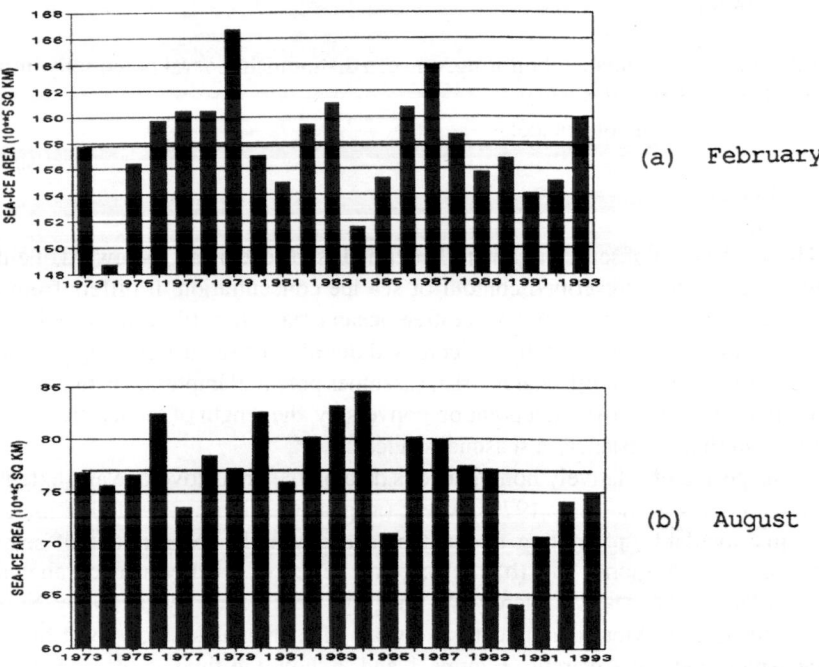

Fig. 1. Time series of Arctic sea ice extent (10^5 km^2) for: (a) February; and (b) August. Thin lines denote means for 1973–1993. Note different scales for February and August. (From Halpert *et al.*, 1994, p. 39).

provided single-source data on sea ice. Gloersen and Campbell (1991) reported a decrease of 2.1% in the extent of Arctic sea ice during this period; this northward migration of the ice margin was accompanied by a reduction in the area of open water within the ice boundary. By contrast, the coverage of sea ice in the Antarctic showed no trends. Parkinson (1992, 1994) used data from the same sensor to address variations in the length of the sea ice season. In both hemispheres, the recent trends of ice season length varied (and changed sign) with longitude. The ice season decreased in the eastern hemisphere of the north polar region but generally increased in western longitudes of the north polar region. The Antarctic data showed spatially coherent trends toward shorter ice seasons in the northern Weddell and Bellingshausen Seas and toward longer ice seasons in the Ross Sea, the East Antarctic offshore waters, and the south central Weddell Sea.

Similar longitudinal dependencies have been rpeorted in recent trends of the areal coverage of sea ice. Chapman and Walsh (1993), for example, show that sea ice coverage increased in the Baffin Bay/Labrador Sea region during 1961–1990,

Antarctic sea ice extent

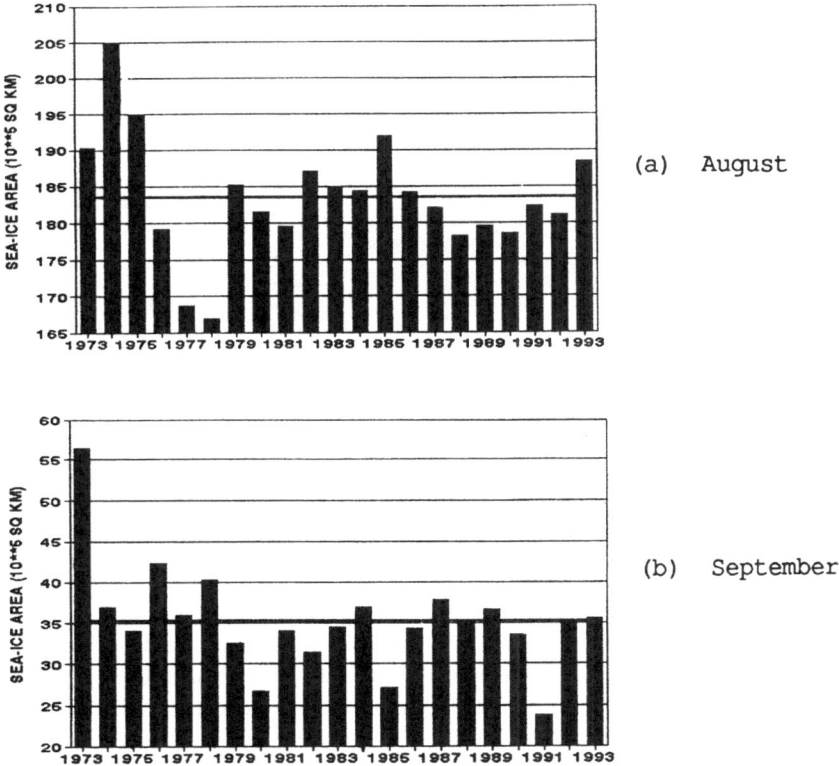

Fig. 2.   Time series of Antarctic sea ice extent (10^5 km^2) for: (a) August; and (b) February. Thin lines denote means for 1973–1993. Note different scales for August and February. (From Halpert *et al.*, 1994, p. 40).

in contrast to the decrease in the region from southern Greenland eastward to the Barents and Kara Seas. Marko *et al.* (1994) and the JSC (1993) also document the increases of sea ice extent over eastern Canada; these variations have a multiyear character that correlates highly with the North Atlantic Oscillation (Figure 3). The longitudinal variations of recent trends of sea ice coverage are generally consistent with temperature data for the past several decades. In particular, the region from eastern Canada to southern Greenland has shown a recent cooling trend, while most of the remainder of the subarctic shows a warming trend (IPCC, 1992, Plate A; Chapman and Walsh, 1993, Figure 1). There is no indication of a recent warming over the central Arctic Ocean, as Kahl *et al.*'s (1993) analysis of ice station and

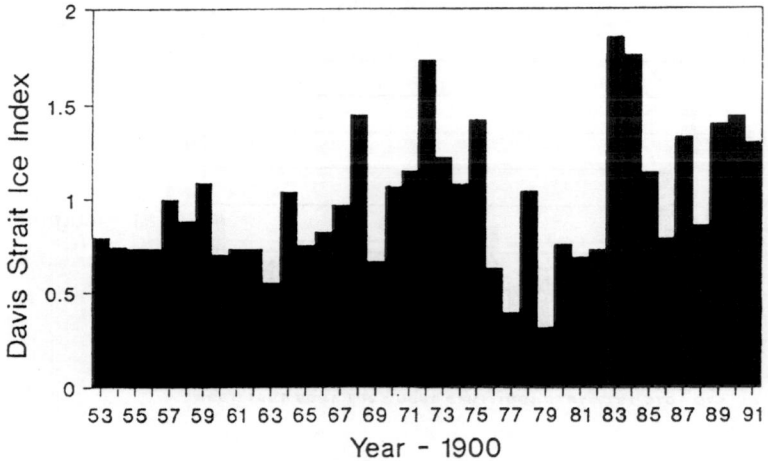

Fig. 3. Normalized extent of sea ice concentrations greater than 2/10 in the Davis Strait region of eastern Canada. (Adapted from Marko *et al.*, 1994, p. 1345).

dropsonde data suggests even a cooling during winter and autumn over the western Arctic Ocean.

The recent variations of sea ice extent and areal coverage may be viewed with confidence because: (a) they are compatible with the available data on air temperature, as summarized above; and (b) these variables have been monitored by satellite with satisfactory accuracy for more than two decades. Indeed, tracking of the sea ice boundary is one of the major successes of remote sensing of the cryosphere. Concentrations of sea ice within the pack are monitored to an estimated accuracy of ±7% (Gloersen *et al.*, 1992), although the uncertainties are larger under conditions of melting ice and new ice formation. If summer is indeed the season in which greenhouse effects on sea ice are earliest and greatest, then these limitations on the accuracy of derived ice concentration may hinder the detection of change in the central Arctic Ocean.

The thickness of sea ice is also a potentially useful indicator of climate change, especially since greenhouse experiments with global climate models indicate that arctic ice thicknesses will decrease substantially if the climate warms. Unfortunately sea ice thickness is a notoriously difficult variable to monitor over large spatial and temporal scales. The most extensive measurements have been obtained from bottom-moored sonars, which provide time series of ice draft at a single point, and from declassified sonar measurements made by nuclear submarines, which provide 'snapshots' of ice draft along a transect. Bottom-moored sonars have provided time series for the Greenland Sea and Alaskan waters for several years, and the network is slowly expanding (WCRP, 1994). The declassified submarine data include measurements from approximately 15 cruises of U.S. submarines (dating

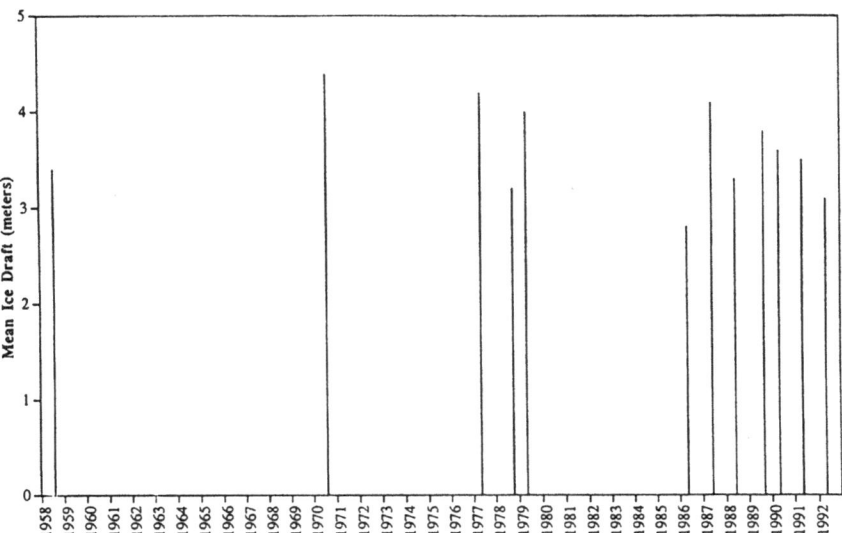

Fig. 4.   Mean draft of sea ice at the North Pole as measured by sonar on twelve submarine cruises. (From McLaren *et al.*, 1994).

back to 1958) and from several cruises of British submarines. Figure 4 (from McLaren *et al.*, 1994) shows the mean ice draft at the North Pole as derived from the sonar measurements of 12 U.S. submarine cruises in the 1958–1992 period. The slight trend toward smaller thicknesses is overwhelmed by the interannual variability and is further complicated by differences in the seasons and tracks of the various cruises. Furthermore, Wadhams (1990) has shown that differences of mean ice draft obtained from two submarine cruises can be consistent with the wind-driven drift of sea ice over short antecedent periods. Thus the monitoring of sea ice thickness requires one or more of the following: (1) more systematic sampling by submarines, which introduce major logistical and cost factors into a monitoring system; (2) a significantly expanded network of bottom-moored sonars, which are presently limited to 1–2 year lifetimes by power requirements; and (3) new developments in satellite remote sensing capabilities for the mapping of sea ice thickness. To date, satellite remote sensing of ice thickness has been limited largely to the discrimination of first-year and multiyear ice, and even this capability is limited to seasons in which ice melt is not occurring.

Data on surface albedo, which is influenced dramatically by cryospheric variations, have long been cited by modelers as a key observational need for model verification. Surface albedo data can also be useful in the monitoring of the surface for possible changes in snow distribution, frequency and timing of melt, sea ice state, etc. Comprehensive datasets on the space-time variations of surface albedo must come from satellites. However, the compilation of such datasets has been hindered by problems of cloud-surface discrimination in visible and infrared wavelengths when the surface is snow- or ice-covered. Serreze *et al.* (1993) have recently compared the monthly surface albedo fields derived from: (1) DMSP (Defense Meteorological Satellite Program) visible and infrared data (Robinson *et al.*, 1992); and (2) ISCCP (International Satellite Climatology Program) analyses (Rossow and Schiffer, 1991). While the spatial patterns obtained by the two tecniques were similar, especially for the key summer months of June–August, systematic differences of 0.05 to 0.10 were found. The ISCCP cloud-clearing algorithm is a likely contributor to these differences, although the technique used to derive surface albedo fields from DMSP visible and IR imagery has its own uncertainties. Nevertheless, the differences of 0.05–0.10 are probably typical of the accuracies of current climatologies of surface albedo in high latitudes. While these uncertainties are comparable to the interannual variations of surface albedo in regions covered by snow and ice, they are considerably smaller than those associated with local transitions from a snow/ice-covered regime to a snow/ice-free regime. Thus the uncertainties in surface albedo do not impose major limitations on the monitoring of the extent and areal coverage of sea ice. The impact of albedo biases or errors of this magnitude in climate models remains to be established.

2.2. CONTINENTAL SNOW COVER

As in the case of sea ice, several properties of continental snow cover are relevant to climate and global change:
  - areal coverage;
  - snowfall;
  - snow depth;
  - liquid water equivalent;
  - albedo;
Robinson (1993) has recently summarized the available databases on these properties; the following survey draws upon Robinson's summary and places the databases into the context of recent trend assessments.

Standardized annual summaries based on data for the entire globe and spanning more than a decade are available only for areal coverage, although useful satellite-derived estimates of snow depth are available for regions in which vegetative masking effects are negligible (see below). In addition, fields of surface albedo of snow-covered regions can be obtained from the ISCCP database, subject to uncertainties arising from cloud-snow discrimination as noted above in the discussion of sea ice albedo.

The primary global dataset depicting snow cover and its temporal variations is the set of weekly charts produced by NOAA/NESDIS and quality-controlled by D. Robinson of Rutgers University (Robinson *et al.*, 1993). The charts are based on the visual interpretation of visible and infrared satellite imagery by trained meteorologists. The weekly series of these operational charts began in 1966 and is the only such continuous hemispheric product. The charts comprise the longest satellite based environmental record available. However, since snow extent was underestimated during the early years (primarily in the autumn season and over Asia), the use of this dataset for monitoring purposes generally begins with the 1972 data. Figure 5 shows the time series of Northern Hemisphere snow-covered area for spring, autumn and winter as derived from this dataset. The outstanding feature of Figure 5 is the decrease of spring and autumn snow cover; the 1972–1981 spring mean is 6.5% smaller than the 1984–1993 mean. The corresponding changes for autumn and winter are −3.7% and +0.5%, respectively. The springtime decrease is apparent in the data for both North America and Eurasia. The reduced extent of snow since the mid-1980s has coincided with some of the warmest surface air temperatures of the past century. The warming has been strongest in spring and over the northern land areas (Jones et al., 1991; Chapman and Walsh, 1993). Groisman et al. (1994) have recently shown that the snow-radiation feedback accounts for a substantial portion of the warming over the northern land areas from the 1970s to the early 1990s. These results are consistent with the fact that the long-term (20th century) increase in surface air temperature has been greatest in spring over the Northern Hemisphere land areas. However, both observational data and model results indicate that the natural variability of surface air temperature is also relatively large over the northern land areas (Stouffer et al., 1994, Figure 4).

A further caution concerning recent trends of snow cover is Brown's (1994) finding that the duration of snow cover over the western Canadian prairies showed an increasing trend over much of the twentieth century, then declined rapidly during the 1970s and 1980s. Although derived from station data, Brown's annual index of western Canadian snow cover correlates very highly with the NOAA/NESDIS total North American snow cover during the 1970s and 1980s.

Additional uses of station data, adjusted for changes of instrumentation and gauge biases, have been the recent assessments of snowfall variations by Karl et al. (1993) and Groisman and Easterling (1994). Although variations of snowfall will differ from variations of snow coverage (and duration), the variables can be expected to correlate positively. Snowfall and its water equivalent (see below) may well be the more appropriate for studies of snow-hydrology linkages. Karl et al.'s analysis of station data for the past four decades (1951–1990) showed an increase of 4–5% per decade of both solid and total precipitation over northern Canada, while the global warmth of the 1980s was accompanied by a 10% increase of annual precipitation in Alaska. As shown in Figure 6, annual precipitation over Canada has been below the 1948–1993 mean only once in the past 21 years (Environment Canada, 1994). However, the increase of precipitation has been accompanied by higher ratios of liquid to solid precipitation as the temperatures have increased, resulting in a slight decrease (−0.7% ± 1.4%) of snowfall in Canada south of 55° N (Groisman and Easterling, 1994). On the century timescale, annual precipitation has increased in this zone by 13%, although the recent decrease of solid precipitation

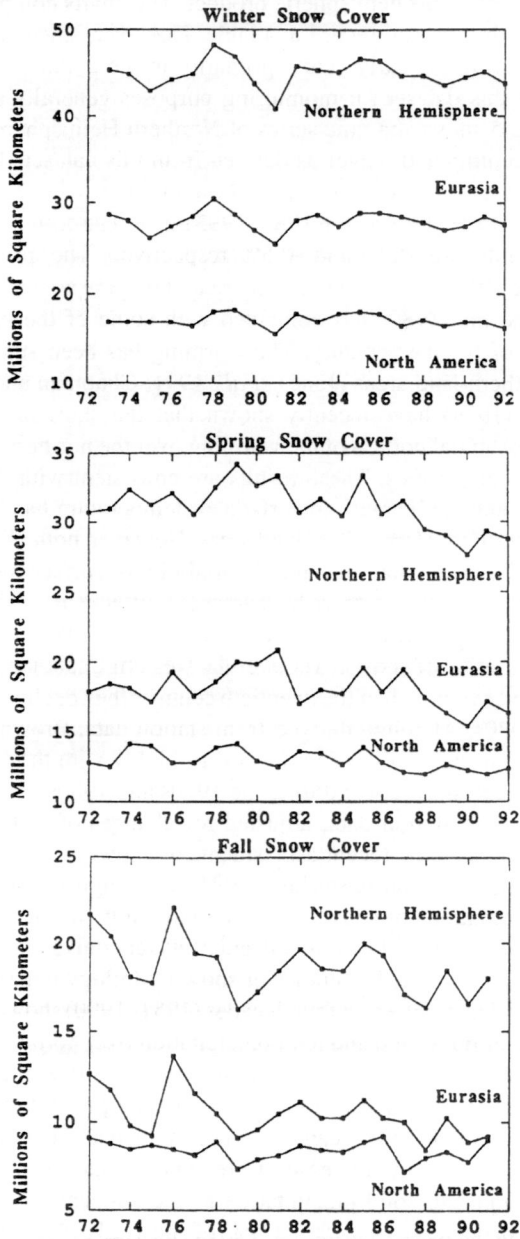

Fig. 5.   Time series of snow cover over Eurasia and North America (Greenland excluded) for: (a) winter; (b) spring; and (c) autumn. Values are derived from NOAA weekly snow charts using the Rutgers Routine. (From Robinson, 1993, p. 12).

is again apparent (Figure 7). Figure 7 also supports Brown's (1994) contention that the recent downward trend of snow over Canada is limited to the past 20 years.

Decadal summaries of snowfall measurements in China were compiled by Li (1987), who found a decrease of snowfall over China during the 1950s followed by an increase during the 1970s and 1980s. An update through the early 1990s would be timely, as would a synthesis of station snow data from western Europe and the former Soviet Union. Such efforts have been initiated at the National Snow and Ice Data Center (Barry *et al.*, 1993, 1994) and more comprehensive studies of high-latitude snowfall and hydrologic linkages have been assigned high priority in the upcoming Arctic Climate System Study of the World Climate Research Program (WCRP, 1994). A key issue in the use of historical snowfall data will continue to be the 'undercatch' of falling snow; the degree of undercatch varies substantially with the type of gauge, with its siting, and with wind (Legates, 1993).

Southern Hemisphere snowfall, limited largely to the Antarctic continent and the surrouding waters, is notoriously difficult to quantify because of the sparseness of station data. The most promising approach seems to be the use of rawinsonde data from the Antarctic periphery to estimate the convergence of the flux of atmospheric water vapor (Bromwich, 1988). Similar computations with fields produced by data assimilation cycles of numerical weather prediction are becoming feasible (Yamazaki, 1992).

Fields of snow depth and liquid water equivalent can, in principle, be derived from two approaches: *in situ* measurements and satellite remote sensing. In practice, each approach presents difficulties that have not yet been surmounted. *In situ* measurements of snow depth are made routinely at thousands of weather observing stations. Unfortunately, the observing sites are often unrepresentative of the surrounding terrain and vegetative regimes. An additional obstacle to the use of such data is that the measurements are buried in diverse and cumbersome synoptic archives, usually on a country-by-country basis. As noted above, attempts to synthesize such data for at least some regions and time periods have been initiated. *In situ* measurements of the liquid water equivalent of snow are generally made in conjunction with snow course surveys. These surveys are driven primarily by hydrologic concerns and are concentrated in the drainage basins of mountainous areas. Thus the database of *in situ* measurements of liquid water equivalent is highly uneven both spatially and temporally.

The most promising approach to large-scale mapping of liquid water equivalent is the application of satellite remote sensing, particularly the passive microwave sensors. Radiances from Nimbus 7's SMMR and the DMSP's SSM/I have been used experimentally to map the snow water equivalent or, assuming a constant density, snow depth. Chang *et al.* (1987) and Hall (1988) have consolidated SMMR-derived fields of snow depths into time series of snow volume; the gridded snow depths have recently become available through the National Snow and Ice Data Center (NSIDC, 1994). The gridded depths correspond reasonably well to actual snow depths in areas that are not heavily vegetated and in situations when the snow is

JOHN E. WALSH

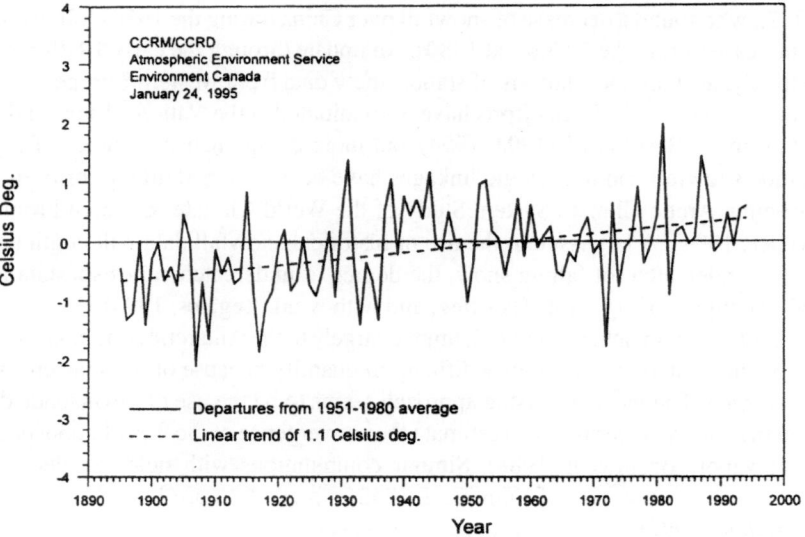

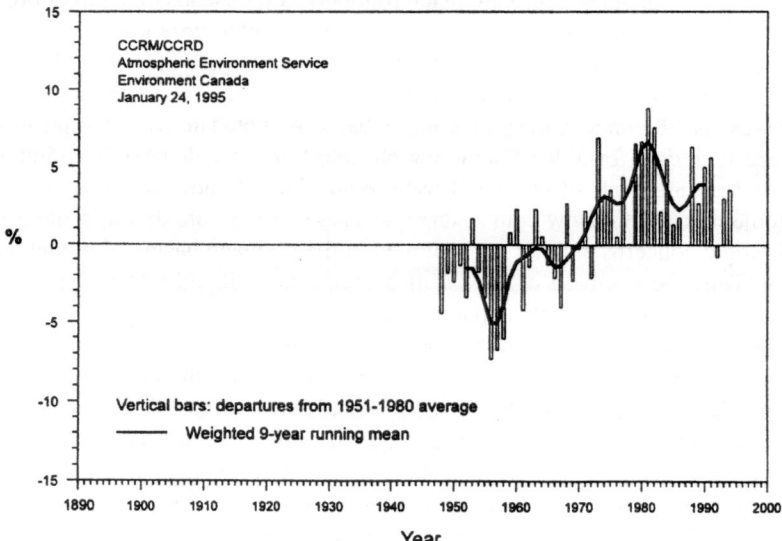

Fig. 6. Annual temperature (upper) and precipitation (lower) averaged over Canada and expressed as departures from the 1951–1980 means. (From Environment Canada, 1994, p. 2).

[250]

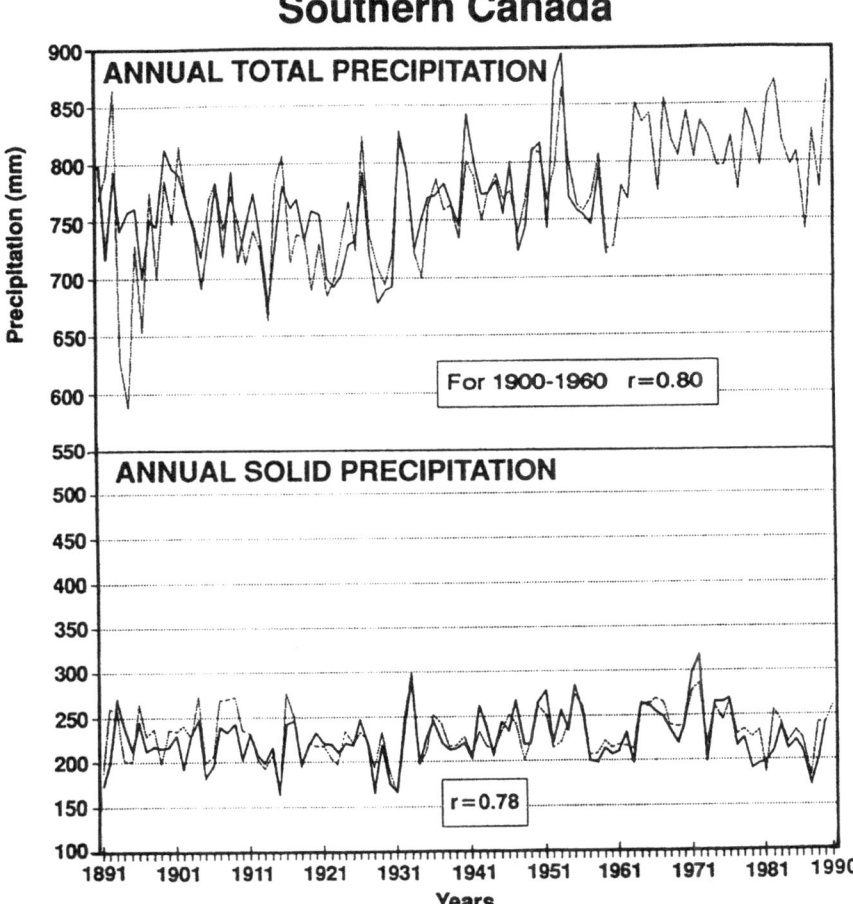

Fig. 7.   Annual snowfall and total precipitation over southern Canada (south of 55° N). Dashed and solid lines represent estimates based on two networks of stations; correlation between each pair of time series is indicated below the time series plots. (From Groisman and Easterling, 1994, p. 191).

not wet (Tait and Armstrong, 1994). The latter two constraints must be overcome before passive microwave mapping can provide the continuous spatial and temporal coverage required for trend determinations. The National Snow and Ice Data Center has developed a prototype snow depth model that uses passive microwave data in conjunction with station reports (Armstromg and Brodzik, 1995). This blend of remotely sensed and *in situ* data is the most promising approach to the monitoring of the hydrologically most important characteristics (depth and water equivalent) of snow cover on the global scale.

2.3. ICE SHEETS AND GLACIERS

The terrestrial ice sheets and glaciers contain the vast majority of the ice that is present in the global system. While both these forms of terrestrial ice are potentially useful indicators of climate, the ice sheets of Greenland and Antarctica contain sufficient ice that changes in their volume could significantly influence sea level. Unfortunately, the current state of glacier and ice sheet mass balances is poorly known (van der Veen, 1991).

Records of the glacier length have been compiled for hundreds of glaciers worldwide by the World Glacier Monitoring Service in Zurich. While the record lengths are highly variable, some of the records now span the past several hundred years. Oerlemans (1994) has recently used data from 48 of these glaciers in an attempt to quantify the global warming of the past 140 years. However, the climatic interpretation of changes in glacier length are complicated by the differences in the response times of different glaciers. As noted by Oerlemans, typical response times of valley glaciers are 10 to 50 years. Longer-term changes depend on the extent to which a glacier's mass balance is in equilibrium with climate. Using the linear component of changes in the length of the 48 glaciers, together with a scaling to allow for differences in glacier geometry and climate sensitivity, Oerlemans found that the retreat of glaciers over the past 100 years appears to be globally coherent (Figure 8). On the basis of a climate sensitivity model of the glaciers, the observed retreat corresponds to a linear warming trend of 0.66 °C per century (Oerlemans, 1994, p. 243). Consistent with this general retreat of glaciers since the mid-to-late 19th century is Meier's (1984) conclusion that glacier melt can account for 30–50% of the observed rise of sea level from 1884 to 1975. More recent calculations summarized by Meier (1990) indicate that, as a result of greenhouse changes, the expected contributions to global sea level rise by the year 2050 are 0.16 ± 0.14 m from small glaciers and ice caps, 0.08 ± 0.12 m from the Greenland ice sheet, and -0.3 ± 0.2 m from the Antarctic ice sheet.

While Oerlemans' trends represent a linear fit to data of the past 140 years, the more recent (post-1960) trends suggest a return to a more 'mixed regime' of glacier advance and retreat (van der Veen, 1991). Wood (1988), for example, found that the percentage of 400 glaciers showing advances increased from 6% in 1960 to more than 50% in 1980; a slowing of the retreat is also apparent in the time series of Figure 8. Additional data on glacier retreat are provided by Barry (1990) and IPCC (1992). However, as stressed by van der Veen, variations in glacier volume are often not proportional to glacier length. Since the mass balance records exist only for selected glaciers and for only the past few decades, and since the flow dynamics vary from glacier to glacier, any generalizations about recent changes in accumulation and ablation rates must be made with caution.

While high-resolution satellite visible imagery (e.g., Landsat) has been and will continue to be useful in the monitoring of alpine glacier extent. The ASTER (Advanced Spaceborne Thermal Emission and Reflection radiometer) planned for

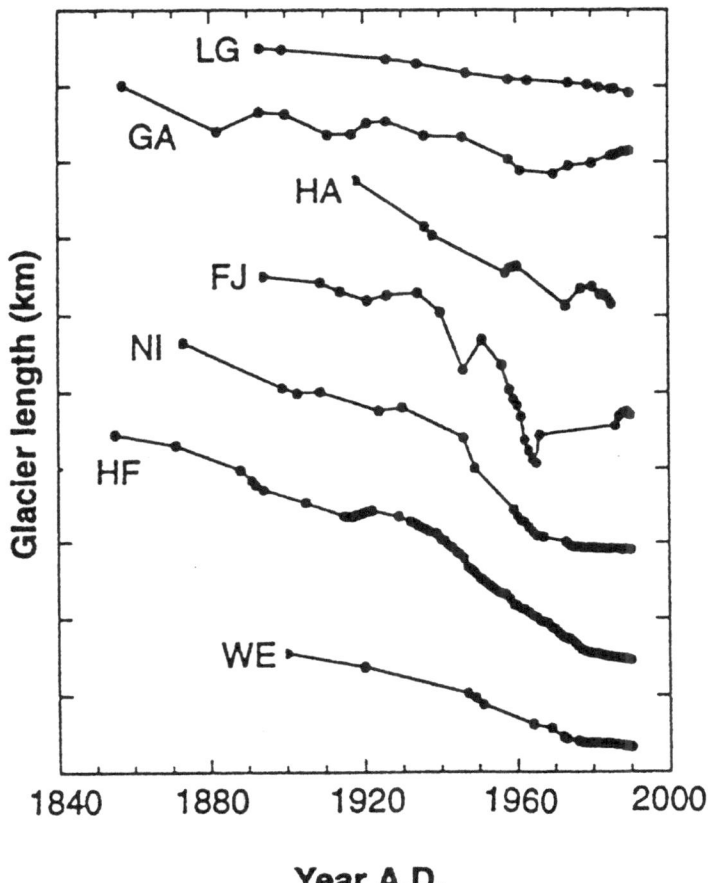

Year A.D.

Fig. 8.   Fluctuations in glacier length based on Oerlemans' (1994) analysis of data compiled by World Glacier Monitoring Service. Each dot represents an observation. Time series are shown for Lewis Glacier (LG, Kenya), Glacier d'Argentiere (GA, France), Hansbreen (HA, Spitsbergen), Franz Joseph Glacier (FJ, New Zealand), Nigardsbreen (NI, Norway), Hintereisferner (HF, Austria), and Wedgemount Glacier (WE, Canada). (From Oerlemans, 1994, p. 243).

the EOS 'AM Platform' in 1998 will permit glacier mapping at 15 m resolution, thereby making feasible a uniform global database on glacier extent (NSIDC, 1994). In addition, satellite altimetry is a potentially important tool in the monitoring of the Greenland and Antarctic ice sheets. Zwally *et al.* (1989) compared radar altimeter measurements from Seasat and Geosat over southern Greenland. The results indicated that the higher-elevation portions of the ice sheet thickened by approximately 1.6 m between 1978 and 1986. Questions have arisen, however, about the effects of orbital uncertainties on such comparisons. Other estimates based on field measurements made during 1959–1968 indicate a lowering of the surface in the ablation zone and a thickening by ~ 0.085 m yr^{-1} in the accumulation zone (Seckel, 1977). Mass balance studies by Reeh and Gundestrup (1985) and Kostecka and Whillans (1988) indicate small thickening rates (< 0.10 m yr^{-1}) in the accumulation zone. The aggregate of the mass balance studies suggest that any present mass imbalance of the Greenland ice sheet is small (van der Veen, 1991, p. 439).

Applications of satellite altimetry to the Antarctic ice sheet have been limited by the fact that the orbital coverage of most of the earlier sensors extended to only 65°–72° latitude. The more recently launched ERS-1 carries an altimeter in an orbit reaching 82° latitude. Mass balance studies of individual drainage basins of the Antarctic ice sheet have indicated positive balances in some regions and negative balances in others, although such estimates are complicated by uncertainties in the locations of the drainage basin boundaries and the fact that accumulation measurements are often made along linear traverses (van der Veen, 1991). Furthermore, accumulation rate data are unavailable for large parts of the interior. On the basis of approximately ten mass balance studies surveyed by Meier (1983), it appears that the mass balance of the entire ice sheet is positive, corresponding to a lowering of global sea level by 0 to 1.2 mm yr^{-1}. This large range is a measure of the uncertainty in the mass balance estimates.

Alternative approaches to the mass balance of the ice sheets include the analysis of accumulation rates from ice cores and the estimation of areally-averaged precipitation from the convergence of the atmospheric moisture influx. An application of the former approach is Morgan et al.'s (1991) study of annual accumulation layers since 1806 in ice cores along a 700-km segment of East Antarctica. The authors find a significant increase in the accumulation rate following a relative minimum, leading to recent values that are about 20% above the long-term mean (Figure 9). Morgan et al. estimate that this increase represents a positive imbalance of 5–25% of the mass input and may correspond to a lowering of sea level by 1.0–1.2 mm yr^{-1}. This increase of accumulation is consistent with the Stark's (1994) recent analysis of Antarctic temperatures, which show a statistically significant warming over the past several decades in the vicinity of the Antarctic peninsula. However, the trends of air temperature over other portions of the Antarctic continent appear to be much smaller (Samson, 1989).

Recent trends of precipitation over Antarctica and Greenland have been examined by Bromwich and Robasky (1993) using estimates of atmospheric water vapor flux convergence (cf. Section 2.2). In agreement with the Morgan et al.'s (1991) results from the Antarctic ice cores, Bromwich and Robasky find that Antarctic precipitation appears to have increased by about 5% over the past several decades. Corresponding estimates for Greenland show a decrease of about 15% from 1963 to 1977, followed by an increase over the southern part of the ice sheet (Robasky and Bromwich, 1994).

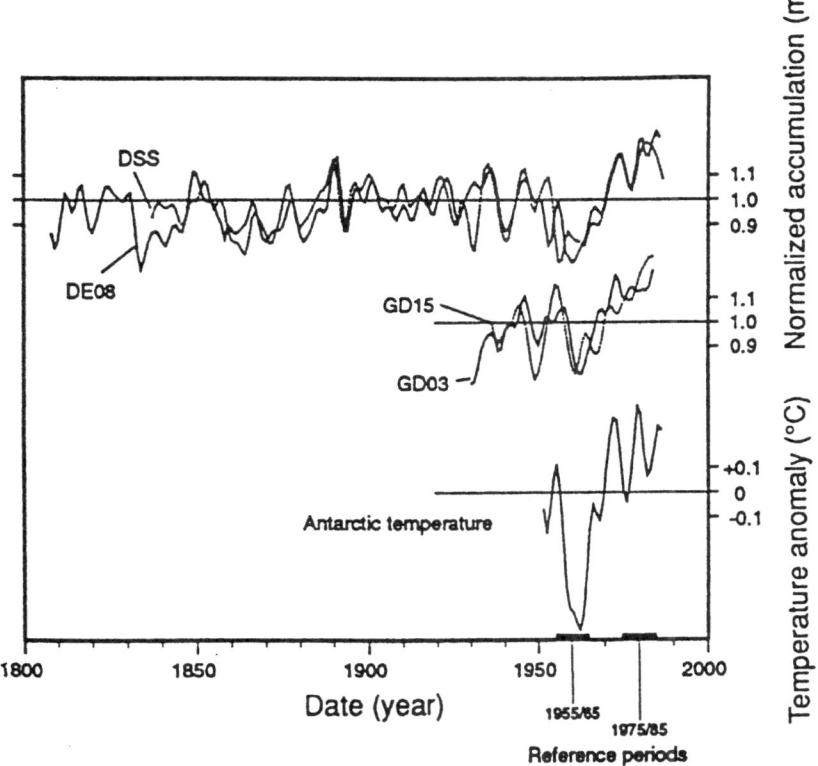

Fig. 9.   Accumulation data for four Antarctic ice cores, as presented by Morgan *et al.* (1991). Values have been divided by mean values for 1930–1985 and smoothed by a Gaussian filter (3-year bandwidth). Also shown are Antarctic temperature anomalies smoothed by the same filter. (From Morgan *et al.*, 1991, p. 58).

There is a need for systematic intercomparisons of ice sheet mass imbalances deduced from satellite altimetry, from the synthesis of ice core data, and from computations of atmospheric moisture flux convergence. While satellite altimetry clearly offers the greatest potential for routine monitoring in the future, it will be necessary that satellite orbits extend to the poles if the Antarctic ice sheet is to be monitored in its entirety. Since ice cores and atmospheric wind/moisture profiles offer the only means to assess the historical trends of the past few decades to centuries, intercomparisons of changes derived from satellite altimeters and from *in situ* dates will be necessary in order to merge the products of future monitoring with the variations deduced for the past decades to centuries.

[255]

2.4. PERMAFROST

Permafrost (a layer in the ground that remains frozen throughout the year) generally occurs in areas in which the annual mean air temperature is below about –1 °C (Barry, 1985; Lachenbruch *et al.*, 1988). In most northern land areas, permafrost underlies an 'active layer' that thaws in summer for periods of weeks to months. Continuous permafrost occurs over most of the land area north of 65° N, while discontinuous permafrost extends southward to approximately 50° N in eastern Canada and to approximately 47° N in eastern Asia. Permafrost also occurs off-shore of some Arctic coasts. The total areal coverage (excluding Antarctica) is approximately 7.6×10^6 km^2 of continuous permafrost and 17.3×10^6 km^2 of discontinuous permafrost (Barry, 1985), although the boundaries in many regions have not been mapped precisely. The volume of frozen water in permafrost is equivalent to a sea level change of 0.08–0.18 m (Hollin and Barry, 1979).

Since the increase of surface air temperature over the past several decades has been largest (~ 0.5 °C per decade) over the northern land areas, the stability of permafrost has become an important issue. Much of Canada, for example, appears to be underlain by permafrost that is classified as 'unstable' relative to present air temperatures (Harriss, 1986). Kwong and Gan (1994) have indeed shown that the southern limit of the discontinuous permafrost zone has migrated northward in the region south of Great Slave Lake, Canada, since the early 1960s. Kwong and Gan also show that the annual air temperatures in this region increased by 0.02 to 0.05 °C yr^{-1} during the period 1949–1989. Thus the northward migration is consistent with that expected from the average latitudinal gradient of air temperature: 150 ± 50 km per 1 °C warming (Barry, 1985). There is also some evidence that the area underlain by permafrost is advancing in eastern Canada (Levesque *et al.*, 1988), where air temperatures have decreased and sea ice coverage has increased in the past few decades (see Section 2.1).

Determinations of permafrost temperature changes over the century timescale come primarily from data obtained through mining exploration, construction, and geophysical surveys that involve sampling to depths of tens of meters. Lachen-bruch and Marshall (1986), for example, used borehole measurements from north-ern Alaska to infer surface temperature increases of 2–4 °C in this region over the past century. Osterkamp (1988) subsequently showed that this warming extended from the Chukchi coast eastward to the Arctic National Wildlife Refuge, although permafrost data for the past decade suggest a short-term cooling in the mid-1980s. Haberli *et al.* (1993) summarize data that show recent increases of alpine per-mafrost temperatures in several regions. However, the interpretation of the deeper permafrost temperatures is complicated by possible changes in snowfall, as increas-es of snow cover will tend to damp the effects of changes in air temperatures on permafrost temperatures. The compounding of the temperature signal by changes of snow depth is noted by Zhang and Osterkamp (1993). The seasonality of changes in air temperature and precipitation can also exert major controls on the permafrost regime. For example, permafrost degradation in many regions will be enhanced not only by summer warmth but also by summer dryness, which will favor higher soil temperatures and greater thawing of the active layer (Barry, 1985).

Satellite remote sensing applications to permafrost mapping and monitoring have thus far been minimal in comparison with the other cryospheric elements. It is possible that wintertime SAR (Synthetic Aperture Radar) data may provide an indirect indication of permafrost extent, but such applications are still in the development phase. In the meantime, syntheses of existing permafrost data (especially data from Asia) and systematic surveys of permafrost are needed if permafrost monitoring is to play a key role in the detection of global change.

2.5. LAKE ICE

In terms of both its mean seasonal cycle and its interannual variability, freshwater ice is an indicator of local atmospheric forcing. Since freshwater ice is a readily observable quantity, it has considerable potential for climate monitoring. The primary variables affecting the duration and thickness of freshwater ice on lakes and rivers are air temperature, snow depth, and wind, as well as the heat storage and impurity content of the body of water. In addition, the rate and temperature of the inflow water affects the ice regime of lakes and rivers. Because inflow (and human) effects are generally more significant in rivers than in lakes, data on river ice are less useful as climate indicators than are data on lake ice (Barry, 1985).

As indicated by the surveys of Robertson (1988) and Barry and Maslanik (1993), a 4–6 day change in the mean freeze-up or break-up date for each 1 °C change in autumn or spring air temperature is typical of lakes in middle latitudes of the Northern Hemisphere. The relationship with air temperature is generally stronger in colder climates and in relation to freeze-up, since snow depth can have a significant influence on the timing of break-up (Barry, 1993). Palecki and Barry (1986) demonstrate the use of lake freeze-up and break-up dates as an index of temperature changes in Finland during autumn and spring.

Using data for 1955–1989, Hanson *et al.* (1992) have shown that the break-up of Great Lakes ice has come significantly earlier in the spring over the past four decades. At some observing sites, the earlier end of the ice season corresponds to a change of 10–15 days over the period of record, approximately three decades. Schindler *et al.* (1990) found that the length of the ice-free season of lakes in northwestern Ontario increased by approximately 3 weeks over a 20-year period beginning in the 1960s; corresponding increases of air temperature were about 2 °C. These changes correlate well with the recent increase of springtime temperatures over much of central and western Canada, and with the earlier retreat of springtime snow cover over the past 20 years (Groisman *et al.*, 1994).

On the century timescale, North American lakes showed a trend toward later break-up from about 1870 to 1940 (Hall, 1993), although there was little change between 1940 and 1970 (Williams, 1971). Long-term records of lake ice cover in

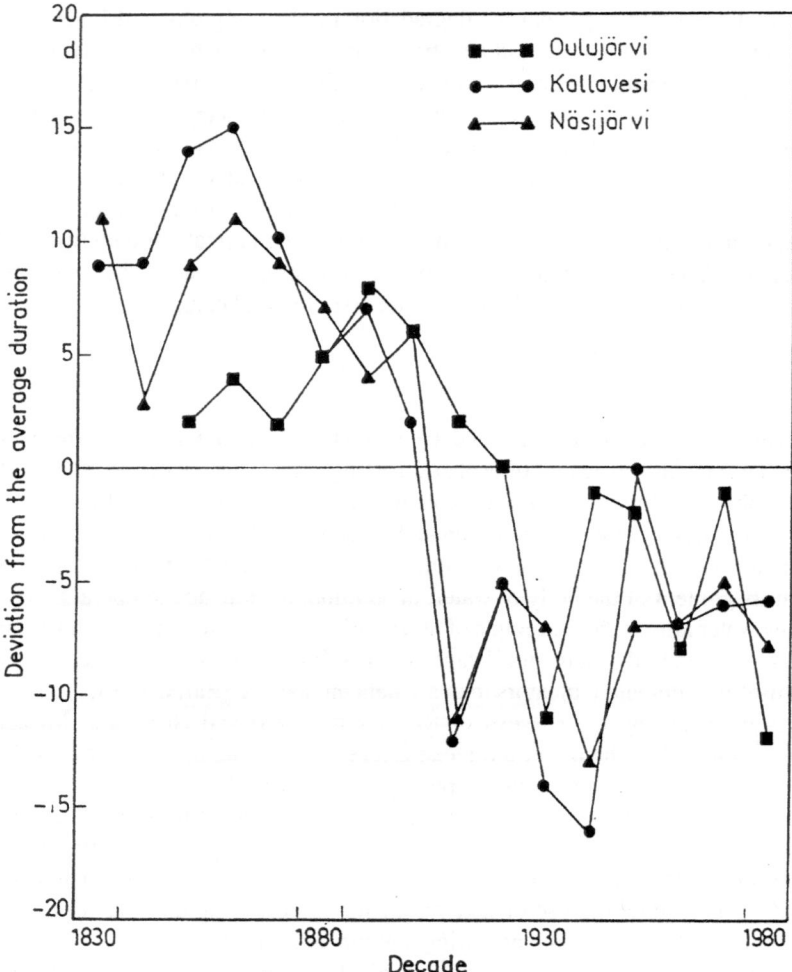

Fig. 10. Decadal mean durations of ice cover on three lakes in Finland; durations are plotted as departures from the means for the entire period. (From Kuusisto, 1993, p. 59).

Finland show a substantial decrease of ice cover duration in the early 1900s (Figure 10), followed by 5–6 decades of less change (Kuusisto, 1993). Break-up dates in recent decades have nevertheless been 1–3 weeks earlier than in the middle 1800s. Records of river ice (River Tornionjoki, 1690–1990) and of Baltic Sea ice in the Helsinki harbor (1850–1980) show similar decreases of ice duration.

Although the data for North America and Europe demonstrate that lake ice is a useful proxy indicator of climate variations, the potential of lake ice in monitoring large-scale climate has not been fully realized. One reason why this potential is untapped is that there is presently no global inventory or archive of data on freshwater ice formation and disappearance. Data are generally scattered among national agencies, whose holdings, data formats and storage media are not well known (Barry, 1993). The nature and status of records from the former Soviet Union are particular uncertain. The volume of lake ice data is sufficiently modest that the establishment of a single archive at an existing data center should be a high priority. Some efforts in this direction have begun at the National Snow and Ice Data Center.

Since the number of unmonitored lakes in high latitudes far exceeds the number of monitored lakes, the use of lake ice to monitor climate in a global context will clearly require the application of satellite remote sensing. The reliance on satellites is likely to increase further as manned stations in northern latitudes are eliminated. As noted by Barry (1993), effective monitoring of lake ice by satellites requires: (a) orbital coverage that permits sampling at frequent ($\sim$daily) intervals in order to caputure the relatively rapid freeze-up and break-up; and (b) sufficiently fine spatial resolution ($\lesssim$ 1 km) to detect changes in the small- to medium-sized lakes that predominate in high latitudes. To date, the instruments with the all-weather and day/night capabilities required for (a) have been the passive microwave sensors, which have spatial resolutions ($\sim$ 25 km) that are far too coarse for the monitoring of most lakes. Synthetic aperture radar (SAR) offers the all-weather capability and adequate spatial resolution, but the frequency of coverage and ambiguities in the interpretation of the backscattered signal are issues requiring attention. Satellites planned for the Earth Observing System will carry sensors having improved spatial as well as spectral resolution. Prior to the EOS era, the selection of a 'baseline network' of lakes for monitoring by satellites appears to be a prerequisite for the extension of lake ice monitoring from the scale of local to global climate. This selection should be based both on satellite capabilities (resolution, coverage, spectral capabilities, etc.) and climate monitoring priorities.

3. Synthesis and Recommendations

In many respects, the cryospheric data surveyed here present a consistent picture of recent climate variability over the past few decades to a century or longer. The warming of the northern land areas over the past 30 years is apparent in the reduced continental snow cover, and the warming even appears to have been enhanced by a feedback from the reduced snow cover during spring. The duration of lake ice has decreased, at least over North America and Europe, and permafrost measurements indicate a retreat over northwestern Canada. Over the past 100 years, the retreat of mountain glaciers is consistent with a warming of 0.5°–1.0 °C. Although sea ice coverage has changed little in the winter portion of the year, there are indications of recent decreases of summertime ice extent in both the Arctic and the Antarctic. However, the interannual variability of ice coverage is sufficiently large and the period of homogeneous sea ice data is sufficiently short (~ 20 years) that the summertime decrease is not yet statistically significant.

Several subregions appear to be exceptions to the above generalizations, especially in the context of model projections of greenhouse warming. There is no evidence of a recent warming over the central Arctic Ocean; the data even suggest a cooling in some seasons. Eastern Canada, including the Baffin Bay-Labrador Sea region, has experienced colder conditions and an increase of sea ice coverage during the past few decades. There are also indications of an increased extent of permafrost in this region. By contrast, the Antarctic Peninsula has shown a particularly strong and statistically significant warming over the past four decades. Since the Peninsula is in the marginal sea ice zone, its climatic variations bear close watching over the next few decades.

Temporal variations of precipitation, which contributes directly to snow cover and to the mass balance of the ice sheets, also appear to be generally coherent over high latitudes during the past several decades to a century. Snowfall data show an increase over northern Canada and Alaska, although the higher temperatures over much of Canada have decreased the percentage of precipitation that falls as snow. In this respect, the recent decrease of Northern Hemisphere snow coverage is not inconsistent with the trend toward greater precipitation in high latitudes. In addition, the bulk of the evidence indicates that Antarctica's precipitation has increased in recent decades and that the mass balance of the Antarctic ice sheet is currently positive (accumulation exceeds ablation). Satellite altimeter data suggest a recent thickening of the southern part of the Greenland ice sheet, although the sign of the mass balance of the entire Greenland Ice Sheet is still open to question.

The above assessment clearly contains uncertainties, and these uncertainties are key considerations in the setting of priorities for long-term monitoring of the cryosphere. Other considerations are: (a) the extent to which existing data have been utilized; and (b) the capabilities of current and anticipated remote sensing tools. One may argue that the extents of sea ice and snow cover are presently observed adequately for monitoring purposes. However, it is essential that the 20–25 year databases be continued into the future through passive microwave sensing of sea ice and the weekly charting (by NOAA/NESDIS) of snow cover based on visible and infrared imagery. These databases are the 'flagships' of the snow and ice monitoring efforts. Priority should also be assigned to development of monitoring capabilities for sea ice thickness and snow depth or liquid water equivalent. Coordinated measurements by underice sonar will likely be the most promising approach to sea ice thickness monitoring, while a blend of quality-controlled *in situ* data and satellite passive microwave measurements will likely provide the optimal determinations of snow depth on a routine basis.

Ice sheets and glaciers also appear to be amenable to routine monitoring by satellite. Satellite-borne altimeters are clearly a top priority in the monitoring of ice sheets, while high-resolution visible imagery is the most promising approach to the systematic monitoring of alpine glaciers on a world-wide basis. Since high-resolution imagery is also needed for the monitoring of lake ice, several motivations exist for such sensors on satellites with orbits reaching the northernmost land areas.

Several non-satellite approaches also deserve mention in the context of long-term monitoring of the cryosphere. First, areal means of snowfall may be deduced from the convergence of water vapor fluxes in the atmosphere (e.g., Bromwich *et al.*, 1993). As data assimilation cycles of numerical weather prediction models produce increasingly accurate analyses (and 're-analyses' from ongoing projects), gridded precipitation 'histories' spanning multiyear to multidecadal periods will become available. While validation efforts will clearly be required, these 'histories' have the potential to augment substantially the databases for the monitoring of snowfall over high-latitude land areas, sea ice and, perhaps most importantly, the Greenland and Antarctic ice sheets. Second, the survey of cryospheric variability in Section 2 has pointed to the need for syntheses of existing data on several cryospheric variables: station data on snow depth and/or water equivalent, lake ice data, and permafrost temperature measurements. Such data exist in a variety of locations, formats and storage media. The collection, quality-control and synthesis of such data into centralized archives will provide valuable historical databases on which future monitoring efforts can build.

Acknowledgements

Preparation of this paper was supported by the National Science Foundation through Grants DPP-9214793 and ATM-9319952. Thanks are due Norene McGhiey for word-processing the manuscript and William Chapman for the computational analysis of the sea ice data.

References

Armstrong, R. L. and Brodzik, M. J.: 1995, 'An Earth-Gridded SSM/I Data Set for Cryospheric Studies and Global Change Monitoring', COSPAR 30th Scientific Assembly (July 11–21, 1994, Hamburg, Germany), *Advances in Space Research*, (in press).

Barry, R. G.: 1985, 'The Cryosphere and Climate Change', in MacCracken, M. C. and Luther, F. M. (eds.), *Detecting the Climatic Effects of Increasing Carbon Dioxide*, DOE/ER-0235, U.S. Dept. of Energy, Washington, D.C., pp. 109–148.

Barry, R. G.: 1990, 'Observational Evidence of Changes in Global Snow and Ice Cover', in Parker, D. (ed.), *Contributions in Support of Section 7 of the 1990 IPCC Scientific Assessment*, Intergovernmental Panel on Climate Change, World Meteorological Organization/United Nations Environmental Programme, pp. I.1–I.20.

Barry, R. G.: 1993, 'Lake Ice Cover as a Climatic Indicator', Report of Working Group #3 in Barry, R. G., Goodison, B. E., and LeDrew, E. F. (eds.), *Snow Watch '92: Detection Strategies for Snow and Ice, Glaciological Data* **GD-25**, pp. 264–265.

Barry, R. G., Armstrong, R. L., and Krenke, A. N.: 1993, 'An Approach to Assessing Changes in Snow Cover: An Example from the Former Soviet Union', *Proceedings, 50th Annual Eastern Snow Conference*, Quebec City, 25–34.

Barry, R. G., Fallot, J.-M., and Armstrong, R. L.: 1994, 'Assessing Decadal Scale Changes in the Cryosphere: Eurasian Snow Cover', *Proceedings, Fifth Symposium on Global Change Studies*, Nashville, TN, American Meteorological Society, 148–155.

Barry, R. G. and Maslanik, J. A.: 1993, 'Monitoring Lake Freeze-up/Break-up as a Climatic Index', in Barry, R. G., Goodison, B. E., and LeDrew, E. F. (eds.), *Snow Watch '92: Detection Strategies for Snow and Ice, Glaciological Data* **GD-25**, pp. 66–79.

Bromwich, D. H.: 1988, 'Estimates of Antarctic Precipitation', *Nature* **343**, 627–629.

Bromwich, D. H. and Robasky, F. M.: 1993, 'Recent Precipitation Variations over the Polar Ice Sheets', *Meteor. Atmos. Phys.* **51**, 259–274.

Bromwich, D. H., Robasky, F. M., Keen, R. A., and Bolzan, J. F.: 1993, 'Modeled Variations of Precipitation over the Greenland Ice Sheet', *J. Clim.* **6**, 1253–1268.

Brown, R.: 1994, 'Long-Term Variability in Canadian Snow Cover', *CMOS Bulletin* (Canadian Meteorological and Oceanographic Society) **22 (2)**, 10–11.

Chang, A. T. C., Foster, J. L., and Hall, D. K.: 1987, 'Nimbus-7 SMMR Derived Global Snow Cover Parameters', *Ann. Glaciol.* **9**, 39–45.

Chapman, W. L. and Walsh, J. E.: 1993, 'Recent Variations of Sea Ice and Air Temperature in High Latitudes', *Bull. Amer. Meteor. Soc.* **74**, 33–47.

D'Arrigo, R. D. and Jacoby, G. C.: 1993, 'Secular Trends in High Northern Latitude Temperature Reconstructions Based on Tree Rings', *Clim. Change* **25**, 163–177.

Environment Canada: 1994, 'Annual Review (1994)', *Clim. Perspectives* **16**, 1–10.

Gloersen, P. and Campbell, W. J.: 1991, 'Recent Variations in Arctic and Antarctic Sea Ice Covers', *Nature* **352**, 33–36.

Gloersen, P., Campbell, W. J., Cavalieri, D. J., Comiso, J. C., Parkinson, C. L., and Zwally, H. J.: 1992, 'Arctic and Antarctic Sea Ice, 1978–1987: Satellite Passive-Microwave Observations and Analysis', NASA SP-511, National Aeronautics and Space Administration, Washington, D.C., 290 pp.

Groisman, P. Y. and Easterling, D. R.: 1994, 'Variability and Trends of Total Precipitation and Snowfall over the United States and Canada', *J. Clim.* **7**, 184–205.

Groisman, P. Y., Karl, T. R., and Knight, R. W.: 1994, 'Observed Impact of Snow Cover on the Heat Balance and the Rise of Continental Spring Temperatures', *Science* **263**, 198–200.

Haberli, W., Chang, C., Gorbunov, A. P., and Harriss, S. A.: 1993, 'Mountain Permafrost and Climatic Change', *Permafrost and Periglacial Process.* **4**, 165–174.

Hall, D. K.: 1988, 'Assessment of Polar Climate Change Using Satellite Technology', *Rev. Geophys.* **26**, 26–39.

Hall, D. K.: 1993, 'Active and Passive Microwave Remote Sensing of Frozen Lakes for Regional Climate Studies', in Barry, R. G., Goodison, B. E., and LeDrew, E. F. (eds.), *Snow Watch '92: Detection Strategies for Snow and Ice, Glaciological Data* **GD-25**, pp. 80–85.

Halpert, M. S., Bell, G. D., Kousky, V. E., and Ropelewski, C. F.: 1994, *Fifth Annual Climate Assessment 1993*, U.S. Dept. of Commerce, Climate Analysis Center, Camp Springs, MD, 111 pp.

Hanson, H. P., Hanson, C. S., and Yoo, B. H.: 1992, 'Recent Great Lakes Ice Trends', *Bull. Amer. Meteor. Soc.* **73**, 577–584.

Harriss, S. A.: 1986, 'Permafrost Distribution, Zonation and Stability along the Eastern Ranges of the Cordillera of North America', *Arctic* **39**, 29–38.

Hollin, J. and Barry, R. G.: 1979, 'Empirical and Theoretical Evidence Concerning the Response of the Earth's Ice and Snow Cover to a Global Temperature Increase', *Environm. Internat.* **2**, 437–444.

IPCC: 1990, *Climate Change: The IPCC Scientific Assessment*, in Houghton, J. T., Jenkins, G. J., and Ephraums, J. J. (eds.), Intergovernmental Panel on Climate Change, Cambridge Univ. Press, Cambridge, 365 pp.

IPCC: 1992, *Climate Change 1992: The Supplementary Report to the IPCC Scientific Assessment*, in Houghton, J. T., Callander, B. A., and Varney, S. K., (eds.), Intergovernmental Panel on Climate Change, Cambridge Univ. Press, Cambridge, 200 pp.

Jones, P. D., Wigley, T. M. L., and Farmer, G.: 1991, 'Marine and Land Temperature Data Sets: A Comparison and a Look at Recent Trends', in Schlesinger, M. E. (ed.), *Greenhouse-Gas-Induced Climatic Change: A Critical Appraisal of Simulations and Observations*, Elsevier, pp. 153–172.

JSC: 1993, *JSC Ocean Observing System Development Panel (OOSDP), Eighth Session*, Joint Scientific Committee, Intergovernmental Oceanographic Commission, UNESCO, 28 pp.

Kahl, J. D., Charlevoix, D. J., Zaitseva, N. A., Schnell, R. C., and Serreze, M. C.: 1993, 'Absence of Evidence for Greenhouse Warming over the Arctic Ocean in the Past 40 Years', *Nature* **361**, 335–337.

Karl, T. R., Groisman, P. Ya., Heim, R. R., Jr., Knight, R. W.: 1993, 'Recent Variations of Snow Cover and Snowfall in North America and their Relation to Precipitation and Temperature Variations', *J. Clim.* **6**, 1327–1344.

Kostecka, J. M. and Whillans, I. M.: 1988, 'Mass Balance along Two Transects of the West Side of the Greenland Ice Sheet', *J. Glaciol.* **34**, 31–39.

Kuusisto, E.: 1993, 'Lake Ice Observations in Finland in the 19th and 20th Century: Any Message for the 21st?', in Barry, R. G., Goodison, B. E., and Le Drew, E. F. (eds.), *Snow Watch '92: Detection Strategies for Snow and Ice, Glaciological Data* **GD-25**, pp. 57–65.

Kwong, Y. T. J. and Gan, T. Y.: 1994, 'Northward Migration of Permafrost along the Mackenzie Highway and Climate Warming', *Clim. Change* **26**, 399–419.

Lachenbruch, A. H., Cladouhos, T. T., and Saltus, R. W.: 1988, 'Permafrost Temperature and the Changing Climate', in Senneset, K. (ed.), *Proceedings: Fifth International Conference on Permafrost, Vol. 3*, Trondheim, Norway (August 1988), Tapir Publishing, pp. 9–18.

Lachenbruch, A. H. and Marshall, B. V.: 1986, 'Changing Climate: Geothermal Evidence from Permafrost in the Alaskan Arctic', *Science* **234**, 689–696.

Legates, D. R.: 1993, 'The Need for Removing Biases from Rain and Snowgage Measurements', in Barry, R. G., Goodison, B. E., and LeDrew, E. F. (eds.), *Snow Watch '92: Detection Strategies for Snow and Ice, Glaciological Data* **GD-25**, pp. 164–171.

Levesque, R., Allard, M., and Seguin, M. K.: 1988, 'Regional Factors of Permafrost Distribution and Thickness, Hudson Bay Coast, Quebec, Canada', in Senneset, K. (ed.), *Proceedings: Fifth International Conference on Permafrost, Vol. 2*, Trondheim, Norway, (August 1988), Tapir Publishing, pp. 199–204.

Li, P.: 1987, 'Seasonal Snow Resources and their Fluctuations in China', in Goodison, B. E., Barry, R. G., and Dozier, J. (eds.), *Large-Scale Effects of Seasonal Snow Cover*, IAHS Press, Wallingford, U.K., pp. 93–104.

Manabe, S., Spelman, M. J., and Stouffer, R. J.: 1992, 'Transient Response of a Coupled Ocean-Atmosphere Model to Gradual Changes of Atmospheric CO_2. Part II: Seasonal Response', *J. Clim.* **5**, 105–126.

Marko, J. R., Fissel, D. B., Wadhams, P., Kelly, P. M., and Brown, R. D.: 1994, 'Iceberg Severity off Eastern North America: Its Relationship to Sea Ice Variability and Climate Change', *J. Clim.* **7**, 1335–1351.

McLaren, A. S., Bourke, R. H., Walsh, J. E., and Weaver, R. L.: 1994, 'Variability in Sea-Ice Thickness over the North Pole from 1958 to 1992', *Geophys. Monogr.* **85**, Amer. Geophys. Union, 363–371.

Meier, M. F.: 1983, 'Snow and Ice in a Changing Hydrological World', *Hydrol. Sci. J.* **28**, 3–22.

Meier, M. F.: 1984, 'Contribution of Small Glaciers to Global Sea Level', *Science* **226**, 1418–1421.

Meier, M. F.: 1990, 'Reduced Rise in Sea Level', *Nature* **343**, 115–116.

Morgan, V. I., Goodwin, I. D., Etheridge, D. M., and Wookay, C. W.: 1991, 'Evidence from Antarctic Ice Cores for Recent Increases in Snow Accumulation', *Nature* **354**, 58–60.

NSIDC: 1994, *NSIDC Notes*, Issue No. **10**, National Snow and Ice Data Center, Boulder, CO, 8 pp.

Oerlemans, J.: 1994, 'Quantifying Global Warming from the Retreat of Glaciers', *Science* **264**, 243–245.

Osterkamp, T. E.: 1988, 'Permafrost Temperatures in the Arctic National Wildlife Range', *Cold Regions Sci. and Technology* **15(2)**, 191–193.

Osterkamp, T. E., Zhang, T., and Romanovsky, V. E.: 1994, 'Evidence for a Cyclic Variation of Permafrost Temperatures in Northern Alaska', *Permafrost and Periglacial Processes*, (in press).

Palecki, M. A. and Barry, R. G.: 1986, 'Freeze-up and Break-up of Lakes as an Index of Temperature Changes During the Transition Seasons: A Case Study for Finland', *J. Clim. Appl. Meteor.* **25**, 893–902.

Parkinson, C. L.: 1992, 'Spatial Patterns of Increases and Decreases in the Length of the Sea Ice Season in the North Polar Region, 1979–1986', *J. Geophys. Res.* **97**, 14,377–14,388.

Parkinson, C. L.: 1994, 'Spatial Patterns in the Length of the Sea Ice Season in the Southern Ocean, 1979–1986', *J. Geophys. Res.* **99**, 16,327–16,339.

Reeh, N. and Gundestrup, N. S.: 1985, 'Mass Balance of the Greenland Ice Sheet at Dye 3', *J. Glaciol.* **31**, 198–200.

Robasky, F. M. and Bromwich, D. H.: 1994, 'Greenland Precipitation Estimates from the Atmospheric Moisture Budget', *Geophys. Res. Lett.* **21**, (in press).

Robertson, D. M.: 1988, 'Lakes as Indicators of and Responders to Climate Change', in Greenland, D. and Swift, L. W., Jr. (eds.), *Climte Variability and Ecosystem Response*, Southeastern Forest Exp. Station, General Tech. Rept. SE-65, Forest Service, U.S. Dept. of Agriculture, pp. 38–46.

Robinson, D. A.: 1993, 'Monitoring Northern Hemisphere Snow Cover', in Barry, R. G., Goodison, B. E., and LeDrew, E. F. (eds.), *Snow Watch '92: Detection Strategies for Snow and Ice, Glaciological Data* **GD-25**, pp. 1–25.

Robinson, D. A., Dewey, K. F., and Heim, R. R., Jr.: 1993, 'Global Snow Cover Monitoring an Update', *Bull. Amer. Meteor. Soc.* **74**, 1689–1696.

Robinson, D. A., Serreze, M. C., Barry, R. G., Scharfen, G., and Kukla, G.: 1992, 'Interannual Variability of Snow Melt and Surface Albedo in the Arctic Basin', *J. Clim.* **5**, 1109–1119.

Rossow, W. B. and Schiffer, R. A.: 1991, 'ISCCP Cloud Data Products', *Bull. Amer. Meteor. Soc.* **72**, 2–20.

Samson, J.: 1989, 'Antarctic Surface Temperature Time Series', *J. Clim.* **2**, 1164–1172.

Schindler, D. W., Beaty, K. G., Fee, E. J., Cruikshank, D. R., De Bruyn, E. R., Findlay, D. L., Linsey, G. A., Shearer, J. A., Stainton, M. P., and Turner, M. A.: 1990, 'Effects of Climatic Warming on Lakes of the Central Boreal Forest', *Science* **250**, 967–970.

Seckel, H.: 1977, 'Das Geometrische Nivellement über das Grönländische Inlandeis der Gruppe Nivellement A der Internationalen Glaziologischen Grönland Expedition 1967–68', *Medd. Gronl.* **187** (3), 86 pp.

Serreze, M. C., Schweiger, A. J., and Key, J. R.: 1993, 'Comparison of Two Satellite-Derived Albedo Data Sets for the Arctic Ocean', in Barry, R. G., Goodison, B. E., and LeDrew, E. F. (eds.), *Snow Watch '92: Detection Strategies for Snow and Ice, Glaciological Data*, **GD-25**, 172–187.

Stark, P.: 1994, 'Climatic Warming in the Central Antarctic Peninsula Area', *Weather* **49**, 215–220.

Stouffer, R. J., Manabe, S., and Vinnikov, K. Ya.: 1994, 'Model Assessment of the Role of Natural Variability in Recent Global Warming', *Nature* **367**, 634–636.

Tait, A. B. and Armstrong, R. L.: 1994, 'Validation of SMMR Satellite-Derived Snow Depth with Ground-Based Measurements', *Internat. J. Rem. Sensing*, (in press).

van der Veen, C. J.: 1991, 'State of Balance of the Cryosphere', *Rev. Geophys.* **29**, 433–455.

Wadhams, P.: 1990, 'Evidence of Thinning of the Arctic Ice Cover North of Greenland', *Nature* **345**, 795–797.

WCRP: 1994, *Arctic Climate System Study (ACSYS) Initial Implementation Plan*, WCRP-85, WMO/TD-No. 627, World Climate Research Programme, Geneva, 66 pp.

Williams, G. P.: 1971, 'Predicting the Date of Lake Ice Breakup', *Water Resourc. Res.* **7**, 323–333.

Wood, F. B.: 1988, 'Global Alpine Glacier Trends, 1960s to 1980s', *Arctic Alpine Res.* **20**, 404–413.

Yamazaki, K.: 1992, 'Moisture Budget in the Antarctic Atmosphere', *proc. NIPR Symp. Polar Meteorol. Glaciol.* **6**, 36–45.

Zhang, T. and Osterkamp, T. E.: 1993, 'Changing Climte and Permafrost Temperatures in the Alaskan Arctic', in *Proceedings: Sixth International Conference on Permafrost, Vol. 1*, Beijing, China (July 1993), South China Univ. of Technology Press, Wushan, China, pp. 783–788.

Zwally, H. J., Brenner, A. C., Major, J. A., Bindschadler, R. A., and Marsh, J. G.: 1989, 'Growth of Greenland Ice Sheet: Measurement', *Science* **246**, 1587–1589.

(Received 23 January, 1995; in revised form 7 July, 1995)

SATELLITE MONITORING OF GLOBAL LAND COVER CHANGES AND THEIR IMPACT ON CLIMATE

RAMAKRISHNA R. NEMANI and STEVEN W. RUNNING

School of Forestry, University of Montana, Missoula, MT 59812, U.S.A.

Abstract. Land cover is a crucial, spatially and temporally varying component of global carbon and climate systems. Therefore accurate estimation and monitoring of land cover changes is important in global change research. Although, land cover has dramatically changed over the last few centuries, until now there has been no consistent way of quantifying the changes globally.

In this study we used long-term climate, soils data along with coarse resolution satellite observations to quantify the magnitude and spatial extent of global land cover changes due to anthropogenic processes. Differences between potential leaf area index, derived from climate-soil-leaf area equilibrium and actual leaf area index obtained from satellite data were used to estimate changes in land cover.

Forest clearing for agriculture and irrigated farming in arid and semi-arid lands are found to be two major sources of climatically important land cover changes. Satellite derived Spectral Vegetation indices (SVI) and surface temperatures (Ts) show strong impact of land cover changes on climatic processes. Irrigated agriculture in dry areas increased energy absorption and evapotranspiration (ET) compared to natural vegetation. On the other hand, forest clearing for crops decreased energy absorption and ET.

A land cover classification and monitoring system is proposed using satellite derived SVI and Ts that simultaneously characterize energy absorption and exchange processes. This completely remote sensing based approach is useful for monitoring land cover changes as well as their impacts on climate. Monitoring the spatio-temporal dynamics of land cover is possible with current operational satellites, and could be substantially improved with the Earth Observing System (EOS) era satellite sensors.

Introduction

The role of land cover has been recognized as a crucial, temporally and spatially varying component of the climate system (Dickinson *et al.*, 1986). Recent numerical simulation experiments suggest that changes in land cover could be as important as the increase in atmospheric greenhouse gases in climatic change processes, particularly at local to regional scales (Shukla *et al.*, 1990). Several studies dealing with impacts of deforestation on climate reported increased surface temperatures and reduced amounts of rainfall following land clearing, suggesting warmer and drier climate (Shukla *et al.*, 1990; Lean and Warilow, 1989; Dickinson and Henderson-Sellers, 1988). On a global scale, however, climatic impacts of land cover changes have yet to be quantified accurately. Although it is clearly evident that the global land cover has been dramatically altered by human habitation, forest clearing and agriculture etc., until now there was no means of quantifying these changes globally. Anthropogenic processes, deforestation, shifting cultivation, expansion and contraction of croplands and pastures, land degradation and human settlements,

Climatic Change **31**: 395–413, 1995.

[265]

that have significant climatic impacts operate at various spatio-temporal scales (Skole, 1993). Consistent methodology and data to quantify these changes are not yet available. Skole (1993) suggested that land cover changes can be better studied by first reconstructing a historical land cover map and use advanced techniques to estimate current rates of land cover change. While historical data is difficult to assemble (Houghton, 1994; Skole, 1993), various techniques exist for quantifying current land cover changes: aerial photography, imagery from air or space borne sensors at various spatial resolutions (Green and Sussman, 1990; Skole and Tucker, 1993). Many of the above methods provide accurate estimates of land cover changes, however they are relatively expensive, time consuming and often limited to local scales. Coarse resolution satellite data from NOAA/AVHRR are now available globally (Goward et al., 1993). Though the coarse resolution of the data severely restricts its usefulness at local scales, it is useful for studying large scale land cover changes (Goward et al., 1993). Two types of land cover changes are predominant and extensive around the globe: (1) forest clearing for agriculture (e.g. eastern U.S., western Europe, India and China); (2) land conversion from grass/shrubland to irrigated/dryland farming (e.g. central valley California, Columbia basin of eastern Washington, Murray Darling basin of Australia and Punjab state of India). Each of the above changes is extensive, climatically significant and amenable for using coarse resolution data.

In this study we used satellite observations to first, present a quantitative evaluation of the magnitudes and spatial extent of land cover changes as a result of anthropogenic processes, and then provide evidence of the impact of land cover changes on surface energy balance. Finally, we outline a methodology for regular satellite monitoring of land cover changes and their impacts on climatic processes.

Background

LAND COVER ANALYSIS

Recognition of the importance of land cover in carbon and climate modeling provided the impetus to quantify the various types and rates of change in land cover (Dickinson et al., 1986; Myers, 1988; Skole and Tucker, 1993). The complexity of the problem is evident from the variations (10–20%) in the amount of global land surface perturbed from its natural vegetation by anthropogenic processes (Matthews, 1983; Townshend et al., 1991). Even the current, increasing rates of change in land cover are unknown for many parts of the world (Myers, 1988; Myer and Turner, 1993). Quantitative analysis of the incredible diversity in global vegetation (species, growth habits and life forms) is a difficult problem (Running et al., 1994a). Many classification efforts have been attempted, and their results show wide variations in current estimates of global land cover (Townshend et al.,

1991). A detailed discussion of various land cover classification schemes and their problems is found in Townshend *et al.* (1991) and Running *et al.* (1994a).

Change detection in land cover is further complicated by the spatio-temporal variations in causal factors. For example, large scale deforestation is easily detectable by various methods compared to selective logging. Similarly human settlements are a rapid phenomenon compared to land degradation. Satellite data has been successfully used to study land cover changes at fine to coarse spatial scales. For example, Thematic Mapper data at 30 m resolution was used in the estimation of urban expansion (Haack *et al.*, 1987) and deforestation (Green and Sussman, 1990; Skole and Tucker, 1993). On the other hand, AVHRR data at 1 and 16 km resolution were used to estimate land cover changes at continental scales (Malingreau and Tucker, 1988; Tucker *et al.*, 1985; Turner *et al.*, 1993; Houghton, 1994). At present all the global satellite databases are derived from the coarse resolution AVHRR, therefore useful only to study large scale changes in land cover (Townshend *et al.*, 1991; Goward *et al.*, 1993).

IMPACT OF LAND COVER CHANGES ON CLIMATE

The influence of vegetation on energy balance and microclimate has been documented in numerous studies (Denmead, 1969; Stanhill *et al.*, 1966). Vegetation plays a crucial role in available energy (A), energy partitioning between sensible (H) and latent heat (LE) fluxes through changes in albedo, roughness and stomatal control of water losses.

Land cover changes, for example, from forests to crops affect A through changes in albedo (0.12–0.18 for forests compared to 0.22–0.26 for crops), H and LE through changes in aerodynamic resistance, ra (10 s/m vs. 100 s/m for crops and forests), and stomatal resistance, rs (500 vs. 100 s/m for crops and forests) (Gash and Shuttleworth, 1991). Increase in LAI from grass (1–3) to forest (3–10) also contributes to lower canopy resistance and higher LE (Jarvis and McNaughton, 1986). The magnitude of A and its partitioning into latent (LE) and sensible heat (H) fluxes has important implications for hydrology and atmospheric dynamics. For example, forests with higher A together with the permanency of the ecosystem enables them to evaporate more water than any other vegetation. The net effect of the above changes is usually reflected in canopy temperatures (Denmead, 1969). Satellite observations show that grass and crop canopies are 10–20 °C warmer than forest canopies (Nemani and Running, 1989b; Goward *et al.*, 1994; Smith and Choudhury, 1990). Although changes in energy balance as a result of land cover changes are well documented at local scales, the spatial extent and temporal dynamics of such changes at larger scales are not well known.

[267]

SATELLITE REMOTE SENSING

Earlier studies used satellite data primarily for land cover classification. However recent work has shown spectral vegetation indices (SVI) and surface temperature (Ts) observations from NOAA/AVHRR to be useful for quantifying the energy absorption and exchange processes (Goward et al., 1994; Choudhury, 1991; Nemani et al., 1993). SVIs combine reflectances in the red (chlorophyll absorption) and near-infrared (multiple scattering in leaves) wavelengths, and are widely used to quantify vegetation extent and activity (Tucker et al., 1985). Theoretical and experimental studies have shown SVIs to be related to various vegetation characteristics such as biomass, canopy cover, leaf area index and absorbed photosynthetically active radiation (Asrar et al., 1992; Goward et al., 1994; Sellers et al., 1994). SVIs are found to be negatively related to ground heat flux (G) (Choudhury, 1991) and positively related to unstressed ET (Sellers, 1985). On the other hand, satellite derived land surface temperatures are a function of energy exchange processes that are controlled by the sources of net radiation (soil vs. vegetation) and water availability for evapotranspiration (Goward et al., 1985; Nemani et al., 1993; Price, 1989). Under dry surface conditions, surface temperatures are linearly related to canopy density across different vegetation types (Nemani et al., 1993). Lower surface temperatures generally mean higher evapotranspiration either due to higher canopy density or high surface moisture content (Nemani and Running, 1989a; Choudhury, 1991; Goward et al., 1994; Nemani et al., 1993).

Methods

In this study we used satellite observations to address two important aspects of land cover: (1) quantifying the spatial changes in land cover from anthropogenic processes; (2) analysing impacts of land cover changes on climatic processes.

LAND COVER CHANGES

To allow for global comparability and satellite evaluation, we define vegetation only in terms of biome type and leaf area index, two of the most important characteristics used to represent vegetation in current climate and carbon models. Biome type differentiates among forest, shrubs and grass etc. Leaf area index (area of leaves per unit ground area, LAI) provides a simple measure of plant canopy density that ignores the complexities of canopy geometry, but quantifies canopy energy and mass exchange processes between plants and the atmosphere (Woodward, 1987; Running and Hunt, 1993). We define land cover changes in terms of changes in leaf area index between potential natural vegetation, vegetation that would have existed prior to anthropogenic disturbances, (Potential LAI) and current vegetation, (Actual LAI). Potential and Actual LAIs are calculated as follows:

Fig. 1.   Global land cover changes expressed as differences in leaf area index between potential and actual vegetation.

Potential LAI (PLAI): To present vegetation that would have existed before large scale human influences began, we used a potential biome map, using biogeographic principles associating vegetation with long-term climate (Leemans and Cramer, 1991) and soil characteristics (Zobler, 1986), produced at 0.5×0.5 degree resolution. This geographic biome distribution is based on specific physiological responses of different plant types to cold tolerance, growing season heat sums and drought stress (Prentice *et al.*, 1993). Using the same climate and soils data, we also computed geographic variations in potential maximum leaf area index based on well established principles of climate-soil-leaf area hydrologic equilibrium (Neilson, 1995; Woodward, 1987; Nemani and Running, 1989a). The hydrologic equilibrium theory suggests that plants adjust their leaf area to optimize the use of climate and soil resources. For example areas with longer growing seasons of optimum temperature and water balance conditions support higher leaf area. This definition of vegetation does not account for natural disturbances such as fire, insect and diseases, and wind extremes on biome distribution.

We used BIOME-BGC, an ecosystem simulation model, to estimate potential all-sided leaf area index for each 0.5×0.5 grid cell (Running and Hunt, 1993). The model uses climate (Solar radiation, humidity, air temperature and rainfall), soil (texture and depth) and vegetation (biome, LAI) information to compute carbon (photosynthesis and respiration) and hydrologic budgets (interception, evaporation, transpiration, outflow). Snowpack dynamics are simulated using air temperatures and rainfall. For each grid cell the model starts with an absolute maximum LAI value of 12. Then LAI is optimized through an iterative process until the peak canopy water stress is below critical level (pre-dawn leaf water potential < -2.0 M Pa) and positive carbon balance (assimilation $>$ respiration) is maintained at the end of the growing season (Woodward, 1987; Neilson, 1995).

Actual LAI: We used the NOAA/Global Vegetation Index (GVI) data collected during 1985–90 mapped to a 10 minute grid. This data was calibrated by K. Gallo (NOAA) and distributed by National Geophysical Research Center, Boulder, CO. As we are interested in comparing only the maximum LAI that a particular area can achieve under potential versus actual conditions, we first computed a yearly maximum $NDVI$ for the six years and then averaged the six maximum values. We assume this six year average maximum value to minimize the impact of cloud contamination, atmopsheric influence and inter-annual variability in climate. Finally, $NDVI$ for each 0.5×0.5 grid cell was produced by taking an average of all the 10 minute grid cells within each 0.5×0.5 area.

To account for the differences in structural and optical properties among different vegetation types, we used separate $NDVI - LAI$ relations for grass ($LAI = NDVI * 1.71 + 0.48$, Asrar *et al.*, 1985), needle leaf ($LAI = (NDVI/0.31) * 0.26$, Spanner *et al.*, 1990; Nemani and Running, 1989a) and broadleaf canopies ($LAI = (NDVI/0.26) * 2$, Pierce *et al.*, 1993). We used Olson *et al.* (1983) vegetation map to represent actual vegetation in terms of grass, needle and broadleaf canopies. The empirical relations based on field studies were found to corroborate theoretically derived forms of $NDVI - LAI$ relations (Asrar *et al.*, 1992; Sellers *et al.*, 1994). For example, because of their canopy structure needle leaf canopies have higher LAI per $NDVI$ than broafleaf canopies (Sellers *et al.*, 1994). We acknowledge that the simple empirical relations do not account for problems like variations in background, atmospheric influences and viewing geometry (Myneni *et al.*, 1990; Asrar *et al.*, 1992). However we believe that using only maximum $NDVI$ values would reduce the impact of many of the above problems.

To quantify the changes in land cover we assumed an accuracy of ± 1 for LAI in both cases (Woodward, 1987; Nemani and Running, 1989a; Pierce *et al.*, 1993). Further we also used a 20% of maximum LAI (5 for grass, shrubs and 10 for forests) as a threshold for detecting changes. Grid cells with LAI differences less than 20% maximum for each biome are considered unchanged.

IMPACT OF LAND COVER CHANGES ON CLIMATE

Micrometeorological observations as well as large scale numerical simulation experiments have shown the potential impacts of land cover changes on energy balance and surface temperature (Denmead, 1969; Lean and Warilow, 1989; Shukla *et al.*, 1990). We believe high resolution (space and time) satellite observations provide a rapid means of testing such predictions under real world conditions.

In order to study the changes in seasonal energy balance under the two types of land cover changes (deforestation and land conversion in arid and semi-arid areas), we chose Lake states area (type 1) and Columbia basin (type 2) in the U.S. (Figure 2). We had to limit our analysis to the U.S. because of the lack of multi-temporal 1 km AVHRR data for other areas of interest around the globe.

TEST AREAS

Lake States
Area around the Great Lakes in the U.S. was once predominantly covered with deciduous and mixed forests. Agriculture (mainly corn) replaced much of the area under forests in this region (Turner *et al.*, 1993). Annual average rainfall varies between 125–150 cm, more than 50% of the rainfall is received during summer enabling crop production.

Columbia Basin
Area east of the Cascade mountains in Washington state is dry with an average annual rainfall of 25–50 cm. Natural vegetation consists of shrubs and grass. This region underwent dramatic changes in land cover over the last 100 years with dryland wheat replacing much of the natural vegetation. Irrigation from the Columbia river is also used to produce corn, alfalfa etc.

[271]

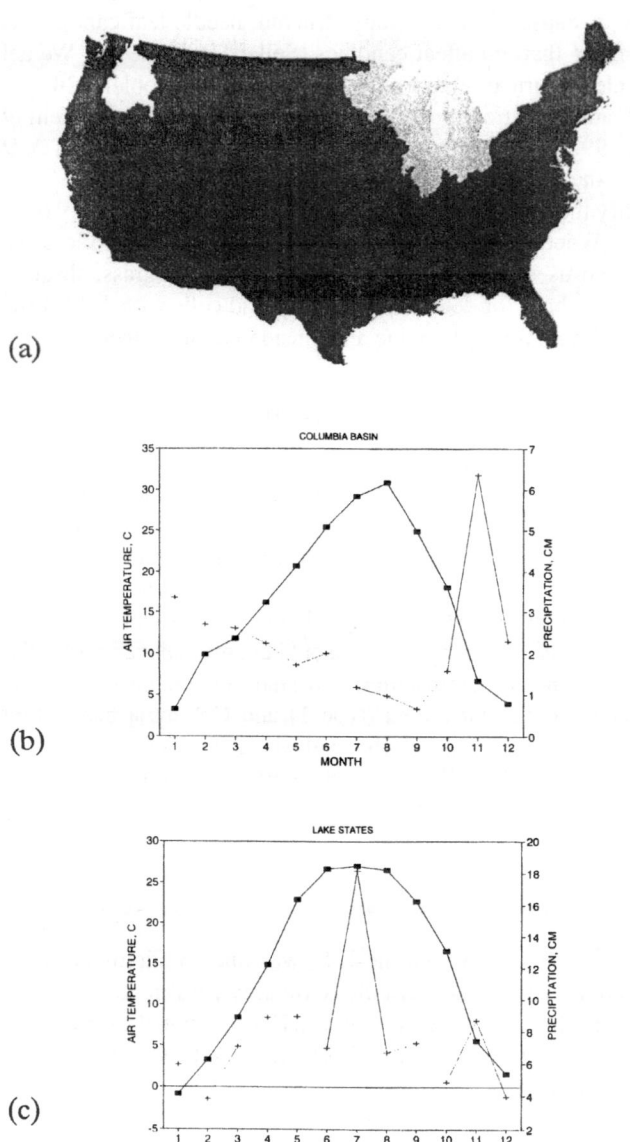

Fig. 2.   (a) Test areas in the U.S. for studying the climatic impacts of two common types of global land cover changes: (1) conversion of natural vegetation in arid and semi-arid areas to dry/irrigated agriculture, Columbia basin in Washington state; (2) forest clearing for agriculture in wet climates, area around the Great lakes; Monthly average air temperature and total precipitation for Columbia basin (b); and Lake states (c).

DATA

We used 1 km NOAA/AVHRR data (channels 1-RED, 2-NIR, TIR-4 and TIR-5) collected and composited bi-weekly during 1991 and 1992 for the conterminous U.S. by the EROS Data Center. This dataset is widely distributed and is similar to the 1990 dataset used for various land cover studies (Eidenshink, 1992; Loveland et al., 1991). For each bi-weekly composite period a 500×500 km area was extracted over the two areas. Then, for each pixel we computed $NDVI$ as $(NIR - RED)/(NIR + RED)$ and land surface brightness temperature as, $Ts = Ts4 + 3.3 * (Ts4 - Ts5)$ (Price, 1984), where $Ts4$ and $Ts5$ are brightness temperatures derived from TIR channels 4 and 5.

We considered only anthropogenic changes involving land conversion from natural vegetation to agriculture. To map locations of various land covers we used a 1 km land cover map of the U.S. (Loveland et al., 1991). Because of the coarse 1 km spatial resolution, landscape mosaics such as forest/crops and shrubs/grass were also included. For each cover class, monthly average maximum $NDVI$ and Ts were computed using data from 1991 and 1992. Digital elevation data were used to normalize Ts using environmental lapse rates (5.5 °C/km).

Results and Discussion

LAND COVER CHANGES

Figure 1 shows changes in global land cover as a result of anthropogenic processes. Agriculture and human settlements replacing native forest vegetation with high potential LAI resulted in a decrease in LAI, as in the case of Southeast Asia, Africa, Madagascar, India and western Europe (Myers, 1988). Tropical forests in South and Southeast Asia, Central America and West Africa appear to have changed considerably, compared to only a few grid cells in the Amazon basin of Brazil (Flint and Richards, 1991; Kummer and Turner, 1994).

On the other hand, crops replacing native vegetation in arid and semi-arid areas, the central valley of California, Australia, midwestern U.S. and parts of northern India, resulted in an increase in LAI over potential conditions. Forest plantations in the Southeast U.S. also show increase in LAI over potential conditions. Few areas showed spuriously high amounts of land cover change. Notable among these are Llanos of South America and York peninsula of Australia. In Llanos climatic potential for vegetation is limited by seasonal flooding arresting plant growth, which is not accounted for in potential LAI estimation. In Australia, we found the extrapolated long-term climate to be wetter than existing conditions, leading to high potential LAI over York peninsula.

Increase in population and the associated pressures on natural resources have been identified as one of several causes of global land cover changes (Kummer and Turner, 1994; Myer and Turner, 1993). Results from this study show that areas with

[273]

high population densities such as India, China and western Europe also exhibited the largest changes in LAI (Myer and Turner, 1993). Change detection is a highly scale dependent process. While top-down approaches such as this study provide rough estimates, high resolution studies along with cause-to-cover relations must be pursued (Skole and Tucker, 1993).

Finally, we would like to emphasize that land cover changes from our approach should be viewed only as gross features useful for carbon and climate models. Accurate estimation of LAI at 0.5×0.5 degree resolution is not an easy task even with ground based method. While the techniques for computing potential and actual LAI are validated at stand level, their global application relies heavily on quality of input data. In spite of the constraints with climate extrapolation, soils data and spatial resolution associated with our Potential LAI and Actual LAI estimates, our analysis seems to provide a relatively accurate picture of land cover changes. Significant improvements in the analysis can be made with better global datasets. For example, AVHRR data from the Pathfinder effort is better calibrated and comes at 8 km resolution compared to the 16 km GVI. The proposed algorithms for the Earth Observing System era sensors such as MODIS would account for various problems that limit application of current AVHRR products (Running et al., 1994b).

IMPACT OF LAND COVER CHANGES ON CLIMATE

Figure 3 shows seasonal changes in energy absorption ($NDVI$) and exchange processes (Ts) as a result of land cover changes in dry (Columbia basin) and wet (Lake states) climates. In Columbia basin land use changes resulted in an increase in energy absorption and evapotranspiration (ET) with maximum changes in summer, while in the wet Lake states forest clearing for agriculture resulted in a decrease in energy absorption and ET with maximum changes in spring-to-early summer.

Differences of $NDVI$ and Ts between dryland agriculture and native vegetation are found to be negligible in the Columbia basin. In both cases water stored in soil from spring snowmelt provides the only source of water for plant growth. When soil moisture is depleted by early summer both systems end their growth cycle. Irrigated crops, though constitute $< 5\%$ of the area, continue to maintain high $NDVI$ and low Ts through the summer. On average Ts of irrigated crops is 7.15 °C cooler than dry crops and native vegetation. Maximum differences in $NDVI$ and Ts are found during August: 0.41 and 34.3 for irrigated crops, 0.19 and 44.3 for dry crops and 0.2 and 45.1 °C for grass/shrub respectively.

Differences in phenology and growth rates between forests and crops are mainly responsible for the observed changes in $NDVI$ and Ts in the Lake states area. The changes are prominent in the spring and early summer when forests leaf-out and quickly achieve full canopy cover compared to slow growing crops. By achieving full canopy cover, forests maintain high ET rates and low Ts (Schwartz and Karl, 1990). Average Ts for crops is approximately 5 °C warmer than for forests.

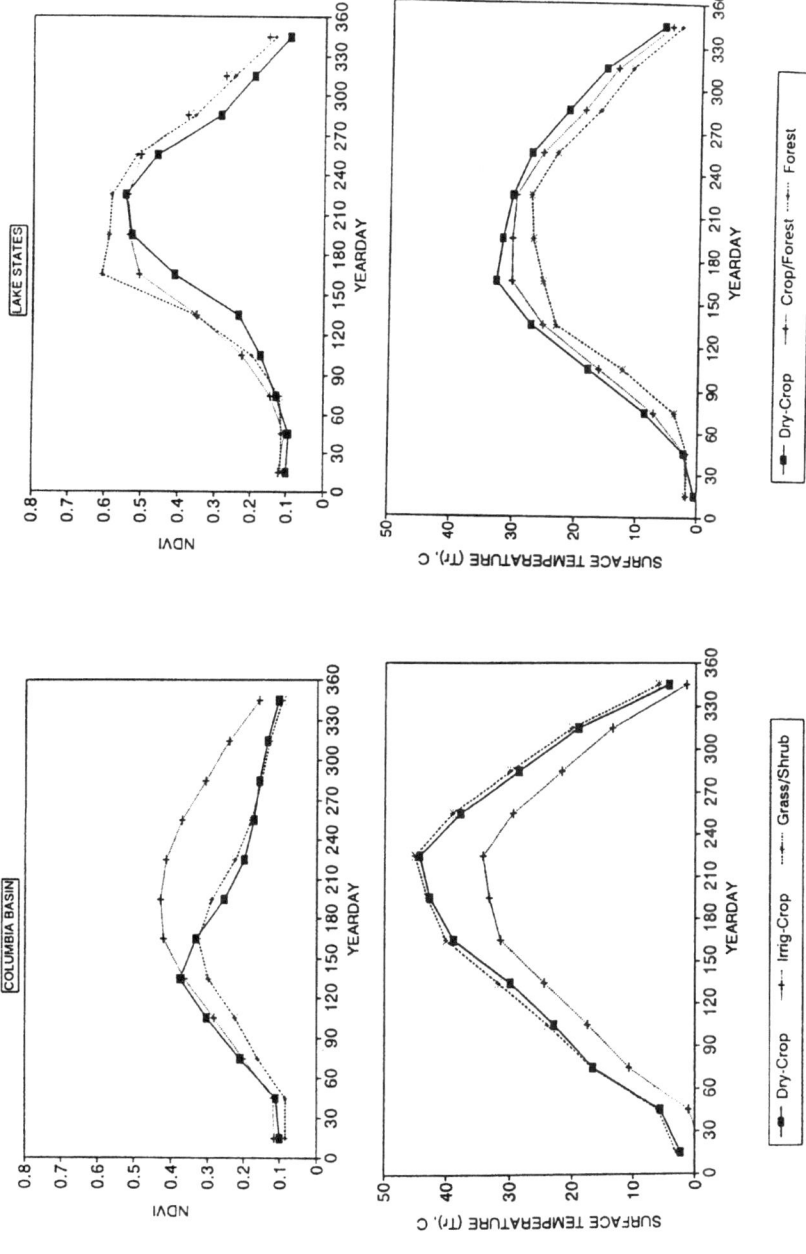

Fig. 3. Average monthly $NDVI$ and Ts for agricultural and natural vegetation over the dry Columbia basin and wet Lake states area.

Maximum differences in $NDVI$ and Ts are found during June: 0.61 and 25.3 for forests, and 0.4, 32.8 °C for crops respectively.

Changes in our satellite derived temperatures as a result of land cover changes confirm results from numerical experiments discussed earlier (Shukla *et al.*, 1990). Limited land based observations also provide evidence of the impact of land cover changes on climate. Overgrazing has been found to increase air temperatures by 3–4 °C along the Mexican border of Sonoran desert (Balling, 1988). Bore hole temperatures in Cuba also indicate a warming of 0.9 °C per century as a result of changes in energy balance due to land cover changes (Cermak *et al.*, 1992).

In addition, recent research has shown that the strong thermal contrasts generated by land cover changes could be sufficient to trigger circulation patterns leading to severe thunder storm activity (Anthes, 1984; Pielke and Avissar, 1990; Segal and Aritt, 1992). Chu *et al.* (1994) reported increased convective activity over the Amazon basin following deforestation. Extrapolating the results from our two scenarios to areas of similar land cover changes shown in Figure 1, we may conclude that spatial heterogeneity in heat fluxes has increased from pre-agricultural times (Li and Avissar, 1994).

On a global scale, the net effect of land cover changes seems to be less available energy for latent and sensible heat fluxes, because crops replacing forests is more predominant compared to irrigated agriculture in arid and semi-arid areas. The predominantly local to regional scale changes in vegetation roughness, albedo and wetness from irrigation, and the corresponding changes in surface fluxes are not well represented in current coarse resolution General Circulation Models (GCM) (Li and Avissar, 1994; Pielke and Avissar, 1990). Advanced GCMs with appropriate spatial resolution are needed to study the impact of land cover changes on global climate.

SATELLITE MONITORING OF GLOBAL LAND COVER

Multi-spectral remote sensing data offers an excellent opportunity to simultaneously monitor land cover changes and their influence on climate. Satellite derived $NDVI$ and Ts observations are useful for monitoring energy absorption and exchange processes, which in turn could be used to characterize land cover (Achard and Blasco, 1990; Running *et al.*, 1994b). Idealized seasonal trajectories of $NDVI - Ts$ for various vegetation types in a temperate climate show the logical basis for a land cover monitoring system (Figure 4a) The $NDVI - Ts$ space is be divided into four groups representing: (1) water limited (grass, shrub); (2) Energy limited (snow, water, wetlands); (3) Atmospherically decoupled (crops with higher aerodynamic resistance); and (4) Atmospherically coupled (forests with low aerodynamic resistance). Robust thresholds for $NDVI$ representing canopy cover and energy absorption, applicable globally, are possible with better algorithms for atmosphere and background corrections (Running *et al.*, 1994b). Thresholds for Ts should be based on climate, particularly maximum air temperatures. Seasonal

$NDVI - Ts$ profiles computed from AVHRR data over different land covers confirm the utility of our logic (Figure 4b). Further separation in each group is achieved using growing season average $NDVI$ ($NDVI_{gs}$) for shrub and grass, seasonal $NDVI$ amplitude ($NDVI_{amp}$) for deciduous vs. evergreen, and Near-IR (NIR) reflectance for broadleaf vs. needle leaf vegetation (Figure 5a). A first application of this logic over the continential U.S. produced spatial patterns of land cover that are similar to the ones derived from satellite data and an extensive amount of anciallary information (Loveland et al., 1991; Running et al., 1994a). The logic presented in Figure 5a is completely remote sensing based, yet produces land cover classes that are compatible with current global carbon and climate models (Running et al., 1994a).

The Earth Observing System (EOS) era satellite sensors (MODIS, MSIR, ASTR etc.) will significantly improve our ability to monitor land cover at various spatio-temporal scales. For example, the MODIS sensor will produce radiances from more spectral channels at higher spatial resolution (500 m and 250 m) than any previous globally operating satellite sensor (Running et al., 1994b). Incorporation of multi-directional reflectances for deriving land cover, albedo and leaf area index would result in better monitoring of land cover. Similarly, corrections for the atmospheric absorption and surface emissivity will improve the retrieval of land surface temperatures (Wan and Dozier, 1989). Regular global coverage of variables such as leaf area index, surface temperature allow accurate surface change detection which is fundamental determinant of global change.

Conclusions

Consistent and accurate monitoring of global land cover changes required in many global change issues is possible using satellite data. Changes in surface temperature resulting from land cover changes are an important aspect of climate change that need to be better quantified. Land cover changes also produce several indirect effects on climatic processes through release of CO_2 from deforestation, biomass burning, trace gas emissions, land degradation leading to an increase in albedo and changing hydrologic retention. Better monitoring of land cover changes and resulting biophysical responses is necessary to understand the true impacts of land cover changes on global environment and habitability.

Acknowledgements

Funding for this research was provided by NASA contract NAS5-31368. We wish to acknowledge Dr. John Townshend and two anonymous reviewers for their comments.

Dynamics of Ts-NDVI for Various Vegetation Types

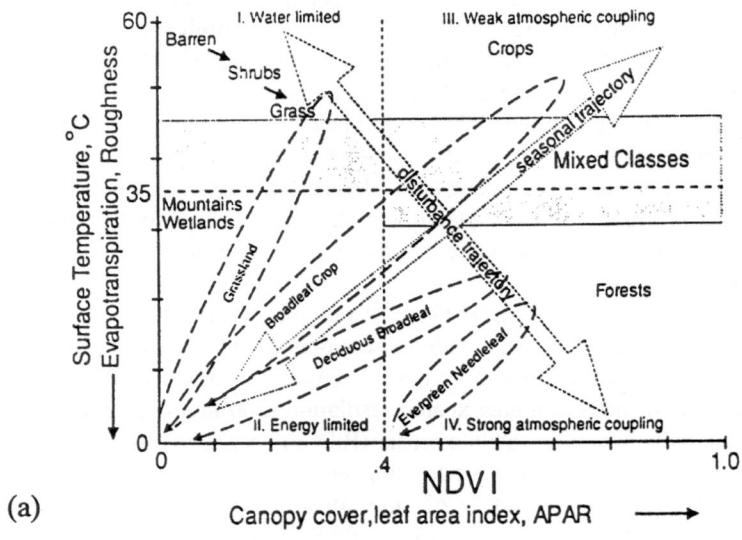

(a)

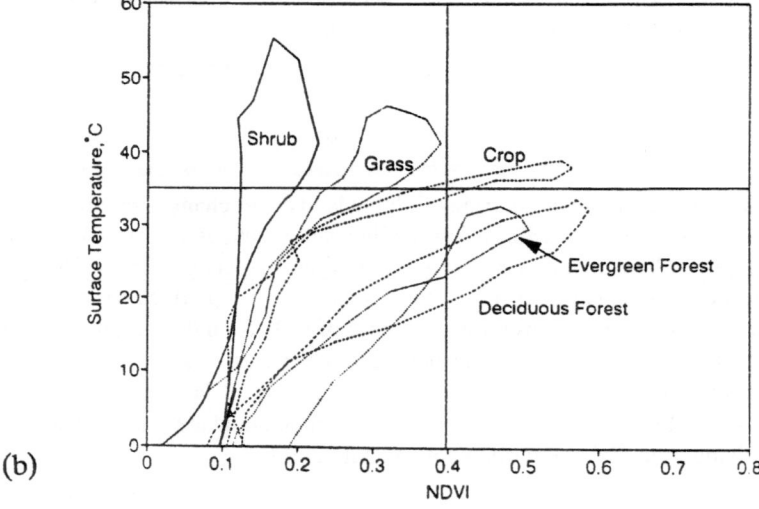

(b)

Fig. 4. (a) A conceptual diagram showing the seasonal trajectories of different land cover types in the $Ts - NDVI$ space. The $Ts - NDVI$ space is divided into four simple groups with thresholds for $NDVI$ (0.4) and Ts (35 °C) chosen to represent energy absorption and exchange characteristics of various land cover types. While the seasonal trajectories indicate phenological evolution of vegetation, the disturbance trajectory is useful for change detection over time. Domains of mixed landscapes are identified as hatched areas; (b) Examples of NOAA/AVHRR derived seasonal trajectories of $NDVI - Ts$ for shrub, grass and conifer forests in the Columbia basin, and crop, deciduous forests in the Lake states area. The clear separation between various land cover types in the $NDVI - Ts$ space provides the logical basis for a global land cover classification and monitoring scheme.

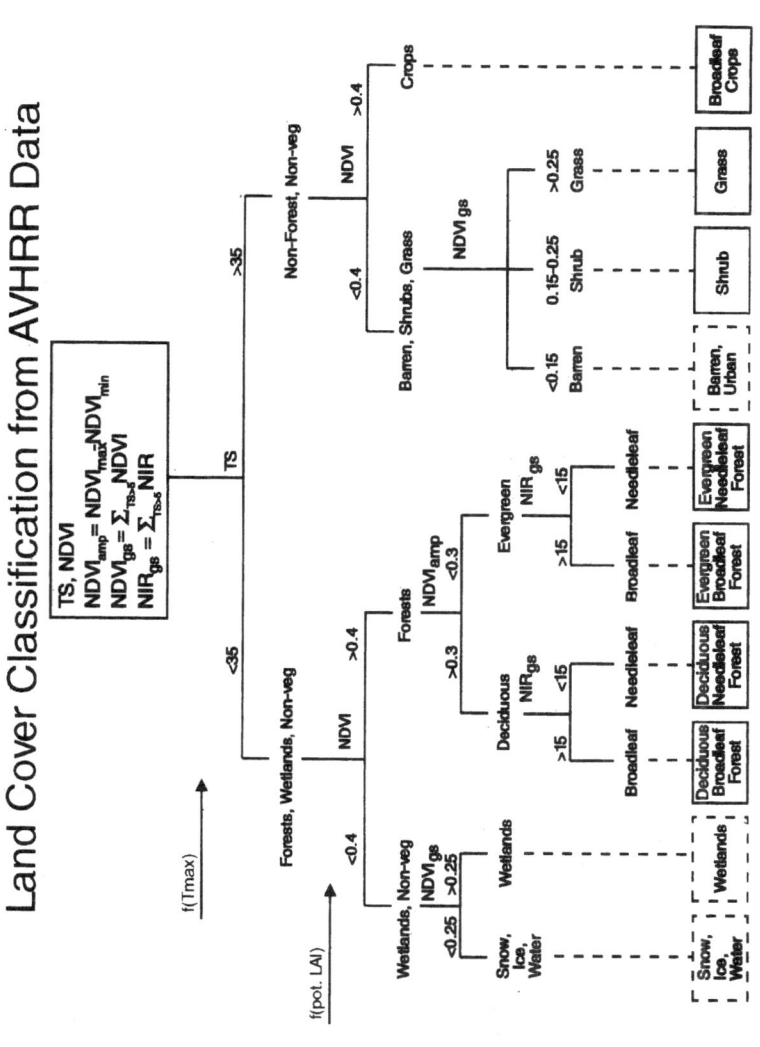

Fig. 5. (top) A flowchart of our land cover classification logic showing the required variables and thresholds. Initial separation into the four simple groups is accomplished by the seasonal trajectory of $Ts - NDVI$ observations. The six vegetation classes defined in Running *et al.* (1994a) are shown in bold. Climate is excluded from these definitions.

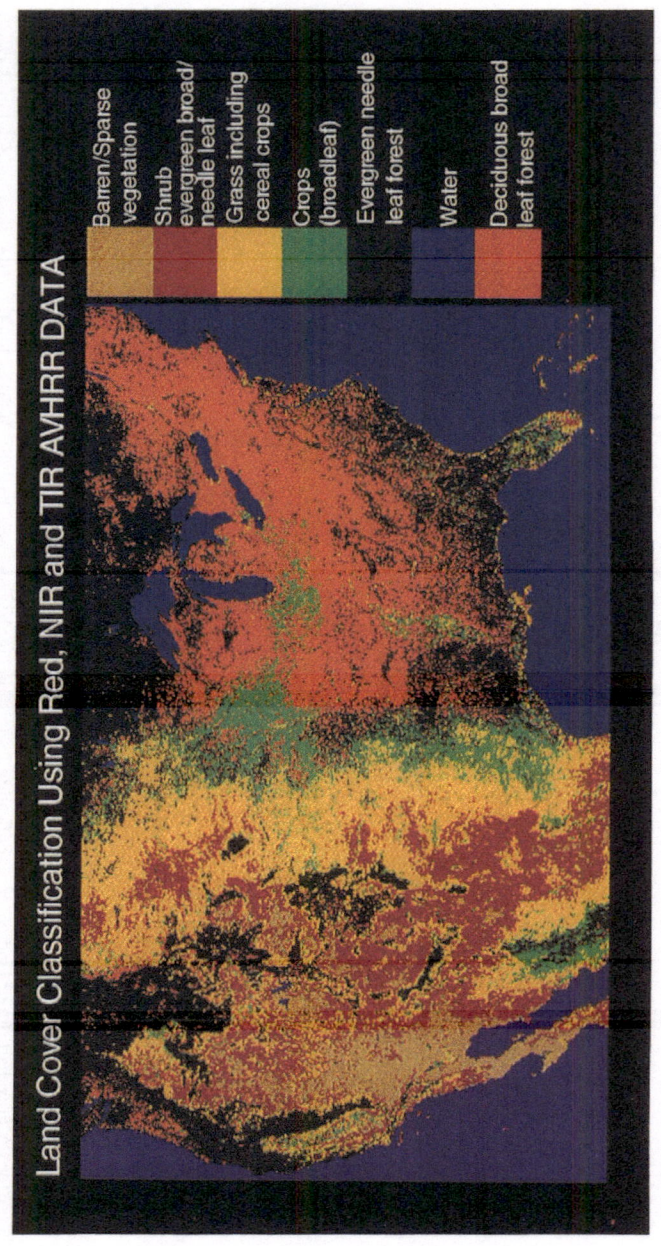

Fig. 5. (bottom) A map showing various land cover classes for the conterminous U.S. derived using only remotely sensed data and classification rules presented above.

References

Achard, F. and Blasco, F.: 1990, 'Analysis of Vegetation Seasonal Evolution and Mapping of Forest Cover in West Africa with the Use of NOAA AVHRR HRPT Data', *Photogram. Eng. Remote Sensing* **56**, 1359–1365.

Anthes, R. A.: 1984, 'Enhancement of Convective Precipitation by Mesoscale Variations in Vegetative Covering in Semi-Arid Regions', *J. Clim. Appl. Meteor.* **23**, 540–553.

Asrar, G., Kanemasu, E. T., and Yoshida, M.: 1985, 'Estimates of Leaf Area Index from Spectral Reflectance of Wheat under Different Cultural Practices and Solar Angle', *Remote Sensing Environm.* **17**, 1–11.

Asrar, G., Myneni, R. B., and Choudhury, B. J.: 1992, 'Spatial Heterogeneity in Vegetation Canopies and Remote Sensing of Absorbed Photosynthetically Active Radiation: A Modeling Study', *Remote Sensing Environm.* **41**, 85–101.

Balling, R. C.: 1988, 'The Climatic Impact of a Sonoran Vegetation Discontinuity', *Clim. Change* **13**, 99–109.

Cermak, V., Bodri, L., and Safanda, J.: 1992, 'Recent Climate Change Recorded in the Underground: Evidence from Cuba', *Paleogeog. Paleoecoclim. Paleoeco.* **98**, 219–223.

Choudhury, B. J.: 1991, 'Multispectral Satellite Data in the Context of Land Surface Heat Balance', *Rev. Geophys.* **29** (2), 217–236.

Chu, P-S., Yu, Z-P., and Hastenrath, S.: 1994, 'Detecting Climate Change Concurrent with Deforestation in the Amazon Basin: Which Way Has It Gone?', *Bull. Amer. Meteor. Soc.* **75**, 579–583.

Denmead, O. T.: 1969, 'Comparative Micrometeorology of a Wheat Field and a Forest of *Pinus radiata*', *Agcl. Meteor.* **6**, 357–371.

Dickinson, R. E. and Henderson-Sellers, A.: 1988, 'Modeling Tropical Deforestation: A Study of GCM Land-Surface Parameterization', *Q. J. R. Meteor. Soc.* **114**, 439–462.

Dickinson, R. E., Henderson-Sellers, A., Kennedy, P. J., and Wilson, M. R.: 1986, 'Biosphere-Atmosphere Transfer Scheme (BATS) for the NCAR Community Climate Model', Tech. Note TN-275+STR, Nat. Center for Atm. Research, Boulder, CO.

Eidenshink, J. C.: 1992, 'The 1990 Conterminous U.S. AVHRR Data Set', *Photogramm. Engineer. Remote Sensing* **58**, 809–813.

Flint, E. P. and Richards, J. F.: 1991, 'Historical Analysis of Changes in Land Use and Carbon Stock of Vegetation in South and Southeast Asia', *Can. J. For. Res.* **21**, 91–110.

Gash, J. and Shuttleworth, W.: 1991, 'Tropical Deforestation: Albedo and the Surface Energy Balance', *Clim. Change* **19**, 123–133.

Goward, S. N., Cruickshanks, G. D., and Hope, A. S.: 1985, 'Observed Relation between Thermal Emission and Reflected Spectral Radiance of a Complex Vegetated Landscape', *Remote Sensing Environ.* **18**, 137–146.

Goward, S. N., Dye, D. G., Turner, S., and Yang, J.: 1993, 'Objective Assessment of the NOAA Global Vegetation Index Data Product', *Int. J. Remote Sensing* **14**, 3365–3394.

Goward, S. N., Waring, R. H., Dye, D. G., and Yang, J.: 1994, 'Ecological Remote Sensing at OTTER: Satellite Macroscale Observations', *Ecol. Applic.* **4** (2), 322–343.

Green, G. M. and Sussman, R. W.: 1990, 'Deforestation History of the Eastern Rain Forests of Madagascar from Satellite Images', *Science* **248**, 212–215.

Haack, B., Bryant, N., and Adams, S.: 1987, 'An Assessment of Landsat MSS and TM Data for Urban and Near-Urban Land-Cover Digital Classification', *Remote Sensing Environm.* **21**, 201–203.

Henderson-Sellers, A., Dickinson, R. E., Burbidge, T., Kennedy, P., McGuffe, K., and Pitman, A.: 1993, 'Tropical Deforestation: Modeling Local to Regional Scale Climate Change', *J. Geophys. Res.* **98**, 7289–7315.

Houghton, R.: 1994, 'The World-Wide Extent of Land-Use Change', *Bioscience* **44**, 305–313.

Jarvis, P. G. and McNaughton, K. G.: 1986, 'Stomatal Control of Transpiration: Scaling Up from Leaf to Region', *Adv. Eco. Res.* **15**, 1–49.

Kummer, D. M. and Turner, B. L. II: 1994, 'The Human Causes of Deforestation in Southeast Asia', *Bioscience* **44**, 323–328.

Lean, J. and Warilow, D. A.: 1989, 'Simulation of the Regional Climatic Impact of Amazon Deforestation', *Nature* **342**, 411–413.

Leemans, R. and Cramer, W.: 1991, *The IIISA Climate Data Base for Land Areas on a Grid with 0.5 Degree Resolution*, WP-90-41, Int. Institute for Applied Systems Analysis, Laxenburg, Austria.

Li, B. and Avissar, R.: 1994, 'The Impact of Spatial Variability of Land Surface Characteristics on Land Surface Heat Fluxes', *J. Clim.* **7**, 527–537.

Loveland, T. R., Merchant, J. W., Ohlen, D. O., and Brown, J. F.: 1991, 'Development of a Land-Cover Characteristics Database for the Conterminous U.S.', *Photogramm. Engineer Remote Sensing* **57**, 1453–1463.

Malingreau, J. and Tucker, C.: 1988, 'Large Scale Deforestation in the Southeastern Amazon Basin of Brazil', *Ambio* **17**, 49–55.

Matthews, E.: 1983, 'Global Vegetation and Land Use: New High-Resolution Data Bases for Climate Studies', *J. Clim. Appl. Meteorol.* **22**, 474–487.

Myer, W. B. and Turner, B. L. (eds.): 1993, *Global Land-Use/Land-Cover Change: A Global Perspective*, Cambridge University Press, New York.

Myers, N.: 1988, 'Tropical Deforestation and Climate Change', *Environ. Cons.* **15**, 293–298.

Myneni, R., Asrar, G., and Gerstl, S.: 1990, 'Radiative Transfer in Three-Dimensional Leaf Canopies', *Transp. Theory Stat. Phys.* **19**, 205–250.

Neilson, R. P.: 1995, 'A Model for Predicting Continental Scale Vegetation Distribution and Water Balance', *Ecol. Appl.*, (in press).

Nemani, R., Pierce, L. L., Running, S. W., and Goward, S. N.: 1993, 'Developing Satellite Derived Estimates of Surface Moisture Status', *J. Appl. Meteorol.* **32**, 548–557.

Nemani, R. and Running, S. W.: 1989a, 'Testing a Theoretical Climate-Soil-Leaf Area Hydrologic Equilibrium of Forests Using Satellite Data and Ecosystem Simulation', *Agric. Forest Meteorol.* **44**, 245–260.

Nemani, R. and Running S. W.: 1989b, 'Estimation of Regional Surface Resistance to Evapotranspiration from NDVI and Thermal-IR AVHRR Data', *J. Appl. Met.* **28**, 276–284.

Olson, J. S., Watts, J. A., and Allison, L. J.: 1983, *Carbon in Life Vegetation of Major World Ecosystems*, TR004, U.S. Dept. of Energy, Washington, D.C.

Pielke, R. A. and Avissar, R.: 1990, 'Influence of Landscape Structure on Local and Regional Climate', *Landscape Ecol.* **4**, 133–155.

Pierce, L. L., Walker, J., Dowling, T., Mcvicar, T., Hatton, T., Running, S., and Coughlan, J.: 1993, 'Ecohydrological Changes in Murray-Darling Basin III. A Simulation of Regional Hydrological Changes', *J. Appl. Ecol.* **30**, 283–294.

Prentice, C., Cramer, W., Harrison, S., Leemans, R., Monserud, R., Solomon, R.: 1993, 'A Global Biome Model Based on Plant Physiology and Dominance, Soil Properties and Climate', *J. Biogeogr.* **19**, 117–134.

Price, J. C.: 1984, 'Land Surface Measurements from the Split Window Channels of NOAA-7 AVHRR', *J. Geophys. Res.* **89**, 7231–7237.

Price, J. C.: 1989, 'Using the Spatial Context, in Satellite Data to Infer Regional Scale Evapotranspiration', *IEEE Trans. Geosci. and Remote Sensing* **28** (5), 940–948.

Running, S. W. and Hunt, R. E. Jr.: 1993, 'Generalization of a Forest Ecosystem Process Model, for Other Biomes, BIOME BGC, and an Application for Global-Scale Models', in Ehleringer, J. R. and Field, C. (eds.), *Scaling Processes between Leaf and Landscape Levels*, Academic Press.

Running, S. W., Loveland, T., and Pierce, L. L.: 1994a, 'A Vegetation Classification Logic Based on Remote Sensing for Use in Global Biogeochemical Models', *Ambio* **23**, 77–81.

Running, S. W., Justice, C. O., Salomonson, V., Hall, D., Barker, J., Kaufmann, Y. J., Strahler, A. H., Huete, A. R., Muller, J-P., Vanderbilt, V., Teillet, P., and Carneggie, D.: 1994b, 'Terrestrial Remote Sensing Science and Algorithms Planned for EOS/MODIS', *Int. J. Remote Sensing* **15**, 3587–3620.

Schwartz, M. D. and Karl, T.: 1990, 'Spring Phenology: Nature's Experiment to Detect the Effect of "Green Up" on Surface Maximum Temperatures', *Mon. Wea. Rev.* **118**, 883–890.

Segal, M. and Aritt, R.: 1992, 'Nonclassical Mesoscale Circulations Caused by Surface Sensible Heat Flux Gradients', *Bull. Am. Met. Soc.* **73**, 1593–1604.

Sellers, P. J.: 1985, 'Canopy Reflectance, Photosynthesis and Transpiration', *Int. J. Remote Sensing* **6** (8), 1335–1372.

Sellers, P. J., Los, S., Tucker, C., Justice, C., Dazlich, D., Collatz, G., and Randall, D.: 1994, 'A Revised Land Surface Parameterization (SiB2) for Atmospheric GCMs. Part 2: The Generation of global Fields of Terrestrial Biophysical Parameters from Satellite Data', *Int. J. Remote Sensing* **15**, 3519–3546.

Shukla, J., Nobre, C., and Sellers, P.: 1990, 'Amazon Deforestation and Climate Change', *Science* **247**, 1322–1325.

Skole, D.: 1993, 'Data on Global Land Cover Change: Acquisition, Assessment and Analysis', in Myer, W. and Turner, B. L. (eds.), *Changes in Land Use and Land Cover: A Global Perspective*, Cambridge University Press, New York, pp. 437–471.

Skole, D. and Tucker, C. J.: 1993, 'Tropical Deforestation and Habitat Fragmentation in the Amazon: Satellite Data from 1978–1988', *Science* **260**, 1905–1910.

Smith, R. C. G. and Choudhury, B. J.: 1990, 'On the Correlation of Indices of Vegetation and Surface Temperature over South-Eastern Australia', *Int. J. Remote Sensing* **11**, 2113–2120.

Spanner, M. A., Pierce, L. L., Running, S. W., and Peterson, D. L.: 1990, 'The Seasonality of AVHRR Data of Temperate Coniferous Forests: Relation to Leaf Area Index', *Remote Sensing Environm.* **33**, 97–112.

Stanhill, G., Hofstede, G. J., and Kalma, J. D.: 1966, 'Radiation Balance of Natural and Agricultural Vegetation', *Q. J. R. Meteor. Soc.* **92**, 128–140.

Townshend, J. R. G., Justice, C. O., Li, W., Gurney, C., and McManus, J.: 1991, 'Global Land Cover Classification by Remote Sensing: Present Capabilities and Future Possibilities', *Remote Sensing Environm.* **35**, 243–255.

Tucker, C. J., Townshend, J. R. G., and Goff, T. E.: 1985, 'African Landcover Classification Using Satellite Data', *Science* **227**, 369–375.

Turner, D. P., Koerper, G., Gucinski, H., and Peterson, C.: 1993, 'Monitoring Global Change: Comparison of Forest Cover Estimates Using Remote Sensing and Inventory Approaches', *Environ. Monotor. Mng.* **26**, 295–305.

Wan, Z. and Dozier, J.: 1989, 'Land-Surface Temperature Measurement from Space: Physical Principles and Inverse Modeling', *IEEE Tran. Geosci. Remote Sensing* **27**, 268–278.

Woodward, F. I.: 1987, *Climate and Plant Distribution*, Cambridge University Press, Cambridge, U.K.

Zobler, L.: 1986, *A World Soil File for Global Climate Modeling*, NASA Tech. memo. 87802.

(Received 23 January, 1995; in revised form 19 June, 1995)

LONG-TERM OBSERVATIONS OF LAND SURFACE CHARACTERISTICS

CHESTER F. ROPELEWSKI

Climate Analysis Center, National Meteorological Center, Washington, D.C. 20233, U.S.A.

Abstract. There is a new awareness of the potential importance of land surface observations in climate studies. This paper suggests a three-pronged, (*in situ,* satellite, and numerical model-based) observation strategy for land surface observations in the Global Climate Observing System (GCOS). This strategy is based on the premise that the land surface's importance lies in its links to the atmosphere and, in particular, in its role in determining and modulating the fluxes of momentum, sensible and latent heat and in the surface radiation budget. Specific suggestions include: (a) harnessing the potential of the WMO's World Weather Watch to provide regular observations of key land surface properties i.e., soil moisture and soil temperature; (b) utilization of satellites to provide land surface 'reference' data sets and encourage development of stable satellite data sets for land surface monitoring; and (c) use of numerical-model-based data assimilation systems (DAS) to augment *in situ* and satellite observations.

1. Introduction

This brief paper attempts to review the current status of land surface characteristics observations and to identify deficiencie in the context of designing a Global Climate Observing System (GCOS). Such an observation system should support monitoring interannual climate variability as well as global change issues. Care has been taken to focus the discussion on observations of land surface characteristics that are directly related to climate and climate change and to specifically exclude those observations that do not directly contribute to monitoring climate on a global basis. Thus, the paper does not address land surface observations which may be critical for many micro-meteorological, micro-climatological, boundary layer and other studies. This paper also does not address the cryosphere which is being treated in a separate review (Walsh, 1995) nor does it specifically address land chemistry and soil composition.

The focus of the discussion is on climate monitoring of the land surface as it influences the fluxes of: (1) moisture; (2) heat; (3) momentum; and the (4) surface radiation budget.

Distinctions are made between land surface characteristics which are not expected to vary or are expected to vary on temporal scales of decades or longer, e.g., soil type, as contrasted with land surface characteristics which may exhibit substanital *seasonal to interannual variability* such as soil moisture.

Climatic Change **31**: 415–425, 1995.

2. Land Surface Characteristics

2.1. REFERENCE LAND CHARACTERISTIC DATA

General land surface characteristics, e.g., elevation, soil type and soil depth, are required to set realistic boundary conditions for numerical models and benchmark data for empirical studies. These characteristics are expected to be relatively stable for several years of even decades. For purposes of this paper these surface land characteristics are referred to as 'Reference' data. Because of their static nature or very slow variation it seems clear that most, if not all, of these characteristics (Table I) should be included in a GCOS reference data base but not be included in the general design of a land observations component for GCOS. The properties listed in Table I are only a sample of the total information potentially available. While we refer to some properties as 'static' land characteristics, it is crucial that GCOS have complete, timely, and accurate descriptions of any changes in these reference land surface characteristics at specific observationv sites. Because of the extreme heterogeneity of the land surface and its effects on the surface flux-es these reference metadata are essential for proper interpretation of land surface measurements. The set of reference land surface characteristics suggests the estab-lishment of a mechanism, within GCOS, for the documentation of *changes* in these so-called 'static' land characteristics but not continuous monitoring. Thus GCOS should forge links with the segments of the land science community responsible for monitoring and observations of these quantities and should look to documented sources of this information, e.g. existing Geographical Information Systems.

2.2. GCOS SCALE (SEASONAL/INTERANNUAL THORUGH MULTI-YEAR) DATA

This class of observations needs to be designed so it can support studies and monitoring efforts which attempt to differentiate climate 'change', from 'normal' climate variability, i.e. climate variability associated with seasonal through inter-annual fluctuations about an unchanging climate mean. The spectrum of climate variability in most measurements is fairly 'red' at frequencies from several seasons through several years. Thus, it is difficult to differentiate chance 'runs' of climate extremes and climate change. Still more difficult, is the unambiguous identification of changes, or trends, in higher order statistics, e.g., changes in the frequency of storms, rain, or drought. However, since society is extremely sensitive to changes in these higher order statistics and how they affect the land surface, GCOS needs to addresss the high frequency variability even though its goal is the study of climate scale variations.

TABLE I

Examples of 'reference' land surface characteristics

Station elevation
Slope or grade
Aspect
Land type (e.g. rock, soil, marsh)
Soil properties
Type
Soil depth
Composition
Porosity
Water holding capacity
Mean depth of the water table
Land use
Urban
Cultivated
Natural forest
Pasture
Vegetation
Types
Mean percent coverage
Rooting depth

3. Observational Requirements

For purposes of this paper, the land surface characteristics include land surface properties that change and/or modulate the climate-scale surface fluxes of heat, moisture, and momentum and the surface radiation budget. These include land properties that have an annual cycle, exhibit interannual variability and are generally included in detailed boundary layer, biosphere (e.g., Xue *et al.*, 1991) or land surface models (Johnson *et al.*, 1993; Henderson-Sellers *et al.*, 1995) listed in Table II. There are several valid and interesting scientific reasons to study and observe the micro-meteorology and micro-climate over specific small regions with spatial scales ranging from meters to tens of kilometers. However, because of the large spatial scales and long temporal scales inherent in climate, and GCOS, one faces the fundamental challenge of scaling (see e.g., Li and Avissar, 1994) the local measurements listed in Table II to a *global* observing system representative of *climate*.

Since land surface characteristics may vary down to extremely small scales e.g., less than a few meters, it seems unlikely that a global network of surface land

TABLE II

Land characteristics – physical quantities GCOS

Quantity	Primary measurement requirement	Supporting measurement requirement
Sfc momentum flux	Roughness	Changes in roughness, winds, stability
Sensible heat flux	Surface temperature soil temperature through the active soil column (1–2 m)	Winds, stability, shelter temperature
Sfc moisture flux	Surface moisture, soil moisture in the active soil column (1–2 m)	Winds, stability, shelter temperature and moisture
Latent heat flux	All of the above	All of the above
Surface radiation budget	Incoming, outgoing, net, shortwave and longwave radiation, surface albedo	clouds, humidity, aerosols, optical depth

observations will be sufficient to capture all of the heterogeneity inherent in land surface measurements. This suggests a three-pronged strategy for observations of the land surface for GCOS; (a) the establishment and maintenance of detailed land surface observations at a limited number of 'representative' operational sites; (b) the use of remotely-sensed (satellite) observations; and (c) the use of analyses from data assimilation systems of numerical models.

3.1. REPRESENTATIVE IN SITU OBSERVATIONS

The World Meteorological Organization (WMO) first order stations were established to provide real-time weather observations in support of operational meteorology. However, these 'weather' observations have also become invaluable for climate studies (e.g., Ropelewski et al., 1985). While many modeling studies (e.g., Milly and Dunne, 1994) have established the crucial role of soil properties to climate there is no global observational network for soil moisture and soil temperature. GCOS should encourage the expansion of the WMO global land observation network for in situ observations of surface land properties and components of the surface radiation budget. The WMO Guide to Agricultural Meteorological Practices (WMO, 1981) provides detailed guidelines for the observation of many land-surface properties including soil moisture and depth of groundwater. These observations are regularly taken at some sites but are not generally included in the international data exchange in the WMO's World Weather Watch (WWW). In principle, data from these sites could provide a source of land surface observations for GCOS.

A number of international field experiments (e.g., FIFE, HAPEX, FIRE) have been at the forefront of the activities to increase scientific understanding of land surface processes. One common thread of these field experiments was the goal of scaling local observations to spatial scales comparable to numerical models (tens of hundreds of kilometers). GCOS needs to develop a mechanism to maintain awareness of the findings from these field studies and to use them to improve and guide the GCOS land surface monitoring efforts. It is not clear whether the maintenance of a few highly instrumented land surface observation sites, similar to the CART/ARM (Cloud and Radiation Test Bed/Atmospheric Radiation Measurement) sites in the United States, should be part of GCOS or whether they should be part of some other, perhaps, international effort. However, the need and requirement for a small number of continuous highly instrumented land observation sites for climate could be argued strenuously and should be investigated. If 'representative' sites are identified they could serve as benchmark stations for numerical model and satellite studies as well as providing stable land-surface measurements for climate change studies.

3.2. SATELLITE OBSERVATIONS

Satellite observations have great potenital for land surface observations in GCOS because they can provide global coverage with fairly high temporal resolution using the same instruments e.g., the suite of Pathfinder Products (NOAA/NASA Pathfinder, 1994; Townshend *et al.*, 1994). However, there is a fairly wide gap between the required observations (either the surface fluxes or the quantities needed to derive them, see Table II) and measurements provided by operational satellites. The discussion here is limited to operational satellites since, almost by definition, research satellites will not provide the temporal continuity and long record required for GCOS. In addition, for climate studies we presumably do not wish to ignore the observations from the operational satellites over the past 20 years (NOAA/NASA, 1994). On the other hand, it would be difficult to ignore enhanced capabilities for satellite monitoring of the land surface that may result from programs such as NASA Earth Observing System (EOS). *Thus, GCOS will require careful intercomparisons between the satellite observations of the same or similar quantities taken from satellites that precede and follow EOS if satellites observations from that system are to be useful for climate monitoring.*

Operational satellite land surface observations are currently available from the Advanced Very High Resolution Radiometer (AVHRR) carried aboard the NOAA series of polar orbiting satellites since late in 1978 and from the Special Sensor Microwave/Imager (SSM/I) flown aboard Defense Meteorological Satellite Program (DMSP) satellites since 1987.

The AVHRR provides estimates of surface 'skin' temperature, (Tarpley, 1988) and of vegetation state (e.g., Goward *et al.*, 1991; Gutman, 1991; Townshend *et al.*, 1987) at spatial scales ranging from 1 km to 8 km as well as an average estimate on 2.5° × 2.5° latitude/longitude grids. The surface (skin) temperature has not been examined as an independent variable for climate change detection (even though it has been used in conjunction with simple models and surface observations to estimate monthly evapotranspiration, Tarpley, 1994). However, it has been used in studies to identify the urban heat island (Gallo *et al.*, 1993). The potential use of the satellite surface skin temperature as a climate monitoring tool needs to be investigated and, if found useful, included in the GCOS.

The AVHRR NDVI (Normalized Difference Vegetation Index), derived from the near-infrared and visible channels, is related to vegetation or 'greenness' and can be related to net primary productivity over some areas (Prince, 1991; Tucker *et al.*, 1986). A more detailed discussion of this index can be found in Asrar *et al.*, 1984; and Sellers, 1985. The NDVI has been successfully used in a number of climate related studies. Some examples include; the determination of land cover characteristics over the United States (Loveland *et al.*, 1991), estimation of vegetation properties for use in atmospheric Global Circulation Models (Los *et al.*, 1994; Sellers *et al.*, 1994; Gutman, 1994), investigation of interaction with atmospheric carbon-dioxide (Fung *et al.*, 1986) and for documentation of the mean annual cycle of vegetation (e.g., Schultz and Halpert, 1995). The NDVI has also been used to study interannual variability in Saharan desert boundaries (Tucker *et al.*, 1994) and for the study of droughts in North America (Kogan, 1995). In addition, the relatively high spatial resolution of these data i.e., as high as 1 km, can be used to monitor some sorts of land cover change caused by human disturbance e.g., in conjunction with studies of deforestation (Skole and Tucker, 1993).

However, the NDVI has been found to be very sensitive to atmospheric aerosols (Kogan *et al.*, 1994; Schultz and Halpert, 1995), to atmospheric water vapor, to residual clouds and to satellite viewing geometry as well as to sensor degradation and to drifts in the satellite orbit, e.g., Goward *et al.*, 1991. The orbital drift results in the local satellite observation time steadily moving away from local noon towards late afternoon. Changes in observation time are associated with variations in the vegetation index that are comparable to the interannual variability, severely limiting the usefulness of this index for the detection of climate change. Perhaps even a more severe restriction for the climate change detection problem is that the effects of stratospheric aerosols on the NDVI are also comparable to, or greater than, interannual variability e.g., the aerosols associated with the Mount Pinatubo eruption in 1991 produced the largest variations in the NDVI in the record. More complete discussions of these limitations and attempts to minimize their impact can be found e.g., in Goward *et al.*, 1994; and Eidenshink and Faundeen, 1994. In spite of these limitations for climate change detection, the NDVI and similar vegetation indices, may be useful in monitoring the length of the growing season,

season when vegetation is 'active', and their variations (Gallo and Flesch, 1989; Kogan, 1990; Schultz and Halpert, 1995).

The SSM/I has great potential for monitoring snow cover and perhaps snow depth and/or liquid water equivalent. (See the paper by Walsh, 1995 in this issue). The SSM/I also has potential for monitoring soil 'wetness' or standing water over usually dry land areas. Satellites, in general, and the SSM/I in particular appears to be limited to detecting moisture only on the surface (which may be land or vegetation or a mixture of both) or, at best, to depths of only a few centimeters (over bare land or sparsely vegetated areas). There is still considerable developmental work to be performed before these estimates of 'wetness' can be made useful for estimating of the surface flux quantities listed in Table II. The best use of the soil wetness observtions may be developed in conjunction with the numerical data assimilation schemes, discussed below, cross-validated with the *in-situ* measurements, discussed above.

The use of other satellite observations for land surface monitoring has yet to be developed. If GCOS is to be a dynamic system, it will have to develop a mechanism for monitoring advances in satellite observations and for including appropriate satellite data for climate monitoring as they become available.

3.3. MODEL DATA ASSIMILATION SYSTEMS (DAS)

Several weather modeling centers have initiated projects to reanalyze past data using a state of the art DAS to provide physically consistent gridded analyses e.g., Kalnay *et al.*, 1995, that may be useful for monitoring changes in atmospheric circulation features. These are discussed in more detail elsewhere in this issue, Trenberth (1995). The following brief discussion focuses on aspects of the reanalyses related to surface land characteristics. Several of the reanalyses will provide twice daily gridded atmospheric fields for ten or more years and at least one, the NMC reanalysis, will attempt to produce reanalyses back to 1958. These assimilation systems were originally designed to provide the best gridded atmospheric analyses possible for use in conjunction with operational numerical weather models for day-to-day operational weather forecasts. While these systems are mainly aimed at atmospheric analyses they also produce a suite of land surface analyses, e.g., Table III. It should be noted at the outset that most, if not all, of these quantities listed in Table III are derived from model-based parameterization schemes and thus are subject to the limitations of those schemes.

Given the limitations of the model land-surface parameterizations it is doubtful that the absolute values of these quantities will be adequate for climate monitoring and climate change detection. Nonetheless, year-to-year variations in some of the DAS-derived land surface quantities may be useful for studies of inter-annual variability through the use of differences. In addition the DAS-based land surface quantities may be useful in conjunction with station observations and satellite estimates to develop improved parameterizations and algorithms.

TABLE III

Typical land surface related output, e.g., from the NMC Reanalysis and
Climate Data Assimilation System (CDAS), Janowiak et al., 1994

Surface flux quantities	Related quantities
Zonal wind stress	Surface pressure
Meridional wind stress	Zonal wind
Sensible heat flux	Meridional wind
Latent heat flux	2 meter temperature
Upward and downward longwave	Specific humidity
Radiation flux	Soil moisture
Upward and downward shortwave	Snow depth
Radiation flux	Total precipitation (24 hr)
Heat flux into the ground	High, middle, low cloud amount

There is also a potential use of DAS in conjunction with regional, boundary
layer, and land surface models. The DAS analyses may provide the large scale
boundary and initial conditions for these smaller scale models in conjunction with
climate change studies that attempt to scale global change to the local level.

The potential usefulness of a DAS for climate monitoring in GCOS is enhanced
by the fact that some of the reanalysis projects will continue indefinitely into
the future (NOAA's Climate Data Assimilation System (CDAS) for certain and
perhaps also at the European Center for Medium-Range Weather Forecasting,
ECMWF. Thus by early in the next century there may be a sufficient length of
record in global gridded analyses to examine low frequency (multi-year to decadal)
variability in the atmospheric circulation and possible interactions with the land
surface.

4. Discussion

Global observations of the land surface characteristics present a daunting challenge
for GCOS. The heterogeneity inherent in the land surface calls to question the
representativeness of *in situ* observations at any particular location. The same issue
of heterogeneity calls into question the usefulness of areal averaged observations
from satellites or from model assimilation systems. However, the problems of
going from local observations to the larger scale and visa-versa shouldn't prevent
GCOS from including land surface observations. In fact, there are some studies
which suggest that land surface processes may scale more simply than initially
indicated (Robock et al., 1994). Given all of the uncertainties in the present state
of land surface observations for climate studies, for *in situ* observations GCOS
should:

1. Encourage the WMO:

    (a) To increase the number of locations where land surface characteristics (e.g., soil moisture, soil temperature) are regularly observed and develop a network of 'representative' observation sites for these data built on the existing WWW;

    (b) To provide for the regular exchange of *in situ* land surface observations;

2. Consider support for a small number of 'benchmark' land observation sites which are fully instrumented to measure all of the fluxes at the land/atmosphere interface.

For **satellite** data, GCOS needs to clearly define the climate related observational requirements. These include:

1. The production of stable, unbiased, estimates of the quantities directly measured by satellites that are directly related to the fluxes at the land/atmosphere interface. In addition to the fluxes themselves these could be satellite observations of the quantities listed in Table II;

2. Land surface observations from different satellite instruments and different satellites must be inter-compared through analyses of data from overlapping periods e.g., as was done for the analysis of the Microwave Sounding Unit (MSU) mean tropospheric temperature data (Spencer *et al.*, 1990);

3. Until the satellite 'drift' problem is properly addressed the use of satellites for climate studies will be extremely limited. Operational satellites should be made to operate in stable orbits;

4. Routine satellite observations of the albedo and components of the surface radiation budget are required.

In terms of **gridded global** land surface data from the DAS of reanalyses and real time monitoring systems GCOS should:

1. encourage the archival of these data;

2. actively pursue the use of these data for analysis of climate variability;

No doubt many readers will be disappointed that this paper does not include lists of land surface parameters with accompanying desired accuracies, precision and spatial and temporal scales. At times such specifications become counterproductive in that they are focused on only a few narrow aspects of the problem, on the one hand, or they become so encyclopedic that they lose any practical usefulness on the other. A good starting place, however, may be twice daily observations, on atmospheric GCM spatial scales, of the quantities needed to compute the fluxes listed in Table II.

In terms of global monitoring for **climate** the status of land surface observations is not as mature as for other components of the climate system. The broad outline presented here is meant to stimulate some deeper discussion and thought on developing a comprehensive monitoring system for the land surface and to simulate the expansion of the few observations that are now being made.

Acknowledgements

This brief review of land surface characteristics for GCOS is made from the very limited perspective of an atmospheric scientist. I claim no special insight nor expertise this area and apologize in advance for any glaring omissions. To the extent that this review may be useful it has benefited from discussions and input from a number of people. These include Drs. John Townshend, Piers Sellers, Anne Henderson-Sellers, Dan Tarpley, Garick Gutman, Kevin Gallo, and Ken Mitchell. My thanks to them all.

References

Asrar, G., Fuchs, M., Kanemasu, E. T., and Hatfield, J. L.: 1984, 'Estimating Absorbed Photosynthetic Radiation and Leaf Area Index from Spectral Reflectance in Wheat', *Agronomy J.* **76**, 300–306.
Eidenshink, J. C. and Faundeen, J. L.: 1994, 'The 1 km AVHRR Global Land Data Set: First Stages in Implementation', *Int. J. Remote Sens.* **15**, 3443–3462.
Fung, I. Y., Tucker, C. J., and Prentice, K. C.: 1986, 'On the Applicability of the AVHRR Vegetation Index to Study the Atmospheric-Biosphere Exchange of CO$_2$', *J. Geophys. Res.* **92**, 2999–3015.
Gallo, K. P., McNab, A. L., Karl, T. R., Brown, J. F., Hood, J. J., and Tarpley, J. D.: 1993, 'The Use of a Vegetation Index for Assessment of the Urban Heat Island Effect', *Int. J. Remote Sens.* **14**, 2223–2230.
Gallo, K. P. and Flesch, T. K.: 1993, 'Large-Area Crop Monitoring the NOAA AVHRR: Estimating the Silking Stage of Corn Development', *Remote Sens. Environm.* **27**, 73–80.
Goward, S. N., Turner, S., Dye, D. G., and Liang, S.: 1994, 'The University of Maryland Improved Global Vegetation Index Product', *Int. J. Remote Sens.* **15**, 3365–3395.
Goward, S. N., Dye, D. G., Turner, S., and Yang, J.: 1991, 'Objective Assessment of the NOAA Global Vegetation Index Data Product', *Int. J. Remote Sens.* **13**, xxx–xxxx.
Gutman, G.: 1994, 'Global Data on Land Surface Parameters from NOAA AVHRR for Use in Numerical Climate Models', *J. Clim.* **7**, 669–680.
Gutman, G.: 1991, 'Vegetation Indices from AVHRR: An Update and Future Prospects', *Remote Sens. Env.* **35**, 121–138.
Henderson-Sellers, A., Pitman, A. J., Love, P. K., Irannejad, P., and Chen, T. H.: 1995, 'The Project for Intercomparison of Land Surface Parameterization Schemes (PILPS): Phases 2 and 3', *Bull. Amer. Met. Soc.* **76**, 489–503.
Janowiak, J. E., Ebisuzaki, W., Chelliah, M., Saha, S., Kistler, R., Kanamitsu, M., and White, G.: 1994, 'Reanalysis Archives', *Proc. of the 10th AMS Conf. on Numerical Weather Prediction*, Portland Oregon, July 18–22, 1994.
Johnson, K. D., Entekhabi, D., and Eagleson, P. S.: 1993, 'The Implementation and Validation of Improved Land-Surface Hydrology in an Atmospheric General Circulation Model', *J. Clim.* **6**, 1009–1026.
Kalnay, E., Kanamitsu, M., Kistler, R., Collins, W., Deaven, D., Gandin, L., Irdell, M., Saha, S., White, G., Woollen, J., Zhu, Y., Chelliah, M., Ebisuzaki, W., Higgins, W., Janowiak, J., Mo, K. C., Ropelewski, C., Wang, J., Leetmaa, A., Reynolds, R., Jenne, R., and Dennis Joseph: 1995, 'The NMC/NCAR 40-year Reanalysis Project', *Bull. Amer. Met. Soc.* **76**, (in press).
Kogan, F.: 1990, 'Remote Sensing of Weather Impacts on Vegetation in Non-Homogeneous Areas', *Int. J. Remote Sens.* **11**, 1405–1419.
Kogan, F. N.: 1995, 'Droughts of the Late 1980's in the United States as Derived from the NOAA Polar-Orbiting Satellite Data', *Bull. Amer. Met. Soc.* **76**, 653–667.
Kogan, F., Sullivan, J., Carey, R., and Tarpley, D.: 1994, 'Post-Pinatubo Vegetation Index in Central Africa', *Geocarto Internat.* **3**, 63–66.
Li, B. and Avissar, R.: 1994, 'The Impact of Spatial Variability of Land-Surface Characteristics on Land-Surface Heat Fluxes', *J. Clim.* **7**, 527–537.

Los, S., Justice, C. O., and Tucker, C. J.: 1994, 'A Global 1 Degree by 1 Degree NDVI Data Set for Climate Studies. Part 1', *Int. J. Remote Sens.* **15**, 3493–3518.

Loveland, T. R. J., Merchant, J. W., Ohlen, D. O., and Brown, J. F.: 1991, 'Development of Land-Cover Data Characteristics Database for the Coterminous U.S.', *Photogramm. Engineer. Remote Sens.* **12**, 1313–1330.

Milly, P. C. D. and Dunne, K. A.: 1994, 'Sensitivity of the Global Water Cycle to the Water-Holding Capacity of Land', *J. Clim.* **7**, 506–526.

NOAA/NASA: 1994, *The NOAA-NASA Pathfinder Program*, published by the University Corporation for Atmospheric Research (UCAR), 22 pp.

Prince, S. D.: 1991, 'A Model of Regional Primary Production for Use with Coarse-Resolution Satellite Data', *Int. J. Remote Sens.* **12**, 1313–1330.

Robock, A., Vinnikov, K. Ya., Schlosser, D. A., Speranskaya, N. A., and Xue, Y.: 1994, 'Use of Mid-latitude Soil Moisture and Meteorological Observations to Validate Soil MOisture Simulations with Biosphere and Bucket Models', *J. Clim.* **8**, 15–35.

Ropelewski, C. F., Janowiak, J. E., and Halpert, M. S.: 1985, " he Analysis and Display of Real Time Surface Climate Data', *Mon. Wea. Rev.* **113**, 1101–1106.

Schultz, P. A., Halpert, M. S.: 1995, 'Global Analysis of the Relationships among a Vegetation Index, Precipitation, and Land Surface Temperature', *Int. J. Remote Sens.* **16**, (in press).

Sellers, P. J., Los, S. O., Tucker, C. J., Justice, C. O., Collatz, G. J., and Randall, D. A.: 1994, 'A Global 1 Degree by 1 Degree NDVI Data Set for Climate Studies. Part 2: The Adjustment of the NDVI and Generation of Global Fields of Terrestrial Biophysical Parameters', *Int. J. Remote Sens.* **15**, 3519–3546.

Sellers, P. J.: 1985, 'Canopy Reflectance, Photosynthesis, and Transpiration', *Int. J. Remote Sens.* **6**, 1335–1371.

Skole, D. L. and Tucker, C. J.: 1993, 'Tropical Deforestation and Habitat Fragmentation', *Science* **260**, 1905–1910.

Spencer, R., Christy, J., and Grody, N.: 1990, 'Global Atmospheric Temperature Monitoring with Satellite Microwve Instruments: Methods and Results', *J. Clim.* **3**, 111–128.

Tarpley, J. D.: 1994, 'Monthly Evapotranspiration from Satellite and Conventional Meteorological Observations', *J. Clim.* **7**, 704–713.

Tarpley, J. D.: 1991, 'The NOAA Global Vegetation Index Product, A Review', *Paleogeogr. Paleoclimatol. Paleoecol. (Global and Planetary Change Section)* **90**, 189–194.

Tarpley, J. D.: 1988, 'Some Climatological Aspects of Satellite-Observed Surface Heating in Kansas', *J. Appl. Meteor.* **27**, 20–29.

Townshend, J. R., James, M. E., Liang, S., Goward, S. N.: 1994, 'A Long Term Data Set for Global Terrestrial Observations', NOAA/NASA Pathfinder Land Surface Working Group Report NASA, (in press).

Townshend, J. R., Justice, C. O., and Kalb, V.: 1987, 'Characterization and Classification of South American Land Cover Types Using Satellite Data', *Int. J. Remote Sens.* **8**, 1189–1207.

Tucker, C. J., Newcomb, W. W., and Drenge, H. E.: 1994, 'Satellite Determination of Desert Spatial Extent', *Int. J. Remote Sens.* **15**, 3547–3566.

Tucker, C. J., Newcomb, W. W., and Prince, S. D.: 1986, 'Monitoring the Grasslands of the Sahel 1984–1985', *Int. J. Remote Sens.* **7**, 1171–1581.

Walsh, J.: 1995, 'Long-Term Observations for Monitoring of the Cryosphere', *Clim. Change* **31**, (this issue).

WMO: 1981, *Guide to Agricultural Meteorological Practices*, WMO – No. 134, (Available from the World Meteorological Organization, Case postale No 2300, 1211 Geneva, Switzerland).

Xue, Y., Sellers, P. J., Kinter, J. L., and Shukla, J.: 1991, 'A Simplified Biosphere Model for Global Climate Studies', *J. Clim.* **4**, 345–364.

(Received 23 January, 1995; in revised form 7 July, 1995)

ATMOSPHERIC CIRCULATION CLIMATE CHANGES

KEVIN E. TRENBERTH

National Center for Atmospheric Research*, P.O. Box 3000, Boulder, CO 80307-3000, U.S.A.

Abstract. The role of the atmospheric circulation in climate change is examined. A review is given of the information available in the past record on the atmosheric circulation and its role in climate change, firstly at the surface via sea level pressure in both the northern and southern hemispheres and secondly for the free atmosphere. As with most climate information, the climate record is compromised by non-physical inhomogeneities arising from changes in observing and analyzing techniques and changes in data coverage. Problems with and threats to the rawinsonde network are discussed. Global analyses produced by the operational centers, U.S. National Meteorological Center (NMC) and the European Centre for Medium Range Weather Forecasts (ECMWF), for weather forecasting purposes contain many discontinuous changes in the analyses arising from improvements in the system used to produce them. A discussion is given of the prospects for and motivation behind an activity known as 'reanalysis' in which the historical data are reanalyzed using a state-of-the-art system that is held constant for the entire record. The only sources of spurious change then are the changes in the observing system, such as the introduction of space-based observations. Recommendations are made on needed actions for better understanding and monitoring climate change.

The role of the atmospheric circulation and the strong links to other variables such as temperature, precipitation and wind are established and illustrated with a survey of decadal variability, the evidence for it, and the way in which the observed atmospheric circulation is involved in the Pacific and Atlantic sectors. The importance of teleconnections is stressed, especially in the winter half year, for understanding local climate change. The likelihood that changes will be manifested in the frequency and intensity of preferred modes of behavior in the atmosphere, such as the El Niño-Southern Oscillation and Pacific-North American teleconnection patterns, rather than in changes in the modes is also emphasized. The recently observed climate changes and the tendency for an unprecedented prolonged El Niño are interpreted in this framework. The key coupled atmosphere-ocean character of decadal variability is noted with the atmosphere providing the spatial scales, the ocean the memory, but also with the need for collaborative, as opposed to destructive, interactions through the atmospheric circulation.

1. Introduction

Changes in climate, whether from natural variability or anthropogenic causes, and whether interannual or decadal in time scale have various manifestations in different climate variables. Because humans live in and breath the atmosphere it is natural to focus on the atmosheric changes. But for the atmosphere, we need to recognize that phenomena and events are loosely divided into the realms of 'weather' and 'climate'.

The atmospheric part of the climate system varies rapidly as weather systems develop, evolve, mature and decay as turbulent instabilities in the flow and the atmosphere is subject to the phenomenon known as chaos. We normally think of

---

* The National Center for Atmospheric Research is sponsored by the National Science Foundation.

Climatic Change **31**: 427–453, 1995.

the very large fluctuations in the atmosphere from hour-to-hour or day-to-day as weather. The changes occur primarily because of the passage of weather systems and the development of various weather phenomena. These arise mainly from several atmospheric instabilities and, as such, they are not predictable in any deterministic sense beyond a week or two.

Climate is usually defined to be average weather, although it is essential to realize that 'average' refers to the averages of all statistics. Thus, not only means but also measures of variability and sequences as well as covariability must be included in any description of climate. It is also important to realize that climate deals with variations in which the atmosphere is influenced by and interacts with other parts of the climate system as well as the 'external' forcings. The internal interactive components in the climate system include the atmosphere, the oceans, sea ice, the land and its features (including the vegetation, albedo, biomass, and ecosystems), snow cover, land ice (including the semipermanent ice caps of Antarctica and Greenland and glaciers), and hydrology (including rivers, lakes and surface and subsurface water).

The climate change anticipated in association with the observed increases in greenhouse gases, in particular carbon dioxide, is widely referred to as 'global warming'. But the potential exists for complex climate change involving all aspects of the climate system and global warming is just one aspect. Changes in the atmospheric circulation are potentially very important even if they do not have as direct an impact on human endeavors, because the atmospheric circulation forms the main link between regional changes in wind, temperatures, and precipitation in the atmosphere and other climatic variables such as ocean currents and sea surface temperatures (SSTs) through changes in surface fluxes of heat, moisture and momentum. While this statement is strictly true only at a given time, there is likely to be a reasonably strong relationship between these variables even on monthly or longer time scales. Moreover, internal consistency among analyzed changes in the variables can add substantial confidence to results and provides the physical setting for understanding the changes taking place. Indeed, a strong case can be made that local climate change can only be understood if the changes in the atmospheric circulation are fully factored in. Abrupt climate change at one site, such as the location of an ice core used for assessing paleoclimate, probably arose mainly because of an atmospheric circulation change. While measuring changes in circulation is probably less important for detecting climate change, understanding those changes is essential for attributing the climate change to particular causes.

In this paper, we begin by briefly surveying the information available in the past record on the atmospheric circulation and its role in climate change, firstly at the surface and secondly for the free atmosphere. As with most climate information, the climate record is compromised by non-physical inhomogeneities arising from changes in observing and analyzing techniques and changes in data coverage. In particular, global analyses produced by operational centers such as the U.S. National Meteorological Center (NMC) and the European Centre for Medium

MAJOR ANALYSIS/NWP MODEL CHANGES AT ECMWF

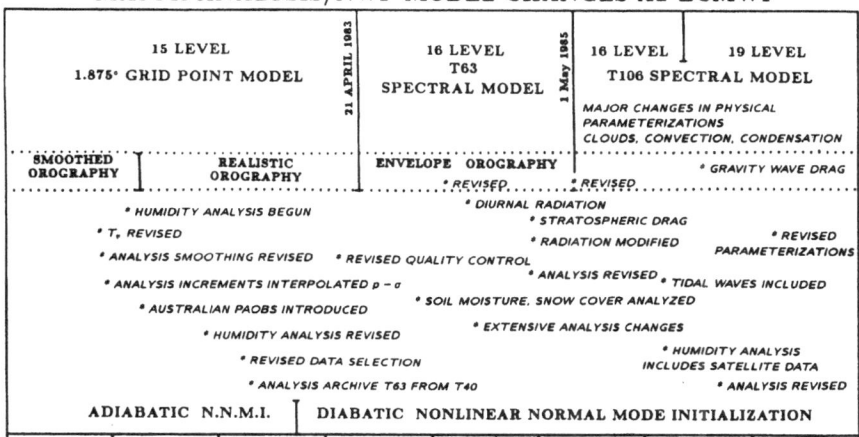

Fig. 1. Time sequence of changes in the ECMWF four dimensional data assimilation system from 1979 to 1987. From Trenberth and Olson (1988).

Range Weather Forecasts (ECMWF) for weather forecasting purposes contain many discontinuous changes in the analyses arising from improvements in the system used to produce them, see Figure 1 (from Trenberth and Olson, 1988). As a result, an activity known as *reanalysis* has grown up at the prompting of the Tropical Oceans Global Atmosphere (TOGA) Programme (see Bengtsson and Shukla, 1988; Trenberth and Olson, 1988) in which the historical data are reanalyzed using a state-of-the-art system that is held constant for the entire record. The only sources of spurious change then are the changes in the observing system, such as the introduction of space-based observations. A short review is given of the reanalysis activities underway. We then briefly discuss decadal variability, the evidence for it, and the way in which the atmospheric circulation is involved in the Pacific and Atlantic sectors. Finally, the roles of the atmospheric circulation and the ocean in climate change are considered.

2. Sea Level Pressure

Observations of the atmospheric circulation have long been recognized as important. Taking advantage of the geostrophic relation, the primary proxy for the atmospheric flow at the surface has been maps of sea level pressure, and thus these have historically been constructed from combined observations of wind and pressure. In fact the sea level pressure maps have always served as a vehicle for integrating all the surface observations, and information on temperature, dew point, cloud cover amount, type and height, and current and past weather, as well as wind speed and direction, sea level pressure and pressure tendency, is plotted. Careful attention is given to analysis of frontal systems, which are marked on the maps, and so the sea level pressure map has historically been a synthesis map of conditions throughout the troposphere, long before upper air observations were available.

The first charts of the global distribution of sea level pressure were published by Buchan (1869) and, in retrospect, were require remarkable in their fidelity. The earliest circulation maps are those of Lamb and Johnson (1966) in which they have reconstructed monthly means based on instrumental data for January and July back to 1750. The domain covered was just a very small part of Europe in the early years, and this gradually expanded to encompass parts of North and South America, the Atlantic and Asia by the late 1800s. The relationships between monthly mean temperatures and precipitation are exploited in more extensive reconstructions of sea level pressures by Jones *et al.* (1983, 1986).

In the United States, daily sea level pressure maps of the region north of 20° N exist since 1899, and monthly means of these from NCAR have been evaluated extensively by Trenberth and Paolino (1980). The United Kingdom Meteorological Office (UKMO) has an alternative version of monthly mean maps dating from 1873 north of 15° N (see Jones, 1987). No analyses are available, however, across the Pacific Ocean between 1881 and 1898, and the UKMO and NCAR versions are essentially the same from 1899–1939 (Williams and van Loon, 1976).

Changes in coverage of observations have impacted some areas and changes in procedures, especially corrections from surface to sea level pressures in regions of high topography, have also had a substantial impact on the homogeneity of the record. Problems are especially apparent over the Arctic region before about 1931 where analysts widely assumed the existence of a high pressure system that was not in fact present (Madden, 1976; Trenberth and Paolino, 1980; Jones, 1987) and Jones has devised corrections for these analyses. For the NCAR dataset, Trenberth and Paolino (1980) devised a number of corrections over the Himalayan-Tibetan Plateau complex that coincided with known changes in procedures to adjust for the main discontinuities. Detailed evaluations including comparisons with many station records led to the conclusion that the analyses were most reliable only after about 1924. Nevertheless, even in recent times there are substantial differences in analyses from different Centers in areas of high topography owing to differing procedures for dealing with the regions below ground (e.g., Trenberth *et al.*, 1993).

A number of regional pressure analyses have been performed in the Southern Hemisphere (SH) but the first extensive series of daily hemispheric maps were prepared leading up to and following the International Geophysical Year (IGY) (1957–58) when there was a major expansion in the observing network. A series of these maps were published by the South African Weather Bureau in *Notos* for 1951–62, and these formed the basis for many studies that culminated in the papers published in the *Southern Hemisphere Monograph*, see especially Taljaard (1972) and van Loon (1972). More recently, gridded hemispheric sea level pressure maps have been made available by the World Meteorological Centre in Melbourne beginning in 1972, and these have been evaluated and used by Trenberth (1979, 1984) and Swanson and Trenberth (1981). Further evaluation and extensions of both sets of analyses has been carried out by Jones and Wigley (1988) for the Antarctic region and Jones (1991) for the region south of 15° S.

In the SH spatial data gaps are a substantial problem in producing reliable analyses as the frequency of ship reports over the southern oceans is usually too few to allow reliable analyses to be constructed. In earlier years in the summer half year, whaling and sealing fishing vessels improved the data coverage. More recently, following the very positive experience during the Global Weather Experiment in 1979, drifting buoys have been used over the southern oceans to provide a basic network of surface observations.

3. Upper Air Observations

Upper level rawinsonde observations began in the 1940s but were not established in many places until about the IGY. The rawinsonde, as an expendable instrument, has to be manufactured cheaply, and it has served weather forecasting well. But it was not designed or operated in a way that guarantees a homogeneous climate record. Histories of station changes, which sondes were used, their calibration and characteristics, and the impacts of changes has been difficult to come by. But changes and shortcomings in the instrumentation are rife. Moreover, the use of different sondes by different countries creates artificial gradients in measurements.

Attempts to improve our knowledge of the station histories of rawinsondes have been made by Elliot and Gaffen (1991) and Parker and Cox (1995). Along with a number of other recent studies (Schwartz and Doswell, 1991; Garand *et al.*, 1992; Elliot and Gaffen, 1993; Gaffen, 1994) they have highlighted a number of problems with the rawinsonde record. These problems have been especially pronounced in the moisture measurements (Elliot and Gaffen, 1991, 1993), for instance in the U.S. as seen in the change from lithium chloride to carbon humidity element sensors in 1965, and in the needs for solar radiation corrections. A new housing for the instrument introduced in 1972 (adopted in 1973) removed a bias that had been introduced in 1965. A more complete listing of changes is given in Table I (adapted and updated from Elliot and Gaffen, 1991).

Note one major change in Table I affects all analyses of geopotential height. Geopotential height Z is defined as

$$g_c Z = \int_0^z g dz$$

where g_c is a constant value of gravity, and up until about 1990 this was taken as $g_c = 9.8$ m s^{-2}, but in recent years* there has been a transition to $g_c = 9.80665$

---

* Different countries have implemented the changeover at different times, in the United States the change was implemented on 1 October, 1993.

TABLE I

Chronology of changes in U.S. radiosonde

Date	Change
1943	Lithium chloride humidity element replaced hair hygrometer
1943	Dark ceramic resistance sensor replaced glass-tube electrolytic temperature element
1948	Relative humidity computed using saturation relative to water instead of ice
1948	Change observing times from 2300, 1100 UTC to 0300, 1500 UTC
1949	Smaller temperature sensor to reduce response time
1950	Correction for solar radiation introduced (until 1960)
1957	Change observing times from 0300, 1500 UTC to 0000, 1200 UTC
1960	Introduced white-coated temperature elements
1965	Carbon humidity element, began reporting low humidities
1972	Redesigned humidity ducts to reduce solar effects
1973	Stopped reporting relative humidity below 20%
1980	New carbon hygristors, new relative humidity transfer equation
1988	Precalibrated hygristor replaced type requiring preflight calibration
1988	New VIZ sonde with new humidity duct
late-	New Space Data Division (SDD) manufactured radiosonde at some stations.
1980's	Differences between VIZ and SDD noted
1993	Relative humidity to be reported over broader range, values $<20\%$ and up to 98% (instead of 95%) in cloud
1993	Gravitational constant used to define geopotential height from geopotential changed from 9.8 to 9.80665 m s^{-2}

m s^{-2}. The latter is the standard value of gravity at $45°$ as used by the World Meteorological Organisation to calibrate barometers. To add to the confusion, this corresponds to an older value and the actual value at $45°$ is now recognized to be 9.80616 m s^{-2}. The differences from these factors are especially significant in the stratosphere.

A substantial problem in the SH has been missing data. Over Antarctica missing data often arise from the hostile environment: the difficulty in successfully releasing balloons and their premature bursting (Trenberth and Olson, 1989). In addition, communications of Antarctic data are often disrupted by auroral phenomena, leading to loss of data entering daily operational analyses. In other continents (Africa and South America), upper air data are often missing for reasons related to cost and availability of expendables. Consequently, monthly mean station data can be corrupted by missing data and it is not certain how much this has affected the reconstructions of Jones and Wigley (1988), for example.

The upper air rawinsonde observing network reached a peak in soundings during the Global Weather Experiment and the number of soundings has since declined by about a third, mostly because of a reduction from two to one sounding per day. A

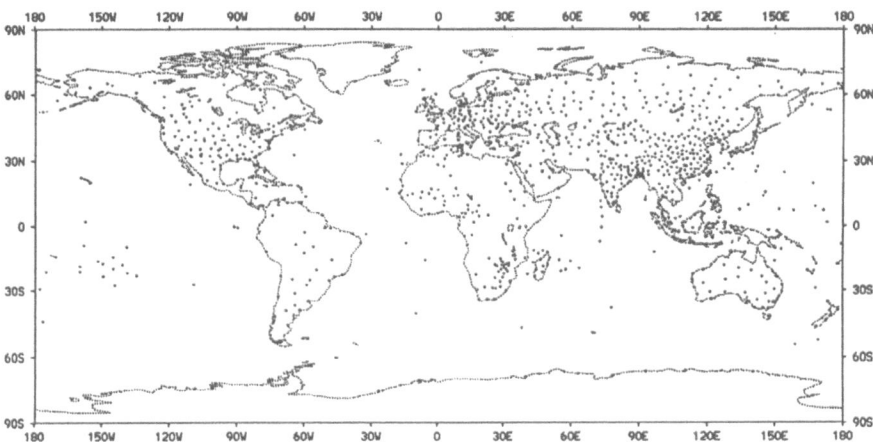

Fig. 2. Distribution of rawinsonde upper air stations reporting for 18 April 1992. (Courtesy D. Shea).

key issue is the cost of expendables. The current network is illustrated in Figure 2. However, the quality of the rawinsondes generally used has improved, although there is scope for further improvements in humidity measurements. Moreover, the rawinsonde network is under severe threat because the Omega and Loran-C navigation systems, on which most rawinsonde systems depend, are being phased out (possibly by 1996) and replaced with the Global Positioning System (GPS) utilizing GPS satellites. No commercially available GPS rawinsondes are yet available and those being tested are expensive. As a consequence, there is a need for the development of low-cost but high-quality expendable sondes that utilize GPS for obtaining winds.

4. Global Analyses

Global analyses are produced using a four-dimensional data assimilation (4DDA) system in which multivariate observed data are combined with the 'first guess' using a statistically optimum scheme. The first guess is the best estimate of the current state of the atmosphere from previous analyses produced using a numerical weather prediction (NWP) model.

The global atmospheric analyses produced as a result of 4DDA operationally consist of global fields of northward and eastward wind components (u, v), geopotential height (z), temperature (T), and relative humidity (RH) or, equivalently, specific humidity (q) each of which are a function of pressure (p). ω (vertical p-velocity) fields are produced diagnostically from the equation of continuity. In recent times, these quantities have been analyzed on the levels of the NWP model used in the 4DDA to provide the first guess for the analyses. Generally, these are σ levels where $\sigma = p/p_s$, and p_s is the surface pressure defined on the model surface topography. Alternatively, a hybrid between σ and pressure coordinates is used which typically reverts to constant pressure above about 100 mb. Analyzed fields on standard constant pressure levels are produced by interpolation. In practice, the changes in the analyses from one synoptic observation time to the next are interpolated to update the standard pressure level fields (after November, 1984 at ECMWF), though the details as to how this has been done have changed with time. Once the fields have been analyzed, they are typically initialized to bring the mass and temperature fields into a dynamical balance with the velocity fields consistent with the predominant low frequency motions in the atmosphere. Recently, with introduction of three dimensional variational techniques (called spectral statistical interpolation (SSI) and introduced 25 June, 1991 at NMC), the initialization step has not been needed.

For many years global analyses were not practicable, partly because the observing systems were not adequate, and partly because the 4DDA systems had not yet been developed to properly utilize the information in observations of different kinds, in particular asynoptic data. The intensive Global Weather Experiment was another major effort, in addition to the IGY, to advance the observing systems and it was only then, in 1979, that global analyses became truly viable. By that time, satellite soundings and miscellaneous other data were also incorporated into the analyses. Since the Global Weather Experiment, the in situ observing network has declined, but advances in 4DDA systems and possibly satellite data have made up for the losses so that the quality of analyses and NWP forecasts have continued to improve (e.g., Trenberth and Olson, 1988; Bengtsson and Shukla, 1988).

It must also be emphasized that the operational analyses are performed under time constraints for weather forecasting purposes and not for climate purposes. Changes in the NWP model, data handling techniques, initialization, and so on, (Figure 1) which are implemented to improve the weather forecasts, may disrupt the continuity of the analyses (Trenberth and Olson, 1988; Trenberth, 1992). The result is spurious trends in the data that can lead to false interpretations as to what the anomaly fields are. The main quantities affected are the divergent wind component and associated vertical motion fields, and the moisture fields. Time series plots reveal the main characteristics of the data. Figures 3, 4 and 5 show some plots from the ECMWF analyses (Trenberth, 1992) for certain variables, and the main changes revealed here are all spurious and can be identified with the dates noted in Figure 1.

Another factor in utilizing operational analyses for climate purposes is that some aspects, such as detailed analyses of the conditions at the surface of the earth, may be of less importance for weather forecasting while of great importance for climate and diagnostic studies. Analyses are not made at the true surface of the earth.

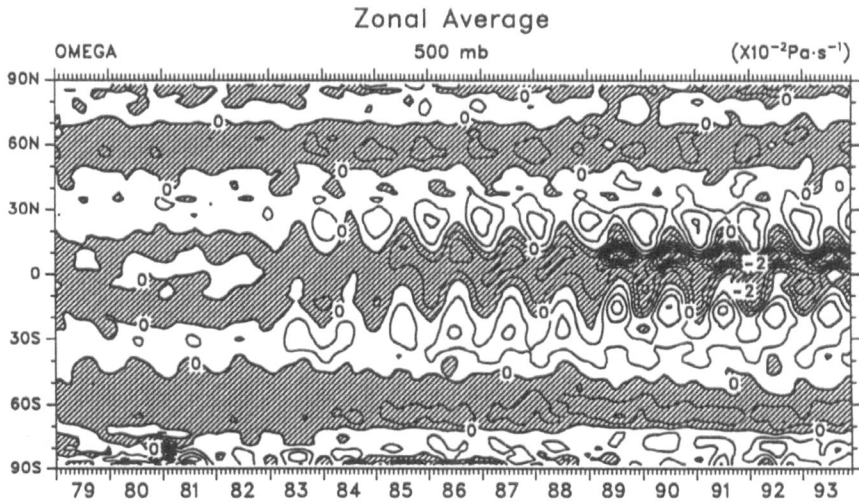

Fig. 3. Latitude-time sections of ω at 500 mb from ECMWF analyses. Negative values are shaded. The trend for increasing amplitude to the annual cycle with time in the tropics arises from the improvements in the 4DDA system shown in Figure 1. Adapted from Trenberth (1992).

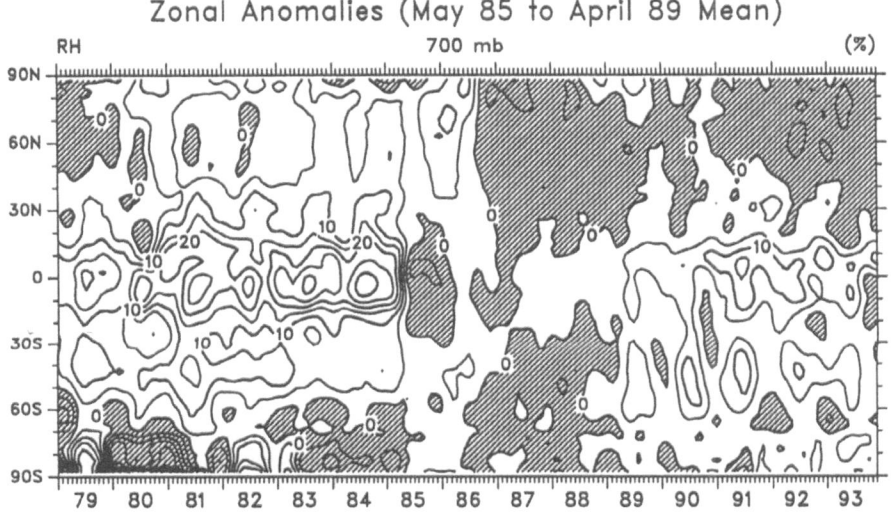

Fig. 4. Latitude-time sequence of departures from the mean annual cycle for May 1985–April 1989 for relative humidity at 850 mb from ECMWF analyses. Almost all changes shown are spurious and arise from changes in Figure 1. Adapted from Trenberth (1992).

Zonal Mean at 1.4S

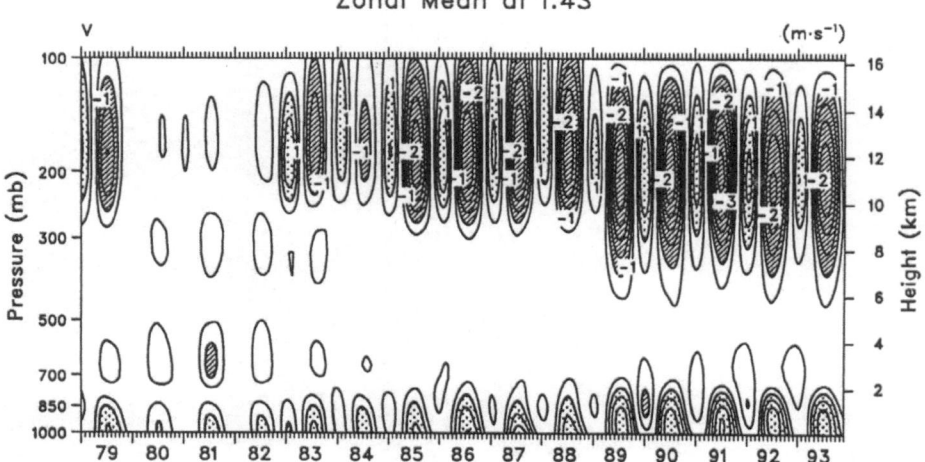

Fig. 5.   Height-time sequence of the zonally averaged v field at 1.4° S. This shows the branches of
the Hadley Circulation as a function of time as seen through the ECMWF analyses and the influences
of the changes in Figure 1.

5. Reanalysis

Because of the spurious changes that disrupt the climate record, a strong case has
been made for *reanalysis* of the observations using a state-of-the-art 4DDA system
(Trenberth and Olson, 1988; Bengtsson and Shukla, 1988). Such retrospective
analysis raises the prospects for a number of improvements:

1. The observations from all sources can be gathered together to incorporate
   into the analyses. This includes not only the observations which made it
   into the original operational analyses but also many which did not. In some
   cases this was because the observations were delayed in transmittal, in other
   cases the observations are simply not available in real time. The latter include
   certain kinds of satellite data as well as observations from fishing fleets that
   are recorded in log books but not sent over the Global Telecommunications
   System (GTS).
2. The analyses have always been performed for NWP which requires tight
   processing schedules. While past data are fully utilized by bringing the information forward in time using the NWP model, future observations are not
   available. Thus although referred to as four-dimensional data assimilation it
   has always been 3 1/2 dimensional in practice. Reanalysis allows the possibility for a true four dimensional system, one that allows the state of the
   atmosphere at a particular time to also take account of the future observations.
   For NWP there has been little incentive to develop such systems. However,

for other reasons, so-called 'adjoint methods' have been under development (see Courtier *et al.*, 1993 for a survey) which provide the framework for incorporating future data. Strictly speaking, while the direct equations describe the evolution forward in time, the adjoint set of equations describe the evolution of sensitivity backward in time.

At NMC, plans are underway to implement a dual NWP and climate analysis system. The latter is referred to as CDAS (climate data assimilation system) in which the cut off for receipt of observations is delayed for seven days before a 'final' analysis is performed. In this situation some look ahead aspects of an analysis system could exploit the new observations of the future atmospheric state taken after the nominal analysis time. Moreover it potentially allows the weather services an opportunity to become proactive in the collection of data, by actively seeking out vital missing observations, by checking up on observations that are flagged in quality control checks as possibly in error, by providing feedback to those responsible for making the observations and reminding them how valuable the data are, and possibly encouraging additional efforts to take data under adverse conditions (such as relaunching a rawinsonde when the balloon carrying the first burst prematurely). In particular, the old time deadlines for taking, transmitting and receiving data can be relaxed for climate purposes. Thus although the relaunch of a balloon may not be justified because of time constraints for NWP, the observations can be fully utilized as part of the climate assimilation system.

3. A state-of-the-art 4DDA system can be exploited and held fixed for the entire period reanalyzed, removing changes in the analysis system as a source of spurious changes. Changes in the data base will still be present, so it is essential for numerical experiments to be carried out to assess their impact. For example, satellite data have had a substantial positive impact on SH analyses, so we can ask whether it is viable to produce SH analyses in the absence of space-based observations, or what is the impact on the quality of the analyses if we do so. It seems desirable to explore strategies where, at least for limited periods (say 5 years), the database is held fixed. Then overlaps can be used to assess the impact of the database changes.

4. An important basic concept inherent in reanalysis is that it should not be done just once. The CDAS-type activity will help create a climate record that is reasonably consistent with the past record as constructed by the reanalysis because the CDAS will be carried out with the same system. But having a system frozen in time means that any bad points or shortcomings are also frozen in time. In addition, it is widely recognized that not everything will go completely smoothly, some data will not be properly assimilated, treatment of future data will be rudimentary, treatment of the earth's surface will be unrefined, the NWP model will improve, and so on. Therefore, it is expected that reanalysis should become a routine activity that would start over again

at regular intervals, perhaps every 5 to 10 years, and in this way could take advantage of the latest state-of-the-art system.

5. One of the largest activities in reanalysis is the organization of the input data base, which is mostly being done at NCAR. Aside from the merits of reanalysis, the assembling and organization of the data itself is an extremely valuable activity for climate studies. Complementary efforts are being made with the upper air data base at the National Climate Data Center under a program called CARDS (Comprehensive Aerological Reference Data Set). These provide an organizing principle for the future archival of data that is incorporated into the operational suites such as in CDAS. Moreover, the task has to be done only once – although no doubt the data base can always be improved – so that future reanalyses will be much easier and will be largely a question of computer time.

   Another substantial activity that has become a part of reanalysis is the development of improved automated quality control procedures (e.g., Gandin, 1988). These are desirable so that the system can run for the most part without human intervention. A major side benefit of these developments has acrued to operational NWP.

6. Because the reanalyses are designed for climate uses, there is an incentive to develop a number of products as it procedes or as part of the process. In particular, monthly means of various circulation statistics and areal averages can be computed on the fly. Also, by utilizing the short (usually 6 hour) forecast that is part of the update cycle many fields not otherwise available can be computed. These include surface fluxes and precipitation, for instance. Periodically, longer forecasts will also be made as part of the reanalysis, and these will prove valuable for validating and evaluating the system, comparing with the operational products, and thus helping to improve the system.

Currently reanalysis activities are underway at a number of organizations. At the Center for Ocean-Land-Atmosphere Studies (COLA) reanalysis of an 18 month period during the 1982–83 El Niño has been performed as a prototype activity (Paolino *et al.*, 1994). Other activities are underway at NASA-Goddard, the Naval Research Laboratory (NRL) in Monterey, NMC/NCAR and ECMWF, and these are briefly described below.

At NASA Goddard, the reanalysis activity is described by Schubert *et al.* (1993). The initial plans called for a 6 year reanalysis of the period 1985–90 and this has been completed. It is likely to be extended in time to encompass more recent years and it may be redone in total or in part. It uses a grid point NWP model with a 2° latitude by 2.5° longitude resolution and 20 levels. The system is unique in that it uses a nudging scheme to obtain an analysis at a particular time by first determining the discrepancy between observations and the model forecast then inserts the differences gradually into the model by rerunning it and adding a fraction of the increment at each time step. In this way, a smooth evolution is achieved in time, there are no spin up problems and initialization is not required.

NASA used the operational data base and it remains for other data to be fully exploited.

At NRL, the focus of the reanalysis effort is to provide better forcings for ocean models. It uses the operationally available observations and tropical cyclone bogus data in the current operational model which is a T79 spectral model with 18 levels. The period 1985 to 1989 has been completed.

At ECMWF (see Gibson *et al.*, 1994), the reanalysis period will be from 1979 to 1993 using a reduced resolution version of their model at T106 resolution with 31 levels. Some new data will be included through collaboration with NCAR and NMC, in particular the COADS (Comprehensive Ocean Atmosphere Data Set) ship observations. A new feature will be the direct assimilation of cloud-cleared satellite radiances using a one-dimensional variational (1D-VAR) scheme to enhance the optimum interpolation system, and this is expected to greatly benefit the moisture field, in particular. Production began early 1995. About eleven days of reanalysis per day can be achieved and processing should be completed by early to mid 1996.

NMC has the most ambitious reanalysis plans and these were in part developed through a community workshop (Kalnay and Jenne, 1991). The system is based on an NWP model with T62 spectral resolution and 28 levels and the SSI scheme is employed with complex quality control. Unique to this effort is the extensive development at NCAR of the most comprehensive data base possible. Moreover, the plans are to carry out the analysis in several stages, eventually back to 1956. The first stage began with analyses in 1985, then 1982–85, with 1982–present expected by the end of 1995. A special challenge will be to deal with the degradation of satellite soundings in the 1970s and then absence of satellite data for 1958 to early 1970s. The system produces one month per day and it began in June 1994, although with several restarts after some initial problems had been ironed out. Over 2000 Gigabytes of data are expected from the project.

All of the above plans are developing rapidly, and may change, but this outline provides an idea of the datasets that might be expected. It is an exciting prospect for many researchers in the climate community.

6. Atmospheric Circulation Links to Temperature and Precipitation

A basic precept in extended range forecasting for many years (Namias, 1948) has been that monthly mean temperature and precipitation anomalies are, to a large extent, determined by the monthly mean geopotential height field for mid-tropospheric levels, and the latter undergoes an orderly evolution and development which can be forecast from physical and kinematic principles and empirical relations. In the United States, the first step in making seasonal forecasts was to determine the expected geopotential height field and anomalies using a variety of methods (including persistence, other statistics, synoptic climatology, analogues,

teleconnections, and SST, snow cover, and other surface anomalies); then the surface temperatures and precipitation are inferred (Gilman, 1983; Epstein, 1988). The
basic guidelines relating surface temperature and precipitation to the circulation in
the mid-troposphere were developed by Martin and Hawkins (1950) for the United States. More recent relationships using pointwise screening regression (Klein,
1983, 1985) were found more skilful than using empirical orthogonal functions
as predictors (Klein and Walsh, 1983). As well as for the United States, quantitative specifications of surface temperatures have been made for Europe and Asia
(Klein and Yang, 1986) and Canada (Klein *et al.*, 1989). Mean annual reduction
of variance in regression equations for monthly mean surface temperature using
the 500 mb height field as predictors for 68 stations in Canada and Alaska is 68%.
Recently, the procedures used for making long lead-time predictions has changed
(O'Lenic, 1995).

It is therefore well established that time-averaged surface temperatures and precipitation changes are empirically related to changes in circulation, and the relationship is extensively exploited in extended range forecasting. However, many
studies of long-term climate trends of temperature or precipitation have not examined parallel changes in the circulation and thereby fully synthesized the available
information about global climate change.

Local climates at the same latitude vary considerably around the globe because
of the distribution of land and sea, the topography over land, and the associated
planetary waves in the atmosphere. Variations in local temperatures on decadal time
scales, or from year to year, are far from uniform but occur in distinctive large-scale
patterns. Studies of circulation changes and their relationships to other variables
often relate to particulare phenomena, such as the Southern Oscillation, the North
Atlantic Oscillation, or to zonal and meridional indices perhaps as part of an index
cycle (see Trenberth, 1993 for a review). General relationships between sea level
pressures or the frequency of synoptic systems and temperature and precipitation
have been outlined using various methods.

The large-scale coherence in these climate variations arises because they are
associated with changes in the quasi-stationary planetary waves and other factors,
including the heat capacity of the underlying surface, in particular whether it is
land or ocean, the role of advection by the mean flow and the planetary waves
which is strong in wintertime but much weaker in summer, and teleconnections, as
discussed by Trenberth (1993). For instance, in winter positive correlations between
precipitation and temperature arise from the dynamical association between warm
advection, increased moisture influx, and vertical motion ahead of a trough, while
negative correlations between precipitation and temperature in summer arise from
the impact of local solar heating which depends greatly on the availability of surface
moisture through the partitioning of heat into sensible and latent components. Thus
heat waves are associated with droughts.

Because of the tendency for atmospheric motions to conserve absolute vorticity,
atmospheric motions occur as Rossby waves. A consequence of this is that local

forcing of such waves sets up a wavetrain of disturbances and 'teleconnections' downstream. In the context of the current topic, teleconnections are important because the anomalies in the circulation and associated temperature and precipitation anomalies may arise from forcing in remote regions. The best known examples of global impacts of local forcing are with changes in SSTs such as the El Niño-Southern Oscillation (ENSO) phenomenon, whereby coupling occurs between the tropical Pacific to higher latitudes via teleconnections.

6.1. DECADAL VARIATIONS IN THE PACIFIC

The winter of December 1993–February 1994 was very cold and snowy with many more than normal winter storms in the northeastern part of the United States. Many questions were raised by the media on how this jibes with expectations of global warming. The exceptionally wintry weather continued for several months, long enough to create a pattern of behavior. However, as part of this pattern there were often mild and sunny conditions inthe western half of the United States and Canada. Also, temperatures were above or much above normal over much of the rest of the hemisphere and the hemispheric mean was +0.2 °C, so the global implications are much less clear.

Moreover, the pattern of change that has been occurring since the late 1970s has been one favoring warming in the western part of North America but little warming, or even slight cooling, in the eastern part. This distinctive pattern of change involves the ocean, for instance as seen through sea surface temperatures, and it has occurred throughout the Pacific and is linked to changes in the tropics. While not predicted by models, this observed teleconnection pattern is a part of an ongoing climate change and it complicates the interpretation of the role of the greenhouse gas forcing.

The decadal variability in the Pacific exists strongly in winter with the period of the fluctuations exceeding 20 years and the details are described by Trenberth (1990) and Trenberth and Hurrell (1994). Considerable evidence has emerged of a substantial at least decade-long change in the North Pacific atmosphere and ocean beginning about 1976 (Figure 6). Observed significant changes in the winter atmospheric circulation involve the so-called Pacific-North American (PNA) tele-connection patterns, so that for 1976–88, throughout the troposphere there was a deeper and eastward-shifted Aleutian low pressure system in the winter half year which advected warmer and moister air along the west coast of North America and into Alaska and colder air over the North Pacific, see Figure 7. Consequently, there were increases in temperatures and sea surface temperatures (SSTs) along the west coast of North America and Alaska but decreases in SSTs over the central North Pacific (Trenberth and Hurrell, 1994; Kawamura, 1994), see Figure 8, as well as changes in coastal rainfall and streamflow and decreases in sea ice in the Bering Sea. Associated changes occurred in the surface wind stress, and, by inference, in the Sverdrup transport in the North Pacific Ocean that have been directly measured

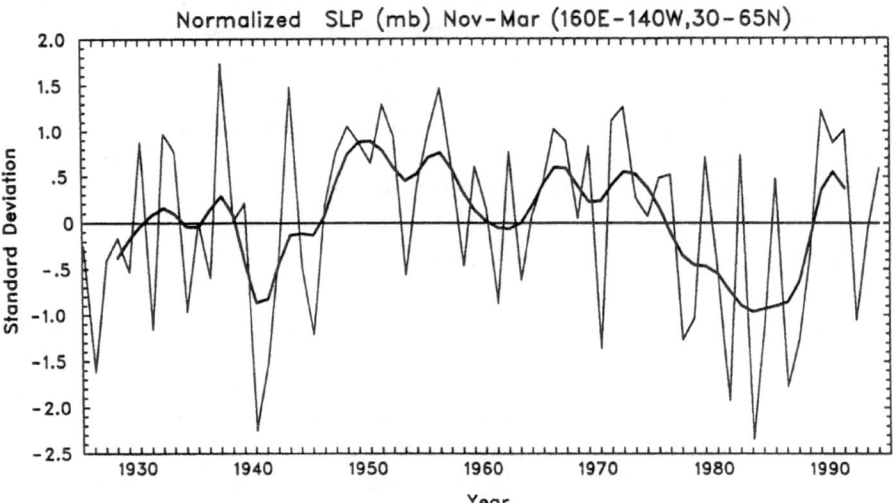

Fig. 6. Time series of mean North Pacific sea level pressures averaged over 30 to 65° N, 160° E to 140° W for the months November through March beginning in 1925 and smoothed with the low pass filter. Updated from Trenberth and Hurrell (1994).

in the Gulf of Alaska. There is also emerging evidence for increased tropospheric temperatures and water vapor in the western tropical Pacific after 1976 (Gaffen *et al.*, 1991).

Changes in the mean flow were accompanied by a southward shift in the storm tracks and associated synoptic eddy activity and in the surface ocean sensible and latent heat fluxes. Also accompanying the changes in the Pacific were higher incidences of cold outbreaks across the plains of North America ultimately leading to major freezes affecting the Florida citrus crop.

In addition to the changes in the physical environment, the deeper Aleutian low increased the nutrient supply as seen through increases in total chlorophyll in the water column, phytoplankton and zooplankton. These changes, along with the altered ocean currents and temperatures, changed the migration patterns and increased the stock of many fish species (McFarlane and Beamish, 1992; Beamish and Bouillon, 1993; Brodeur and Ware, 1995; Trenberth and Hurrell, 1995; Polovina *et al.*, 1994).

The dominant atmosphere-ocean relation in the North Pacific is one where atmospheric changes lead SSTs by one to two months, apparently because of the changes in surface sensible and latent heat fluxes combined with mixing in the ocean and entrainment. Changes in the storm tracks in the North Pacific help to reinforce and maintain the anomalous circulation in the upper troposphere. Trenberth and Hurrell (1994) have also shown that strong ties exist between the North Pacific changes and events in the tropical Pacific, with changes in tropical Pacific SSTs

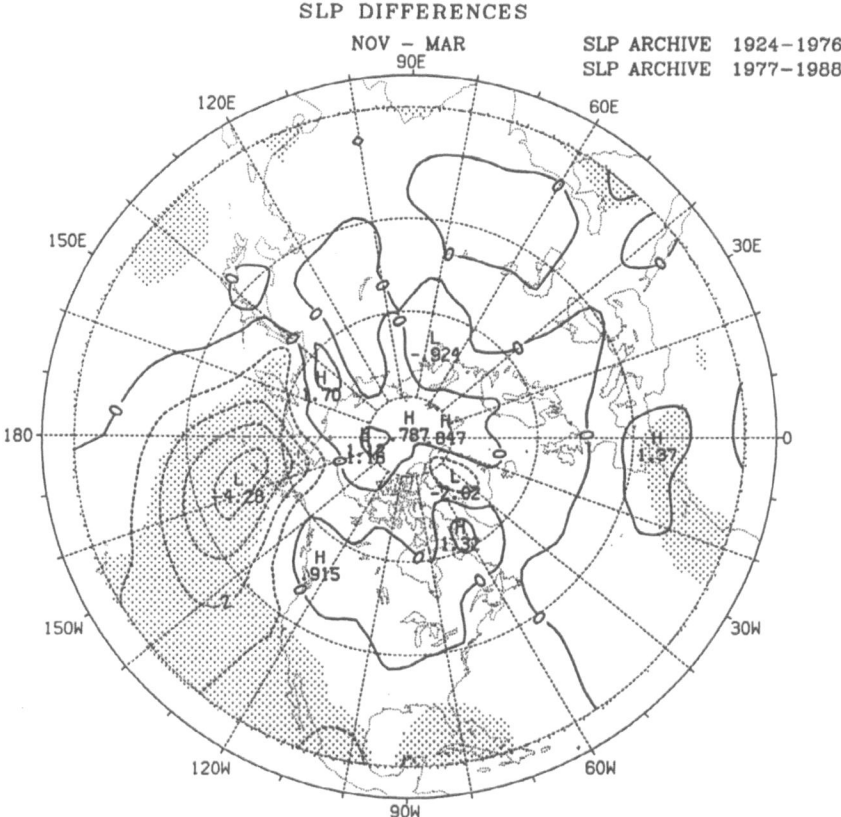

Fig. 7.   The difference in mean sea level pressures from 1977 to 1988 for November through March versus 1924 to 1976 (mb). Stippling indicates statistical significance at 5%. From Trenberth (1990).

leading SSTs in the North Pacific by three months. The Pacific decadal timescale variations have been linked to recent changes in the frequency and intensity of El Niño versus La Niña events and it has been hypothesized that the decadal variation has its origin in the tropics (Trenberth and Hurrell, 1994). Observational studies by Kawamura (1994) and Lau and Nath (1994) have shown that the decadal variation in the extratropics of the Pacific is closely tied to tropical SSTs in the Pacific and Indian Ocean, see Figure 9. This reveals the close association between the PNA pattern and the SSTs throughout the North and tropical Pacific, and the decadal time scale change after the winter of 1975–76. An update of the time series of the Southern Oscillation index is presented in Figure 10. Note that superposed on the warm El Niño and cold La Niña events is a distinct decadal variability and a

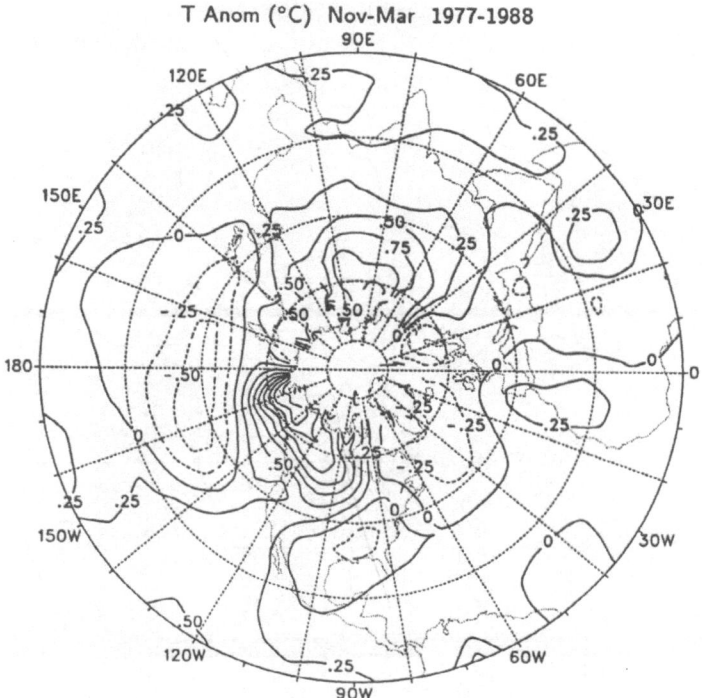

Fig. 8. Twelve year (1977–1988) average surface temperature or sea surface temperature anomalies as departures from the 1951–1980 mean. Contours every 0.25 °C. Shown are the anomalies for the 5 winter months (November to March). Negative values are dashed. From Trenberth and Hurrell (1994).

pronounced tendency toward negative SOI values (El Niño conditions) in recent years.

Several aspects of the decadal-scale change beginning around 1976 have been simulated with atmospheric models using specified SSTs (Kitoh, 1991; Chen *et al.*, 1992; Lau and Nath, 1994; Miller *et al.*, 1994; Graham *et al.*, 1994). These studies confirm that the atmospheric changes are tied to the changes in SSTs, and that the changes over the North Pacific are substantially controlled by the anomalous SST forcing from the tropical Pacific. Recent modeling studies by Kumar *et al.* (1994) and Graham (1995) show that seasonal variations and patterns of surface temperature change observed in recent years over land can be reproduced from models forced solely with SSTs, and that the observed global warming is also reproduced.

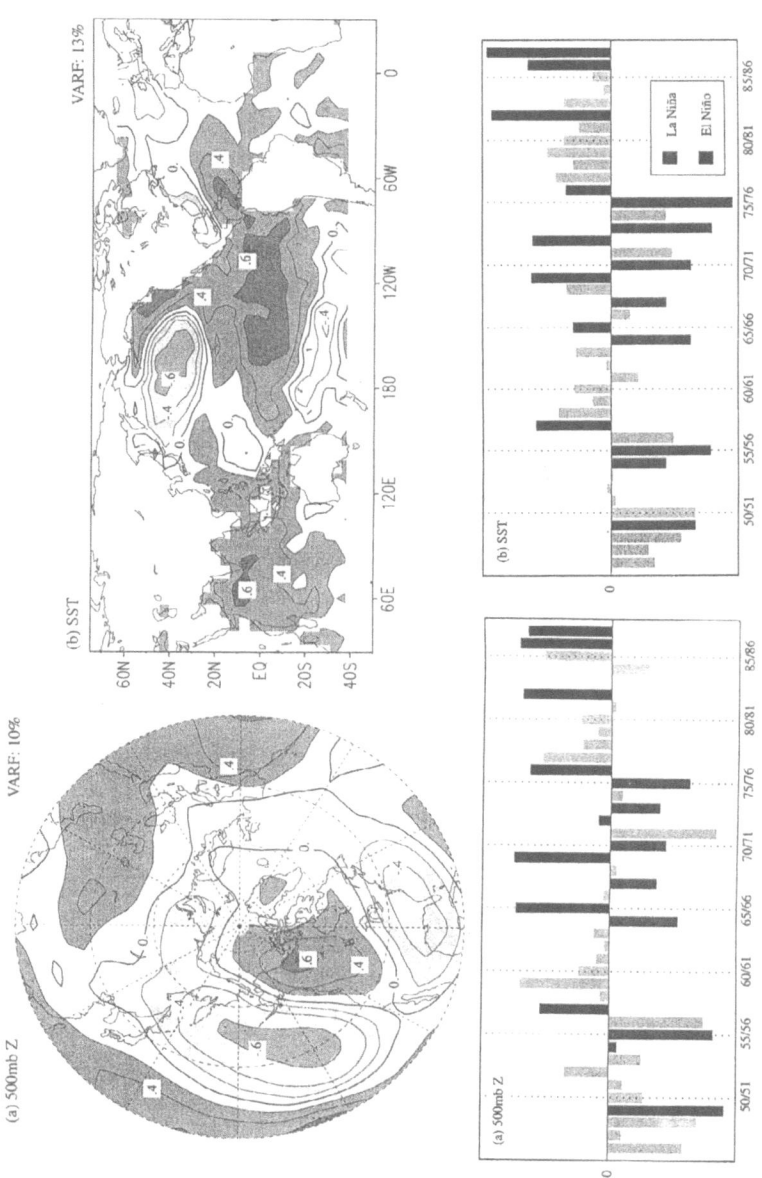

Fig. 9. The first mode of a singular value decomposition of the northern winter 500 mb height field over the extratropical NH and the near global SST field (north of 40° S). Shown are the heterogeneous correlation patterns of the height field (a); and the SST field (b), along with the corresponding time series. In the latter the El Niño and La Niña events are indicated by shading. The correlation is 0.83 and the two patterns account for 10% and 13% of the variance of each field, respectively. This reveals the close association between the PNA pattern and the SSTs throughout the North and tropical Pacific, and the decadal time scale change after the winter of 1975–76. From Lau and Nath (1994).

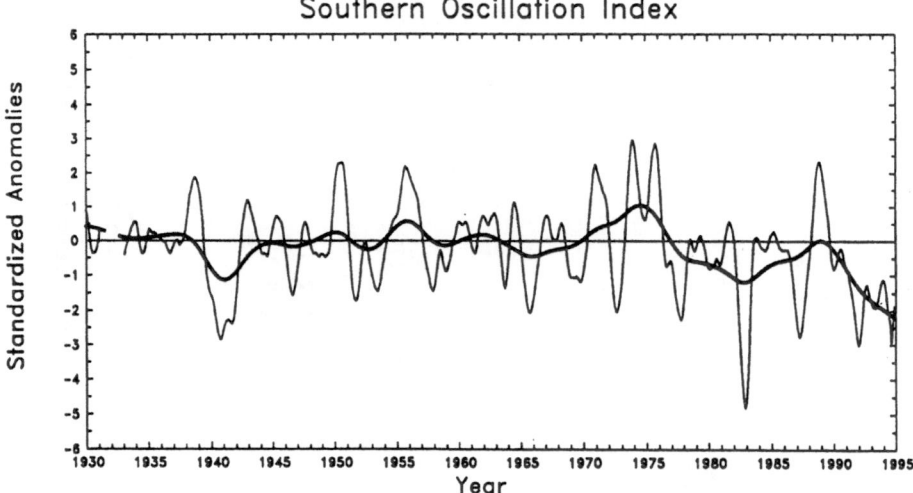

Fig. 10. Time series of the Southern Oscillation Index, sea level pressure anomalies at Tahiti minus Darwin normalized. The dark curve is a smoothing spline designed to show the decadal time scale variations.

The changes in lower tropospheric temperatures in the North Pacific winter do not have much influence on the hemispheric or global mean because there is strong cancellation by regional anomalies of opposite sign (Wallace *et al.*, 1993), and instead the global mean is associated with a different mode which is present in all seasons and which is also present in the Atlantic sector. However, lower tropospheric temperatures, such as given by the 1000–500 mb thickness, differ somewhat from surface temperatures (Trenberth *et al.*, 1992) because of the land-ocean differences in the underlying surface in regions of surface temperature anomalies of opposite sign. Differences in heat capacity amplify temperature anomalies over land.

Over the SH, careful analysis of station data together with the available analyses have revealed decadal-scale climate changes that are not an artifice arising from changes in data or analyses (van Loon *et al.*, 1993; Hurrell and van Loon, 1994). A change in the atmospheric circulation is most evident in the late 1970s with sea level pressures in the circumpolar trough generally lower in the 1980s than in the 1970s and with the changes most pronounced in the second half of the year, so that the tropospheric polar vortex remained strong into November and was associated with a delayed breakdown in the stratospheric polar vortex and the beginning of the ozone deficit in the Antarctic spring (Hurrell and van Loon, 1994). It is not believed that the ozone depletion was responsible for the change but it does highlight a dynamical component to the ozone hole problem. Instead it seems likely that the changes are associated with the increase in tropical SSTs and associated changes noted for the North Pacific.

6.2. DECADAL VARIATIONS IN THE NORTH ATLANTIC

In the North Atlantic, decadal period fluctuations have been found superposed on a longer period (about 70 year) variation (Deser and Blackmon, 1993; Kushnir, 1994; Hurrell and van Loon, 1993; Schlesinger and Ramankutty, 1994; Kawamura, 1994, Hurrell, 1995) which is reflected in the global mean temperature variations. Earlier studies of the North Atlantic are reviewed by Trenberth (1993).

In the Atlantic, one mode of decadal variability is shorter in timescale than that in the Pacific and the mode is found to also vary on quasibiennial timescales. The irregular fluctuations average about 9 years in length before about 1945 and about 12 years thereafter. The mode is characterized by a dipole pattern in SSTs and surface air temperatures, with anomalies of one sign east of Newfoundland, and anomalies of opposite polarity off the southeast coast of the United States in the Atlantic. Distinctive sea level pressure and wind patterns have been identified with this mode and seem to be of the nature that the SSTs result largely from the winds, with cooler-than-normal SSTs accompanied by stronger-than-normal winds (Cayan, 1992; Deser and Blackmon, 1993; Kushnir, 1994). The SST changes are closely associated with variations in sea ice in the Labrador Sea, with heavy sea ice preceding winters of colder than normal SSTs east of Newfoundland by about two years. The mode is identified with an atmospheric teleconnection pattern known as the North Atlantic Oscillation (Hurrell and van Loon, 1993; Hurrell, 1995) and the 'West Atlantic' pattern at 500 mb (Deser and Blackmon, 1993).

The lower frequency mode is identified as a temperature fluctuation with period of about 65–70 years in the global mean temperatures but with largest amplitude in the North Atlantic and surrounding continents (Schlesinger and Ramankutty, 1994) where the period appears to vary somewhat regionally from 50 to 88 years. Relative to the record from 1850 on, it corresponds to a much colder period from the late 1800s to about 1920, much warmer than normal from about 1930 to 1960, and another cooler period in the 1970s and 1980s (Kushnir, 1994; Schlesinger and Ramankutty, 1994). The sea level pressures and surface winds associated with this pattern have been explored by Deser and Blackmon (1993) and Kushnir (1994) who find an anomalous cyclonic circulation during the warmer years and anticyclonic during the cooler years.

The nature of the relationships among the atmosphere and ocean variables is not fully understood although it seems likely that it signifies the patterns of dynamical coupling between the atmosphere and ocean, which may extend outside the Atlantic domain. Because the surface changes are also reflected at depth in the ocean (Levitus, 1989) and changes in the Gulf Stream (Deser and Blackmon, 1993), the ocean circulation is clearly involved. Thus it is likely that there are changes in the ocean thermohaline circulation (Deser and Blackmon, 1993; Kushnir, 1994; Schlesinger and Ramankutty, 1994) such as those seen in models with time scales of roughly 50 years (Manabe and Stouffer, 1988; Delworth et al., 1993) in which an intensified circulation is associated with surface warming at high latitudes of the North Atlantic and a poleward-shifted and intensified Gulf Stream. The thermohaline circulation is greatly influenced by density fluctuations at high latitudes which apparently arises mainly in the fresh water availability which alters salinity, and originates from changes in precipitation, runoff, and evaporation. These kinds of fluctuations may also be related to the abrupt climate changes seen in recent Greenland ice-core analysis (Taylor *et al.*, 1993; Dansgaard *et al.*, 1993).

7. Discussion

Because the atmosphere has a very short memory while the ocean has enormous thermal inertia it is widely assumed that it is the ocean that provides the important memory for decadal and longer-term climate variations. However, it is the collaborative interaction between the atmosphere and the ocean and other parts of the climate system that gives rise to the important climate variations. It is argued that constructive feedbacks between climate system components must be preferred over destructive interference modes (Trenberth and Hurrell, 1994). As the observed variations reveal the very large scales that are apparently set by the scales of the atmosheric planetary-scale waves, the coupled nature of the variations is emphasized. The system appears to be one where considerable chaotic noise arising from weather systems is a prominent factor in the extratropics but where the patterns at least in the Pacific, as seen in decadal climate variations, are moulded by the tropical influences with the ocean selectively bringing out the long time scales. The ocean is never in equilibrium.

Analysis of the changes in surface temperatures, winds and the various measures of the atmospheric circulation has shown coherent coupled variations on decadal and longer times scales that are especially important regionally.The regionalization appears to be more evident in the winter half year and is probably associated with the stronger dynamics operating then, while thermodynamic forcing can emerge more strongly in the summertime (Wallace *et al.*, 1993). As a result, teleconnections are present in the winter extratropics and the changes seen at one location can be caused by local changes in quite remote regions, and especially from the tropics and subtropics.

Of particular note throughout the Pacific has been the changes that occurred in the late 1970s, and it is difficult to assess whether these are merely a part of natural variability or whether they are a complex response to anthropogenic-induced changes such as the increases in greenhouse gases. Palmer (1993) has emphasized that nonlinear dynamics may well change the frequency distribution of weather regimes in the atmosphere, rather than changing the regimes, so that counter-intuitive changes can occur. Thus there is a real prospect that ENSO events may be modified by global climate change through changes in frequency, intensity or by skewing the distribution between warm El Niño and cold La Niña events (Meehl *et al.*, 1993). Indeed the observed increased tendency for the tropics to be preferentially in an El Niño mode may well be one of the primary manifestations of anthropogenic global warming. A consequence is the changes throughout the Pacific involving the PNA teleconnection pattern in the Northern Hemisphere, which has resulted in a particular surface temperature pattern change. The latter may have amplified the global surface temperature changes because of its interaction with land and ocean as temperature changes tend to be amplified over land and more subdued over the ocean because of the large heat capacity of the mixed layer of the latter (e.g., see Karl *et al.*, 1994). Because ENSO events cause very large and important systematic variations in weather patterns on interannual time scales and are responsible for many floods and droughts throughout the world, these changes are probably a factor in the wide spread perception of increasing extremes in weather and climate in recent years.

8. Recommendations

As discussed above, the atmospheric circulation is changing on decadal time scales and these changes have had major impacts on climate locally. It is important to understand the origin of the changes and attempt to sort out the extent to which they might be due to anthropogenic effects. We have noted the decline of the World Weather Watch and the reduction in the number of rawinsonde observations taken since the Global Weather Experiment in 1979. Moreover, there is an imminent threat to the conventional sounding data as the challenge of switching to a GPS-based system is faced. For monitoring climate changes in the atmospheric circulation it is vital that:

1. A new low-cost rawinsonde must be developed. If a design can be made widely available, then multiple manufacturers can compete for contracts with the national weather services. Moreover, the formation of consortia of smaller nations may be beneficial by capitalizing on bulk-buying power in keeping costs of expendables under control.

2. As recommended by the Global Climate Observing System, networks of baseline reference stations should be established both at the surface and for the upper air network. Nations should be encouraged to make special efforts to ensure the integrity and continuity of these designated stations so that a skeletal network will accurately depict the changes occurring. Such a network also provides needed ground truth for satellite-based observations. The climate record should be a factor in any changes made in the observing system, and it is desirable to have a period of overlap when changes are introduced so that the impact on the climate record can be assessed.

3. All observations made should be communicated, processed and archived with the expectation that they will be reused multiple times as part of reanalysis of the past record.

4. Continued efforts should be made to improve the quality and volume of the historical data base, such as the CARDS project. This includes the need for data 'archeology'. Such efforts can help future reanalyses but are also useful for documenting long term circulation changes.
5. Reanalysis activities should be encouraged and should occur at regular intervals, every 5 to 10 years, as data assimilation systems improve.
6. A comprehensive view of the observed climate change must be pursued by analyzing all climate variables and accounting for relationships among them wherever possible. The atmospheric circulation provides a vital link in building this understanding.

Acknowledgements

I thank Jim Hurrell for providing Figure 6 and Dennis Shea for providing Figures 2 and 10, and Walt Dabberdt for information on rawinsondes. The ECMWF data used were provided by ECMWF. This research is partly sponsored by the Tropical Oceans Global Atmosphere Project Office under grant NA87AANRG0208, and NASA under NASA Order No. W-18,077.

References

Beamish, R. J. and Bouillon, D. R.: 1993, 'Pacific Salmon Production Trends in Relation to Climate', *Can. J. Fish. Aquat. Sci.* **50**, 1002–1016.
Bengtsson, L. and Shukla, J.: 1988, 'Integration of Space and In Situ Observations to Study Global Climate Change', *Bull. Amer. Meteor. Soc.* **69**, 1130–1143.
Brodeur, R. D. and Ware, D. M.: 1995, 'Interdecadal Variability in Distribution and Catch Rates of Epipelagic Nekton in the Northeast Pacific Ocean', in R. J. Beamish (ed.) *Climate Change and Northern Fish Populations. Canad. Spec. Publ. Fisheries Aquatic Sci.* **121**, 329–356.
Buchan, A.: 1869, 'The Mean Pressure of the Atmosphere and the Prevailing Winds over the Globe', *Trans. Roy. Soc. Edinb.* **25**, 575–637.
Cayan, D. R.: 1992, 'Latent and Sensible Heat Flux Anomalies over the Northern Oceans: The Connection to Monthly Atmospheric Circulation', *J. Clim.* **5**, 354–369.
Chen, T.-C., van Loon, H., Wu, K.-D., and Yen, M.-C.: 1992, 'Changes in the Atmospheric Circulation over the North Pacific-North America Area since 1950', *J. Met. Soc. Japan* **70**, 1137–1146.
Courtier, P., Derber, J., Errico, R., Louis, J-F., and Vukićevic, T.: 1993, 'Important Literature on the Use of Adjoint, Variational Methods and the Kalman Filter in Meteorology', *Tellus* **45A**, 342–357.
Dansgaard, W., Johnsen, S. J., Clausen, H. B., Dahl-Jensen, D., Gundestrup, N. S., Hammer, C. U., Hvidberg, C. S., Steffensen, J. P., Sveinbjornsdottir, A. E., Jouzel, J., and Bond, G.: 1993, 'Evidence for General Instability of Past Climate from a 250-kyr Ice-Record', *Nature* **364**, 218–220.
Delworth, T., Manabe, S., and Stouffer, R. J.: 1993, 'Interdecadal Variations of the Thermohaline Circulation in a Coupled Ocean-Atmosphere Model', *J. Clim.* **6**, 1993–2011.
Deser, C. and Blackmon, M. L.: 1993, 'Surface Climae Variations over the North Atlantic Ocean During Winter: 1900–1989', *J. Clim.* **6**, 1743–1753.
Elliot, W. P. and Gaffen, D. J.: 1991, 'On the Utility of Radiosonde Humidity Archives for Climate Studies', *Bull. Amer. Meteor. Soc.* **72**, 1507–1520.

Elliot, W. P. and Gaffen, D. J.: 1993, 'Effects of Conversion Algorithms on Reported Upper-Air Dewpoint Depressions', *Bull. Amer. Meteor. Soc.* **74**, 1323–1325.

Epstein, E. S.: 1988, 'Long-Range Weather Prediction: Limits of Predictability and Beyond', *Wea. Forecast.* **3**, 69–75.

Gaffen, D. J.: 1994, 'Temporal Inhomogeneities in Radiosonde Temperature Records', *J. Geophys. Res.* **99**, 3667–3676.

Gaffen, D. J., Barnett, T. P., and Elliott, W. P.: 1991, 'Space and Time Scales of Global Tropospheric Moisture', *J. Clim.* **4**, 989–1008.

Gandin, L. S.: 1988, 'Complex Quality Control of Meteorological Observations', *Mon. Wea. Rev.* **116**, 1138–1156.

Garand, L., Grassotti, C., Hallé, J., and Klein, G. L.: 1992, 'On Differences in Radiosonde Humidity-Resporting Practices and Their Implications for Numerical Weather Prediction and Remote Sensing', *Bull. Amer. Meteor. Soc.* **73**, 1417–1423.

Gibson, J. K., Källberg, P., Nomura, A., and Uppala, S.: 1994, *The ECMWF Reanalysis (ERA) Project – Plans and Current Status*, Preprints, Tenth International Conference on Interactive Information and Processing Systems for Meteorology, Oceanography, and Hydrology, American Meteorological Society, pp. 164–167.

Gilman, D. L.: 1983, 'Predicting the Weather for the Long Term', *Weatherwise* **36**, 290–297.

Graham, N. E.: 1995, 'Simulation of Recent Global Temperature Trends', *Science* **267**, 666–671.

Graham, N. E., Barnett, T. P., Wilde, R., Schlese, U., and Bengtsson, L.: 1994, 'On the Roles of Tropical and Midlatitude SSTs in Forcing Interannual and Interdecadal Variability in the Winter Northern Hemisphere Circulation', *J. Clim.* **7**, 1416–1441.

Hurrell, J. W.: 1995, 'Decadal Trends in the North Atlantic Oscillation: Regional Temperatures and Precipitation', *Science* **269**, 676–679.

Hurrell, J. W. and van Loon, H.: 1993, *Analysis of Low-Frequency Climate Variations over the Northern Hemisphere Using Historical Atmospheric Data*, Fourth Symposium on Global Change Studies, 17–22 January 1993, Anaheim, CA, pp. 355–360.

Hurrell, J. W. and van Loon, H.: 1994, 'A Modulation of the Atmospheric Annual Cycle in the Southern Hemisphere', *Tellus* **46A**, 325–338.

Jones, P. D.: 1987, 'The Early Twentieth Century Arctic High – Factor or Fiction', *Clim. Dyn.* **1**, 63–75.

Jones, P. D.: 1991, 'Southern Hemisphere Sea Level Pressure Data: An Analysis and Reconstructions back to 1951 and 1911', *Intl. J. Climatol.* **11**, 585–607.

Jones, P. D. and Wigley, T. M. L.: 1988, 'Antarctic Gridded Sea Level Pressure Data: An Analysis and Reconstruction back to 1957', *J. Clim.* **1**, 1199–1220.

Jones, P. D., Wigley, T. M. L., and Briffa, K. R.: 1983, 'Reconstructing Surface Pressure Patterns Using Principal Components Regression on Temperature and Precipitation Data', Proc. Intl. Mtg. Statistical Climatol., Inst. Nacional de Met. Geofisica, Lisbon, 4.2.1–4.2.8.

Jones, P. D., Wigley, T. M. L., and Briffa, K. R.: 1986, *Monthly Mean Pressure Reconstructions for Europe (to 1780) and North America (back to 1858)*, DoE Tech. Rep TR0, U.S. Dept. Energy, Washington, D.C.

Kalnay, E. and Jenne, R.: 1991, 'Summary of the NMC/NCAR Reanalysis Workshop of April 1991', *Bull. Amer. Meteor. Soc.* **72**, 1897–1904.

Karl, T. R., Knight, R. W., and Christy, J. R.: 1994, 'Global and Hemispheric Temperature Trends: Uncertainties Related to Inadequate Spatial Sampling', *J. Clim.* **7**, 1144–1163.

Kawamura, R.: 1994, 'A Rotated EOF Analysis of Global Sea Surface Temperature Variability with Interannual and Interdecadal Scales', *J. Phys. Oceanogr.* **24**, 707–715.

Kitoh, A.: 1991, 'Interannual Variations in an Atmospheric GCM Forced by the 1970–1989 SST. Pt II: Low Frequency Variability of the Wintertime Northern Hemisphere Extratropics', *J. Met. Soc. Japan* **69**, 271–291.

Klein, W. H.: 1983, 'Objective Specification of Monthly Mean Surface Temperature from Mean 700 mb Heights in Winter', *Mon. Wea. Rev.* **111**, 674–691.

Klein, W. H.: 1985, 'Space and Time Variations in Specifying Monthly Mean Surface Temperature from the 700 mb Height Field', *Mon. Wea. Rev.* **113**, 277–290.

Klein, W. H., Shabbar, A., and Yang, R.: 1989, 'Specifying Monthly Mean Surface Temperatures in Canada and Alaska from the 500 mb Height Field', *J. Clim.* **2**, 631–638.

Klein, W. H. and Walsh, J. E.: 1983, 'A Comparison of Pointwise Screening and Empirical Orthogonal Functions in Specifying Monthly Surface Temperature from 700 mb Data', *Mon. Wea. Rev.* **111**, 669–673.

Klein, W. H. and Yang, R.: 1986, 'Specifications of Monthly Mean Surface Temperature Anomalies in Europe and Asia from Concurrent 700 mb Monthly Mean Height Anomalies over the Northern Hemisphere', *J. Climatol.* **6**, 463–474.

Kumar, A., Leetmaa, A., and Ji, M.: 1994, 'Simulations of Atmospheric Variability Induced by Sea Surface Temperatures and Implications for Global Warming', *Science* **266**, 632–634.

Kushnir, Y.: 1994, 'Interdecadal Variations in North Atlantic Sea Surface Temperature and Associated Atmospheric Conditions', *J. Clim.* **7**, 141–157.

Lamb, H. H. and Johnson, A. I.: 1966, 'Secular Variations of the Atmospheric Circulation since 1750', *Geophys. Mem.* **110**, London.

Lau, N.-C. and Nath, M. J.: 1994, 'A Modeling Study of the Relative Roles of Tropical and Extratropical SST Anomalies in the Variability of the Global Atmosphere-Ocean System', *J. Clim.* **7**, 1184–1207.

Levitus, S.: 1989, 'Interpentadal Variability of Temperature and Salinity at Intermediate Depths of the North Atlantic Ocean, 1970–1974 versus 1955–1959', *J. Geophys. Res.* **94**, 6091–6131.

Madden, R. A.: 1976, 'Estimates of the Natural Variability of Time-Averaged Sea-Level Pressure', *Mon. Wea. Rev.* **104**, 942–952.

Manabe, S. and Stouffer, R. J.: 1988, 'Two Stable Equilibria of a Coupled Ocean-Atmosphere Model', *J. Clim.* **1**, 841–866.

Martin, D. E. and Hawkins, H. F. Jr.: 1950, 'Forecasting the Weather – the Relationship of Temperature and Precipitation over the United States to the Circulation aloft', *Weatherwise* **3**, 16–19, 40–43, 65–67.

McFarlane, G. A. and Beamish, R. J.: 1992, 'Climatic Influence Linking Copepod Production with Strong Year-Classes in Sablefish, Anoplopoma fimbria', *Can. J. Fish Aquat. Sci.* **49**, 743–753.

Meehl, G. A., Branstator, G. W., and Washington, W. M.: 1993, 'Tropical Pacific Interannual Variability and CO_2 Climate Change', *J. Clim.* **6**, 42–63.

Miller, A. J., Cayan, D. R., Barnett, T. P., Graham, N. E., and Oberhuber, J. M.: 1994, 'Interdecadal Variability of the Pacific Ocean: Model Response to Observed Heat Flux and Wind Stress Anomalies', *Clim. Dyn.* **9**, 287–302.

Namias, J.: 1948, 'Evolution of Monthly Mean Circulation and Weather Patterns', *Trans. Amer. Geophys. U.* **29**, 777–788.

O'Lenic, E.: 1995, *A New Paradigm for Production and Dissemination of the NWS's Long Lead-Time Seasonal Climate Outlooks*, Proc. 19th Climate Diag. Workshop, pp. 408–411.

Palmer, T. N.: 1993, 'A Nonlinear Dynamical Perspective on Climate Change', *Weather* **48**, 314–326.

Paolino, D. A., Yang, Q., Doty, B., Kinter, J. L. III, Shukla, J., and Straus, D.: 1994, *A Pilot Reanalysis Project at COLA*, COLA Rep. No. 5, Center for Ocean-Land-Atmosphere studies, 48 pp.

Parker, D. E. and Cox, D. I.: 1995, 'Towards a Consistent Global Climatological Rawinsonde Data-Base', *Int. J. Climatol.* **15**, 473–496.

Polovina, J. J., Mitchum, G. T., Graham, N. E., Craig, M. P., DeMartini, E. E., and Flint, E. N.: 1994, 'Physical and Biological Consequences of a Climate Event in the Central North Pacific', *Fish. Oceanogr.* **3**, 15–21.

Schlesinger, M. E. and Ramankutty, N.: 1994, 'An Oscillation in the Global Climate System of Period 65–70 Years', *Nature* **367**, 723–726.

Schubert, S. D., Rood, R. B., and Pfaendtner, J.: 1993, 'An Assimilated Dataset for Earth Science Applications', *Bull. Amer. Meteor. Soc.* **74**, 2331–2342.

Schwartz, B. E. and Doswell, C. A. III: 1991, 'North American Rawinsonde Observations: Problems, Concerns and a Call to Action', *Bull. Amer. Meteor. Soc.* **72**, 1885–1896.

Swanson, G. S. and Trenberth, K. E.: 1981, 'Trends in the Southern Hemisphere Tropospheric Circulation', *Mon. Wea. Rev.* **109**, 1879–1889.

Taljaard, J. J.: 1972, 'Synoptic Meteorology in the Southern Hemisphere', in Newton, C. W. (ed.), *Meteorology of the Southern Hemisphere*, Met. Monogr. **13**, Amer. Meteor. Soc., 139–213.

Taylor, K. C., Lamorey, G. W., Doyle, G. A., Alley, R. B., Grootes, P. M., Mayewski, P. A., White, J. W. C., and Barlow, L. K.: 1993, 'The "Flickering Switch" of Late Pleistocene Climate Change', *Nature* **361**, 432–436.

Trenberth, K. E.: 1979, 'Interannual Variability of the 500 mb Zonal Mean Flow in the Southern Hemisphere', *Mon. Wea. Rev.* **107**, 1515–1524.

Trenberth, K. E.: 1984, 'Interannual Variability of the southern Hemisphere Circulation: Representativeness of the Year of the Global Weather Experiment', *Mon. Wea. Rev.* **112**, 108–123.

Trenberth, K. E.: 1990, 'Recent Observed Interdecadal Climate Changes in the Northern Hemisphere', *Bull. Amer. Meteor. Soc.* **71**, 988–993.

Trenberth, K. E.: 1992, *Global Analyses from ECMWF and Atlas of 1000 to 10 mb Circulation Statistics*, NCAR Tech. Note NCAR/TN-373+STR, 191 pp. (plus 24 fiche).

Trenberth, K. E.: 1993, 'Northern Hemisphere Climate Change: Physical Processes and Observed Changes', in Mooney, H. A., Fuentes, E. R., and Kronberg, B. I. (eds.), chapter 3 of *Earth System Responses to Global Change: Contrasts between North and South America*, Academic Press, 35–59.

Trenberth, K. E., Berry, J. C., and Buja, L. E.: 1993, 'Vertical Interpolation and Truncation of Model-Coordinate Data', NCAR Technical Note NCAR/TN-396+STR, 54 pp.

Trenberth, K. E., Christy, J. R., and Hurrell, J. H.: 1992, 'Monitoring Global Monthly Mean Surface Temperatures', *J. Clim.* **5**, 1405–1423.

Trenberth, K. E. and Hurrell, J. W.: 1994, 'Decadal Atmosphere-Ocean Variations in the Pacific', *Clim. Dyn.* **9**, 303–319.

Trenberth, K. E. and Hurrell, J. W.: 1995, 'Decadal Coupled Atmosphere-Ocean Variations in the North Pacific Ocean', in R. J. Beamish (ed.) *Climate Change and Northern Fish Populations. Canad. Spec. Publ. Fisheries Aquat. Sci.* **121**, 15–24.

Trenberth, K. E. and Olson, J. G.: 1988, 'An Evaluation and Intercomparison of Global Analysis from NMC and ECMWF', *Bull. Amer. Meteor. Soc.* **69**, 1047–1057.

Trenberth, K. E. and Olson, J. G.: 1989, 'Temperature Trends at the South Pole and McMurdo Sound', *J. Clim.* **2**, 1172–1182.

Trenberth, K. E. and Paolino, D. A.: 1980, 'The Northern Hemisphere Sea-Level Pressure Data Set: Trends, Errors, and Discontinuities', *Mon. Wea. Rev.* **108**, 855–872.

van Loon, H.: 1972, 'Pressure in the Southern Hemisphere', in Newton, C. W. (ed.), *Meteorology of the Southern Hemisphere, Met. Monogr.* **13**, Amer. Meteor. Soc., 59–86.

van Loon, H., Kidson, J. W., and Mullan, A. B.: 1993, 'Decadal Variation of the Annual Cycle in the Australian Dataset', *J. Clim.* **6**, 1227–1231.

Wallace, J. M., Zhang, Y., and Lau, K.-H.: 1993, 'Structure and Seasonality of Interannual and Interdecadal Variability of the Geopotential Height and Temperature Fields in the Northern Hemisphere Troposphere', *J. Clim.* **6**, 2063–2082.

Williams, J. and van Loon, H.: 1976, 'An Examination of the Northern Hemisphere Sea Level Pressure Dataset', *Mon. Wea. Rev.* **104**, 1354–1361.

(Received 23 January, 1995; in revised form 29 June, 1995)

TEMPERATURE ABOVE THE SURFACE LAYER

JOHN R. CHRISTY

Earth System Science Laboratory, University of Alabama in Huntsville, Huntsville AL 35899, U.S.A.

Abstract. Three published data sets of upper-air global temperatures, two from radiosondes and one from satellites, are examined and compared for the lower stratosphere and troposphere.

The global lower stratosphere exhibits a downward trend for the past 16+ years of –0.53 °C (–0.33 °C per decade). Since the 1960's (using radiosondes before 1979 which are subject to known and unknown inhomogeneities) it is likely that there has been a downward trend of about the same magnitude. Significant issues of the stratospheric radiosonde data are: (1) that the long-term time series is biased toward spurious cooling; and (2) the earliest years of Angell display unrealistic variability. Inhomogeneities in satellite data due to orbit drifting and instrument calibration are examined.

The tropospheric temperature has shown a downward trend of –0.11 °C since 1979 (–0.07 °C per decade). Beginning in earlier years, (relying only on radiosonde data before 1979) the estimated warming trend since the late 1950's is +0.07 to +0.11 °C per decade.

Tropospheric and surface temperature anomalies are compared. There is concern that the disproportionate representation of extratropical continents, with their high temperature variance, may bias any long term 'global' surface trend toward a maximum-possible value than would be calculated had all regions (including those with much lower responsiveness) been monitored.

1. Importance of Upper-Air Measurements

Monitoring the earth system has traditionally focused on measurements at the earth's surface because this is where we live and perform our most obvious life sustaining functions. In just the past few decades, however, efforts have been initiated to assess the character of the atmosphere above the surface. Regular upper-air measurements by balloon ascents in scattered locations began in the 1940's and observations from satellites generally began in 1979.

The upper-air is important in the climate variations context because changes in these regions may offer clearly discernible relationships to such phenomena as global warming due to the enhanced greenhouse effect. Barnett and Schlesinger (1987) point out the probability that climate-change trends in the mid-troposphere (5–8 km) will be more clearly evident against the background of natural variability than at other levels. Indeed, results from coupled ocean-atmosphere models indicate that in terms of a *global mean quantity* the troposphere will actually warm at a rate greater than that of the surface temperature (e.g. Manabe *et al.*, 1991, Boer *et al.*, 1992). At higher elevations, above 12 km, the IPCC (1990) indicates there is a high degree of certainty that a greenhouse-gas induced decrease in stratospheric temperatures will occur. The measure of change in these upper atmospheric layers, in combination with quantities observed at the surface, provide an ensemble of information to give the most robust opportunity for climate change detection.

Climatic Change **31**: 455–474, 1995.
© 1995 *Kluwer Academic Publishers.*

In this paper we shall examine three published data sets of upper-air temperatures. Two are derived from radiosonde (balloon) releases and one from satellite measurements.* Since each has different characteristics in the method of construction and in the vertical layers which are measured we note here that we cannot expect perfect agreement among them. Because of the differences, each set brings something unique in our effort to monitor the upper-air temperature.

2. Methods of Measurement

2.1. RADIOSONDE

As mentioned above, the longest record of upper-air measurements comes from radiosonde releases in which a balloon carries an instrument package aloft which transmits data back to a ground station. These observations are made once or twice per day (00 to 12 UT) and are, of course, confined to the narrow column of air through which the balloon ascends. Variables which are recorded are pressure, temperature and humidity.

Fortunately, the pattern of monthly-averaged variation in temperature of the upper atmosphere is characterized by large-scale coherency. As a result, if stations are properly distributed around the globe, a relatively small number may provide a reasonably accurate sampling of the global atmosphere. For example, a few more than 200 stations would be sufficient to observe the global atmosphere in cells of 750 km radius. It is unfortunate however, that vast regions of the global atmosphere, areas over the Southern Oceans for example, contain no radiosonde stations, and so variations in those regions have not been directly observed.

2.1.1. *Angell*

The two main sources of global temperature data derived from radiosondes employ different methodologies. Angell (1988; see also Angell and Korshover, 1983) has selected a set of 63 radiosonde sites that are somewhat widely distributed and that have fairly continuous records since the late 1950's. The physical quantity utilized by Angell to deduce the mean temperature of a layer is geopotential thickness which is the vertical distance between two levels of prescribed pressure, i.e. the volume of the atmospheric column. Volume and temperature are directly proportional between prescribed pressure levels for conditions in which the density (or composition) of the column is invariant through time. Since changes in atmospheric density due to water vapor fluctuations are usually small, this calculation is normally suitable and is known as virtual temperature (Tv) rather than actual temperature.

The accuracy of the calculation of Tv from thickness (i.e. volume) data depends on the original temperature and humidity measurements, since both are needed

---

* We note that a third global radiosonde data set is being constructed and validated, though data were not available for this report (Parker and Cox, 1994).

to determine air density and thus the depth for each pressure layer. Unfortunately, there have been inhomogeneities in the measurement of both quantities, particularly in regards to humidity, so that spurious trends may be introduced into the final calculation of Tv (see below).

Once Tv is determined in Angell's method, the seasonal average for seven broad latitudinal bands is calculated as the average Tv of those radiosondes which lie in the particular band. A simple calculation is then performed in which the bands are averaged into hemispheric and global mean seasonal quantities. The data set begins in 1958.

2.1.2. *Oort and Liu*

The data sets of Oort and Liu (1993) rely on as many radiosonde sites as possible to produce an objective global analysis of the temperature on several pressure levels. The key point here is that the emphasis changes from a few robust stations (as in Angell) to many stations whose combined information is intended to lead to a more robust product. However, the availability and quality of the data for many of these extra sites are limited as many records are discontinuous, contain inhomogeneities or have considerable missing data. Another point to remember is that the increase in station number from 63 to 800+ should not be viewed as a proportional increase in areal coverage. Since the vast majority of additional stations are in North America, Europe and Asia, these areas are, in effect, oversampled for the large-scale features of the atmosphere, while large regions over oceans are still undersampled. For example, there are only 12 stations in the latitudinal band of 40–60° S in Oort and Liu, of which six are within 20° longitude of New Zealand, three on the tip of South America, and none between 176° W and 73° W.

Oort and Liu computed monthly mean statistics of the radiosonde station data, then applied an objective analysis scheme (Conditional Relaxation Analysis Method; Harris *et al.*, 1966) to generate global fields on a 2.5° lat. by 5° long. grid. Hemispheric and global mean values are then calculated from the gridded data which are areally biased only in the sense that each gridpoint has its own, varying level of confidence. It is important to note here that the layer-mean temperatures of Oort and Liu are calculated as the average of temperatures analyzed at several intervening levels. For example, the 850–300 hPa mean temperature from Oort and Liu is calculated from temperatures analyzed at 850, 700, 500, 400 and 300 hPa. Spurious changes due to changing humidity sensors will not affect the data of Oort and Liu as is possible in Angell. Both groups have reported on variations in the temperatures of three layers: 850–300, 300–100 and 100–50 hPa.

2.2. MICROWAVE SOUNDING UNIT

The second method used in monitoring the temperature of the upper-air was developed by Spencer and Christy (1992a, b) in which the intensity of upwelling microwave radiation as measured by NOAA polar orbiting satellites is converted to temperature.* Oxygen molecules function as blackbody radiators near the 60 GHz frequency (about 5 mm wavelength) and their emissions are recorded by the Microwave Sounding Units (MSUs). Since oxygen is an abundant, well-mixed gas in the air, this radiation serves as an indicator of overall atmospheric temperature. Monitoring the intensity of frequencies slightly lower than 60 GHz, the MSU takes advantage of pressure-broadening effects so that more oxygen is available for observation. In other words, the instrument 'sees' deeper into the atmosphere.

Emission from oxygen at all levels is observed by the instrument, but by selecting the appropriate near-60 GHz frequency, one may pre-determine the vertical profile of measured emissions. For purposes here we shall report on results from channel 4 ($T4$), or lower stratosphere in which the bulk of the emission comes from the 120–40 hPa layer which is comparable with the 100–50 hPa layer measured by radiosondes. For tropospheric comparisons, we shall use a synthetic MSU channel 2R ($T2R$) which peaks at 700 hPa and whose emission is primarily in the 1000–400 hPa layer, or somewhat lower than the 850–300 layer of radiosondes. Though these two MSU layers do not coincide exactly with the 100–50 and 850–300 hPa layers reported from radiosonde measurements, there is sufficient dependency in the overlapping layers to provide remarkable agreement. Each satellite (there are usually two in operation at any given time) will produce about 30,000 readings as it travels over 14 north-south oriented orbits per day (Spencer and Christy 1992a, b, 1993).

The emission that is observed by the satellite instrument is almost, but not entirely, that produced by oxygen molecules in the air. In reality emissions from other constituents in the air and emissions from the surface ($T2R$ only) contribute to the signal. Some of these emissions may not be as strongly related to the temperature of the air as is oxygen and so have the potential for contaminating the desired temperature value. In large-scale thunderstorm complexes for example, the microwave radiation received by the satellite for $T2R$ will include a portion which began above 400 hPa (therefore colder) altitudes and has been reflected upward by the largest precipitation particles. In these systems, the temperature viewed by the satellite is colder than that of the layer of interest. Such cold biases are screened out of the MSU data set. The non-thermometric but systematic contribution of surface emissions is removed by considering only the anomalies from the background mean (Spencer and Christy, 1992a).

To summarize, there are two main categories of upper-air temperature measurements, one in which radiosondes provide the basic data and one in which MSUs serve as the thermometers. To understand the reliability of these very different data sets, we shall briefly report on recent research into concerns of homogeneity and long-term stability.

---

* In an earlier, independent study, Nash and Edge (1989) demonstrated the utility of microwave sounders for monitoring global temperatures.

3. Recent Reports Concerning Data Reliability

3.1. RADIOSONDE

Compilations of the radiosonde temperature of upper-air layers for studies of long-term changes of global averages have been carried out by the two groups identified above, though Oort and Liu has not been updated since reported on in IPCC (1992) in which values were calculated only through 1989. We show in Figures 1 and 2 the seasonal anomalies of global temperature for the troposphere and lower stratosphere from all three datasets. These anomalies have been adjusted to be departures from an average for 1980–1989. We shall refer to the data of Angell as *ANG* and to Oort and Liu as *OL*.

3.1.1. *Spatial Coverage and Other Problems*

An aspect of uncertainty in the *ANG* data set, and to a lesser extent that of *OL*, is related to the non-uniform geographic distribution of stations. Vast areas over oceans, particularly in the Southern Hemisphere, are not monitored at all so there is always a component of the global mean temperature which is missing (Trenberth and Olson, 1991; Karl *et al.*, 1994). The magnitude of the variance in seasonal anomalies in *ANG* was demonstrated to be greater than actual simply because the spatial sampling was not complete and because the latitudinal weighting coefficients over-emphasized the contribution from higher latitudes. Because values in the higher levels (i.e. lower stratosphere) were even less well-sampled in the early years, variations there are highly suspicious (see below).

A further problem was suggested by Parker and Cox (1994) concerning the biasing of earlier reports toward warmer temperatures in the lower stratosphere. It seems that the quality of the balloons was poorer in the first part of the record, and there was a tendency for the balloons to burst on the days which were colder at the higher elevations. In addition, problems with equipment and/or observers generally occurred on the coldest days. As a result, seasonal averages in the early record contained a disproportionate number of warmer observations. Therefore, the overall downward trend in lower stratospheric temperatures in *ANG* and to some extent *OL* contains a spurious component of unknown magnitude.

A comparison with the *T4* over the last 16+ years indicates that the lower stratospheric trend in *ANG* is about 0.7 °C per decade colder than observed by the MSU (Figure 1). This large and recent discrepancy in trends appears to be related to many factors, including discontinuities in a specific subset of the radiosonde stations. Evidently, stations in the tropics which were formerly French-operated experienced major changes through the years in equipment or procedures and stand as outliers on the negative side of the distribution of individual station trends (J. Angell, personal communication). Two other stations, Bombay and Calcutta, have experienced a similar downward trend relative to nearby stations. In addition, a recent investigation has revealed that the phase-in of sondes built by a new

[329]

manufacturer for the Australian network in 1988–89 produced a clear downward displacement of 1 °C when compared with *T4* temperatures (D. Parker, personal communication). Finally, as mentioned above, the difference in trends is influenced by the weighting scheme applied to the latitudinal bands of *ANG* when computing the global mean. Added together, these various factors have caused the *ANG* radiosonde stratospheric trend to be more negative than is observed by the MSU. Methods to adjust the radiosonde time series for those specific stations known to be problematic are currently under investigation.

3.1.2. *Sensor Changes*

The devices which measure temperature and humidity in these radiosondes have undergone many changes in the past 40 years which have had significant effects on long-term trends (IPCC, 1992). Recent studies report on the effects of humidity and temperature sensor changes.

3.1.2.1. Humidity Sensor. As mentioned, the fundamental quantity reported by *ANG* is *Tv* which is derived from the atmospheric thickness between two levels of constant pressure. The calculation of *Tv* assumes that the column has the density of dry air. Since water vapor molecules have less mass than average dry air molecules but exert the same pressure, *Tv* is always warmer than actual temperature for a column containing moisture. Therefore, any changes in moisture content will be viewed as changes in *Tv* in which a moistening of say 10% would be viewed as a small increase in *Tv* of about 0.02 °C.

Since the most troublesome discontinuities and accuracy problems concern the moisture sensors, one must keep in mind that this impacts the derivation of *Tv* from pressure-level heights (Gaffen *et al.*, 1992; Gutzler, 1993). Elliott *et al.* (1994) calculate that the 'drying' of the 850–300 hPa layer due to improved radiosonde humidity sensors has lead to a spurious cooling since 1958 of 0.014 to 0.028 °C per decade in the data set of *ANG*. This cooling is primarily related to sensor changes prior to the 1980's. The temperatures in *OL* are not affected by humidity sensor inhomogeneities.

Of concern for long-term trend calculation is the fact that the troposphere, particularly in the tropics, may be undergoing an increase in humidity (Elliott *et al.*, 1991; Gaffen *et al.*, 1992; Gutzler, 1993). As long as there are no changes in moisture content or in humidity sensors, the *Tv* of *ANG* are quite useful for calculating trends in actual temperature. An increase in moisture, however, would be viewed in *ANG* as an increase in *Tv* while the actual temperature may not have risen (Gutzler, 1993). Changes in either humidity or temperature are equally important, but may have different impacts on the climate system. A trend in *Tv* does not tell us whether it is humidity, temperature or both which may be changing.

3.1.2.2. Temperature Sensor. While information on changes in radiosonde sensors is rarely complete, Gaffen (1994) reported on the effects of documented

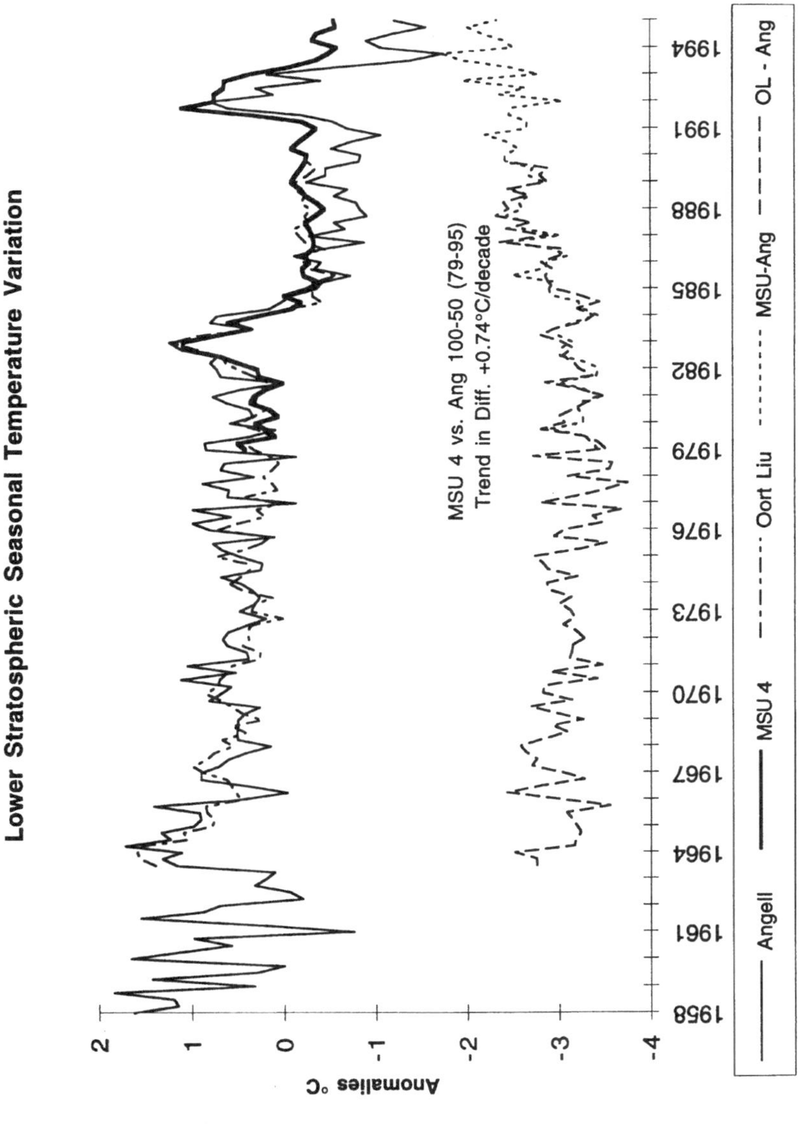

Fig. 1. Top curves: seasonal global temperature anomalies of 100–50 hPa from Angell (thin solid line), Oort and Liu (dashed) and MSU T4 lower stratosphere (Spencer and Christy, thick solid line). Base period is 1980–1989 for all. Lower curves: differences MSU minus Angell (short dashed) and Oort and Liu minus Angell (long dashed).

changes in the temperature sensors and found that at least 43% of the stations used in *ANG* have clear inhomogeneities. (One could easily generalize this result to note that at least this percentage of stations used in *OL* is also inhomogeneous.) The most dramatic inhomogeneities occurred prior to 1979 in several of the stations and involved movement of site, upgrading of equipment (for example, lengthening the cord between the balloon and instrument package), changes in solar radiation-exposure equations and utilization of newer instrumentation. While the complete impact of all inhomogeneities will never be known, Gaffen estimated the net effect as a spurious cooling trend on the temperatures (virtual and actual) since 1958. Though quite large and variable for individual stations, the magnitude of the global tropospheric trend error appears to be small. However, the global lower stratospheric trend was found to have been significantly affected, particularly in the tropics and Southern Hemisphere, producing much colder than realistic trends.

The data of *ANG* and *OL* are not exactly comparable. While *ANG* gives the average Tv between two widely spaced pressure surfaces, *OL* provides the mass-weighted temperature between those surfaces. If there are variations within the column, each data set will be affected in slightly different ways. Consider two cases of fractional warming for the 850–300 hPa layer, firstly that the lowest 100 hPa warms by 1 °C and secondly that the highest 100 hPa warms by 1 °C (all other levels remain the same). In the mass-weighted average temperature of *OL*, there would be essentially no difference between the two 850–300 hPa average temperatures. On the other hand, the thickness change for the lowest 100 hPa would be less than that of the highest 100 hPa, so that the net change of column thickness would be greater in the latter case (over 4 m more). As a result, the Tv value of *ANG* would register a warmer column in the second case. Fortunately, for monthly and seasonal averages, the troposphere is essentially equivalent barotropic so that the temperature of individual layers does not change independently as proposed in the two cases above (Blackmon *et al.*, 1979).

3.2. MICROWAVE SOUNDING UNIT DATA

MSUs on polar orbiting satellites have demonstrated high precision and global coverage for the temperature of deep layers in the troposphere and stratosphere (Spencer and Christy, 1992b, 1993). Monthly (daily) global values are known to within two (five) hundredths of a degree in the lower stratosphere (Spencer and Christy, 1993; Christy and Drouilhet, 1994) and to within four (eight) hundredths in the lower troposphere (Christy *et al.*, 1995).

Because constituents in the atmosphere besides oxygen contribute small amounts of energy to the signal, there is concern that these may contaminate the reported temperature values. Spencer *et al.* (1990) calculated the impact of emissions from the various constituents on the microwave brightness temperature specifically to determine the magnitude of any possible error that might impact the anomalies. The results demonstrated that the air temperature signal is indeed the dominant signal by about two orders of magnitude. As listed below, decreases in MSU global tropospheric temperature measurements of 0.01 °C due to non-thermometric processes would be observed if one of the following conditions occurred:

- a global increase in mid-level cloud amount of 7%
- a global decrease in low clouds of 8%
- a decrease in tropospheric humidity of 15% over all oceans
- an increase of tropospheric humidity of 10% over all land

Global, monthly anomalies have a range of about 0.8 °C, so errors of less than 0.01 °C would be well within the noise level. In any case, none of these changes, to the extent required for influencing the $T2R$ data by 0.01 °C, has been observed. (The stratospheric channel is insensitive to massive changes in these quantities at the 0.01 °C level.)

The time series of MSU data is a combination of data from seven satellites launched at about two-year intervals. Data are received from only two satellites at a time, so as a new satellite is placed in service, an older one is decommissioned. The severe environment of space, particularly for satellites which orbit in the mid-day to mid-night path, causes deterioration of the sensitive components of the system. This was of considerable concern and so tests were devised to validate the MSU data with independent sources of information.

In the first test, MSU data were compared from two concurrently operating satellites (overlaps up to 4 years) and no significant trends in the difference time series for channels 2 and 4 were found through 1991 (Spencer and Christy, 1992a, 1992b, and 1993; Christy and Drouilhet, 1994; Christy and Goodridge, 1994). Because each satellite contained its own independent MSU, a comparison between the two was an excellent validation test of the MSU signal. The results indicate that either the instruments had excellent stability over long time periods, or (highly unlikely) that all sensors drifted spuriously at exactly the same rate. Not only was there no significant drifting observed, but the daily global anomalies were captured by both satellites in which the daily anomaly variance was 30 times that of the noise variance (65 for monthly) in the $T2R$ (Christy et al., 1995). The signal-to-noise ratios for $T4$ were even greater than those of $T2R$.

In a second test of precision, direct ten-year comparisons between gridpoint MSU temperatures and radiosonde-computed temperatures demonstrated the merged-MSU data did not drift relative to many-station averages of the sondes. These comparisons utilized U.S.-controlled radiosondes which had consistent sensors during the 1979–88 validation period. Individual station vs. MSU comparisons usually differed in the trend, but the precision of the radiosonde trends was too poor to make strong conclusions. In some cases, radiosondes that were within 150 km of each other would differ in decadal trend by 1 °C. By taking monthly averages of at least five stations in specified regions and comparing with the corresponding MSU grid cell values, we found the correlations and decadal trends to be within the limit of precision of both radiosondes and MSUs (Spencer and Christy, 1992a, b, and 1993). Once again, in this completely independent test of MSU precision, the satellite data were demonstrated to be accurate representations of deep-layer atmospheric temperatures.

The remarkable record of inter-satellite agreement was interrupted in late 1990. The MSU on NOAA 11 began measuring a tropospheric trend that was warmer than observed from the concurrently operating MSU on NOAA 10. When NOAA 12 was launched to replace NOAA 10, the NOAA 11 MSU continued to show warming relative to that observed on NOAA 12. The health of both NOAA 11 and 12 had been sub-standard so it was difficult to determine whether either or both were spuriously drifting. The subsequent investigation found two problems, one in each satellite, that were of a systematic nature and therefore easily corrected (for details see Christy et al., 1995).

The relative warming in NOAA 11's tropospheric temperature versus the other two MSUs resulted from a time drift of the platform. Since late 1990, NOAA 11 began to loose a few seconds each day in equatorial orbit crossing time so that it observed earth locations at later and later local time. By 1994 the platform was crossing at 12 seconds later each day. So, the original 0130/1330 local crossing times had degraded to 0430/1630 by February 1994. As time went on, the MSU on NOAA 11 was observing a cooler atmosphere as the early morning crossing time drifted from 0130 to 0430 and a warmer atmosphere in the afternoon reading as its crossing time drifted from 1330 to 1630. The net effect of this slippage in the observation time created a spurious warming in the global average temperature of NOAA 11 because the average 1330 to 1630 warming was greater in magnitude than the cooling observed from 0130 to 0430. To compensate for this, the relative trend between NOAA 11 and NOAA 10 & 12 was calculated and removed from NOAA 11 (Christy et al., 1995).

The second problem dealt with the MSU on NOAA 12. A study of the time series of differences between the now-corrected NOAA 11 and NOAA 12 MSUs indicated a weak annual cycle in which NOAA 12 revealed warmer temperatures at the coldest places and cooler temperatures at the warmest locations. Further investigation found that the dynamic range on NOAA 12's MSU was only 98% of normal so that extreme values were being mitigated. The situation was rectified by mathematically expanding the dynamic range to show agreement with the other satellites (Christy et al., 1995).

With these corrections made to the NOAA 11 and NOAA 12 MSUs, the signal-to-noise ratios of the measurements made since 1990 are now at the level observed in the previous 12 years.

4. Assessment of Results Updated to February 1995

4.1. LOWER STRATOSPHERE

The lower stratospheric temperature (16–21 km or 100–50 hPa) has demonstrated dramatic changes over the past decades (Figure 1). Most remarkable are the sudden warmings caused by infrared-absorbing aerosols from volcanoes (Agung, 1963, Nyamuragria, 1981, with El Chichon, 1982, and Mount Pinatubo, 1991). It should be noted that post-eruption temperatures fell to lower levels than pre-eruption values in both recent volcanic episodes, lending support to the idea that aerosols contribute to the depletion of ozone (WMO, 1989).

The cooling in the global stratospheric temperatures of ANG (Figure 1) relative to $T4$ and OL has been discussed above and in the literature (Trenberth and Olson, 1991; IPCC, 1992; Christy and Drouilhet, 1994; Gaffen, 1994; Angell, personal communication). From 1979 through February 1995, the $T4$ global trend is –0.33 °C per decade while of ANG it is –1.08 °C. The trends for 1979–89, a period common to all three data sets, are –0.79, –0.85 and –1.57 °C per decade for $T4$, OL and ANG respectively. Prior to 1979, OL has a trend which is more downward (–0.63 °C per decade for 1964–1978) than does ANG (–0.38 °C). We can see that lower stratospheric trends produced from ANG are indeed too cold for 1979–95, yet for 1964–1978 it is unclear as to the magnitude of any stratospheric trend (IPCC, 1992).

Examination of the anomaly time series (Figure 1) reveals a tendency for dramatic swings in ANG, especially prior to 1964, some of which even exceed the perturbations produced by Mt. Pinatubo aerosols in 1991–92 (see below). The seasonal anomalies of OL and $T4$ show much less inter-seasonal variance and produce between them a correlation of 0.95 (annual anomalies correlate at 0.98, Table I). This is remarkable agreement for two completely independent methods of analysis. We therefore reinforce the conclusions of previous studies that the undersampling of the global stratosphere as exhibited in the ANG data set is a source of considerable error (up to 1 °C) when calculating a single season stratospheric anomaly. In addition, the many issues of inhomogeneity mentioned earlier make it difficult to determine the error characteristics of the data prior to 1979.

Prior to 1964, the global anomalies of ANG seem unrealistic, but have been published to give information on the 1963 Agung eruption. Users have always been cautioned about possible inhomogeneities when drawing conclusions from these stratospheric data, especially for 'global' values which did not include the tropics or southern subtropics before 1964 nor Russia prior to 1970 (Angell, personal communication).

Comparison of all three data sets indicates that the stratosphere has been cooling since 1979 (–0.33 °C per decade in $T4$), and has probably experienced cooling since the 1960's yet with an uncertain magnitude. The 31-year trend calculated by

TABLE I

Correlations between global anomalies of lower stratospheric temperatures where above diagonal values are annual and lower are seasonal. Periods of comparison are different: MSU vs. OL: 1979–1989; MSU vs. ANG: 1979–1995; ANG vs. OL: 1964–1989

	MSU 4	Oort and Liu	Angell
MSU 4		0.98	0.87
Oort and Liu	0.95		0.90
Angell	0.85	0.84	

TABLE II

As above for tropospheric temperatures

	MSU 2R	Oort and Liu	Angell
MSU 2R		0.96	0.95
Oort and Liu	0.92		0.97
Angell	0.78	0.92	

appending *T4* for 1979–95 to OL 1964–78 is –0.36 °C per decade, about the same as the 1979–95 *T4* trend.

4.2. TROPOSPHERE

The problems which plague the radiosonde stratospheric data set are reduced for the troposphere. Data links between instrument packages and ground stations are more likely to be operating in the tropospheric portion of the ascent, and the problem of bursting balloons is much less an issue for these lower, warmer elevations (Gutzler, 1993). The sampling rate is therefore improved with a high likelihood that no bias exists for underreporting days with certain cold or warm weather regimes. When comparing annual global anomalies between the three data sets, we find correlations of 0.95 and above (Table II) even though the layer measured by *T2R* represents a lower elevation than that of the 850–300 hPa layer reported in *ANG* and *OL*.

We note from Figure 2 what was observed in our stratospheric comparison, namely, that the undersampling of *ANG* is evident in the larger season-to-season variability than is measured by *OL* and *T2R*. What is encouraging is that the magnitudes of the differences (lowest two series) are much less than seen for the stratosphere (where errors in the calculation of pressure-level heights tend to accumulate as elevation increases). We have already pointed out the results of previous investigations concerning inhomogeneities in the data for this layer. They generally found that prior to 1980 radiosonde instrument changes (those known to researchers) have probably reduced the magnitude of the observed warming trend, which as measured for 1964–78 in *ANG* is +0.15 °C and for *OL* is +0.08 °C. (Beginning the *OL* vs. *ANG* comparison in the cold year of 1964 and ending with a relatively warm 1978 is misleading when viewing the entire time series (Figure 2) as the trend in *ANG* for 1958–1978 is actually less than zero at –0.08 °C per decade).

One remarkable result of the intercomparison of tropospheric temperatures is that the 16+ year trends for *T2R* and *ANG* (1979–95) are identical, being –0.07 °C per decade. Though single season comparisons may show differences of more than 0.2 °C, their random nature evidently has balanced out over time so that the longer time series is representative of the true trend. Of course, these two systems monitor different layers of air, and their agreement may not be as pure a validation comparison as it seems. Errors may be present which have conspired to produce the startling agreement. The *T2R* includes information from the boundary layer (i.e. below 850 hPa) as well as concentrating its signal below 500 hPa. Should this lower layer have the same trend as the 850–300 hPa layer? Perhaps changes in procedures or instrumentation for radiosondes contributed small perturbations to the true trend (e.g. new software containing changes in solar radiation corrections) to produce the agreement. The estimated precision of the *T2R* trend is ±0.03 °C over the 16+ year period, or that the trend is –0.08 to –0.14 °C per 16+ years (–0.05 to –0.09 per decade). In any case, we report with high confidence that the global tropospheric temperature has experienced a decline since 1979 of –0.07 °C ± 0.02 °C per decade. A hybrid data set in which values from each data source are averaged into a seasonal anomaly yields a trend since 1958 of +0.085 °C per decade. Considering the possible cool trend-bias of the early radiosonde record we report the trend since 1958 as +0.07 to +0.11 °C per decade.

4.3. COMPARISON OF MSU 2R AND IPCC SURFACE TEMPERATURES

With the lenghtening record of satellite measurements, the opportunity arises to investigate the relationship of temperature variations between the lower troposphere and the surface. Previous studies have performed such comparisons, see for example Trenberth *et al.* (1992) concerning SST vs. MSU 2 (mid-troposphere), Christy (1992) for regional comparisons and IPCC 1992 for global and hemispheric comparisons. These studies have shown that *T2R* has greater variance than *Tsfc* in virtually every case of local and large-scale averaged area anomalies except over the mid-continents of the Northern Hemisphere. In addition, the correlations between *Tsfc* and *T2R* were highest in regions with dense surface monitoring networks and in areas with considerable surface/tropospheric mixing. We can illustrate these results in Table II in which the annual and monthly anomalies of *Tsfc* (described

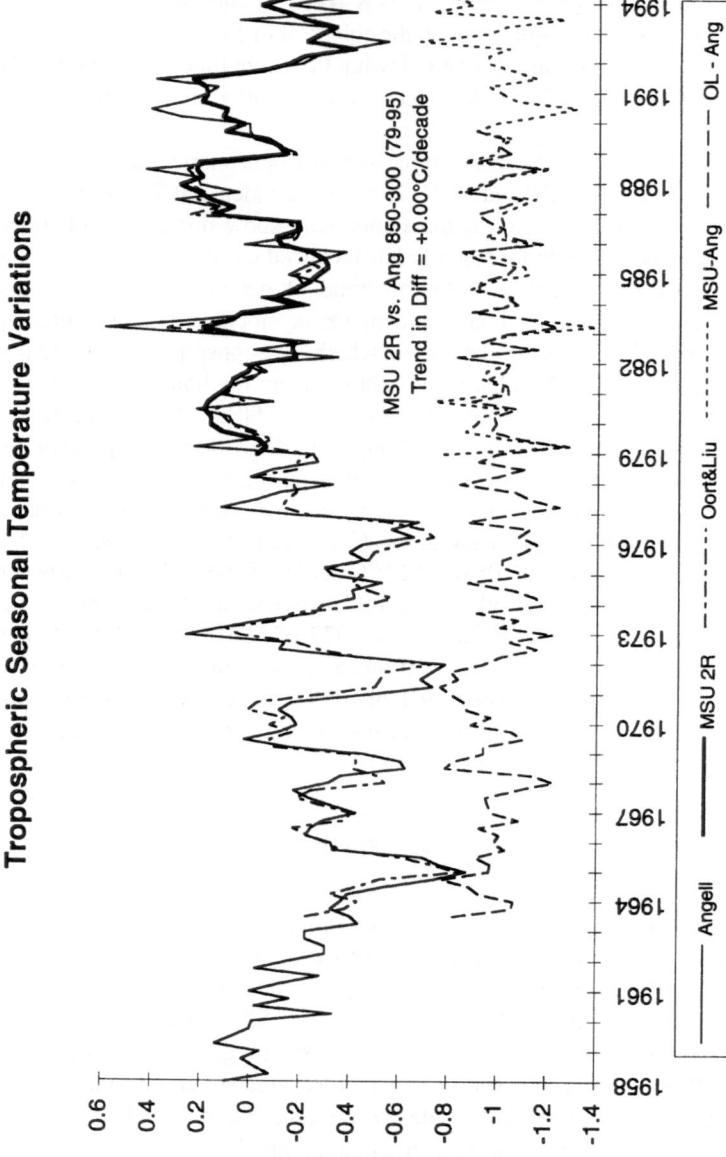

Fig. 2.   As in Figure 1 for 850–300 hPa temperatures for Angell and Oort and Liu, and MSU 2R (lower troposphere).

TABLE III

Statistical, regional comparisons between *T2R* and *Tsfc* (IPCC, 1992, plus updates) 1979–94.
Standard deviations (σ) are in °C

Area	Boundaries				Annual			Monthly		
	W	E	S	N	r	σ T2R	σ Tsfc	r	σ T2R	σ Tsfc
Globe	−180	180	−90	90	**0.68**	0.165	0.106	**0.59**	0.202	0.148
No. Hem.	−180	180	0	90	**0.75**	0.181	0.142	**0.63**	0.230	0.210
So. Hem.	−180	180	−90	0	**0.56**	0.157	0.078	**0.49**	0.221	0.129
Land					**0.71**	0.195	0.151	**0.68**	0.253	0.243
Ocean					**0.63**	0.144	0.081	**0.50**	0.184	0.102
N. America	−125	−75	30	75	**0.95**	0.410	0.512	**0.93**	0.960	1.223
Europe	−10	35	40	70	**0.95**	0.404	0.491	**0.82**	0.958	0.897
Antarctica	−180	180	−90	−65	**0.95**	0.551	0.484	**0.87**	1.227	1.338
N. Trop. Atl.	−80	−15	5	20	**0.90**	0.247	0.220	**0.73**	0.326	0.283
NE Asia	125	150	30	70	**0.89**	0.460	0.400	**0.80**	0.921	0.748
E. Trop. Pac.	−150	−85	−15	10	**0.87**	0.344	0.467	**0.79**	0.444	0.584
N. Cen. Asia	35	125	40	70	**0.86**	0.402	0.611	**0.89**	0.924	1.356
N. Pacific	150	−130	30	50	**0.84**	0.450	0.325	**0.58**	0.706	0.396
S. Pacific	165	−140	−50	−25	**0.79**	0.382	0.232	**0.59**	0.650	0.316
N. Atlantic	−60	−20	25	60	**0.77**	0.254	0.138	**0.59**	0.478	0.220
Australia	115	155	−35	−10	**0.72**	0.305	0.190	**0.69**	0.539	0.435
Argentina	−75	−55	−55	−25	**0.71**	0.279	0.218	**0.69**	0.742	0.545
S. Atlantic	−35	−10	−50	−10	**0.66**	0.175	0.180	**0.46**	0.343	0.248
Indian O.	45	110	−40	0	**0.65**	0.196	0.147	**0.44**	0.283	0.205
W. Trop. Pac.	120	180	−10	10	**0.10**	0.161	0.085	**0.23**	0.232	0.185

and utilized in IPCC (1992) and updated to 1994) and gridded *T2R* are compared
for different 'rectangular' sections of the globe (Figure 3).

The *Tsfc* anomalies are analyzed on a 5° × 5° grid and are not interpolated or
extrapolated to fill in missing areas. The correlation calculations were performed
with a 'common grid' meaning that if one data set reported a missing value, the
other data set was also set to missing for that gridbox and month. The vast majority
of missing grids were due to missing *Tsfc* in oceanic regions. Table III begins with
global, hemispheric, land and ocean correlations, then lists various regions which
are ranked according the magnitude of the annual anomaly correlation. Standard
deviations of the temperature for areal means are also given.

The results verify earlier studies that large continental regions with good monitoring networks and large standard deviations produce excellent agreement between
T2R and *Tsfc*. It was shown in Spencer and Christy (1992b), for example, that for
mid-latitude, mid-continent stations, monthly anomalies for any two layers which
shared a lowest pressure level of 1000 hPa (e.g. 1000–700 vs. 1000–300 hPa) were

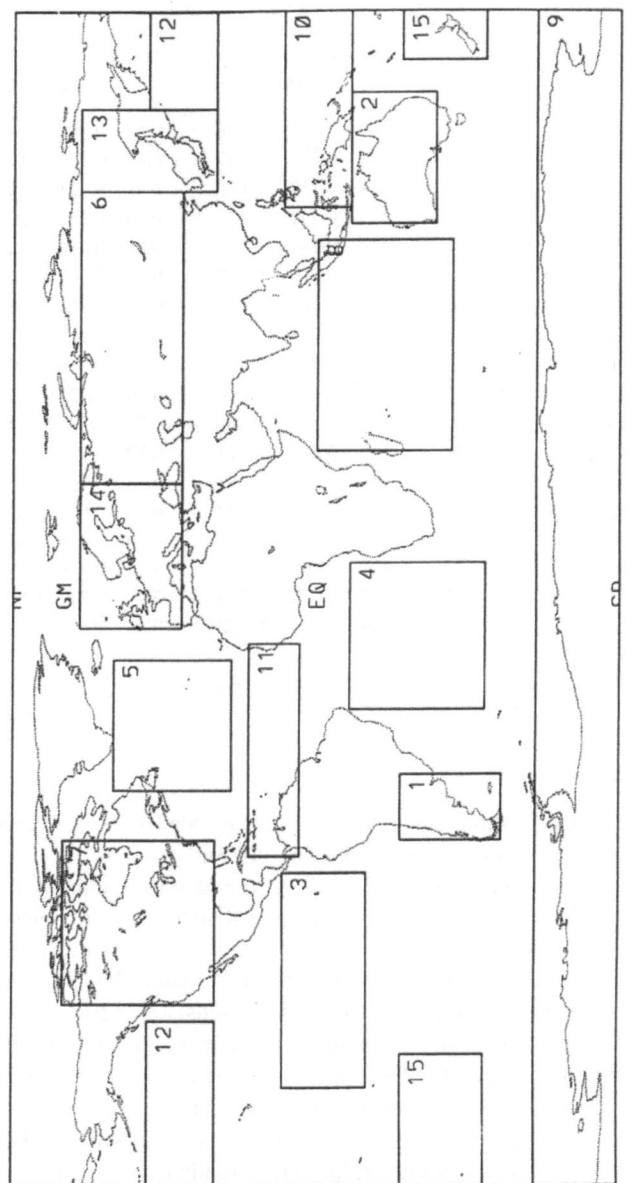

Fig. 3. Geographic areas utilized in the regional correlations of Table III: 1: Argentina; 2: Australia; 3: E. Tropical Pacific; 4: South Atlantic; 5: North Atlantic; 6: N. Central Asia; 7: North America; 8: Indian Ocean; 9: Antarctica; 10: W. Tropical Pacific; 11: N. Tropical Atlantic; 12: North Pacific; 13: NE Asia; 14: Europe; and 15: South Pacific.

intercorrelated at least at the 0.90 level for layers bounded up to 200 hPa. This was true even for winter months when strong inversions were likely to confound the inter-layer correlations. This evidence shows that these inversions have little impact on the interlayer equivalent barotropic nature of the monthly-averaged atmosphere.

Annual correlations above 0.88 are seen for land areas of N. America, Europe, NE Asia, and Antarctica. The lower values found for Australia and Argentina are probably due to the influence of several ocean grid cells which were averaged into the rectangular boxes containing these two regions.

The lower correlations over oceanic areas were examined by Trenberth *et al.* (1992) and found to be due to several factors. One factor is simply that ocean temperatures and air temperatures are not the same physical quantity. Secondly, many ocean grids are not well-monitored, giving rise to larger errors and therefore lower *Tsfc* vs. *T2R* correlations. Finally, the atmosphere over large areas of the ocean is highly stably stratified which allows the cool marine boundary layer to be decoupled from the variations of the free troposphere above. This is mainly true for the subtropical and tropical oceans. The combined effect of uncertainty in the measurements and the real physical difference of temperature between the ocean and troposphere leads to much lower correlations overall. The extreme cases of decoupling and coupling are seen in the tropical Pacific where in the West the correlation is not significantly different from zero while in the East it is 0.87.

Correlations of *Tsfc* vs. *T2R* for mid-latitude oceanic regions are relatively high (0.77 for N. Atlantic and 0.84 for N. Pacific) though the poorer quality of S. Atlantic measurements and the inclusion of equatorial latitudes are probably factors in the 0.66 value there. We note that in the northern midlatitude oceans, sufficient ship traffic has been maintained since 1979 to provide good sampling of the water temperatures.

In Table III we see that the surface temperatures over extratropical continents have a greater impact on global averages than their areal coverage warrants due to their large signal variance in comparison with temperature fluctuations elsewhere. Near-surface air over the oceans and over tropical land has much less temperature variance than air over extratropical land. However, in comparison to extratropical land, these other areas represent much more geographic coverage so that accurate ocean and tropical land measurements are still necessary for the determination of a global surface temperature anomaly.

One contribution to the uncertainty in trends of near-surface air temperatures over the past 140 years is not simply the lack of global coverage in the measurements, but that regions of high temporal variance may be overrepresented. If there has been a global warming trend in the past century, it is certainly plausible that the trend is greater over extratropical continents where variance (i.e. thermal response) is higher than experienced over the remainder of the globe. Unfortunately, these extratropical continents have always had a greater proportion of representation in the 'global' average than regions of less variation. As such, long time series of 'global' *Tsfc* with bracketed error estimates on the trend, may be more representative of the maximum-possible trend than a representation of the true global trend. To a lesser extent this may effect trends from radiosonde networks that are biased toward land stations.

5. Conclusions

We have shown that upper-air temperatures are an important source of information in the determination of climate variations. Three data sets of global temperature have been published for the upper-air and we have examined the character of the temperature trends for the lower stratosphere (100–50 hPa) and troposphere (850–300 hPa, 1000–400 hPa for MSU).

Our assessment for the lower stratosphere is that there has been a downward trend in temperatures for the 16+ years of January 1979 to March 1995 of –0.53 °C (–0.33 °C per decade). Trends calculated from earlier starting dates require data which prior to 1979 are based on radiosondes only and are therefore subject to known and unknown inhomogeneities. It is likely, however, that there has been a downward trend of about the same magnitude (–0.3 to –0.4 °C per decade) since the 1960's. Volcanic aerosols have had a significant impact on the temperature of the lower stratosphere.

We are highly confident that the temperature of the troposphere has shown a downward trend of about –0.11 °C since 1979 (–0.07 °C per decade). The cooling influence of stratospheric aerosols from Mt. Pinatubo (1991–1994) has certainly contributed to this trend (Christy and McNider, 1994). Beginning in earlier years, we rely only on radiosonde data before 1979 and estimate a warming trend since the late 1950's to the present of +0.07 to +0.11 °C per decade.

The most significant issues of the radiosonde stratospheric data are: (1) that the long-term time series is biased toward spurious cooling for various reasons (mainly Angell); and (2) the earliest five years of Angell display unrealistic variability. The radiosonde tropospheric temperatures tend to have fewer problems though the issue of virtual temperature and its dependence on accurate humidity measurements is a new concern because significant inhomogeneities exist for this quantity.

Data from surface networks averaged to estimate a 'global' temperature have many issues of inhomogeneity associated with them (see IPCC, 1990, 1992). One point made here is that the disproportionate representation of extratropical continental regions, with their associated high responsiveness to forcing (i.e. high temperature variance), may bias any long term 'global' trend toward the maximum-possible value versus what would have been calculated had all regions (including those with much lower responsiveness) been sampled. This is less of an issue for radiosonde networks as they sample the upper-air which has less variation in the magnitude of variance from region-to-region. For satellites, which essentially observe the entire globe, errors due to spatial coverage are not a problem.

Because of the problems that have been identified and will continue to be discovered in these upper-air data sets, it is important that multiple monitoring capabilities be maintained (radiosonde and satellite). This is critical to issues of independent validation and determination of measurement precision. We encourage continued research into possible inhomogeneities in all data sets so that the best long-term results may be forthcoming and that current data collection and manipulation adhere to international standards. We also recommend that current monitoring capabilities be expanded, especially with regard to the spatial sampling of the radiosonde network. In the case of the MSU, research will be required to merge the current data set with measurements from its successor (Advanced Microwave Sounding Unit) which will monitor slightly different frequencies than observed now.

Acknowledgements

Discussions with D. Parker, K. Trenberth, A. Oort and R. Spencer helped to clarify several sections of the text. The author is particularly indebted to J. Angell who provided considerable information on the radiosonde data set and on current investigations concerning homogeneity. This research was supported by the U.S. Department of Energy's (DOE) National Institute for Global Environmental Change (NIGEC) through the NIGEC Southeast Regional Center at The University of Alabama, Tuscaloosa (DOE Cooperative Agreement No. DE-FC03-90ER61010). Financial support does not constitute an endorsement by DOE of the views expressed in this article.

References

Angell, J. K.: 1988, 'Variations and Trends in Tropospheric and Stratospheric Global Temperatures, 1958–87', *J. Clim.* **1**, 1296–1313.
Angell, J. K. and Korshover, J.: 1983, 'Global Temperature Variations in the Troposphere and Stratosphere, 1958–82', *Mon. Wea. Rev.* **111**, 901–921.
Barnett, T. P. and Schlesinger, M. E.: 1987, 'Detecting Changes in Global Climate Induced by Greenhouse Gases', *J. Geophys. Res.* **92**, 14,722–14,780.
Blackmon, M. L., Madden, R. A., Wallace, J. M., and Gutzler, D. S.: 1979, 'Geographic Variations in the Vertical Structure of Geopotenital Height Fluctuations', *J. Atmos. Sci.* **36**, 2450–2466.
Boer, G. J., McFarlance, N. A., and Lazare, M.: 1992, 'Greenhouse Gas-Induced Climate Change Simulated with the CCC Second-Generation General Circulation Model', *J. Clim.* **5**, 1045–1077.
Christy, J. R.: 1992, 'Monitoring Global Temperature Changes from Satellites', in Majumdar, S. K., Kalkstein, L. S., Yarnal, B. M., Miller, E. W., and Rosenfeld, L. M. (eds.), *Global Climate Change: Implications, Challenges and Mitigation Measures*, The Pennsylvania Academy of Science, 566 pp.
Christy, J. R. and Drouilhet, S. J.: 1994, 'Variability in Daily, Zonal Mean Lower-Stratospheric Temperatures', *J. Clim.* **7**, 106–120.
Christy, J. R. and Goodridge, J. D.: 1994, 'Precision Global Temperatures from Satellites', *Atmos. Env.*, (in press).
Christy, J. R. and McNider, R. T.: 1994, 'Satellite Greenhouse Warming', *Nature* **367**, 325.

Christy, J. R., Spencer, R. W., and McNider, R. T.: 1995, 'Reducing Noise in the Daily MSU Lower Tropospheric Temperature Data Set', *J. Clim.* **8**, 888–896.

Elliott, W. P., Gaffen, D. J., Kahl, J. D. W., and Angell, J. K.: 1994, 'The Effect of Moisture on Layer Thicknesses Used to Monitor Global Temperatures', *J. Clim.* **7**, 304–308.

Elliott, W. P., Smith, M. E., and Angell, J. K.: 1991, 'On Monitoring Tropospheric Water Vapor Changes Using Radiosonde Data', in M. E. Schlesinger (ed.), *Greenhouse-Gas-Induced Climatic Change: A Critical Appraisal of Simulations and Observations*, Elsevier, pp. 311–328.

Gaffen, D. J.: 1994, 'Temporal Inhomogeneities in Radiosonde Temperature Records', *J. Geophys. Res.* **99-D2**, 3667–3676.

Gaffen, D. J., Elliott, W. P., and Robock, A.: 1992, 'Relationships between Tropospheric Water Vapor and Surface Temperature as Observed by Radiosondes', *Geophys. Res. Lett.* **19**, 1839–1842.

Gutzler, D. S.: 1993, 'Uncertainties in Climatological Tropical Humidity Profiles: Some Implications for Estimating the Greenhouse Effect', *J. Clim.* **6**, 978–982.

Harris, R. G., Thomasell, A., Jr., and Welsh, J. G.: 1966, 'Studies of Techniques for the Analysis and Prediction of Temperature in the Ocean, Part III: Automated Analysis and Prediction', Interim Report, Travelers Research Center, Inc., for U.S. Naval Oceanographic Office, 97 pp.

IPCC: 1990, *Scientific Assessment of Climate Change*, IPCC WG I, WMO, UNEP, Houghton, J. T., Jenkins, G. J., and Ephraums, J. J. (eds.), U. Cambridge Press, 365 pp.

IPCC: 1992, *Climate Change 1992, The Supplementary Report to the IPCC Scientific Assessment*, IPCC WG I, WMO, UNEP, Houghton, J. T., Callander, B. A., and Varney, S. K. (eds.), U. Cambridge Press, 200 pp.

Karl, T. R., Knight, R. W., and Christy, J. R.: 1994, 'Global and Hemispheric Temperature Trends: Uncertainties Related to Inadequate Spatial Sampling', *J. Clim.* **7**, 1144–1163.

Manabe, S., Stouffer, R. J., Spelman, M. J., and Bryan, K.: 1991, 'Transient Responses of a Coupled Ocean-Atmosphere Model to Gradual Changes of Atmospheric CO_2. Part I: Annual Mean Response', *J. clim.* **4**, 785–818.

Nash, J. and Edge, P. R.: 1989, 'Temperature Changes in the Stratosphere and Lower Mesosphere 1979–1988 Inferred from TOVS Radiance Observations', *Adv. Space Res.* **9**, 333–341.

Oort, A. H. and Liu, H.: 1993, 'Upper-Air Temperature Trends over the Globe, 1958–1989', *J. Clim.* **6**, 292–307.

Parker, D. E. and Cox, D. I.: 1994, 'Towards a Consistent Global Climatological Rawinsonde Data-Base', *Int. J. Climatol.*, (in press).

Spencer, R. W. and Christy, J. R.: 1992a, 'Precision and Radiosonde Validation of Satellite Gridpoint Temperature Anomalies, Part I: MSU Channel 2', *J. Clim.* **5**, 847–857.

Spencer, R. W. and Christy, J. R.: 1992b, 'Precision and Radiosonde Validation of Satellite Gridpoint Temperature Anomalies, Part II: A Tropospheric Retrieval and Trends During 1979–90', *J. Clim.* **5**, 858–866.

Spencer, R. W. and Christy, J. R.: 1993, 'Precision Lower Stratospheric Temperature Monitoring with the MSU: Validation and Results 1979–91', *J. Clim.* **6**, 1194–1204.

Spencer, R. W., Christy, J. R., and Grody, N. C.: 1990, 'Global Atmospheric Temperature Monitoring with Satellite Microwave Measurements: Method and Results 1979–84', *J. Clim.* **3**, 1111–1128.

Trenberth, K. E., Christy, J. R., and Hurrell, J. W.: 1992, 'Monitoring Global Monthly Mean Surface Temperatures', *J. Clim.* **5**, 1405–1423.

Trenberth, K. E. and Olson, J. G.: 1991, 'Representativeness of a 63-Stations Network for Depicting Climate Changes', in Schlesinger, M. E. (ed.), *Greenhouse-Gas-Induced Climate Change: A Critical Appraisal of Simulations and Observations*, Elsevier Science Publishers B.V., Amsterdam, pp. 249–259.

World Meteorological Organization: 1989, *Scientific Assessment of Stratospheric Ozone, Vol. 2*, 469 pp.

(Received 23 January, 1995; in revised form 7 June, 1995)

AN OCEAN OBSERVING SYSTEM FOR CLIMATE

The Conceptual Design

NEVILLE R. SMITH[1], GEORGE T. NEEDLER[2] and
THE OCEAN OBSERVING SYSTEM DEVELOPMENT PANEL *

[1] *Bureau of Meteorology Research Centre, Melbourne, Australia,* [2] *Bedford Institute of
Oceanography, Dartmouth, Nova Scotia, Canada*

Abstract. The conceptual design of an ocean observing system for the routine, long-term gathering and processing of ocean data useful for monitoring, describing and predicting ocean climate and its variability is discussed. The ultimate aim of the system is represented by four application areas; atmospheric prediction; ocean and coupled ocean-atmosphere climate prediction; state-of-the-art ocean climate assessment; and model validation. Models are presented as the unifying glue for the system, providing a means for exploiting observed information in many different ways as well as a means for processing complicated and diverse data sets into a form which has practical applications. Monitoring, description and prediction require different supporting environments in order to exploit this potential. The overall objective of the system is broken down into a series of goals and sub-goals roughly aligned with surface, upper ocean and deep ocean applications and with modelling and information management requirements. A prioritization of these goals is presented and it is shown that ordered implementation of the elements supporting these goals will lead to a sensible, staged implementation of the observing system. The research, development and exploitation of appropriate technology is emphasised. Trade-offs and rationalisation across the elements of the ocean observing system for climate and between other climate components and ocean modules is central to the development and successful implementation. The design is presented as the first stage in a constantly evolving and maturing ocean observing system for climate applications.

1. Introduction

In December 1994 the Ocean Observing System Development Panel finished its work on the conceptual design for an ocean observing system for climate. This article provides one view of the approach taken in that endeavour, concentrating on the conceptual design and how it might deliver useful products for understanding and predicting ocean and general climate variability. The conceptual design is built on the premise that it is now feasible to contemplate a permanent, long-term system for ocean climate observations, serving both the needs of the Global Ocean Observing System (GOOS) and the Global Climate Observing System (GCOS)** and delivering benefits to the nations commensurate with the considerable investment of resources that is required to sustain such an enterprise. This premise is based partly on the considerable advances in scientific understanding over recent

---

* The Ocean Observing System Development Panel was sponsored by the Joint Scientific Committee. Its members were W. Nowlin (Chair), A. Alexiov, M. McPhaden, L. Merlivat, G. Needler, R. Schmitt, N. Smith, P. Taylor, A. Vezina, M. Wakatsuchi, and R. Weller.

** GOOS and GCOS agreed to adopt the Ocean Observing System Development Panel for scientific advice on their common interest – the climate module of GOOS and the ocean component of GCOS.

Climatic Change **31**: 475–494, 1995.
© 1995 *Kluwer Academic Publishers.*

[345]

years, partly on advances in technology such as the satellite altimeter, and partly on the recognition that useful, practical products can now be generated from ocean data.

It is instructive to dwell just a little on the keywords that now constitute the names of the parent groups as well as those most often associated with activities of an ocean observing system for climate-monitoring, detection, description, understanding and prediction. The *Global* part can mean several things but here it is interpreted as encapsulating the intended scope of the system, but not necessarily of individual elements. *Ocean* and *climate* together imply all the world's oceans and regional seas are relevant (the non-land and non-atmospheric parts of the earth system), but that small space and short time scales would not be targeted (the cut-off is around the Rossby radius in space and about a month in time). We interpret *observing* to mean the detailed examination of the state of the ocean upon which further diagnoses, analyses and predictions (that is, products) are based. It is not just the act of measurement, though of course that is fundamental, but includes all the stages of data manipulation, management and interpretation that enable thorough examination. *System* is used to imply several things; it means that a group of interrelated and/or interacting elements are being brought together to form a collective entity, the implication being that this unified entity can achieve what the individual elements could not. *System* is also used to imply a methodical approach to the task; previously *ad hoc* arrangements for collection of information are made methodical and routine.

Having now defined what we mean by an ocean observing system for climate, it is useful to consider the main purposes/applications of such a system.

Monitoring is perhaps the most obvious application, though the benefits may be neither obvious nor tangible. It usually implies methodical collection of data over an extended period, with little extra interpretation or extrapolation of the information. A long time series at a particular location or regularly produced analyses (e.g., of sea surface temperature) are examples of monitoring.

Detection, on the other hand, usually requires specialised analysis and interpretation in order to reveal the important signal. For example, Warwick and Oerlemans (1990) discuss the problems of detecting sea level rise from tide gauge measurements while Karl *et al.* (1989) discuss the limitations on detection of, for example, sea surface and air temperature changes, from the recent climate record. Detection of change requires adequate sampling in time (though the length of the record may be beyond control) and space, where adequacy is determined, among other things, by the relative magnitude of the signal and noise, the sampling rate and the quality of the samples.

Description, for our intents and purposes, is inextricably linked to *understanding*. Continued progress in our understanding of the ocean and its natural variability is foremost the task of science; there is, however, a continuing need for baseline information, spanning all relevant research programs, upon which more focussed and intensive ocean research can build. Science will forever remain a key client of the observing system and the observing system will, in turn, provide a raison d'etre and catalyst for ocean research.

Prediction represents the exciting potential of a routine observing effort. Various ocean and ocean-atmosphere models are demonstrating ability for extrapolating information forward in time; it is now feasible to initialise such models based on contemporary observations and thus produce practical climate forecasts.

The Ocean Observing System Development Panel Report (hereafter referred to as OOSDP, 1995) discusses in considerable detail the various issues that must be taken into account in setting down the design of a long-term observing system, including scales of variability of physical and biogeochemical fields and the methods used to observe them. We will not attempt to summarise those discussions here, nor attempt to discuss in detail the recommendations for individual elements of the system. Instead, we will present a topdown view of the concept and the design, working downwards from the general objectives, through various goals and subgoals, emphasising the cohesive nature of the plan and the central role of models in achieving this cohesion. The following section looks at some of the principal products of the observing system and argues that dedicated processing facilities are integral to the development, implementation and sustained strength of the system. The next section looks at the goals of the system, framed specifically to provide the information and data processing and management facilities needed to generate the targeted products. A prioritisation is suggested, reflecting both the immediate prospect of delivering practical benefits and the feasibility of implementing the necessary observing elements. This section is followed by a more detailed examination of two of the highest priority goals and examines how various elements might be implemented in combination to achieve these goals. In particular, the relative impact and feasibility of individual contributions is discussed. The concluding section discusses some of the general issues, including enabling technology and research, trade-offs between elements within the system and between this system and other modules and components of GOOS, GCOS and operational systems such as the World Weather Watch. The evolution of the system is also discussed.

2. Processing and Delivering Benefits

The effectiveness of the observing system is inextricably linked to its ability to deliver real, practical benefits to those nations who support the system. This is a very different measure of success to that used in ocean research where improved understanding is the overriding imperative. The principal beneficiaries include the agriculture and energy industries (via improved seasonal forecasts), managers of coastal and blue water environments (better, routine description of the larger scale evolving climate) and decision makers concerned with possible impacts of an enhanced greenhouse effect (data for climate assessment and model simulation validation). There are also considerable user communities for routine analyses of, for example, sea surface temperature and surface currents. We will avoid a detailed discussion of these user groups here and concentrate instead on the processing and applications upon which such groups will rely.

While in some rare cases useful products can be generated almost directly from observations, it is usually the case that sophisticated processing chains are required before the raw information can be transformed into a 'consumable' product. The OOSDP (1995) design calls for a distributed suite of processing centres dedicated to delivering practical and useful products based on the information derived from the observation network. A single centre may provide processing for several applications or several centres may combine to deliver the product. There may be one or more centres for each application, perhaps organised by nation, as is commonly the case in meteorology.

Modelling is envisaged as a fundamental component of all such application centres since models provide a framework for interpretation, interpolation and extrapolation of raw information, subject to physical and dynamical laws and various conservation constraints. Figure 1 attempts to show schematically several such frameworks for the interfacing of models and data. At the primary level (Figure 1a) the data and modelling streams are essentially independent with the only connectivity being through intermittent model, validation. This category is perhaps representative of the state of the art in climate change detection and coupled model simulation. At the next level observed information is ingested into the model thourgh corrections to the model solution (Figure 1b); this corrected solution (analysis) is used as the initial condition for a subsequent forecast. Many numerical weather prediction systems and some ocean data assimilation schemes operate in this manner. Alternatively, the observed information may be ingested in a closed loop (steady) model solution so that the final solution represents a consistent fit to both the data and model equations (often referred to as an inversion). Such techniques are central to understanding and describing the global ocean thermohaline circulation. Finally, the model-data interface may allow feedback to the observing network and to the assessment (quality control) of the data set (Figure 1c). This represents the ultimate synergism of the measurement and modelling streams.

The OOSDP (1995) design envisages such models being applied in a distributed suite of centres. We will discuss four such applications here.

1. *Atmospheric Prediction.* The observing system is both a provider of information to, and customer for, numerical weather prediction products. Sea surface temperature, surface wind and surface humidity are all important parameters for weather prediction; in turn, numerical weather prediction centres provide consistent, regular analyses of surface wind stress and air-sea heat and water flux. These application centres are operated somewhat independently of the requirements of ocean climate, a state of affairs that has been changing in recent times. The re-analysis projects at the U.S. National Center for Environmental Prediction (Kalnay and Jenne, 1991), and elsewhere, will be

(a) Interpolation and Simulation

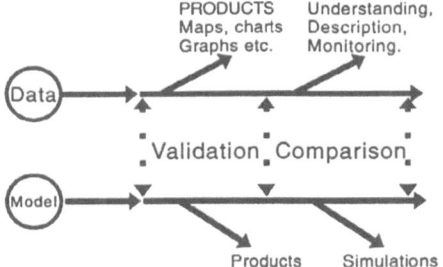

(b) Assimilation and Prediction

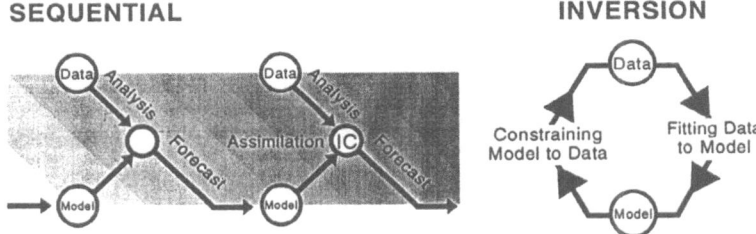

(c) An Integrated Data and Model OOS

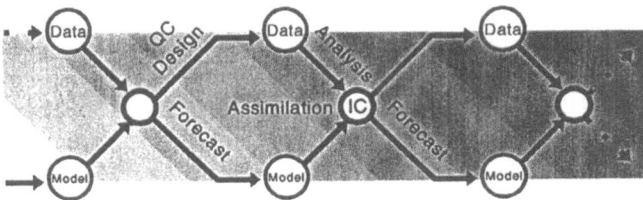

Fig. 1. A schematic of three methods for interfacing observations and models, in order of increasing sophistication: (a) The data are interpreted without the aid of a model, and the model is run without observed information. The interface is provided by series of validation steps; (b) The data and model fields are merged in an analysis (left), which then provides the basis for a sequential model prediction scheme, or (right) the analysis field is obtained from an equilibrium calculation of the circulation based on observation and model constraints (inversions); (c) As (b) but now the analysis also affects the data base, first through quality control, and later through modifications to the observation network design based on the joint model-data analysis and forecast.

an important experience in utilising weather prediction models for obtaining improved, consistent surface flux analyses useful for climate (Trenberth, 1995).

2. *Climate Assessment.* The present concern with climate change and climate variability, perhaps as a result of an enhanced greenhouse effect, has placed increased emphasis on the adequacy of the ocean data base for addressing such issues. The Intergovernmental Panel on Climate Change (IPCC) has sought regular assessments of the state of the earth's climate (Houghton *et al.*, 1990; Houghton *et al.*, 1992), relying in part on various ocean data sets to build this assessment. The severe limitations of this data base are obvious and will not be discussed here. However, it is important to recognise that such assessments and updates are not simply a revision based on new observations but involve the application of improved understanding, analysis procedures and perhaps consideration of rehabilitated data. The processing involved is not trivial and requires expert scientific guidance, relying in part on the novelty and challenge of the process to enlist scientific involvement. The OOSDP (1995) report suggests that consideration should be given to making such processing activities, or at least some part of them, a routine, long-term (operational) element of the observing system. Such a centre (or centres) would continually be reassessing and analysing the ocean data base, attempting to refine and maximise the useful information (the climate signal) available from the data base.

3. *Model Validation.* Models are not a substitute for measurement but rather add value to data sets by interpreting and exporting unevenly distributed information into regions of space and time for which direct sampling is either not available or not possible, subject to dynamical and physical constraints. In some cases, the model simulation may not involve any ocean information at all but will be forced according to some prescribed scenario (e.g., a coupled model forced with transient increases in CO_2 concentration). In all cases it is important that the models faithfully represent, as far as is practical, the actual dynamical, physical, chemical and biological processes of the ocean. The observing system design must provide for observations to detect and quantify model systematic errors and to validate the simulated spectrum of variability. Model validation is essential if the observing system is to evolve and progress. Only by testing models against actual data can we ensure that the methods used to provide products are continually improving. This will not happen as a matter of course but will require positive actions to render the raw information into a form that is easy to use and understand by the modelling groups.

4. *Numerical Ocean and Climate Prediction.* This title is chosen deliberately to draw a parallel with numerical weather prediction and thus invoke a vision of regular and routine data analysis, assimilation and model initialisation for ocean and coupled ocean-atmosphere climate predictions. In reality this is

only feasible, at present, for seasonal to interannual climate variations associated with the El Niño-Southern Oscillation (ENSO) phenomenon. Prototypes of such activities exist at many places already (e.g., Ji *et al.*, 1995; and Smith, 1995) but the scope of the problem suggests further progress may depend on dedicated, combined research and application centres (the European Centre for Medium Range Weather Forecasting is often suggested as a model). This concept has been explored in a proposal for an international research institute for seasonal to interannual prediction (IRICP, 1992) whereby one or several centres would provide core facilities for assembling, analysing and assimilating all information needed to generate seasonal to interannual climate forecasts. The products of this core facility would then be provided to regional application centres who would tailor this information for their regional user groups. No matter how the problem is approached, there is clearly a fundamental role for facilities which routinely process and apply information for forecasting climate change.

While each of these individual activities provides justification for maintaining an ocean observing system, it is the collective operation of the various elements which provides the true strength of the observing system. Numerical ocean prediction will rely on numerical weather prediction for wind stress while the ocean data used for numerical ocean prediction are also used in numerical weather prediction. The processing of data for climate assessments provides a product which is useful for model validation. The data sets used in climate assessment may also be used in numerical ocean prediction. The platforms used to provide measurements for one objective may also be used to provide measurements for another purpose. This cross-utilisation of information and multi-purpose measurement platforms makes clinical design of the observing system impossible but, paradoxically, also provide its fundamental strength.

3. System Goals and Priority Areas

The applications of observing system data provide one approach to the discussion of elements and methods. Another approach is to recast the system design into a set of goals and subgoals which, together, accomplish the overall objective. OOSDP (1995) breaks the objectives into four areas: (i) the surface; (ii) the upper ocean; (iii) the interior ocean: and (iv) processing, which encompasses modelling, analysis, climatologies and information management. Figure 2 presents this approach schematically. The goals of (iv) have been discussed to some extent above and will not be pursued in detail any further.

Figure 2 also shows the different levels of information associated with each task, starting at the integrated measurement platforms (Level 1), progressing through stages of quality control and elementary processing to get oceanographic variables (Level 2), and ending with analyses and data assimilation to produce gridded

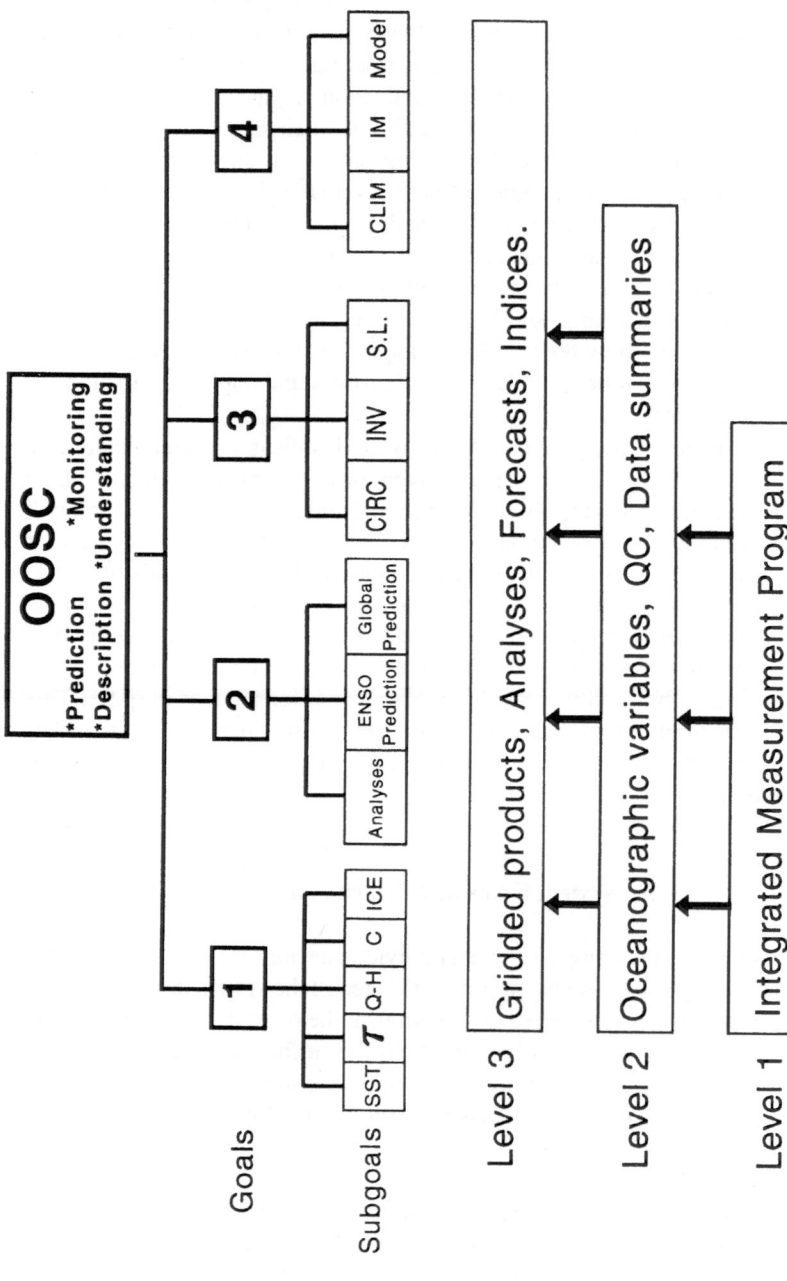

Fig. 2 Schematic of the objectives, goals and sub-goals of the ocean observing system for climate (OOSC) design. See text for further explanation. Also shown are the different levels of information in the observing system, graduating from unprocessed data at Level 1, through to oceanographic variables, quality control and data summaries at Level 2, and analysed fields, forecasts and indices at Level 3.

fields and forecasts. This processing chain in effect overlays the goals. There is an important final step which is not shown in either part of the diagram. The outcome from the system must be a suite of products which offer practical benefits to the customers and users of the system. This 'applications' stage (discussed above) effectively overlays and connects all the sub-goals. Alternatively, we add a further level (Level 4) which represents the added-value processing which is required to turn the scientific products (Level 3) into products which are useful to the customers. This application level provides the coherency in the design. The remainder of this section will concentrate on (i), (ii) and (iii) and discuss an appropriate prioritisation.

    (i) *Surface fields and surface fluxes.* Almost all the information that is needed to determine the ocean's circulation is originally communicated via the air-sea interface, so the estimation of ocean surface fields and air-sea surface fluxes is a fundamental requirement of the ocean observing system. OOSDP (1995) also includes sea ice because its extent, concentration and thickness are intimately related to the fluxes of heat and water to and from the ocean. While our fundamental interest in the surface is because it defines the major sources and sinks of momentum, heat, water, CO_2 and other chemicals to the ocean interior, it is also to be remembered that the ocean surface represents the logical interface between the present observing system and that supporting atmospheric prediction. This means that parameters of the marine boundary layer and ocean surface are of mutual interest and the respective observing systems must be coordinated so as to ensure their respective requirements are met.

    The objective specifically targets estimation of global distributions of sea surface temperature and salinity, wind stress, heat flux and freshwater exchange on monthly and longer time scales. Note that in some cases, such as sea surface temperature, there are strong requirements at both shorter time scales (order of a week) and very long time scales (centuries), while in other cases the emphasis is on the decadal and longer time scales (for example, sea surface salinity). The applications include:

    – data and boundary conditions for NWP;
    – information for climatologies and climate detection;
    – boundary conditions and data for numerical ocean prediction and physical and biogeochemical model simulations; and
    – ocean and coupled ocean-atmosphere model validation.

    Surface wind stress and sea surface temperature represent the highest priority targets for this goal.

    (ii) *The upper ocean.* The upper ocean is a buffer to the exchange of heat and other properties between the atmosphere and the interior of the ocean and thus provides the first level of 'memory' for the ocean-atmosphere system. The upper ocean is characterised by prominent seasonal and interannual thermal signals (see, for example, Levitus and Antonov, 1995) which suggests observation of the upper ocean will be important for prediction over these time scales. As with the surface, our knowledge and experience in observing the upper ocean varies greatly from one parameter to the next and from one region to the next. Note that, for the purposes of this design, deep convection, which involves both the surface and upper ocean, is considered part of the interior ocean goal.

For the global domain in general prediction is not practical, so our first aim is to provide broad-scale data collection which, over time, will result in an accurate determination of the mean state of the temperature and salinity fields and of long-term changes in these fields. This requires relentless, systematic collection of data at modest resolution through many realisations of the annual cycle. For selected regions and, in particular, the tropical Pacific Ocean, climate prediction is feasible so a more ambitious, separate sub-goal is defined which calls for data to initialise and verify ENSO prediction models as well as data for routine monitoring purposes. Applications include numerical ocean and climate prediction but not exclusively. While extension of this activity to regions outside the tropics is not practical at the moment, it is sensible to include a visionary component to the goal which provides for data collection in support of efforts to describe extra-tropical variability and construct useful models for global prediction.

Note that the three strands to this goal are connected and that achieving these goals is not independent of achievements in other goals. ENSO prediction, for example, relies also on the outcomes of the surface and processing goals. While there are no clear demonstrations of the practical utility of observing upper ocean salinity for ENSO prediction, it is taken on faith that improved determination of the seasonal cycle, at least, is important. Note also that for the purposes of this goal, altimeter estimates of surface elevation are highly relevant. Priority is attached to the *in situ* upper ocean measurements for ENSO prediction since the gathering and application of these data has been proven feasible and useful and the data have a proven impact.

(iii) *The Interior Ocean.* The interior ocean is characterised by its capacity to sequester heat, fresh water and chemicals from the surface layers and delay their release from the ocean for long periods of time (from decades to perhaps 1000's of years). Deep ocean observations are essential for monitoring and detection of low frequency natural variations and changes that may be related to anthropogenic forcing of climate. The keywords here are understanding, validation and climate assessment, rather than prediction. The goal is to provide data to describe the mean and variations in the global, large-scale ocean circulation and its transport and storage of heat, freshwater and carbon. This is to be achieved through various methods of deep ocean *in situ* sampling and through observation of sea level (plus fields derived from other goals). More than any other goal, the design here depends on concurrent research programs such as the World Ocean Circulation Experiment and the Joint Global Ocean Flux Study (JGOFS) and on planned research such as the Climate variability program (CLIVAR). The design must recognise the poor state of the baseline data set and adopt a strategy which provides incremental enhancements. Perhaps the exception to this is sea level for which some long records already exist, some of good quality (Gornitz, 1995; Douglas, 1995).

The interpretation of such data is extremely difficult both because of the complexity of the dynamical, physical and biogeochemical processes and because of lack of measurements. Again, models will play a key role, imposing consistency in the interpretation of the data and providing independent estimates of the fields through interpretation and extrapolation of information provided by external boundary conditions.

PRIORITY AREAS

There are many factors that must be considered when deciding the priority/ranking attached to sub-goals including feasibility, technology and logistics, cost, impact, benefits and trade-offs. The prioritisation of OOSDP (1995) was:

1. (a) Estimation of surface wind stress and sea surface temperature effectively impacts on all goals of the observing system and is thus a high priority goal.

   (b) The practicality and feasibility of ENSO prediction has been proven by TOGA. ENSO prediction is possible based on the two data sets of (a) alone but there is now growing evidence that assimilation of subsurface data provides significant improvements. Other upper ocean data (currents, sea level) are also important, principally for model validation.

   (c) Sea level change is accepted as one of the possible consequences of an enhanced greenhouse effect and its estimation is considered important for climate assessments (e.g., Houghton *et al.*, 1990). In view of this fact and the relative ease of taking such measurements, global sea level monitoring is also attached high priority.

2. The second level of priority includes estimation of surface fluxes of heat, freshwater and CO_2 and estimation of the distribution and concentration of sea ice. All these fields are important for model validation and for improved understanding of the ocean. The relentless gathering of broad-scale upper ocean temperature and salinity samples is also accorded second level priority in view of the fundamental need to improve our baseline knowledge of these fields, though the aimed-for global coverage does not look possible with present shipping patterns.

3. Global inventories derived from time series of temperature, salinity and carbon are assigned the third level of priority. These measurements are feasible given present technology and would yield valuable information on the changes in heat, freshwater, and carbon content over the full depth of the ocean on large space and time scales. Estimation of the relevant surface flux fields provides independent and critical supporting information for the fulfilment of this goal. Like (2), it is dependent on long-term, relentless observation.

4. At this time determination of the global ocean circulation and its attendant transports, and global seasonal to interannual climate prediction are assigned lowest priority, basically because there is not yet sufficient understanding nor the relevant cost effective technologies to achieve the goals in the short term. Advances are critically dependent upon on-going research programs.

Existing operational observing elements are effectively restricted to the surface goal. OOSDP (1995) recommends an implementation strategy which would immediately put in place long-term support for the key elements of the highest priority areas/subgoals given in (1) above. Within each subgoal there is also a prioritisation of individual elements, though this does not take account of trade-offs available across the whole observing system (these issues are taken up below). The actions taken in support of the highest priority goals also provide considerable support for lower ranked goals. For example, implementation of the suite of elements proposed for sea surface temperature provides information relevant to most other goals, as well as providing platforms capable of supporting other measurements (e.g., pressure and surface drift from buoys).

While lower priority has been given to other goals, there is nevertheless a compelling case for implementing some elements for other goals immediately, and all other elements as resources and technology will allow. The underlying principal is that each increment in the implementation returns improved benefits not only for the immediately relevant goal but also for all other applications; the staged implementation ensures, on the one hand, a satisfactory, immediate demonstration of the feasibility and practical benefits of the observing system, while at the same time laying the foundations for a comprehensive, multi-faceted long-term observation system. The altimeter rates a particular mention in this respect since it achieves what no other instrument can – global sampling of changes in the ocean interior, admittedly indirect and with some ambiguity in interpretation. Data from this instrument would impact on most goals and so warrants special attention.

4. Elements and Subgoals

In order to illustrate some of the detailed considerations behind the design set out in OOSDP (1995), we turn our attention here to the goals assigned highest priority in the previous section, and discuss the elements of two of these (sea surface temperature and upper ocean measurements for ENSO) in more detail. There are many considerations that must be taken into account when assessing the priority attached to particular elements. One of these is an evaluation of the impact of the data provided by this element in regard to the scientific objective. In all bar a few cases it is extremely difficult to quantify this impact since it depends on the volume, accuracy and quality of the data, the method used to process the data (e.g., a badly biased model may not like any data), the particular application and subjectivity of the method used to 'score' impact (e.g., Pacific surface winds do not score highly for Australian weather prediction but would score very highly if your product was Pacific surface wind stress). Observing system (or system simulation) experiments are often impractical due to the complexity of the data assimilation and modelling processes or proffer ambiguous interpretations of the worth of specific elements (Smith, 1993). In spite of these complexities, it is nevertheless important that at least some attempt is made to provide a relative grading for each element within a subgoal.

A second consideration is loosely described as the feasibility of implementing the subsystem. OOSDP (1995) and others have stressed that implementation and maintenance of an operational observing system places certain constraints on the types of elements and methods that can be considered. The observations must be taken in a methodical manner; must be sustained over the long-term; must be cost-effective; and must be doable in a routine manner (i.e., not require constant scientific oversight and revision). These considerations together, with suitable weighting, effectively define the feasibility. The 'cost-effective' component in this equation is perhaps the most difficult to pin down since in essence it requires knowledge of the benefits that can be attributed uniquely to this measurement. At this stage we must rely on subjective opinions as to the relative feasibility of various elements and accept that it is not a precise or quantitative measure. Such subjectivity, both in respect of feasibility and impact, can and will be criticised, but it should be remembered that such judgements must take place and will never be clinical, and that it is better to force informed scientists to make the first call (and listen to opinion) than to let such assessments be implicit and hidden.

OOSDP (1995) presents a series of feasibility-impact diagrams for each of the subgoals of the observing system. Such feasibility-impact diagrams represent one way of progressing the conceptual design process and assigning priority though, as OOSDP (1995) emphasises, they are clearly not the only way (OOSDP, 1995, includes some alternative views). Naturally, elements with both high feasibility and high impact will normally be rated highly, whereas those with low feasibility and low to medium impact would rate poorly. However there is no set rule, particularly when considering the relative merits of a highly feasible but low impact element against one with the reverse assessment. In regards to implementation, it is also important to consider the relative preparedness of elements for implementation. In some sense this a measure of the research and development required for this element. The first category comprises relevant, *existing* elements; the second category comprises *initial* elements for which technological development is complete and the element is ready now for implementation; the third category includes elements which are potentially useful but require some lead time (referred to as *enhancements*). OOSDP (1995) also discusses areas where research and development is needed.

FEASIBILITY

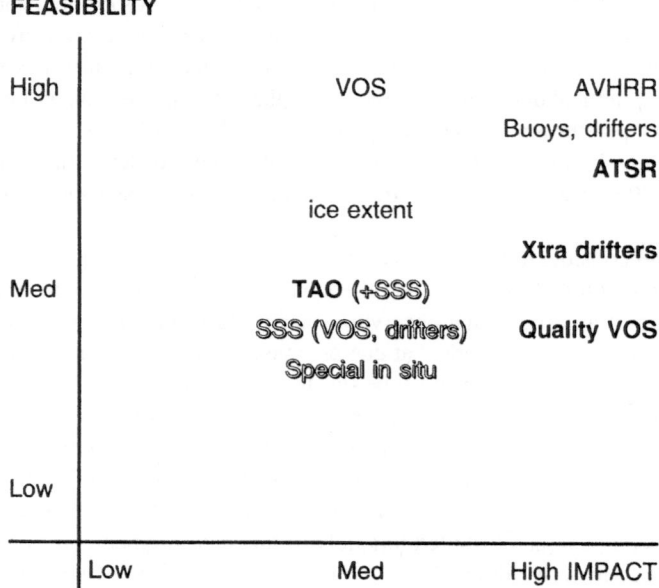

• Exist • **Add now** ○ Enhancements • From other goal

Fig. 3.   Feasibility-impact diagram for the subgoal of estimation of sea surface temperature and sea surface salinity showing the relevant measurement elements. Feasibility and impact are discussed in the text. Elements which exist now are in unaltered font; bold items are elements that should be added for the initial system; shadow font is used for items which are recommended enhancements; a greyshade is used for fields supplied from other subgoals.

EXAMPLE 1. SEA SURFACE TEMPERATURE

Figure 3 shows the feasibility-impact diagram for elements contributing to the observation of sea surface temperature variability on seasonal time scales and longer (sea surface salinity is also included in this goal). Sea surface temperature is a field for which products now exist, many maintained operationally (e.g., Reynolds and Smith, 1994), but for which enhancements are required if the objectives of the ocean observing system for climate are to be met (see also Parker *et al.*, 1995, this issue). The corner stones of this sub-system are the *in situ* data provided by buoys and volunteer observing ships and the satellite estimates provided by Advanced Very High Resolution Radiometers (AVHRR). The sub-system would be improved by more and better *in situ* calibration, for example from drifters in data sparse regions and improved measurement aboard volunteer observing ships; these elements would have high impact, particularly in regard to long-term climate change detection, but require some effort, hence the medium feasibility. Sea surface temperature from subsurface temperature profiles, also taken from volunteer

FEASIBILITY

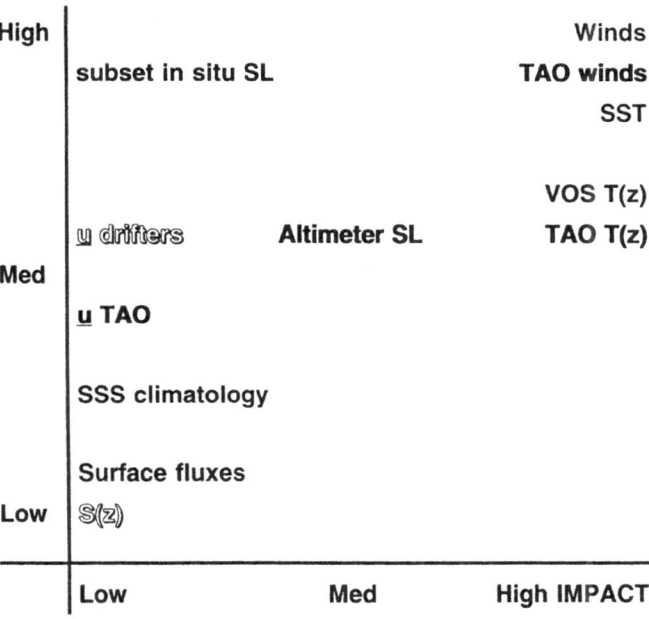

Fig. 4. Feasibility-impact diagram for the subgoal to provide upper ocean data for the initialization and verification of ENSO forecast models.

observing ships, is just as feasible but has less impact because of the quality of the data and lack of coverage. Sea surface temperature measurements from the Along Track Scanning Radiometer aboard the ERS-1 satellite have demonstrated feasibility and the improved resolution would make a significant impact. However, at this time, there is no commitment to an operational (continuing) mission.

None of the methods for sea surface salinity are as well developed as those adopted for temperature, hence the lower feasibility. Measurements from volunteer observing ships would have high impact (quasi-global coverage) but are considered difficult to implement. Measurements are also possible from drifters (more feasible) but would have relatively less impact. Most of the research and development issues are concerned with sea surface salinity rather than sea surface temperature.

EXAMPLE 2. UPPPER OCEAN MEASUREMENTS FOR ENSO PREDICTION

This goal encompasses both measurements needed for the initialisation of models and those needed to validate the models. Figure 4 shows the feasibility-impact diagram, with the same conventions as before. At high feasibility and impact are the observations provided by other sub-systems – sea surface temperature and winds. Winds are absolutely critical for model initialisation and development; sea surface temperature is the critical field for ENSO at the interface between the ocean and atmosphere and has a role in monitoring, detection, model validation and model initialisation. *In situ* sea level is provided by a subset of an existing system and, though it is easy to measure, is now considered of relatively minor importance in ENSO prediction. Sea level from altimeters, on the other hand, may have a significant impact because of its better spatial coverage, is highly feasible, and so is considered an *enhancement* option. Subsurface thermal data from fixed moorings such as the Tropical Ocean Atmosphere (TAO) array and from expendable bathythermographs (XBTs) taken from volunteer observing ships (VOS) are reckoned to have high impact but both require relatively more effort to implement and maintain compared with, say, TAO and VOS SST samples, respectively. The thermal profiles from the equatorial region of the Pacific have far greater impact than those collected for other regions as part of the relentless, but methodical, global survey. Surface velocity from drifters and subsurface velocity from moorings are both feasible but their use is mainly for validation, hence the somewhat lower impact. Salinity, at this point in time, is reckoned to have little impact in the prediction problem, at least relative to the other contributors of information. Sea surface salinity would improve the sub-system. Of the options for improving ENSO prediction, OOSDP (1995) concludes that improvements in processing methodology (assimilation and model skill) and computing offer the best opportunities; data would not be a limiting factor so long as the system was implemented as recommended.

5. Discussion and Conclusions

In this paper we have attempted to convey some of the fundamental concepts in the design of an ocean observing system for climate. We have avoided the detail attached to particular measurement methods and platforms and instead concentrated on the bigger picture and the sub-systems and applications which will deliver practical benefits to the share-holders of the observing system. The immaturity of the methods and technology together with the complexity of the oceanographic and climate processes make clinical design impractical, forcing greater reliance on the subjective assessments and opinions of the oceanographic science community. This will always be the case, in part, since the demands of the system (more informative products) and increased complexity tend to keep pace with improvements in understanding, and vice versa.

While it is probably true that an effective and productive system could be constructed on the basis of technology that exists now, it is also true that the successful evolution and growth of the observing system and its products will depend on technological improvements. These will come in both the measurement and processing sectors.

On the measurement side, satellites and remote sensing offer real potential for measurement networks with global coverage in areas which have hitherto relied on *in situ* data. For example, the TOPEX/POSEIDON altimetric mission, launched in 1992, is now providing data on sea level and its variability, with sampling rates and accuracies capable of defining the upper ocean expression of the world ocean circulation. The accuracy has in fact exceeded design specifications, so much so that validation against independent data sets are now suggesting accuracies of a few centimetres. The reality for GCOS and GOOS is that this mission is experimental, but the potential is obvious, particularly in respect of truly global coverage, so that a concerted effort must be mounted to provide a continuing, operational altimetric system. For biologists, the same potential is seen in ocean colour scanners, such as that planned for SeaWifs. Granted, the interpretation of this data is not straightforward and much research is still required, but in terms of true global sampling, the potential is enormous.

OOSDP (1995) discuss many of the exciting emerging technologies relevant to the observing system, including acoustic thermometry, moored arrays (e.g., TAO), floats, autonomous profilers and sonar methods for ice thickness, as well as some of the needed technologies (e.g., remotely sensed sea surface salinity). In some cases the technologies promise data coverage that is not presently available, in others improved and more efficient methods. Technologies which can sample in presently data sparse regions of the world's oceans are high priority. For example, the ALACE (Autonomous Lagrangian Circulation Explorer) float is an emerging technology which promises a cost-effective method for global hydrography. The cost of developing the technology, the length of the developmental period (typically 10 years), and the projected cost of operation are also significant factors. OOSDP (1995) emphasises the need to be pro-active in identifying and promoting needed technologies.

Technology is also a significant factor in the processing of observational data. OOSDP (1995) places particular emphasis on the need to improve the technology for data telemetry from the instrument site to data assembly points, both to promote more real-time collection of data and to reduce the costs of data telemetry. Many of the applications discussed above rely on the timely and effective delivery of information; in some cases the constraints are imposed by the time windows of numerical weather prediction, in others it is needed to get data to assembly points for quality control and use in numerical ocean prediction. Computing power is another obvious limiting technology. For the ENSO prediction problem and practical forecasts of rainfall, computing resources for processing (data assembly, data assimilation and model forecast) are probably the most significant limiting factor next to lack of scientific understanding. Climate simulations under enhanced greenhouse gas scenarios and inversions of global data sets also require massive computing resources. While much of this work will remain in the province of research communities in the near future, it must be recognised that in the long-term significant resources will need to be devoted to processing information, and that this will need to be balanced against the need to collect information on the actual state of the oceans.

The discussion of new and emerging technologies referred to possible trade-offs through more efficient observing practices. There are also many trade-offs that need to be considered among the methods and platforms underlying the elements of the observation system. Almost without exception, ocean sampling platforms are multi-purpose. For example, a satellite for estimating sea surface temperature may also be used for atmospheric measurements or for estimating sea ice extent; drifters can measure pressure and surface drift as well as sea surface temperature; floats provide subsurface drift as well as temperature (and perhaps salinity) profiles; TAO provides wind, surface pressure, sea surface and sub-surface temperature, and occasionally currents and other marine data. This multi-purpose use of platforms is a significant consideration since extra information can often be obtained with little increment in the cost of the unit,

There are also the trade-offs available between different methods for the same field, and different fields for the same purpose and application. TAO and XBTs from volunteer observing ships both supply tropical ocean sub-surface thermal data, so it is conceivable that this redundancy can be used to reduce the required sampling of one or both systems.

For the ENSO problem, there are a variety of sources of information for model initialisation and validation, including surface wind stress, sea surface temperature, sub-surface thermal profiles and sea level. At this time it is not clear which of these data is most critical for ENSO forecasts or whether some subset of this data combined with a skilful model data assimilation scheme would produce a forecast whose benefit matched that based on the total data set. Observing system simulation experiments have been used in some cases to seek answers to these questions but, in most cases, the complexity of the route from raw data to product is too complex to come up with an unambiguous answer. Finally, we should also consider possible trade-offs and rationalisation between the recommended observations for climate and those maintained and recommended for other modules of GOOS, components of GCOS, and observing sub-systems of the World Weather Watch. There is almost no experience in this area though clearly some fields and platforms will cross observing system boundaries. This has already happened to some extent at the ocean surface. Other modules of GOOS require similar physical and biogeochemical data to that discussed here, though often with different sampling and accuracy requirements.

A facet of the implementation and management of the observing system which deserves further attention is the partnership between 'operational' and 'research' activities. Clearly, the design of the observing system is almost totally reliant on accrued scientific knowledge. The design process will not cease with the publication of OOSDP (1995) but will continue as research learns more about the oceanographic processes and technology which under-pins the system. New demands will also be placed on the system as applications mature and new user communities are created; these demands will in many cases change the goals and priorities which in turn alters the optimal design strategy. It is critical, therefore, that a strong partnership be built between the oceanographic and climate research communities and those agencies and groups with responsibility for the routine operation of the observing system. Indeed, the entire evolution of the system depends on such a partnership being forged. Note that this partnership offers mutual benefit since an effective and strong operational system will open up new and exciting avenues for research, much as occurs in meteorology.

Acknowledgments

The material of this paper is based largely on the work of the Ocean Observing System Development Panel. That Panel also benefited from wise advice in the form of review of earlier reports and several background papers written especially for the Panel. We thank all who have contributed to this effort. We also wish to thank the reviewers for their comments and the organisers of the conference on Long-Term Monitoring of the Global Climate Observing System for their efforts and wisdom in making the conference possible.

References

Douglas, B. C.: 1995, 'Global Sea Level Change: Determination and Interpretation', *Clim. Change* **31** (this issue).

Gornitz, V.: 1995, 'Monitoring Sea Level Changes', *Clim. Change* **31** (this issue).

Houghton, J. T., Callander, B. A., and Varney, S. K. (eds.): 1992, *Climate Change 1992. The Supplementary Report to the IPCC Scientific Assessment*, Cambridge University Press.

Houghton, J. T., Jenkins, G. J., and Ephraums, J. J. (eds.): 1990, *Climate Change. The IPCC Scientific Assessment*, Cambridge University Press.

IRICP: 1992, *International Research Institute for Climate Prediction: A Proposal.* Report by the IRICP Task Group, (A. Moura, Chairman), publ. NOAA Office of Global Programs, Silver Spring, Maryland, USA, 55 pp. + appendices.

Ji, M., Leetmaa, A., and Derber, J.: 1995, 'An Ocean Analysis System for Seasonal to Interannual Climate Studies', *Mon. Wea. Rev.* **123**, (in press).

Kalnay, E. and Jenne, R.: 1991, 'Summary of the NMC/NCAR Reanalysis Workshop of April 1991', *Bull. Amer. Met. Soc.* **72**, 1897–1904.

Karl, T. R., Tarpley, J. D., Quayle, R. G., Diaz, H. F., Robinson, D. A., and Bradley, R. S.: 1989, 'The Recent Climate Record: What It Can and Cannot Tell Us', *Rev. Geophys.* **27**, 405–430.

Levitus, S. and Antonov, J.: 1995, 'Observational Evidence of Interannual to Decadal-Scale Variability of the Temperature-Salinity Structure of the World Ocean', *Clim. Change* **31** (this issue).

OOSDP (Ocean Observing System Development Panel): 1995, *The Scientific Design for the Common Module of the Global Climate Observing System and the Global Ocean Observing System*, Report of the Ocean Observing System Development Panel, publ. U.S. WOCE Office, Texas A&M University, College Station, Texas, 285 pp.

Parker, D. E., Folland, C., and Jackson, M.: 1995, 'Marine Surface Temperature: Observed Variations and Data Requirements', *Clim. Change* **31** (this issue).

Reynolds, R. W. and Smith, T. M.: 1994, 'Improved Global Sea Surface Temperature Analyses Using Optimum Interpolation', *J. Clim.* **7**, 929–948.

Smith. N. R.: 1993, 'Ocean Modelling in a Global Ocean Observing System', *Rev. Geophys.* **31**, 281–317.

Smith, N. R.: 1995, 'The BMRC Ocean Thermal Analysis System', *Aust. Met. Mag.*, (to appear).

Trenberth, K. E.: 1995, 'Atmospheric Circulation Climate Changes', *Clim. Change* **31** (this issue).

Warwick, R. A. and Oerlemans, H.: 1990, 'Sea Level Rise', in *Climate Change. The IPCC Scientific Assessment*, Cambridge University Press, pp. 261–279.

(Received 23 January, 1995; in revised form 27 July, 1995)

OBSERVATIONAL EVIDENCE OF INTERANNUAL TO DECADAL-SCALE VARIABILITY OF THE SUBSURFACE TEMPERATURE-SALINITY STRUCTURE OF THE WORLD OCEAN

SYDNEY LEVITUS and JOHN ANTONOV

NODC/NOAA (E/OC5), Universal Building, 1825 Connecticut Ave., N.W., Washington, D.C. 20235, U.S.A.

Abstract. We present a brief summary of some results describing interannual to decade scale variability of ocean parameters, focusing on subsurface temperature and salinity. We focus attention on the North Atlantic, where it is very clear that a major redistribution of heat and salt has been occurring since 1960, from the sea surface to at least 3000 m depth. We then discuss implications of these examples of historical variability toward development of an ocean monitoring system for the ocean interior. The purpose of this paper reflects our belief that the justification of, and planning for, any such monitoring system for the world ocean, should be based on analyses of historical oceanographic data. These analyses should document the time and space scales associated with variability of the world ocean.

1. Introduction

During the past 20 years, but particularly the last 10 years, the international oceanographic community has presented convincing evidence that the temperature-salinity structure of parts of the world ocean have undergone major changes. These changes have occurred in the deep ocean, as well as in the near surface and intermediate layers. Evidence for such variability is most numerous for the North Atlantic Ocean, which is the most heavily sampled, and perhaps most intensely studied of all oceans. However, evidence of interannual to decadal variability is also accumulating for other regions of the World Ocean.

The purpose of this paper is two-fold. The first purpose is to review observational evidence for interannual to decadal scale variability of subsurface physical properties (temperature and salinity primarily) and circulation of various parts of the world ocean. Non-specialists may not be aware of the accumulating body of evidence indicating large-scale changes occurring in some parts of the world ocean. In addition, the existence of such phenomena provides support for justification of an ocean monitoring system on global spatial scales and decadal time scales. Scientists have long speculated about the possible role of the ocean in determining the earth's atmospheric climate. The first step towards determination of the role of the ocean as it might affect the atmospheric climate, is to document such ocean variability. Any such variability needs to be described, whether natural or anthropogenic in origin. The second purpose of this paper is to broadly discuss implications of the observed variability toward design of an ocean monitoring system for climate change detection.

[365]

Climatic Change **31**: 495–514, 1995.

Only a small portion of this work is devoted to chemical and biological data, in part because of the lack of measurements, and in part because of our desire to limit the scope of this review. We note however that such topics are of importance for the study of the earth's biogeochemical cycles and their temporal variability. The continued increase of atmospheric carbon dioxide as a result of anthropogenic activities demands that any global climate observing system address the monitoring of parameters important to the earth's biogeochemical cycles. We have not included many references to studies describing the variability of sea surface temperature, such as those based on *in situ* measurements (Folland *et al.*, 1984; Parker *et al.*, this volume; Woodruff *et al.*, 1987; The Ocean Observing System development Panel, 1995). We focus this report on the North Atlantic Ocean, where it is clear that at a minimum, a major redistribution of heat and salt has been taking place since 1960. The variability of the North Atlantic is such that we believe large-scale interaction between the ocean and atmosphere must be involved. If this is the case then we may have an example demonstrating the role of the ocean as part of the earth's climate system. In particular, we note the suggestion by Rossby (1959) that the ocean may play a large role in determining the earth's atmospheric climate, by storing heat from the atmosphere for relatively long periods of time.

A review of oceanic decadal-scale variability was published recently by IOC (1992) and provides some examples of variability not presented here.

2. Variability in Individual Ocean Basins

2.1. NORTH ATLANTIC OCEAN

The temperature and salinity distributions of the subarctic and subtropical regions of the North Atlantic have undergone major changes during the last forty years. The work of Lazier (1980, 1988, 1995) has described the changes throughout the entire vertical extent of the water column in the Labrador Sea (0–2000 m).

In Lazier's most recent work he notes that Labrador Sea Deep Water has cooled by more than one degree centigrade from the early 1970's through 1990. Antonov and Groisman (1988), Antonov (1990, 1993), Read and Gould (1992), Koltermann and Sy (1994), and Levitus *et al.* (1989a, 1989b, 1989c, 1994a, 1994b, 1995) further document the changes in the subarctic gyre of the North Atlantic. Their works greatly expand on the earlier observations by Brewer *et al.* (1983), Swift (1984, 1985) that describe the cooling of the subarctic gyre, particularly the deep water in this gyre. Roemmich and Wunsch (1984) and Parilla *et al.* (1994) have documented decadal changes in the subtropical North Atlantic temperature structure from the upper ocean down to 2500 m depth.

Before describing decadal scale variability for two of the longest deep ocean time series, we present Figure 1 which shows the climatological distribution of temperature at 1750 m depth in the North Atlantic Ocean. The two locations

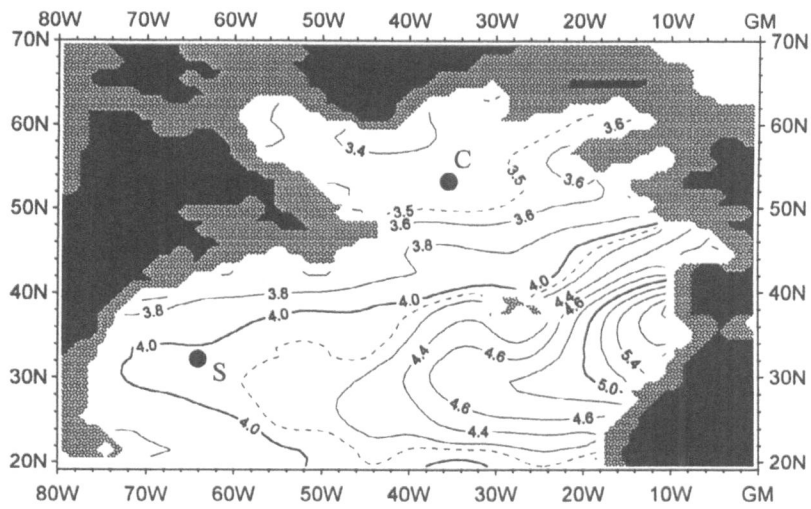

Fig. 1.   Distribution of temperature (°C) at a depth of 1750 m in the North Atlantic Ocean.

for which we have time series are Ocean Weather Station 'C' whose location is marked by the letter 'C' and Station 'S' marked by the letter 'S'. Station 'S' was established in 1954 through the efforts of Henry Stommel (Stommel, 1988) and represents one of the longest time series of subsurface oceanographic measurements of temperature and salinity in existence. To document changes in the deep water temperature structure of the subarctic gyre of the North Atlantic, we show Figure 2 from Levitus *et al.* (1995). The temperature time series in the 1000–1500 m depth range at Ocean Weather Station 'C' clearly show that temperatures reached a maximum around 1973–1974 and then cooled through 1990. In contrast, the deep waters of the subtropical gyre have undergone a significant warming. Figure 3 taken from Levitus *et al.* (1995), shows that temperature at 1750 m depth at Station 'S' located near Bermuda, exhibits a linear warming trend with a magnitude equivalent to 1.0 °C per century for the 1960–1990 period. A warming of deep ocean temperature at this location was first noted by Frankignoul (1981) and Roemmich (1985).

In addition to the warming trend, a decadal-scale oscillation is apparent in this record. Salinity exhibits similar signals in the deep ocean at Station 'S'. It is clear that while the deep water of the North Atlantic subarctic gyre is cooling, the deep water of the subtropical gyre is warming. Thus it is not possible at this time to estimate whether there is a net cooling or warming of the world ocean. There are simply not enough measurements from most of the world ocean to even guess at such a possibility. What is important is that we can now state that at a minimum, a major redistribution of heat and salt has occurred in the North Atlantic since

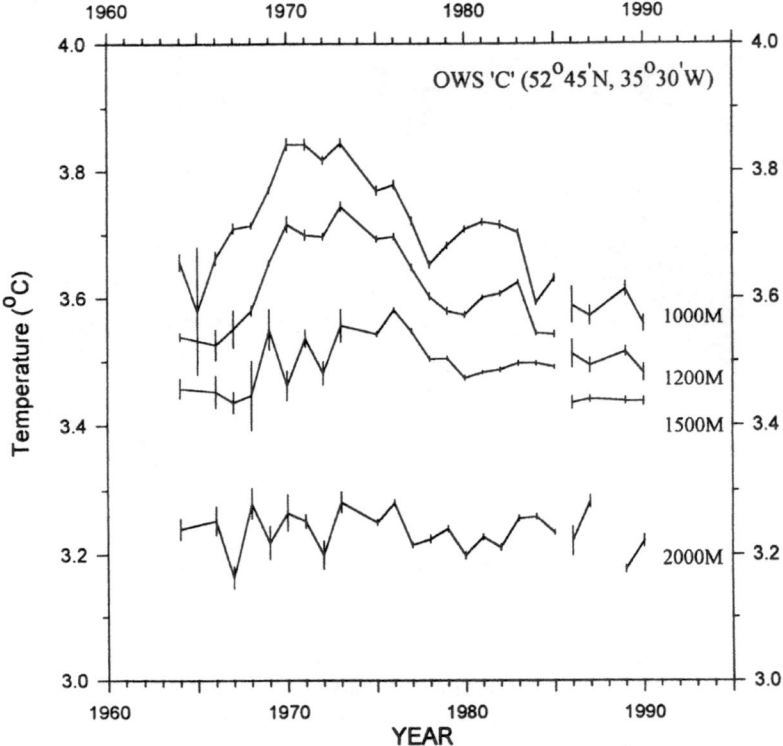

Fig. 2. Time series of annual mean deep (1000 m, 1200 m, 1500 m, 2000 m) ocean temperatures (°C) at OWS 'C'. The annual means are computed from the average of the daily means in each year. The vertical bars extending from each annual mean represent plus and minus one standard error of the daily means about each annual mean.

1960. The warming of the deep waters of the subtropical gyre of the North Atlantic began around 1960 at the Bermuda location. The cooling and freshening of the deep waters of the subarctic gyre of the North Atlantic Ocean began in 1971, when deep convection in the Labrador Sea brought relatively cool, fresh surface waters to the deeper levels of this gyre.

To present further evidence of trends in the upper North Atlantic Ocean temperature fields we present Figures 4 and 5 (based on the analyses by Levitus *et al.*, 1994b). Figure 4 shows the linear trend in temperature for the 1966–1990 period at 100 m depth and the percent variance accounted for by the linear trend. Figure 5 shows these same quantities at 400 m depth. It is clear that there are statistically significant (from results of statistical tests not shown here) temperature trends at 100 m depth in the North Atlantic Ocean with the subarctic gyre cooling, midlatitudes warming, and the subtropics along 15° N–20° N cooling. Figure 6 shows the interannual variability of annual mean temperature at 100 m depth at Ocean

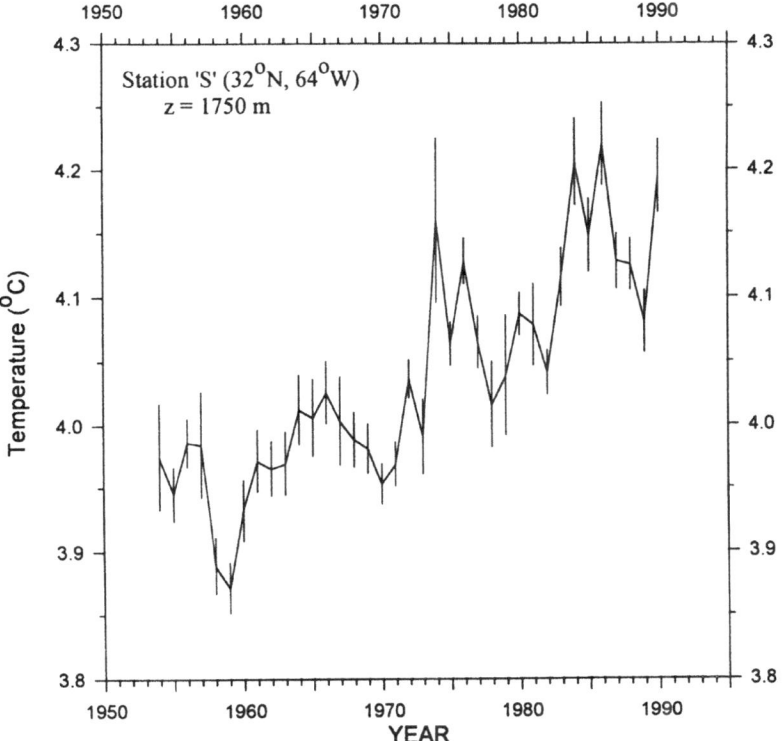

Fig. 3. Time series of annual mean temperature (°C) at 1750 m depth at Station 'S'. The annual means are computed from the average of the monthly means in each year. The vertical bars extending from each annual mean represent plus and minus one standard error of the monthly means about each annual mean.

Weather station 'C' presented by Levitus *et al.* (1995). In addition to the negative linear trend that we expect to see based on Figure 4, a quasi-decadal scale signal exists with a range of about 1.0 °C. This signal is correlated with an oscillation of the atmospherlc sea surface pressure field known as the East Atlantic Oscillation. Cayan (1992) documents some aspects of this oscillation. Deser and Blackmon (1994) analyzed sea surface temperature data which exhibits the same oscillation and noted that the signal exists in the pre-World War II period. At 400 m depth (Figure 5) there are differences in the structure of the linear trend field as compared to the trend field at 100 m depth. The biggest difference being the tongue-shaped cooling trend extending westward from Africa centered along 30° N. A similar tongue-shaped warming pattern was identified at this location at 1750 m depth by Levitus (1989c). These patterns were also found by Antonov and Groisman (1988) and Antonov (1993). It is not known whether these two features are related. The

[369]

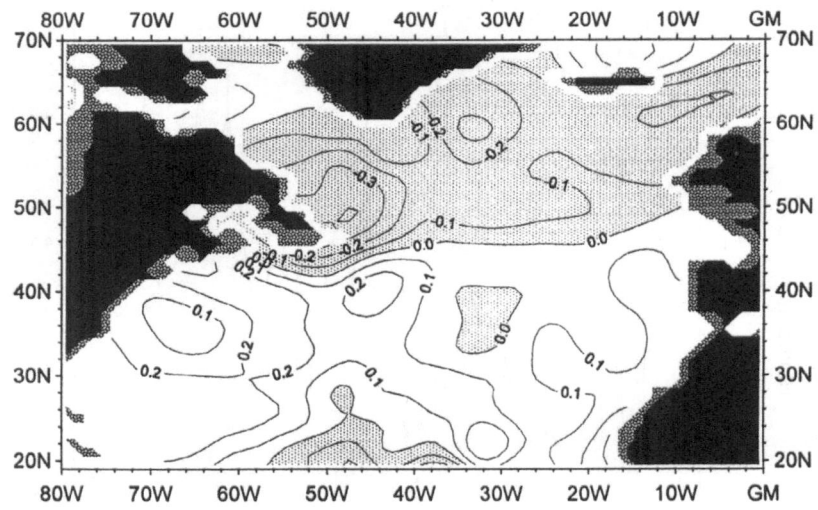

Fig. 4a. Linear temperature trend (°C/10 yr) at 100 m depth in the North Atlantic for the 1966–90 period.

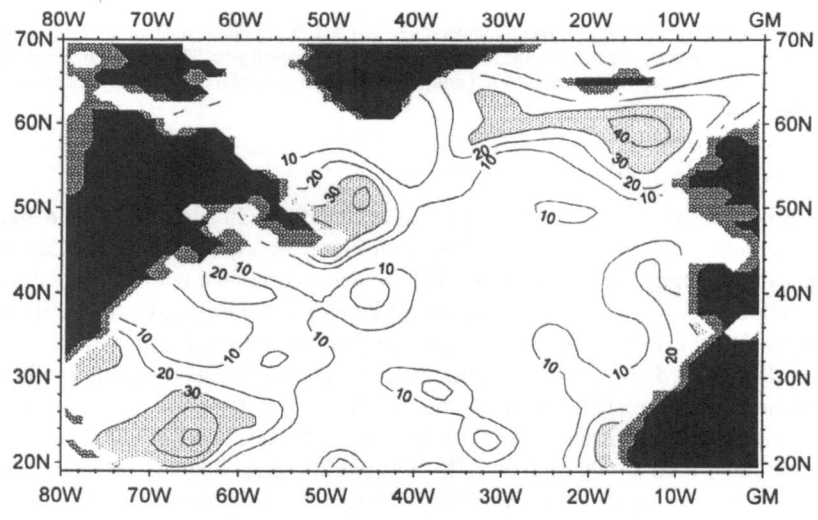

Fig. 4b. Percent variance explained by linear trend (1966–90) at 100 m depth.

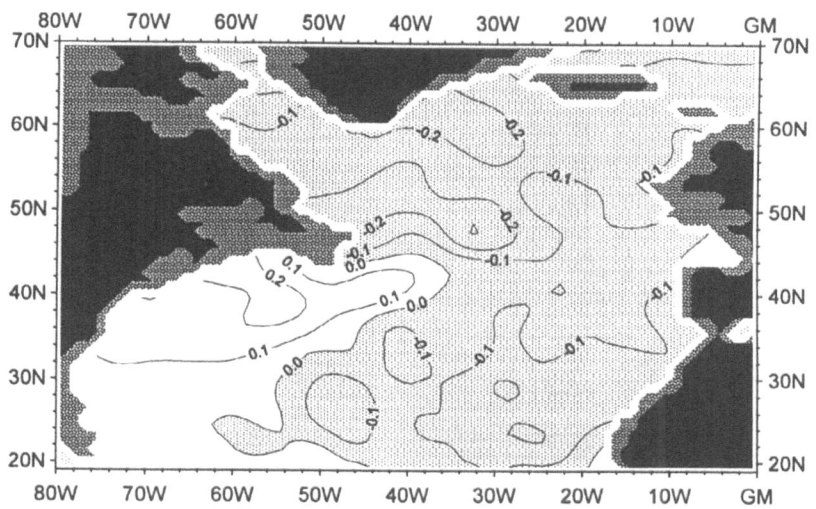

Fig. 5a. Linear temperature trend (°C/10 yr) at 400 m depth in the North Atlantic for the 1966–90 period.

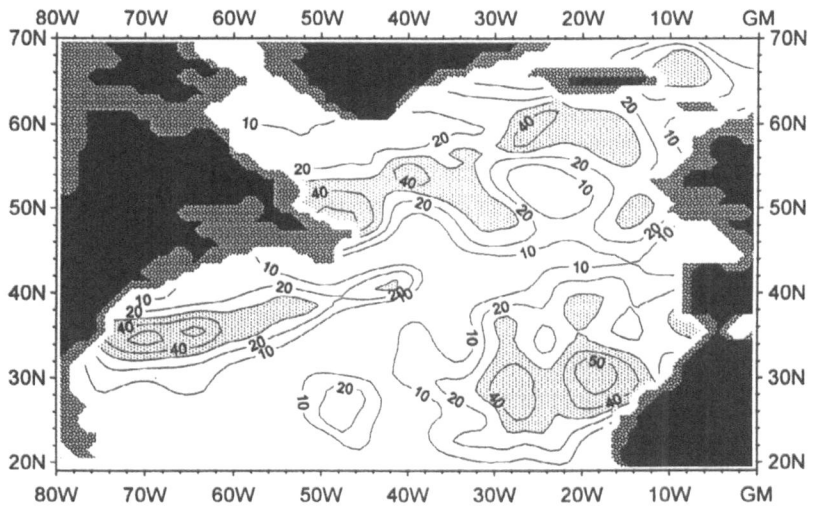

Fig. 5b. Percent variance explained by linear trend (1966–90) at 400 m depth.

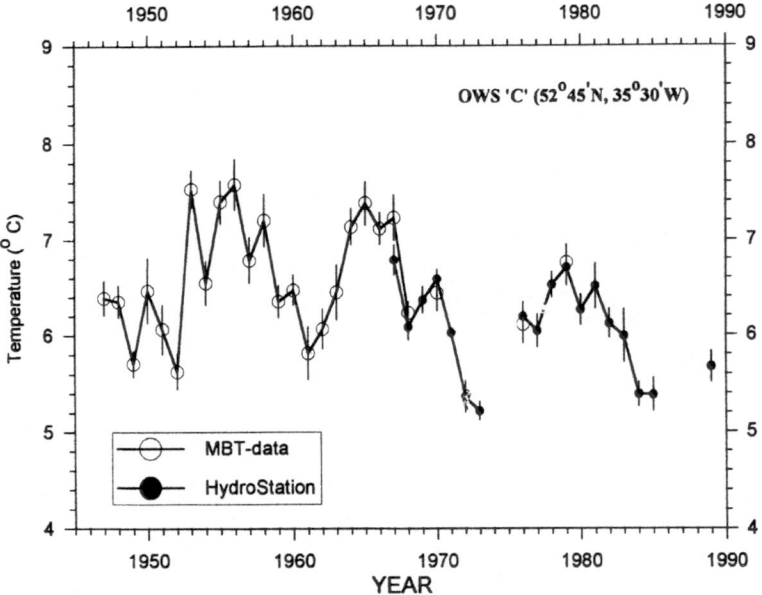

Fig. 6. Interannual variability of annual mean temperature at a depth of 100 m based on data from Ocean Weather Station 'C'. The vertical bars extending from each annual mean represent plus and minus one standard error of the monthly means about each annual mean.

upper boundary and lower boundaries of the warm, saline Mediterranean tongue could be considered to be at approximately 400 m and 2000 m depth.

Petrie and Drinkwater (1993) documented a cooling and freshening of subsurface waters along the western boundary of the subarctic North Atlantic (Scotian Shelf and Gulf of Maine) from 1952 through 1967. They associate these changes with an increase of the westward transport of the Labrador Current from one Sverdrup in the early 1950's to four Sverdrups in the mid-1960's. Morgan *et al.* (1993) document a decrease of surface air temperature at coastal locations around the periphery of the northern North Atlantic that corresponds to the decrease in SST of the subarctic North Atlantic. Dooley *et al.* (1984) presented evidence documenting that not only did surface salinity exhibit a large decrease during the 1968–79 period in the northern North Atlantic (known as the Great Salinity Anomaly, Dickson *et al.*, 1988) but the salinity of the Arctic Intermediate Water in the Rockall Trough decreased.

Evidence for gyre and basin-scale changes in the circulation pattern of the North Atlantic was presented by Levitus (1989a, 1990). In the subtropical gyre, density surfaces were as much as 50 m higher in the 1970–74 period as compared to the 1955–59 period. In the subarctic gyre immediately north of the Gulf Stream the depth of density surfaces exhibited the opposite sign. These changes were quantified by Greatbatch *et al.* (1991) using diagnostic techniques described by Mellor *et al.* (1982). Their results indicate that the net eastward transport between Bermuda and Halifax was approximately thirty Sverdrups less during the 1970–74 period as compared to the 1955–59 period. Confirmation of these changes is presented by Ezer *et al.* (1995) using a different analysis technique. Ezer *et al.* used a numerical model of the Atlantic Ocean to dynamically adjust the density and wind fields for the 1955–59 and 1970–74 pentads with the model bottom topography. As they document in their paper, their results confirm the thirty Sverdrup transport decrease but provide more realistic estimates of meridional heat transport and meridional circulation than can be obtained with purely diagnostic techniques (Ezer and Mellor, 1994). Their model results agree quite well with changes in observed sea level along the east coast of North America that are observed between the two pentads.

2.2. MEDITERRANEAN SEA

For the first half of the twentieth century, temperature and salinity of the deep waters of the Mediterranean Sea were believed to be approximately constant. This view changed, after the report by Lacombe *et al.* (1985). They noted that between 1970 and 1972 the western Mediterranean Deep Water (WMDW) had warmed by as much as 0.05 °C. Additional observations described by Bethoux *et al.* (1990) indicate that the warming of the WMDW accelerated during the 1955–60 period, an observation confirmed by Leaman and Schott (1991) and Rohling and Bryden (1992). Bethoux suggested that global warming might be playing a role in the changes and that salinity had increased by 0.015 parts per thousand. In addition to the changes in WMDW, Lacombe *et al.* (1985), Rohling and Bryden (1992), and Hecht (1992) also described changes that occurred in the temperature-salinity characteristics of Levantine Intermediate Water (LIW) of the Mediterranean Sea. Hecht (1992) documented changes in salinity of not only LIW but also the Atlantic Water found in the Southeastern Levantine Basin. These changes occurred in 1982. The fact that Atlantic Water is changing characteristics may be an indication that changes in the exchange of water between the Mediterranean and Atlantic Ocean are occurring. Rohling and Bryden presented calculations demonstrating that the changes in water mass properties of the LIW and WMDW are consistent with reduction of fresh water inflow to the Mediterranean, due to the damming of rivers in this region. However, Bethoux and Gentili (1994) argue otherwise and suggest that neither the damming of the Nile, nor the damming of rivers flowing into the Black Sea during the 1947–85 period can account for the observed changes in the western Mediterranean. Bethoux and Tailliez (1994) suggest that changes in the properties of WMDW are due to regional changes in precipitation.

2.3. SOUTH PACIFIC OCEAN

Bindoff and Church (1992) described changes in the temperature-salinity structure in the western South Pacific along 28° S and 43° S, between 1967 (Scorpio

Expedition) and 1989–1990. They observed that below the mixed layer, the depth-averaged temperature had warmed by 0.04 °C at 43° S and 0.03 °C at 28°S. They also described changes in the temperature-salinity structure that were not density compensating. Thus an increase in geopotential thickness of 2–3 cm resulted from the changes in the temperature-salinity structure in these regions. Particularly interesting is their finding that the amount of Antarctic Bottom Water in this region decreased by approximately three percent during this time interval.

Roemmich and Cornuelle (1990) used high resolution XBT sections to investigate the structure of the subtropical gyre of the South Pacific during 1986–1990. They found that the gyre appears to exist in two distinct steady states with transitions between the two states occurring very rapidly.

Ridgway and Godfrey (1995) found changes in the Coral Sea that represented mechanical displacements of the thermocline in this area. Changes in the circulation on the order of 10–15 Sverdrups were estimated between 1975–79 and 1985–89. These changes are not related to the ENSO phenomenon, at least not directly.

2.4. NORTH PACIFIC OCEAN

No gyre or basin-scale changes have been detected in the temperature-salinity structure of the deep waters of the North Pacific Ocean. Antonov (1993) has analyzed temperature changes at standard level depths from 300 m to 3000 m for the 1957–1981 period. Basin-mean temperature decreased by 0.1 °C at depths of 300 m and 400 m in the North Pacific Ocean. This is consistent with about 0.3 °C temperature drop at the surface. There were no statistically significant changes of the basin-mean temperature below the depth of 400 m. Lukas (1994) has reported year-to-year variability in the potential temperature of the 3500–3600 db layer (magnitude 0.04 °C) at the Hawaii Ocean Time Series (HOTS) site, but it is not clear what is the scale of the phenomenon responsible for these changes.

Changes in the upper ocean thermal structure of the North Pacific have been documented by Royer (1989), Levitus et al. (1994b) and Watanabe and Mizuno (1994). Watanabe and Mizuno documented changes on the order of 0.5 °C using a different choice of compositing and differencing periods from those of Levitus et al. Levitus et al. presented evidence of gyre and basin-scale temperature changes in the upper ocean characterized by magnitudes exceeding 0.5 °C over periods as short as five years. Their choice of compositing and differencing periods was based on the results of EOF analyses. Certain characteristic patterns appear to be associated with these changes. Generally the interior mid-latitude of the North Pacific Ocean seems to be characterized by an anomaly that is opposite in sign to the anomaly in the region along the periphery of the subtropical gyre. This mode of variability is evidenced by Figure 7 (taken from Levitus et al., 1994b) which shows the mean temperature anomaly fields at 125 m depth for the 1960–62 and 1965–69 periods in the North Pacific Ocean. During the earlier period the central Pacific was characterized by a cold anomaly, whereas during the latter

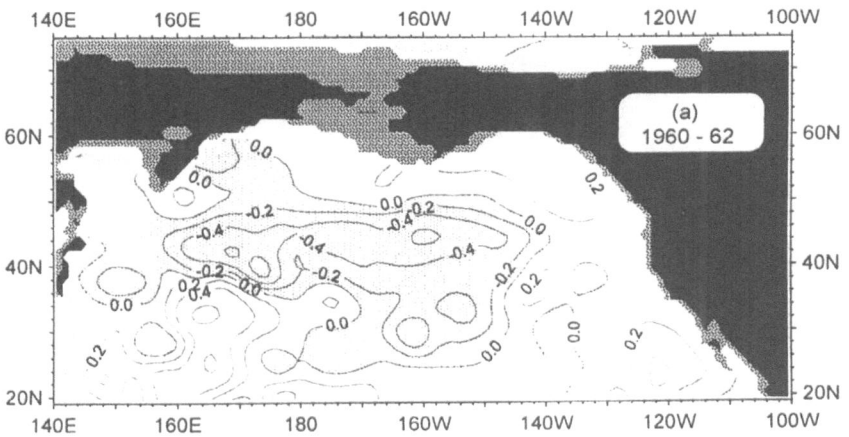

Fig. 7a. Mean temperature anomaly (°C) at 125 m depth in the North Pacific for the 1960–62 period.

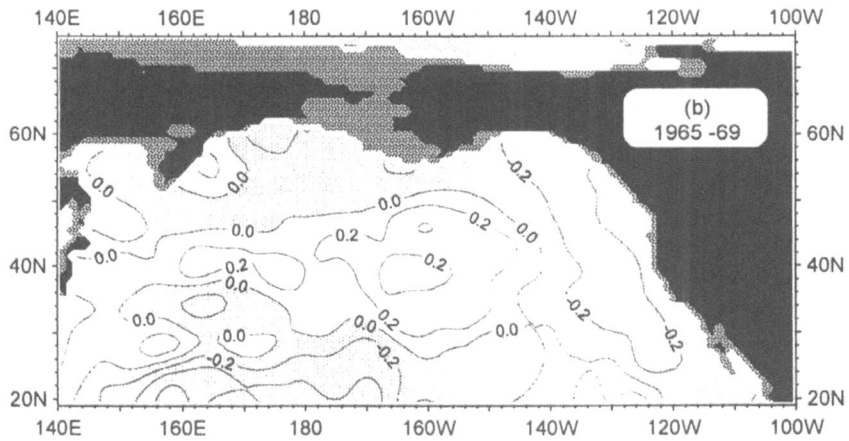

Fig. 7b. Mean temperature anomaly (°C) at 125 m depth in the North Pacific for the 1965–69 period.

period the entire anomaly pattern reversed sign. The temperature anomaly patterns for the periods 1981–85 and 1988–1990 were similar to the 1960–62 and 1965–69 anomaly patterns respectively (Levitus *et al.*, 1994b). Meteorologists have identified a decade-long shift in the atmospheric circulation over the North Pacific that began in 1976 (Douglas *et al.*, 1982; Trenberth, 1990). A natural question to ask is what is the role of the North Pacific Ocean in causing or responding to this phenomenon. The response to this question is complex and will require exhaustive studies. However some characterizations on the nature of gyre and basin scale

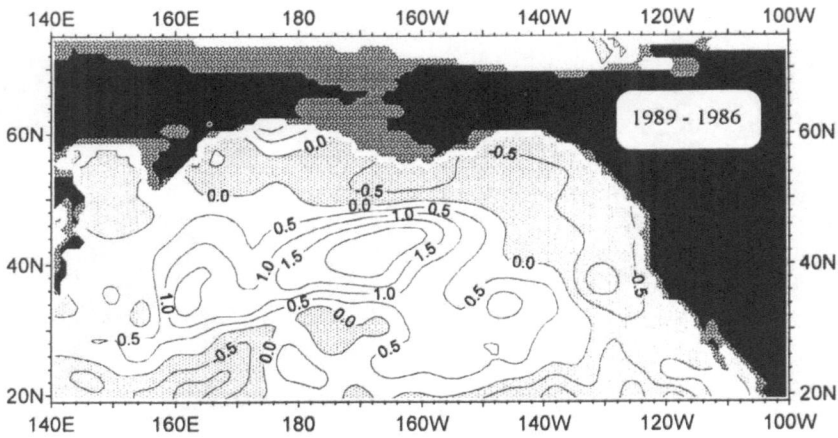

Fig. 8.   Temperature difference (°C) for 1989 minus 1986 at 125 m depth in the North Pacific.

subsurface temperature changes can be made based on analyses such as those shown in Figure 7.

We also want to emphasize that changes similar to those shown in Figure 7 can occur over shorter time periods, as happened at 125 m depth between 1986 and 1989. Figure 8 shows that the temperature change over these three years in the central North Pacific is characterized by warming that exceeds 0.5 °C over a large area. The region outside of the warmed region is characterized by magnitudes exceeding 0.5 °C. Clearly, the upper ocean temperature structure of the North Pacific can undergo what we consider to be relatively large changes in magnitude, over a period as short as three years.

Based on analyses of upper ocean thermal variability such as documented in Figures 7 and 8, the immediate research problem is to determine what, if any, air-sea interactions are involved in such changes on interannual to interdecadal scales. Gridded data fields are becoming available to perform such work. Monthly objective analyses of surface marine meteorological data based on historical observations for the 1945–89 period are presented by Da Silva *et al.* (1994). The patterns of variability described by Figures 7–8 (interior North Pacific opposite in phase from variability along the northern and eastern borders of the North Pacific) have been termed the 'El Niño' mode by Weare *et al.* (1976). We believe that comparison of results shown in Figures 7–8 and Figure H19 from Levitus *et al.* (1994b) with the occurrence of El Niño suggests that more phenomena than El Niño may be involved in this mode of oscillation.

Qiu and Joyce (1992) and Shuto (1994) have described variability along 137° E based on the repeat surveys carried out by the Japanese Meteorological Agency. Bingham *et al.* (1992) have documented changes in upper ocean thermal structure between the late 1930's and the 1970's. Suga and Hanawa (1995) have documented changes in the Subtropical Mode Water of the North Pacific Ocean that they relate to changes in the position of the Kuroshio as well as changes in formation rate. The changes in formation rate were found to be correlated with the intensity of winter cooling.

3. Other Evidences of Long-Term Variability

While we have concentrated on describing changes in the subsurface distributions of temperature and salinity in the world ocean, there are other examples of long-term variability. For example there is evidence accumulating of changes in biochemical and biological parameters. Particularly interesting are the long-term changes in zooplankton productivity at Ocean Weather Station 'P' noted by Brodeur and Ware (1992) as well as a fifty percent decrease in winter nitrate levels along Line P from 1989 to 1994. Venrick et al. (1987) described changes in chlorophyll distribution in the upper Pacific Ocean. Harris et al. (1988) describe interannual variability of climate and fisheries in the South Pacific Ocean. Haake et al. (1993) describe interannual variability in the flux of particles to the deep Arabian Sea. Karl et al. (1995) as well as Polovina et al. (1994) have documented ecosystem changes in the subtropical gyre of the North Pacific that appear to be associated with ENSO phenomena.

Interannual variability of surface salinity has been studied by Delcroix and Henin (1991) and Dessier and Donguy (1994). Delcroix and Henin (1989, 1990) describe variability of subsurface temperature in the tropical Pacific. Variability of the surface salinity of the subarctic gyre of the North Atlantic has been studied by Dickson et al. (1988). The 'Great Salinity Anomaly' as it has been termed occurred during the late 1960's and 1970s in the northern North Atlantic. Surface salinity reached its lowest values in this region since the early 1900's.

4. Requirements for Monitoring and Detecting Long-Term Change in the World Ocean

In order to monitor the world ocean to detect climate change, an observing system must, at a minimum, measure the parameters of state, temperature and salinity, as a function of depth (pressure). Ideally these measurements should be made on a global scale but the expense involved means only a limited system can be deployed. In order to maximize limited resources, careful observing system design is required. Certainly to forecast the future state of the ocean the parameters of state must be specified, particularly if convective activity is to be forecast. An ocean monitoring system must be capable of detecting changes of deep water properties such as those observed in the North Atlantic Ocean. One may also want to monitor the ocean for the purpose of measuring variability of biogeochemical cycles such as the carbon cycle.

The parameters to be measured as part of a Global Ocean Observing System, will in fact, determine the nature of the observing system. For example, some of the systems now in use for monitoring temperature, (e.g. the Volunteer Observing Ship program), could not be used for monitoring subsurface biogeochemical parameters. For example the technology does not exist for measuring nutrients at depth using an expendible instrument. Research ships with scientific leadership and a trained technical staff are required to make such measurements. In a few sentences one can specify requirements for a monitoring system that requires substantial resources to maintain. Choices must be made based on existing knowledge and limited budgets of how to implement such a system. New technology can assist the international research community in establishing such a monitoring system.

The most important aspect of the studies we have reviewed is that interannual-to-decadal scale variability exists in major parts of the world ocean and includes the deep waters of the North Atlantic. An important characteristic of the observed decadal-scale variability in the subtropical gyre of the North Atlantic and South Pacific is that the variability was due to vertical displacements of density surfaces. The displacements in the North Atlantic corresponded to large enough (17 cm) changes so that existing satellite altimeters can accurately estimate this variability. Variability associated with changes in water mass properties on potential density surfaces has also been identified. The separation of variability into these two categories is significant. To the extent that changes on potential density surfaces are density compensating, there will be smaller changes in steric sea level associated with these type of variability. In addition, in the case of nearly compensating density changes of temperature and salinity, advective fluxes of heat and salt can occur with only small corresponding changes in a circulation pattern. Changes over any layer associated purely with mechanical displacements result in distinct changes in steric sea level as we have noted.

5. Components of a Subsurface Ocean Observing System

It is appropriate to list different types of oceanographic observing systems that are being used now or could possibly be used to monitor subsurface conditions in the world ocean. Again we note that an extensive review of technologies has been given by The Ocean Observing System Development Panel (1995) (also see Smith *et al.*, 1995, this volume). Here we simply include a brief description of some systems that are, or may be used, to monitor subsurface temperature and salinity.

5.1. VOLUNTEER OBSERVING SHIP PROGRAMS

Provide coverage along ship-of-opportunity tracks of parameters that can be estimated using expendible instruments. The measurements from this program have
proven to be invaluable for studies describing interannual variability of upper
ocean thermal structure. Rossby *et al.* (1995) review the Volunteer Ship Program and provide information on ways in which vessels can support the task of
ocean monitoring. These include profiling physical properties, chemical sampling
via automated water samplers, optical techniques to measure various biological
parameters, and provision of ground truth measurements for remote sensing from
orbiting and geostationary satellites.

5.2. SATELLITE ALTIMETRY

Provides global coverage of vertically integrated changes of temperature and salinity (via the density of sea water) in the ocean water column. This technology
is already deployed although not on a routine basis (See the special collection of
papers on TOPEX/POSEIDON Geophysical Evaluation that are companion papers
to the mission overview given by Fu *et al.* (1994)). Additional results from this
mission have been reported at various scientific meetings. It is clear that this technology provides the scientific community with an unprecedented ability to monitor
the sea level of the world ocean.

5.3. MOORED INSTRUMENTS

Provides real-time measurements of upper ocean parameters as well as surface
meteorological data.

5.4. POP-UP FLOATS

As described by Davis *et al.* (1992), Autonomous Lagrangian Circulation Explorer
(ALACE) floats can provide vertical profiles of temperature and possibly salinity
at a relatively low cost. Floats can be programmed to cycle up and down through
the water column at predetermined intervals (e.g. monthly). One can 'seed' regions
that are relatively inaccessible and receive data in near real-time via satellite communication links.

5.5. HYDROGRAPHIC SECTIONS

Can provide basin-wide sections of any parameter that can be measured by the
scientific community. The World Ocean Circulation Experiment (WOCE) and
Joint Global Ocean Flux Study (JGOFS) will be providing the most comprehensive
description to date of the distribution of various water mass properties in the world
ocean based on hydrographic sampling. One component of WOCE is designed to

carry out repeat occupations of hydrographic sections to detect variability. It is very expensive to operate ships but there is no other way to make measurements of certain ocean parameters at present.

5.6. ACOUSTIC TOMOGRAPHY

Experimental technology that has potential to provide basin-scale fields of ocean temperature on a routine basis.

6. Conclusion

Each of the systems noted has different costs associated with its operation and can provide different types of data for monitoring the world ocean. It is beyond the scope of this paper to articulate all the characteristics of these systems and weight their relative merits. We believe that an ocean monitoring system will make use of many, if not all, of these systems.

The foremost requirement for an ocean monitoring system is to monitor the temperature/salinity structure of the world ocean in real-time, or nearly real-time (e.g. within weeks of measurement). This requirement is to insure that data are available for prognostic forecasts. The Global Temperature-Salinity Pilot Project (GTSPP) has been successful in improving the international distribution of oceanographic measurements in real-time (Searle, 1992).

Satellite altimeter measurements provide a vertically integrated measure of the state parameters with near-global coverage. However one must use some statistical relation to estimate individual profiles of temperature and/or salinity from these satellite measurements. Acoustic tomography holds the promise of measuring the ocean temperature field on gyre and basin scales but the technology still must be proved before it becomes a routine method of monitoring the ocean.

However it is also important to note that some changes may take place in the form of density compensating changes of temperature and salinity on density surfaces. Such changes are not detectable by a satellite altimeter system since they correspond to zero change in sea level.

Acknowledgements

Our work has been supported from grants provided by the NOAA Climate and Global Change program and the NOAA Earth System Data and Information program. We acknowledge the efforts of our colleagues from the NODC Ocean Climate Laboratory. During this work J.A. was a research associate (National Research Council) at the NODC. The data used in our work has been acquired because of the efforts of scientists and data managers of the international scientific community.

References

Antonov, J.: 1990, 'Recent Climatic Changes of Vertical Thermal Structure of the North Atlantic and the North Pacific Oceans', *Meteorologiya i Gidrologiya* N4, 78–87 (in Russian).

Antonov, J.: 1993, 'Linear Trends of Temperature at Intermediate and Deep Layers of the North Atlantic Ocean and the North Pacific Ocean', *J. Clim.* **6**, 1928–1942.

Antonov, J. and Groisman, P. Ya.: 1988, 'Sea-Water Temperature Changes below the Seasonal Thermocline in the North Atlantic', *Meteorologiya i Gidrologiya* N3, 57–63 (in Russian).

Bethoux , J. P. and Gentili, B.: 1994, 'The Mediterranean Sea, a Test Area for Marine and Climatic Variation', in Malanotte-Rizzoli, P. and Robinson, A. R. (eds.), *Ocean Processes in Climate Dynamics: Global and Mediterranean Examples*, Kluwer Academic Publishers, Boston, pp. 239–254.

Bethoux, J. P., Gentili, B., Raunet, J., and Tailliez, D.: 1990, 'Warming Trend in the Western Mediterranean Deep Water', *Nature* **347**, 660–662.

Bethoux, J. P. and Tailliez, D.: 1994, 'Deep-Water in the Western Mediterranean Sea, Yearly Climatic Signature and Enigmatic Spreading', in Malanotte-Rizzoli, P. and Robinson, A. R. (eds.), *Ocean Processes in Climate Dynamics: Global and Mediterranean Examples*, Kluwer Academic Publishers, Boston, pp. 355–369.

Bindoff, N. L. and Church, J. A.: 1992, 'Warming of the Water Column in the Southwest Pacific Ocean', *Nature* **357**, 59–62.

Bindoff, N. L. and McDougall, T. J.: 1994, 'Diagnosing Climate Change and Ocean Ventilation Using Hydrographic Data', *J. Phys. Oceanogr.* **24**, 1137–1152.

Bingham, F. M., Suga, T., and Hanawa, K.: 1992, 'Comparison of Upper Ocean Thermal Conditions in the Western North Pacific Between Two Pentads: 1938–42 and 1978–92, *J. Oceanogr.* **48**, 404–425.

Brewer, P. G., Broecker, W. S., Jenkins, W. J., Rhines, P. B., Rooth, C. G., Swift, J. H., Takahasi, T., and Williams, R. T.: 1983, 'A Climatic Freshening of the Deep North Atlantic (North of 50° N), over the Past 20 Years', *Science* **222**, 1237–1239.

Brodeur, R. D. and Ware, D. M.: 1992, 'Long-Term Variability in Zooplankton Biomass in the Subarctic Pacific Ocean', *Fisheries Oceanogr.* **1**, 32–38.

Cayan, D. R.: 1992, 'Latent and Sensible Heat Flux Anomalies over the Northern Oceans: Driving the Sea Surface Temperature', *J. Phys. Oceanogr.* **22**, 859.

Da Silva, A., Young, C., and Levitus, S.: 1994, *Atlas of Surface Marine Data 1994, Vol. 1: Algorithms and Procedures. NOAA Atlas NESDIS 6*, (in press).

Davis, R. E., Webb, D. C., Regier, L. A., and Dufour, J.: 1992, 'The Autonomous Lagrangian Circulation Explorer (ALACE)', *J. Ocean and Atmosph. Tech.* **9**, 264–285.

Delcroix, T. and Henin, C.: 1989, 'Mechanisms of Subsurface Thermal Structure and Sea Surface Thermohaline Variabilities in the Southwestern Tropical Pacific During 1975–85', *J. Mar. Res.* **47**, 777–812.

Delcroix, T. and Henin, C.: 1990, 'Sea Surface Thermohaline and Subsurface Thermal Structure Variabilities in the Southwestern Tropical Pacific Ocean During 1979–1985', in *Air-Sea Interaction in Tropical Western Pacific: Proceedings of U.S.-PRC International TOGA Symposium. 1988, Beijing.*

Delcroix, T. and Henin, C.: 1991, 'Seasonal and Interannual Variations of Sea Surface Salinity in the Tropical Pacific Ocean', *J. Geophys. Res.-Oceans* **96**, 22135–22150.

Deser, C. and Blackmon, M.: 1994, 'Surface Climate Variations over the North Atlantic Ocean During Winter: 1900–1989, *J. Clim.* **6**, 1743–1753.

Dessier, A. and Donguy, J. R.: 1994, 'The Sea Surface Salinity in the Tropical Atlantic between 10° S and 30° N Seasonal and Interannual Variations (1977–1989)', *Deep-Sea Res.* **41**.

Dickson, R. R., Meincke, J., Malmberg, S. Aa., and Lee, A. J.: 1988, 'The "Great Salinity Anomaly" in the Northern North Atlantic', *Prog. in Oceanogr.* **20**, 103–151.

Dooley, H. D., Martin, J. H. A., and Ellet, D. J.: 1984, 'Abnormal Hydrographic Conditions in the Northeast Atlantic during the 1970's', *Rapp. P. v. Reun. Cons. int. Explor. Mer* **185**, 179–187.

Douglas, A. V., Cayan, D. R., and Namias, J.: 1982, 'Large-Scale Changes in North Pacific and North American Weather Patterns in Recent Decades', *Mon. Wea. Rev.* **110**, 1851–1862.

Ezer, T. and Mellor, G. L.: 1994, 'Diagnostic and Prognostic Calculations of the North Atlantic Circulation and Sea Level Using a Sigma Coordinate Ocean Model', *J. Geophys. Res.-Oceans* **99**, 14159–14171.

Ezer, T., Mellor, G. L., and Greatbatch, R. J.: 1995, 'On the Interpentadal Variability of the North Atlantic Ocean: Modelling Changes in Transport, Meridional Heat Flux and Coastal Sea Level Between 1955–1959 and 1970–1974', *J. Geophys. Res.-Oceans*, (submitted).

Folland, C. K., Parker, D. E., and Kates, F. E.: 1984, 'Worldwide Marine Temperature Fluctuations 1856–1981', *Nature* **310**, 670–673.

Frankignoul, C.: 1981, 'Low Frequency Temperature Fluctuations off Bermuda', *J. Geophys. Res.* **86**, 6522–6528.

Fu, L., Christensen, E., Yamarone, C., Lefebvre, M., Menard, Dorer, Y., and Escudier, P.: 1994, 'TOPEX/POSEIDON Mission Overview', *J. Geophys. Res.-Oceans* **99(C12)**, 24369–24382.

Greatbatch, R., Fanning, A. F., Goulding, A. D., and Levitus, S.: 1991, 'A Diagnosis of Interpentadal Circulation Changes in the North Atlantic', *J. Geophys. Res.-Oceans* **96**, 22009–220024.

Haake, B., Ittekkot, V., Rixen, T., Ramaswamy, V., Nair, R. R., Curry, W. B.: 1993, 'Seasonal and Interannual Variability of Particle Fluxes to the Deep Arabian Sea', *Deep-Sea Res.* **40**, 1323–1344.

Harris, G. P.: 1988, 'Interannual Variability in Climate and Fisheries in Tasmania', *Nature* **333(6175)**, 754–757.

Hecht, A.: 1992, 'Abrupt Changes in the Characteristics of Atlantic and Levantine Intermediate Waters in the Southeastern Levantine Basin', *Oceanol. Acta* **15**, 25–42.

IOC: 1992, 'Oceanic Interdecadal Climate Variability', *Intergovernm. Oceanogr. Commiss. Techn. Ser.* **40**, 40 pp.

Karl, D. M., Letelier, R., Hebel, D., Tupas, L., Dore, J., Christian, J., Winn, C.: 1995, 'Ecosystem Changes in the North Pacific Subtropical Gyre Attributed to the 1991–92 El Niño', *Nature*, 230–234.

Koltermann, K. P. and Sy, A.: 1994, 'Western North Atlantic Cools at Intermediate Depths', *International WOCE Newsletter* **15**, unpublished manuscript, 5–6.

Lacombe, H., Tchernia, P., and Gamberoni, L.: 1985, 'Variable Bottom Water in the Western Mediterranean Basin', *Prog. in Oceanogr.* **14**, 319–338.

Lazier, J. R. N.: 1980, 'Oceanographic Conditions at O.W.S. Bravo, 1964–1974', *Atmos. Ocean* **18**, 227–238.

Lazier, J. R. N.: 1988, 'Temperature and Salinity Changes in the Deep Labrador Sea, 1962–1986', **35**, 1247–1252.

Lazier, J. R. N.: 1995, 'The Salinity Decrease in the Labrador Sea over the Past Thirty Years', *Natural Climate Variability on Decade-to-Century Time-Scales*, National Academy of Sciences Press, (in press).

Leaman, K. D. and Schott, F. A.: 1991, 'Hydrographic Structure of the Convection Regime in the Gulf of Lions: Winter 1987', *J. Phys. Oceanogr.* **21**, 575–598.

Levitus, S.: 1989a, 'Interpentadal Variability of Temperature and Salinity at Intermediate Depths of the North Atlantic Ocean, 1970–74 versus 1955–59', *J. Geophys. Res.-Oceans* **94**, 6091–6131.

Levitus, S.: 1989b, 'Interpentadal Variability of Salinity in the Upper 150m of the North Atlantic Ocean, 1970–74 versus 1955–59', *J. Geophys. Res.-Oceans* **94**, 9679–9685.

Levitus, S.: 1989c, 'Interpentadal Variability of Temperature and Salinity in the Deep North Atlantic Ocean, 1970–74 versus 1955–59', *J. Geophys. Res.-Oceans* **94**, 16125–16131.

Levitus, S.: 1990, 'Interpentadal Variability of Steric Sea Level and Geopotential Thickness of the North Atlantic Ocean, 1970–74 versus 1955–59', *J. Geophys. Res.-Oceans* **95**, 5233–5238.

Levitus, S., Antonov, J., and Boyer, T. P.: 1994a, 'Interannual Variability of Temperature at 125 m Depth in the North Atlantic Ocean', *Science* **266**, 96–99.

Levitus, S., Boyer, T. P., and Antonov, J.: 1994b, *World Ocean Atlas 1994, Volume 5: Interannual Variability of Upper Ocean Thermal Structure. NOAA NESDIS Atlas series*, 176 pp.

Levitus, S., Antonov, J., Zhou, X., Dooley, H., Selemenov, K., Tereschenkov, V.: 1995, 'Decadal-Scale Variability of the North Atlantic Ocean', *Natural Climate Variability on Decade-to-Century Time Scales*, National Academy of Sciences Press, (in press).

Lukas, R.: 1994, 'HOT Results Show Interannual Variability of Pacific Deep and Bottom waters', *WOCE Notes: The U.S. role in the World Ocean Circulation Experiment* **6**(2), (unpublished manuscript).

Mellor, G. L., Mechoso, C. R., and Keto, E.: 1982, 'A Diagnostic Model of the General Circulation of the Atlantic Ocean', *Deep-Sea Res.* **29**, 1171–1192.

Morgan, M. R., Drinkwater, K. F., and Pocklington, R.: 1993, *Climatol. Bull.* **27**, 135–153.

Parilla, G., Lavin, A., Bryden, H., Garcia, M., and Millard, R.: 1994, 'Rising Temperatures in the Subtropical North Atlantic', *Nature*.

Parker, D. E., Folland, C. K., and Jackson, M.: 1995, 'Marine Surface Temperature: Observed Variations and Data Requirements', *Clim. Change* **31** (this issue).

Petrie, B. and Drinkwater, K. F.: 1993, 'Temperature and Salinity Variability on the Scotian Shelf and in the Gulf of Maine 1945–1990', *J. Geophys. Res.-Oceans*, 20079–20089.

Polovina, J. J., Mitchum, G. T., Graham. N. E., Craig, M. P., Demartini, E. E., and Flint, E. N.: 1994, 'Physical and Biological Consequences of a Climate Event in the Central North Pacific', *Fisher. Oceanogr.* **3**, 15–21.

Qiu, B. and Joyce, T. M.: 1992, 'Interannual Variability in the Mid- and Low-Latitude Western North Pacific', *J. Phys. Oceanogr.* **22**, 1062–1160.

Read, J. F., and Gould, W. J.: 1992, 'Cooling and Freshening of the Subpolar North Atlantic Ocean since the 1960s', *Nature*, 55–57.

Rohling, E. J. and Bryden, H. L.: 1992, 'Man-Induced Salinity and Temperature Increases in the Western Mediterranean Deep Water', *J. Geophys. Res.-Oceans* **97**, 11191–11198.

Ridgway, K. and Godfrey, S.: 1995, *Long Term Temperature and Circulation Changes in the East Australian Current*, (in preparation).

Roemmich, D.: 1985, 'Sea Level and the Variability of the Ocean', in *Glaciers, Ice Sheets, and Sea Level: Effect of a CO₂-induced Climatic Change*, DOE/ER/G0235-1, Dept. of Energy, Washington, D. C., pp. 104–115.

Roemmich, D. and Cornuelle, B.: 1990, 'Observing the Fluctuations of Gyre-Scale Ocean Circulation: A Study of the Subtropical South Pacific', *J. Phys. Oceanogr.* **20**, 1919–1934.

Roemmich, D. and Wunsch, C.: 1984, 'Apparent Changes in the Climatic State of the Deep North Atlantic', *Nature* **307**, 447–450.

Rossby, C. L.: 1959, 'Current Problems in Meteorology', in *The Atmosphere and Sea in Motion*, Rockefeller Institute Press, New York, pp. 9–50.

Rossby, T., Siedler, G., Zenk, W.: 1995, 'The Volunteer Observing Ship and Future Ocean Monitoring', *Bull. Amer. Meteor. Soc.* **76**, 5–11.

Royer, T. C.: 1989, 'Upper Ocean Temperature Variability in the Northeast Pacific Ocean: Is It an Indicator of Global Warming?', *J. Geophys. Res.-Oceans* **94**, 18175–18183.

Searle, B.: 1992, 'Global Temperature and Salinity Pilot Project', in *Proceedings of the Ocean Climate Data Workshop*, Sponsored by NOAA and NAS, Greenbelt, Md, U.S.A., pp. 97–108.

Shuto, K.: 1994, 'Scientific Background of JMA Surveys along PR2', *International WOCE Newsletter* **15**, 6–10, (unpublished manuscript).

Smith, N. R., Needler, G. T., and the Ocean Observing System Development Panel: 1995, *Clim. Change* **31** (this issue).

Stommel, H.: 1988, *Station 'S' off Bermuda: Physical Measurements 1954–1984*, Published by Woods Hole Oceanographic Institution and Bermuda Biological Station Library of Congress Number 88–51552, 189 pp.

Suga, T. and Hanawa, K.: 1995, 'Interannual Variation of North Pacific Subtropical Mode Water in the 137° E Section', *J. Phys. Oceanogr.* **25**, 1012–1017.

Swift, J. H.: 1984, 'A Recent θ-S Shift in the Deep Water of the Northern North Atlantic', in Hansen, J. E. and Takahashi, T. (eds.), *Climatic Processes and Climate Sensitivity, Geophys. Monogr. Series* **29**, AGU, Washington, D.C., pp. 39–47.

Swift, J. H.: 1985, 'A Few Comments on a Recent Deep-Water Freshening', *Glaciers, Ice Sheets, and Sea Level: Effect of a CO₂-induced Climatic Change*, DOE/ER/G0235-1, Dept. of Energy, Washington, D.C., pp. 104–115.

The Ocean Observing System Development Panel: 1995, *Scientific Design for the Common Module of the Global Climate Observing System and the Global Climate Observing System: An Ocean*

Observing System for Climate, Department of Oceanography. Texas A&M University, College Station, Texas, 265 pp.

Trenberth, K. E.: 1990, 'Recent Observed Interdecadal Climate Changes in the Northern Hemisphere', *Bull. Amer. Meteorol. Soc.* **71**, 988–993.

Venrick, E. L., McGowan, J. A., Cayan, D. R., Hatward, T. L.: 1987, 'Climate and Chlorophyll', *Science* **238**, 70–73.

Watanabe, T. and Mizuno, K.: 1994, 'Decadal Changes in the Thermal Structure in the North Pacific', *International WOCE Newsletter* **15**, 10–12, (unpublished manuscript),

Weare, B. C., Navato, A. R., and Newell, R. E.: 1976, 'Empirical Orthogonal Analysis of Pacific Sea Surface Temperatures', *J. Phys. Oceanogr.* **6**, 671–678.

Woodruff, S. D., Slutz, R. J., Jenne, R. J., and Steurer, P. M.: 1987, 'A Comprehensive Ocean-Atmosphere Data Set', *Bull. Amer. Meteorol. Soc.* **68**, 1239–1250.

Received 23 January, 1995; in revised form 28 July, 1995

MONITORING SEA LEVEL CHANGES

VIVIEN GORNITZ

Center for Climate Systems Research, Columbia University and NASA GSFC Institute for Space Studies, 2880 Broadway, New York, NY 10025, U.S.A.

Abstract. Future sea level rise arouses concern because of potentially deleterious impacts to coastal regions. These will stem not only from the loss of land through inundation and erosion, but also from increased frequency of storm floods, with a rising base level, even with no change in storm climatology, and from saltwater intrusion and greater amounts of waterlogging. Current sea level trends are important in formulating an accurate baseline for future projections. Sea level, furthermore, is an important parameter which integrates a number of oceanic and atmospheric processes. The ocean surface demonstrates considerable variability on diurnal, seasonal, and interannual time scales, induced by winds, storm waves, coastal upwelling, and geostrophic currents. Secular trends in sea level arise from changes in global mean temperature and also from crustal deformation on local to regional scales. The challenge facing researchers is how best to extract the climate signal from this noise.

This paper re-examines recent estimates of sea level rise, discusses causes of variability in the sea level records, and describes methods employed to filter out some of these contaminating signals. Evidence for trends in long-term sea level records and in extreme events is investigated. Application of satellite geodesy to sea level research is briefly reviewed.

Introduction

Sea level is a fundamental environmental parameter, which integrates diverse physical processes involving the ocean-atmosphere system. The ocean surface responds to diurnal, seasonal, and interannual variations in atmospheric pressure, winds, storm waves, coastal upwelling, and geostrophic currents. Long-term variations in sea level arise from deformation of the earth's crust as well as eustatic and steric changes that are associated with climate change. An improved understanding of these physical processes will lead to more accurate prediction of future climate-induced sea level rise.

Sea level has varied in response to past global climate changes, ranging from 120 m below present at the peak of the last ice age, around 20,000 years BP (Fairbanks, 1989), to 2–6 m higher than today in many locations, during the last interglacial, around 125,000 years ago (Matthews, 1990). The greenhouse gas-induced climate warming is anticipated to elevate rates of sea level rise by factors of 3–6 within the next 100 years (Warrick and Oerlemans, 1990; Wigley and Raper, 1992). Even at present, 70% of the world's sandy beaches are retreating (Bird, 1985), and projected sea level trends could exacerbate erosion and inundation of low-lying coastal areas, such as major river deltas.

The reported trend of mean global sea-level rise (*SLR*) over the last 100 years ranges between 1–2 mm yr^{-1} (Table I). These rates of sea-level rise are consistently higher than those of the past few thousand years, based on matched tide-gauge and geological or archeological data from widely-separated localities (Table II; Gornitz, 1995a; Gornitz and Seeber, 1990; Shennan and Woodworth, 1992; Varekamp *et al.*, 1992; Flemming and Woodworth, 1988), suggesting a relatively recent climate signal. However, the diverse physical processes that influence sea level on local and even global scales can introduce large uncertainties in the measurements (Table III; Gornitz, 1995b). In addition to coupled atmospheric-oceanic interactions, and crustal movements (including glacio-isostasy, neotectonism, and sediment compaction) accurate measurement of sea level is beset with problems of data quality and geographic distribution, effects of anthropogenic activity (such as water and sediment impoundment behind reservoirs and fluid withdrawal), and river runoff (Meade and Emery, 1971). These effects vary widely in their relative importance. This paper will review the factors contributing to the recent sea level rise, methods of filtering some of these contaminating signals, examine trends in extreme events, and discuss recent advances in sea level measurement technology.

Sea Level Observations

Sea-level observations over the last 100–150 years come from tide-gauge records collected by the Permanent Service for Mean Sea Level (PSMSL), Bidston Observatory, Birkenhead, England (Spencer and Woodworth, 1993), which has compiled records from over 1,400 tide-gauge stations worldwide (Figure 1). However, most tidal records are too short or have too many gaps to be useful. Very few stations have records older than 1880. Most of these are located in northwestern Europe, with a few in the United States (e.g. Key West; Maul and Martin, 1993). A time series of at least 50 years, and preferably longer, should be used to reduce effects of interannual to decadal fluctuations in sea surface temperatures, ocean currents, or long-term tides, such as the 18.6 year lunar nodal cycle. Most records longer than 40 years show confidence levels higher than 0.95 (Emery and Aubrey, 1991, p. 83). However, only 40 stations have nearly continuous sea level records over a 50 year period (Table I; Peltier and Tushingham, 1989, 1991). Only 21 stations survive with time series > 60 years (Table I; Douglas, 1991). Thus the selection of the optimum record length is a trade-off between the longer records required to minimize interdecadal variability and the number of available stations (Figure 1).

The quality of sea level data suffers from a number of problems, including too short, and broken time-series, considerable variations in the number of simultaneous records, and a geographic bias toward stations in the Northern Hemisphere, particularly for those with longer records (Figure 1). A critical assessment of these data limitations has led several researchers (Gröger and Plag, 1993; Pirazzoli, 1993) to conclude that the tide-gauge data set is inadequate for the determination of a global *SLR* trend.

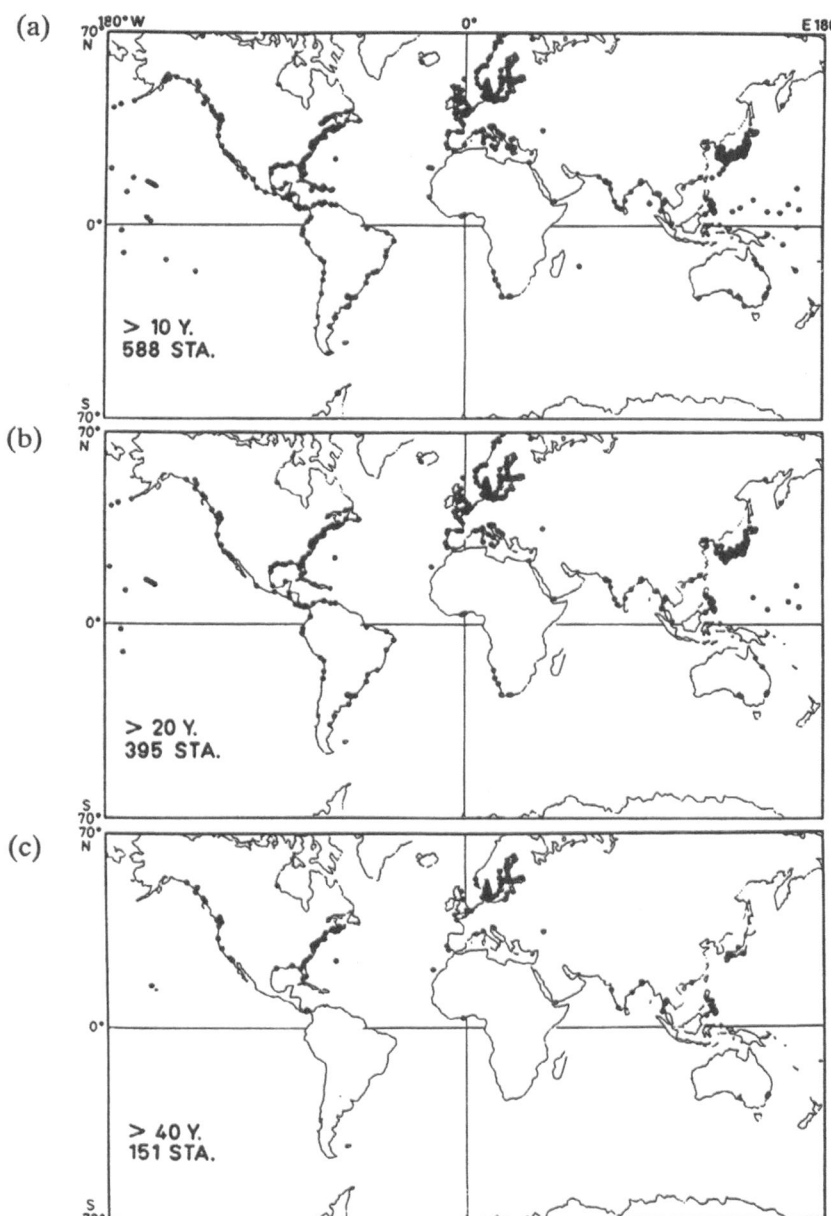

Fig. 1. Distribution of tide-gauge stations from the PSMSL: (a) all stations with record lengths > 10 years; (b) stations with record lengths > 20 years; (c) stations with record lengths > 40 years (after Emery and Aubrey, 1991).

TABLE I

Estimates of sea-level rise (updated from Warrick and Oerlemans, 1990)

Rate (mm/yr)	Comments	References
> 0.5	Cryologic estimate	Thorarinsson (1940)
1.1 ± 0.8	Many stations, 1807–1939	Gutenburg (1941)
1.2–1.4	Combined methods	Kuenen (1950)
1.1 ± 0.4	Six stations, 1807–1943	Lisitzin (1974)
1.2	Selected stations, 1900–1950	Fairbridge and Krebs (1962)
3.0	Many stations, 1935–1975	Emery (1980)
1.2 ± 0.1[†]	193 stations → 14 regions, 1880–1980	Gornitz et al. (1982)
1.5	Many stations, 1900–1975	Klige (1982)
1.5 ± 0.15[†]	Selected stations, 1903–1969	Barnett (1983)
1.4 ± 0.14[†]	Many stations → regions, 1881–1980	Barnett (1984)
1.2 ± 0.3[†]	130 stations, 1880–1982	Gornitz and Lebedeff (1987)*
1.0 ± 0.1[†]	130 stations → 11 regions, 1880–1982	Gornitz and Lebedeff (1987)*
1.15	155 stations, 1880–1986	Barnett (1988)
2.4 ± 0.9[§]	40 stations, 1920–1970	Peltier and Tushingham (1989, 1991)**
1.75 ± 0.13[§]	84 stations, 1900–1986	Trupin and Wahr (1990)**
1.67 ± 0.33	69 stations, 1900–1986	Wahr and Trupin (1990)**
1.8 ± 0.1[•]	21 stations, 1880–1980	Douglas (1991)**
1.15 ± 0.38[§]	655 stations → 10° × 10° blocks, ~1807–1990	Nakiboglu and Lambeck (1991)**

[†] = Value plus 95% confidence interval removed.

[§] = Mean and standard deviation effects removed.

[•] = Standard error.

* Long-term crustal motions.

** Glacio and hydro isostatic.

Physical processes account for most of the spatial and temporal variability in the sea level record (Table III). Atmospheric and oceanographic processes are responsible for a large fraction of the interannual variability. These fluctuations, 1–2 years to a decade in duration, and coherent over long distances, reflect steric changes (in temperature or salinity), currents, and coupled oceanographic-atmospheric forcing, such as the El Niño-Southern Oscillation (ENSO) in the Pacific Ocean (Komar and Enfield, 1987), or the Northern Atlantic Oscillation (Maul and Hanson, 1991).

TABLE II

Recent sea level rise

Region	Number stations	Sea level, recent trend, mm/yr	Sea level, late Holocene trend, mm/yr	Difference mm/yr
Eastern N. America	42	2.48 ± 1.81	0.97 ± 1.94	1.50 ± 0.71
Southeast Australia	5	1.55	0–0.3 (and local land subsidence)	0.95–1.13
New Zealand	4	1.68	0.0–0.15	1.70–1.81
Northwest Europe	18	1.12 ± 0.95	−0.013 ± 0.92	1.13 ± 0.52 (Shennan and Woodworth, 1992)

ENSO events, particularly moderate to strong ones, recur on average every 3–4 years (Diaz and Pulwarty, 1993). Amplitudes of the ENSO signal are between 10–50 cm (Table III). The onset of an ENSO episode is marked by a relaxation of the easterly trade winds, reversing the normal W to E equatorial sea-level gradient. As a result, anomalously warm water from the western equatorial Pacific propagates eastward, splitting into N and S poleward-propagating Kelvin wves, as the water masses approach the eastern Pacific rim (Jacobs et al., 1994; Komar and Enfield, 1987). This causes sea level to rise along the west coasts of both North and South America. The Kelvin wave is reflected from the American continents as a westward-propagating Rossby wave, during and after major ENSO occurrences, such as the 1982–83 event. The Rossby wave following this event has been traced across the Pacific basin for over a decade, using satellite altimetry (Jacobs et al., 1994).

DIRECT ANTHROPOGENIC EFFECTS

Anthropogenic processes over the last 100 years may also affect sea level (Figure 2, Table IV). The direct anthropogenic contribution to SLR over a specified time period can be expressed as:

$$SLR_a = (G \pm D + W) - (R + I),$$

(1)

where:

TABLE III

Summary of processes affecting sea level changes

Process		Rate (mm/yr)	Period (yr)
A	*Glacio-eustasy*	10 (av.)	The first $\sim 7,000$ of deglaciation
		~ 1–2	Last 150
B	*Vertical land movements* Long-wavelength processes 100–1000 km		
	Glacio-isostatic changes	± 1–10	10^4
	Shelf subsidence – ocean lithosphere cooling, sed. and water loading	0.03	10^7–10^8
	Shelf sedimentation Short-wavelength processes < 100 km	0.02–0.05	10^2–10^6
	Neo-tectonic movements	± 10	10^2–10^4
	Deltaic sedimentation	1–5	10–10^4
C	*Anthropogenic activity* Hydrologic cycle changes (see Table IV for details) Subsidence due to ground-water/oil/gas withdrawal	-1.6	< 100
	(local)	2–10	< 100
D	*Ocean-atmosphere effects*	Amplitude (cm)	
	Geostrophic currents Low-frequency atmospheric	1–100	1–10
	forcing	1–4	1–10
	El Niño	10–50	3–8

G = SLR due to groundwater mining;
D = SLR due to deforestation (combustion, oxidation, $\pm$runoff);
W = SLR due to drainage of wetlands;
R = SLR reduction from reservoir impoundment, infiltration, and water vapor storage;
I = SLR reduction due to irrigation (infiltration and water vapor storage).

Sahagian *et al.* (1994) estimate a net positive addition of 0.54 mm/yr to SLR over the last 60 years from such activities as groundwater mining, deforestation, and wetland loss (positive components) and the impounding of water reservoirs (negative component). They claim that this direct anthropogenic contribution to SLR could account for as much as a third of the observed rise, thereby significantly reducing the share attributable to climate warming. However, they have substantially underestimated the volume of water behind dams (Chao, 1994; Rodenburg, 1994). Furthermore, they have overlooked additional sources of water storage (Figure 2). One source is water lost through deep infiltration beneath both reservoirs and irrigated soils. Deep seepage beneath reservoirs recharges the aquifers, which ultimately increases groundwater discharge to the ocean. However, over short (<100 year) periods, the extent of marine discharge from this source may be relatively minor. If average seepage losses from deep percolation under reservoirs represents ~5% of reservoir volume (Gleick, 1992), this could amount to 250 km^3, equivalent to an average rate of 0.69 mm/yr withheld from SLR over the last 60 years since major dam construction began. (A volume of 360 km^3 is equivalent to 1 mm/yr SLR). Similarly, assuming a 5% loss of irrigation water to deep seepage results in a 0.37 mm/yr reduction in SLR (Table IV; Gornitz et al., 1994).

A second potential source comes from enhanced evaporation over reservoirs and evapotranspiration over irrigated soils that could increase the average atmospheric water vapor concentration, at least locally or regionally in arid areas. Although most of this water will return to the ground as precipitation, a portion of which will then flow to the ocean, a small although as yet unquantified amount remains in the atmosphere (shown schematically in Figure 2 as enhanced vapor storage). An accurate assessment of this fraction needs further investigation. If, for purposes of illustration, we assume that only 10% is retained in the atmosphere, there will be a net SLR reduction of 0.57 mm/yr from evapotranspiration over irrigated soils, and another 0.05 mm/yr from evaporation over reservoirs (Table IV).

The impact of deforestation on sea level could be either positive or negative. While combustion and oxidation of forest biomass add water to the ocean via the atmosphere (Figure 2), forest clearing could increase runoff. However, one recent model (Henderson-Sellers et al., 1993) suggests a net decrease in runoff, due to a greater decrease in precipitation than in evapotranspiration. With the given set of assumptions, the net effect of all direct anthropogenic processes would reduce SLR by an amount comparable to that actually observed (Table IV). However, the net contribution to SLR depends on the relative rates of opposing processes that are not yet well-quantified and will require comprehensive global hydrological modeling of the type now under development for use in general circulation models (GCMs).

CRUSTAL DEFORMATION AND SEA LEVEL

Crustal deformation is a major source of spatial variability in tide-gauge records (Emery and Aubrey, 1991; Table III). Post-glacial and hydro-isostatic crustal warping extends over 100–1000's kilometers (Tushingham and Peltier, 1991). Differen-

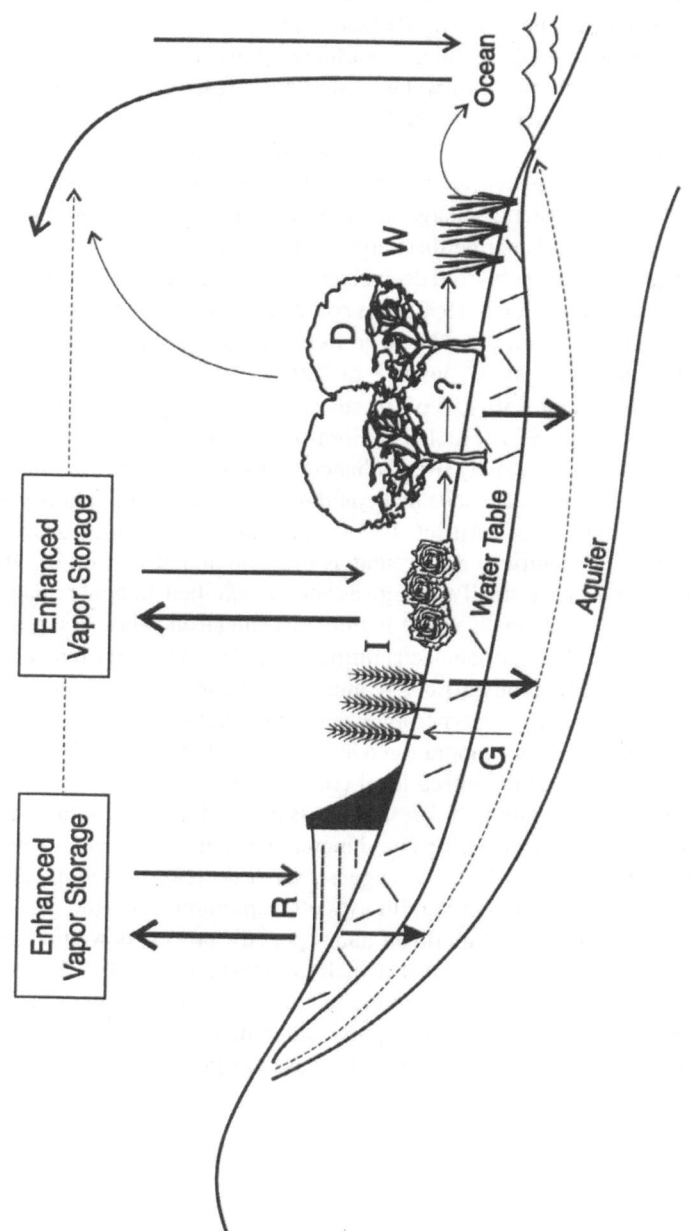

Fig. 2. Schematic diagram illustrating the effects of anthropogenic activities on sea level: R = reservoir impoundment; G = groundwater mining; I = irrigation; D = deforestation; and W = wetland drainage. Thick arrows indicate anthropogenic processes witholding water from the sea; thin arrows show water added to the sea, medium arrows are neutral.

TABLE IV

Anthropogenic contributions to sea level rise over the last 60 years

Process	Sea level rise (mm/yr)	Reference
Groundwater mining (G)	0.39	Sahagian *et al.*
Deforestation (D)		
combustion, oxidation;	0.030	
runoff	−0.15	
Wetlands loss (W)	0.006	Sahagian *et al.*
Reservoirs and dams (R)		
storage	−0.23	
infiltration	−0.69	
evaporation	−0.047	
Irrigation (I)		
infiltration	−0.37	
evapotranspiration	−0.57	
TOTAL	−1.63	

tial vertical movements over shorter distances are associated with tectonic deformation and fault displacements. Locally, heavy sediment loads carried by major rivers and/or compaction of peats in tidal marshes can lead to anomalously high coastal subsidence (for example, rates of 9–12 mm yr^{-1} along the Louisiana coast (Penland and Ramsey, 1990), or up to 5 mm yr^{-1} in the Nile delta (Chen *et al.*, 1992). Withdrawal of subsurface fluids (oil, gas, water) by pumping has led to land subsidence and increased rates of relative SLR at many coastal sites (Chi and Reilinger, 1984; Dolan and Goodell, 1986). An extreme example is Bangkok, Thailand, where the local subsidence had reached as much as 13 cm yr^{-1} (Milliman *et al.*, 1989).

Strategies to reduce the presence of crustal motions in sea level data include selection of a limited number of representative stations, with relatively long, complete records, from tectonically-stable regions (Fairbridge and Krebs, 1962; Barnett, 1983; Table I). Alternatively, a larger number of records have been averaged from non-tectonic or glacially rebounding sites, in the expectation that net residual land movements will tend to cancel (Barnett, 1984, 1988, Table I).

Late Holocene paleosea-level indicators (e.g. molluscs, corals, peats, wood, etc.; Gornitz *et al.*, 1982; Gornitz and Lebedeff, 1987) have been used to filter out long-wavelength crustal motions. The sea level curve recorded on tide-gauges

(2) includes long-term trends of glacio-isostatic, tectonic, and possibly residual eustatic origin.

$$SL_S = I + E_L + T_L + E_S \pm T_S \qquad (2)$$

The late Holocene sea level curve (3) is a composite of these three components.

$$SL_L = I + E_L + T_L \qquad (3)$$

Over the last few thousand years, these longer-term components can be approximated by linear trends. Although each point on the earth's surface has responded differently to the combined effects of glacial rebound and the post-glacial marine transgression, by differencing the sea level curves (2)–(3) from the same area, it is assumed that the long-term components will cancel, leaving only short-term changes (4).

$$SL_S - SL_L = (I + E_L + T_L + E_S \pm T_S) - (I + E_L + T_L) = E_S \pm T_S \qquad (4)$$

where:

SL_S = Short-term sea level curve from tide-gauge data;

SL_L = Long-term sea level curve (Late Holocene ≤ 6000 yr);

I = Glacio-isostatic component (rebound or uplift as in Canada, Scandinavia; subsidence as in U.S. East Coast, Holland);

E_L = Long-term eustatic component, often assumed to be 0, following final melting of the ice sheets by $\sim 6,000$ years BP. Minor residual melting (~ 0.2–0.4 mm/yr) cannot be completely ruled out;

T_L = Long-term tectonic component (either uplift or subsidence). This term includes continental shelf subsidence due to lithospheric cooling and sediment loading (est. ~ 0.03 mm/yr U.S. East coast), as well as more active plate boundary deformation;

T_S = Short-term land movements, including neotectonic motions (co-seismic) and subsurface fluid withdrawal (gas, oil, groundwater);

E_S = Short-term, recent eustatic and steric components (≤ 100–150 years).

Paired tide-gauge and radiocarbon data have been compared from: (a) 42 localities in eastern North America, extending from the Canadian Arctic to Key West, Florida; (b) 4 stations in New Zealand; (c) 4 stations in southeastern Australia (Gornitz, 1995a); and (d) 18 stations in northwestern Europe (Shennan and Woodworth, 1992). The North American and European sites cover both glacio-isostatically rebounding and subsiding terrains. Southeastern Australia has remained relatively geologically stable during the Holocene. Although New Zealand straddles the boundary between the Australian and Pacific Plates, and is therefore subject to active deformation, 3 our of the 4 tide-gauge stations (Auckland, Dunedin, and Christchurch-Lyttleton) are in comparatively stable areas. However, Wellington sits on a potentially unstable transition zone lying between the two plates, marked by strike-slip faulting. The estimated vertical component of motion is indicated in Table II.

The residual trends (Equation 4) from these sites range between 1.0 and 1.8 mm/yr (Table II). These values fall within the range of sea level rise estimates reported elsewhere (Table I). The small range in residual trends from widely-separated regions with contrasting geologic environments supports the hypothesis of a global sea level rise presumably associated with the recent climate warming.

Glacial rebound models have been applied to tide-gauge data in order to filter out elevation changes associated with ice melting and water loading (Peltier and Tushingham, 1989, 1991; Trupin and Wahr, 1990; Douglas, 1991). Radiometrically-dated sea level data are utilized to calibrate geophysical models that incorporate the deglaciation history, mantle visco-elastic and gravitational behavior. These studies have employed older versions of Peltier's model, but have differed in their averaging strategies and station selection criteria. To date, removal of glacial-isostatic effects from tide-gauge records by means of the latest version of Peltier's model, ICE-4G (Peltier, 1994) has not yet been undertaken.

Nakiboglu and Lambeck (1991) find a global mean sea-level rise of 1.15 ± 0.38 mm yr^{-1}, which lies on the lower end of the range in Table I. They have attempted to extract meaningful coefficients from a global fit of trends to a spherical harmonic expansion and have used records as short as 10 years. Their analysis made no attempt to compensate for disparate record lengths and uneven spatial distribution of tide gauges. Thus, their results should not carry as much weight as those of the other studies.

Both geological and geophysical methods are imperfect. The geological method is seriously limited by the spatial and temporal availability of paleosea-level data in the vicinity of a tide-gauge station, as well as the quality of dating these records. While glacial isostatic models are sensitive to assumptions of lithospheric thickness and viscoelastic properties of the earth's interior, particular choice of these parameters may not be too critical to corrections for post-glacial rebound (PGR). Inasmuch as model parameters can be traded off against each other, a more important consideration is whether the model correctly predicts the current amount of PGR. More uncertain are the thickness and melting histories of the Arctic and Antarctic ice sheets. The precise form of the deglaciation history, especially for the two largest ice sheets (North American and Antarctic) is still open to debate (cf. Kerr, 1994; Peltier, 1994; Colhoun et al., 1992; Lambeck, 1990). The most recent version of Peltier's model (ICE-4G; Peltier, 1994) predicts a reduced Laurentide ice sheet thickness of 2 km, in contrast to the earlier CLIMAP reconstruction of close to 3.5 km. Other ice sheets have similarly reduced thickenesses. As a consequence, the post-glacial eustatic SLR is only 105.2 m, which is ~15 m lower than that inferred from the Barbados sea level curve (Fairbanks, 1989). The model predictions, however, do not always accord with geological data. For example, dating of exposed, raised beaches in Antarctica suggests a contribution of only

0.5–2.5 m to post-glacial SLR (Colhoun et al., 1992), as compared to the ICE-4G model prediction of 21.8 m (Peltier, 1994).

Neither methodology takes into consideration local ground subsidence or neo-tectonism, as for example in the Pacific Northwest, where a significant residual spatial variability in sea level is associated with ongoing tectonic deformation (Mitchell et al., 1994; Kelsey et al., 1994). Analysis of the change in uplift rates on different time scales may ultimately contribute to a clearer separation between the tectonic and eustatic signals, by leading to improved models of the earthquake cycle. The uplift recorded by tide-gauges and geodetic leveling measurements over the last ~ 50 years at convergent plate boundaries represents interseismic deformation, most of which is recovered elastically during a major earthquake. The fraction that is not recovered coseismically contributes to the long-term ($\sim 10^4$ to 10^5 yrs) rock uplift, as expressed, for example, in both the raised marine terraces and the mean topography along the Oregon-Washington coast (Kelsey et al., 1994). This long-term uplift represents the integrated effects of numerous earthquake cycles. However, the eustatic signal cannot yet be unambiguously separated from the tectonic signal (both short-term interseismic and long-term integrated components), inasmuch as geodesists usually assume some eustatic trend in order to infer the uplift (Mitchell et al., 1994, for example assume 1.8 mm/yr, after Douglas, 1991). Satellite geodesy may eventually help resolve this issue, by providing an absolute measure of crustal motion (see below), but the geological and geophysical data on several time scales are essential in interpreting the results.

A more direct approach uses hydrodynamic time-series to investigate long-term, upper ocean steric changes off southern California (Roemmich, 1992). A trend of $+0.9 \pm 0.2$ mm yr^{-1} in steric height between 1950–1991 was associated predominantly with water temperature rise. However, this method only yields the steric component of SLR and provides no measure of any possible glacial meltwater contribution.

The mean of the sea-level trends shown in Table I is 1.49 ± 0.53 mm yr^{-1}, excluding Thorarinsson (1940), which was not based on tide-gauge data. Peltier and Tushingham (1991) find that the sea level trends depend strongly on the choice of minimum record length and on the particular time interval selected. Thus, the SLR estimates based on record lengths > 40 years (i.e. Douglas, 1991; Trupin and Wahr, 1990; and Peltier and Tushingham, 1989) tend to lie closer to 2 mm/yr than those based on shorter record lengths (Table I). Since spatial variability of many geophysical phenomena tends to be damped at low frequencies, long time series can partially compensate for the effects of uneven geographic distribution of tide gauges.

Another important source of difference is that Peltier's post-glacial rebound curve produces a much smaller residual sea level trend than does a linear fit to the geological data. Hence the corrected tide-gauge trends using the former (Peltier and Tushingham, 1989, 1991) will be higher than those using the latter (Gornitz and Lebedeff, 1987). Although the post glacial rebound follows an exponential decay law since the last glacial maximum 20,000 yrs BP, a linear interpolation produced a satisfactory fit to the geological data, in most cases, over the last few thousand years (see also discussion in Shennan and Woodworth, 1992).

ACCELERATION OF SEA LEVEL

A weak acceleration in the sea level trend was found for a few of the longest tide-gauge records from northwestern Europe (Woodworth, 1990; Gornitz and Solow, 1991; Figure 3) and Key West, Florida (Maul and Martin, 1993). However, no statistically significant acceleration of SLR was observed for a more globally representative data set of 23 stations in 10 groups, between 1905 and 1985. The acceleration in SLR still remained near zero even for a larger set of 37 stations within the same 10 groups, using a longer mean record length of 92 years over the period 1850–1991 (Douglas, 1992). Thus the instrumental record does not exhibit compelling evidence for any acceleration in SLR since the 1850's.

The fairly consistent 1–2 mm/yr difference between the sea level trend of the past 150 years and that of the last few thousand years from a number of widely separated regions (Table II; Gornitz, 1995a; Shennan and Woodworth, 1992) implies a comparatively recent acceleration of sea level. But the exact timing of the onset of the period of accelerated SLR remains uncertain.

In general, geological and archeolological data from a number of localities suggest that sea level has fluctuated by no more than several decimeters over the past two millennia (Stapor et al., 1991; Hofstede, 1991; Tanner, 1992; Varekamp et al., 1992). Several of the longest known tidal records (e.g. Amsterdam and Stockholm, see Mörner, 1973; Ekman, 1988) show increased rates of SLR over the last 100–150 years relative to the preceeding centuries. Microfaunal and geochemical studies of Connecticut salt marshes suggest a sharper sea level rise (average 2.2 mm/yr) over the last 200–250 years, in contrast to a more 'sluggish' rise (< 1 mm/yr) between 1500–1750 AD (Thomas et al., 1993; Varekamp et al., 1992). In the Chesapeake Bay area, eastern U.S.A., older marsh-stratigraphic studies implied a fairly constant sea level rise of 3.0 mm/yr, close to the present rate, since at least 1650 (Froomer, 1980). As pointed out above, this trend exceeds that of the last few millennia on the U.S. East Coast, and elsewhere. More recent studies around Chesapeake Bay, however, suggest that the apparent acceleration in marsh accretion and land loss rates began around the mid-19th century (Kearney and Stevenson, 1991; Kearney et al., 1994; Downs et al., 1994). The geomorphological changes around Chesapeake Bay are consistent, in part, with an increased rate of SLR since the end of the Little Ice Age around 1850 (Grove, 1988). However, an as yet unquantified portion of that increase may be caused by local anthropogenic factors, such as increased sediment loading due to 19th century land clearance, and more recent groundwater overdraft (Kearney and Stevenson, 1991).

On the other hand, Allen and Rae (1988) determined an accelerating curve of salt-marsh accretion from the Severn estuary, in southwest England, over the last

[397]

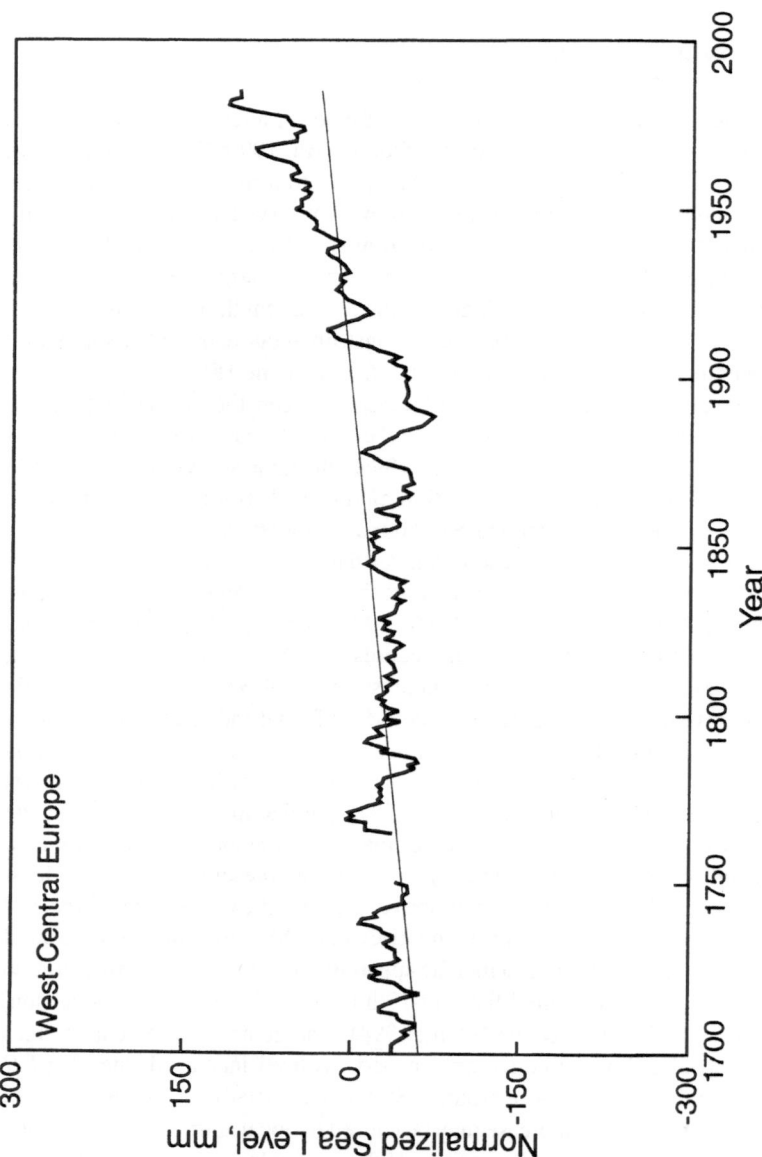

Fig. 3.   Acceleration in sea level from west-central Europe (after Gornitz and Solow, 1991). The curve is a regional composite of the 5-year running mean from Brest, Aberdeen, Amsterdam, and seven other Dutch stations.

2,000 years, presumably linked to the rate of relative sea level rise. Yet, there is no indication of a marked acceleration in secular trends of maximum flood levels over the last few centuries in Germany (Rohde, 1980), nor in England (Horner, 1972) (Figures 5a, b).

The geological and archeological records generally concur that sea level has been rising more rapidly for at least the last 150 years than in previous centuries, but precise dating of the onset of this most recent transgressive phase is not firmly established. Additional studies are clearly needed to resolve this issue.

Trends in Extreme Events

Trends in extreme sea levels, such as produced by storm surges and waves, are of concern because of the flood damage potential to low-lying coastal areas. Even in the absence of a change in storm climatology, rising sea levels will lead to a decrease in the return period of a storm surge of given elevation, because the surge is superposed on a higher base level (Figure 4).

Some historical data exist for Europe. Maximum flood levels for North Sea coastal cities in Germany have shown an average rise of ~ 2.5 mm/yr since 1570 (Figure 5a; Rohde, 1980). This compares with an average trend of 2.6 mm/yr, between 1890 and 1989, for 10 tide-gauge stations along the same coast (Führböter et al., 1990). Similarly, the high water level of the Thames River, in London, has increased linearly at a rate of ~ 7.1 mm/yr since 1791 (Figure 5b; Horner, 1972). Although this exceeds the mean sea level trend from the two nearest tide-gauge stations (1.2 mm/yr, Southend; 2.4 mm/yr Sheerness; Shennan and Woodworth, 1992), the flood data do not show any sign of acceleration over the last 200 years. Trends in extreme sea levels from 62 British stations are similar to mean sea level, both in spatial distribution and in value (Dixon and Tawn, 1992). On the other hand, the dramatic increase in surge levels in Venice, since 1900 is largely due to land subsidence from groundwater pumping and dredging (Camuffo, 1993; Pirazzoli, 1991), although older periods of enhanced surge activity (1500–1550; 1720–1830) may be associated with stormier climate.

An increase in mean significant wave heights in the North Atlantic has been linked to an increase in the annual mean atmospheric gradient (defined as the pressure difference between the Azores High and the Iceland Low divided by their distance – an index of the North Atlantic Oscillation; Bacon and Carter, 1993). The North Atlantic pressure gradient shows an upward trend since 1873, with an anomalous period (persistent NAO reversal with the Iceland Low weaker than normal) during the 1960's. However, analysis of geostrophic wind speeds in the German Bight, as a proxy for storminess, has not shown any significant change in the last 100 years (Figure 5c). Similarly, no trend in mean annual wave heights was observed along the coast of the Netherlands, between 1949 and 1980, nor was there any significant change in storm surge events at the Hook of Holland between

VIVIEN GORNITZ

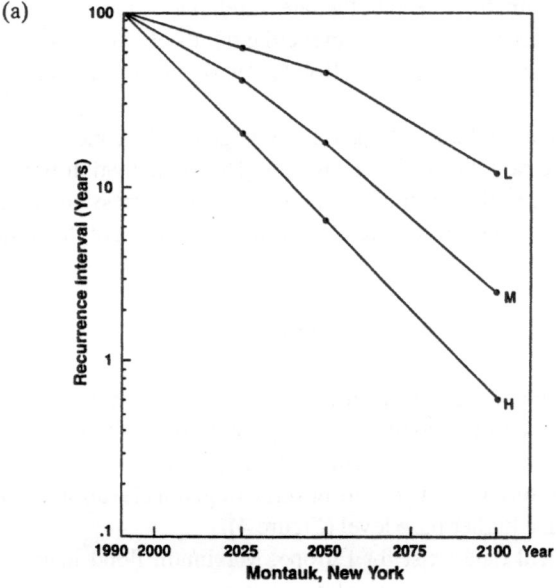

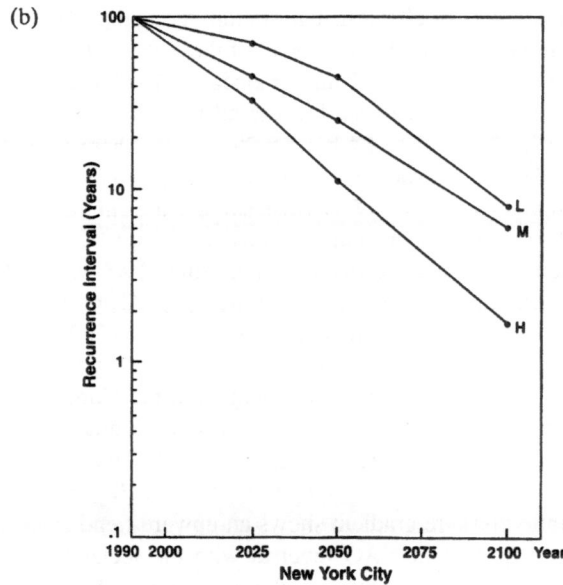

Fig. 4. Decrease in recurrence intervals for the '100-year' surge event for three sea level rise scenarios (modified from Warrick and Oerlemans, 1990; surge data from Ebersole, 1984), at (a) Montauk, New York; and (b) New York City.

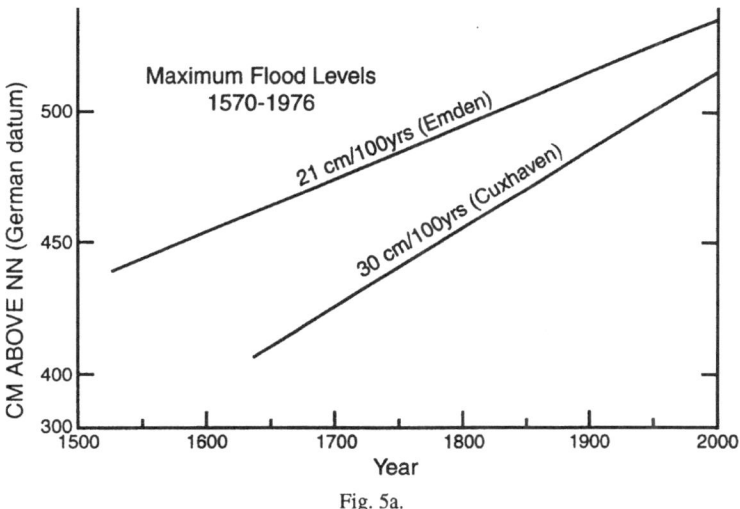

Fig. 5a.

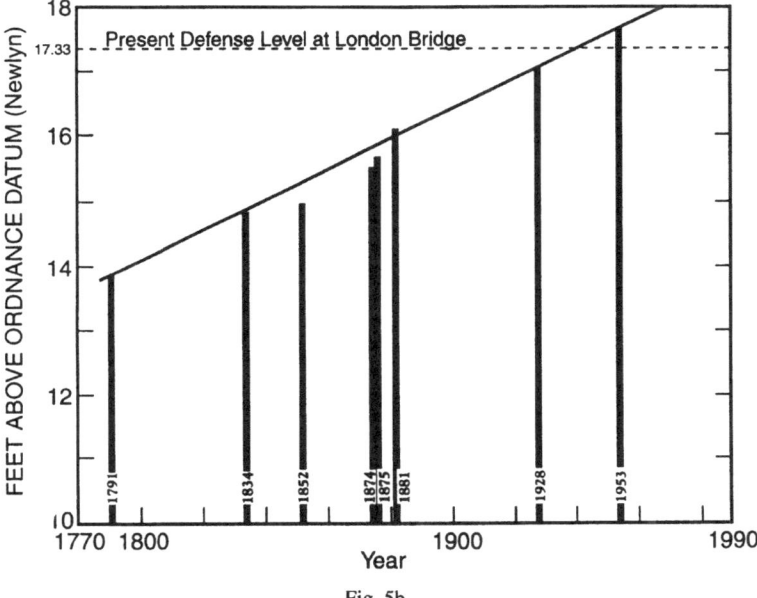

Fig. 5b.

1887 and 1985 (Figure 5d). The absence of a trend in wind velocities and storm surges appears to indicate a relatively constant storm climate in the German Bight over the record period.

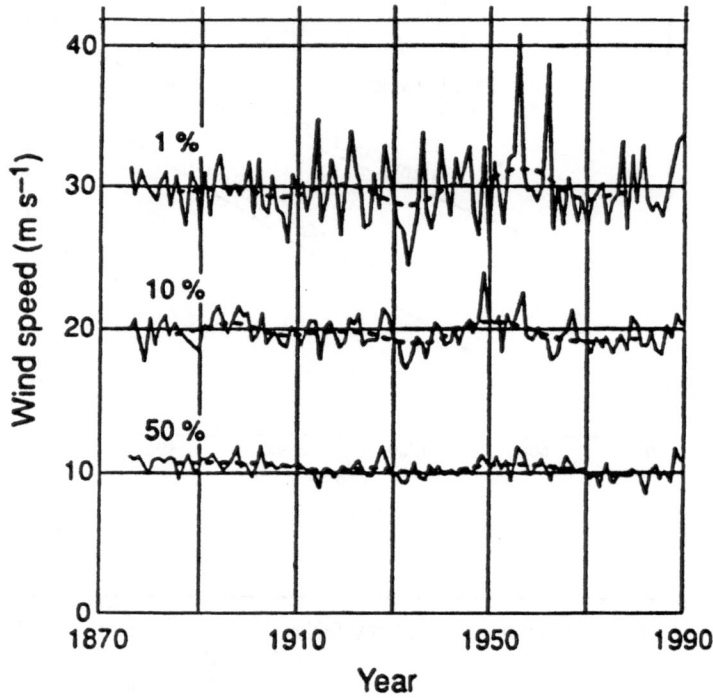

Fig. 5c.

These seemingly contradictory findings can be better understood in terms of the synoptic climatology. The annual average number of deep cyclones (central pressure ≤ 900 hPa) in the north-central Atlantic has increased by around 30% in the 1980's relative to the 1930's (Schinke, 1992). However the increase in storminess has been centered to the southwest of Iceland, whereas the annual number of deep cyclones over the Low Countries has remained relatively constant during this period (Figure 7 in Schinke, 1992).

Damaging waves and storm surges along the U.S. East Coast are largely caused by extratropical cyclones. Coastal cyclone frequencies and intensities increased between 1942 and 1959, peaked in the 1960's, whereas the 1970's and 1980's were characterized by less storms, and of shorter duration (Fenster and Dolan, 1994; Dolan *et al.*, 1988; Hayden, 1981). The peak in storm frequency and intensity during the mid-1960's relative to subsequent decades appears to have led to a systematic response of much of the East Coast toward greater beach erosion since 1967–68 (Fenster and Dolan, 1994).

Dolan *et al.* (1988) have suggested that the number of East Coast extratropical cyclones and the duration of storms with high waves may be associated with winter blocking anticyclones southwest of Greenland (i.e. strong North Atlantic Oscilla-

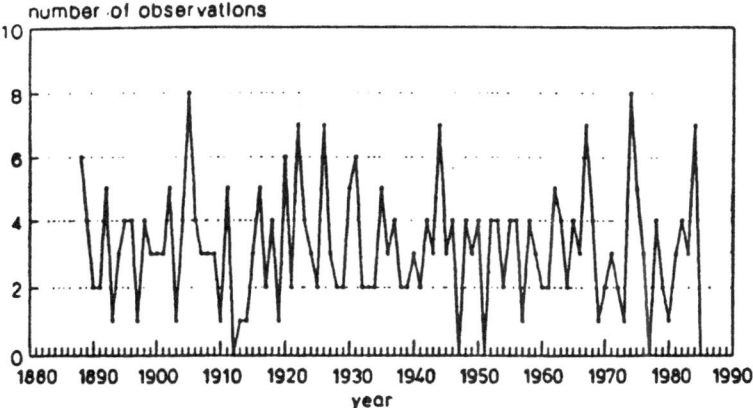

Fig. 5d.

Fig. 5. (a) (Top) Trend of maximum flood levels for two German cities on the North Sea Coast, 1570–1980, in cm/100 yr. These represent the minimum and maximum trends among a sample of 10 German coastal cities. Vertical scale is in cm above NN (current German reference datum; after Rohde, 1980); (b) (Bottom) Trend of high tide levels, London, 1791–1953 (after Horner, 1972); (c) Time series of geostrophic wind speeds from the German North Sea Coast. The percentages are the 1, 10, and 50% exceedance levels derived from the annual distributions of the daily wind speeds (i.e. 1, 10, and 50% of the observations, respectively, in each year, equal or exceed the value indicated). Solid lines are annual averages; broken lines, 30 year low pass (after Schmidt and von Storch, 1993); (d) Number of observations of storm surges greater than 90 cm in each storm season (Oct. 1–Mar. 15) at the Hook of Holland, 1888–1985) (from Hoozemans and Wiersma, 1992).

tion reversals). The persistent NAO reversal of the 1960's (Moses *et al.*, 1987) may have also contributed to a positive sea level anomaly along the southeastern U.S. coast at that time (Maul and Hanson, 1991). This period was also affected by the 'Great Salinity Anomaly', a near-surface, below-normal-salinity water mass that propagated cyclonically within the North Atlantic Ocean (Dickson *et al.*, 1988), and was also associated with expanded sea ice extents (Mysak and Manak, 1989; Mysak *et al.*, 1990).

Causes of Recent Sea Level Rise

The recent rise in sea level is consistent with the historical temperature record (Hansen and Lebedeff, 1988; Folland *et al.*, 1990; IPCC, 1992). Other indicators of recent global warming include the worldwide recession of alpine glaciers (Oerlemans, 1994; Grove, 1988; Meier, 1984), boreal tree-ring data (Jacoby and D'Arrigo, 1989), upward migration of Alpine plant species (Grabherr *et al.*, 1994), and near-surface warming of permafrost within the Arctic circle (Lachenbruch and

TABLE V

Estimated contributions to recent sea-level rise (cm)

Process	Sea level rise	Time interval	Study
Thermal expansion	1.4–4.5	1880–1980	Gornitz et al., 1982
	2.3–4.8	1880–1985	Wigley and Raper, 1987
	2–6	1890–1990	Warrick and Oerlemans, 1990
	5.5	1900–1980	Church et al., 1991
Mountain glaciers	3.7	1900–1980	Meier, 1984
	1.6	1900–1980	Oerlemans and Fortuin, 1992
Greenland	2.3	1880–1980	Warrick and Oerlemans, 1990
	2.5	1890–1990	van de Wal and Oerlemans, 1994
Antarctica	−7.5	100 years	Meier, 1990
	−10.0–12.0	100 years	Morgan et al., 1991
	−1.0–11.0	100 years	Bentley and Giovinetto, 1991
	up to +13.0	100 years	Jacobs, 1992

Marshal, 1986), in unfrozen soils from eastern Canada (Beltrami and Mareschal, 1991) and elsewhere (Lewis, 1992).

Thermal expansion of the upper layers of the oceans could account for 0.14 to 0.45 mm yr^{-1} of the observed SLR (Gornitz et al., 1982), depending on the equilibrium climate sensitivity for doubled CO_2, and diffusion coefficients used in the one-dimensional diffusion model of Hansen et al. (1984; Table V). Wigley and Raper (1987) calculated a thermal expansion of 2.3–4.8 cm between 1880–1985, assuming a global warming of 0.4–0.6 °C over that period, and using a box-upwelling diffusion energy-balance climate model (Table V). The projected thermal expansion decreases somewhat by including plant CO_2 fertilization effects on atmospheric CO_2 levels, possible negative feedbacks from halocarbons on ozone depletion, and sulfate aerosols, and particularly a reduced ocean temperature change from high-latitude sinking water relative to the global mean change (Wigley and Raper, 1992). A more realistic three-dimensional model which tracks thermohaline circulation over time produced a thermal expansion of 55 mm between 1900 and 1980, which is slightly higher than the above-mentioned studies (Church et al., 1991; Table V).

Meltwater from receding mountain glaciers also adds to the observed SLR. A sea-level rise of around 28 mm between 1900 and 1962 (or 0.46 mm yr^{-1}) was estimated using mass-balance data from 25 glaciers with records exceeding 50 years (Meier, 1984). Oerlemans and Fortuin (1992) instead calculated an increased meltwater equivalent to a SLR of only 0.20 mm yr^{-1}, for a temperature increase of 0.35 K over the same period, and a temperature sensitivity of 0.577 mm yr^{-1} °C^{-1}, derived from a glacier mass-balance model applied to 12 glaciers worldwide, with adequately reliable data (Table V). The difference between these two studies may be caused by the large spatial variability in temperature trends, such that the small sample sizes used in each case may not yield globally reliable results. Furthermore, the role of the subpolar ice caps may have been previously underestimated.

Addition of the higher thermal expansion estimates of ~ 5 cm SLR to the 2–4 cm from mountain glacier melting (Table V) accounts for ~ 7–9 cm of the observed SLR of the last 80–100 years, which is slightly below the lower end of the range given in Table I. If, however, the higher SLR values (Table I) are closer to the true value, then an additional source of water is required, presumably derived from the melting of the Greenland and possibly the Antarctic ice sheets.

Satellite radar altimetry suggests that the Greenland ice sheet may have thickened by some 0.23 ± 0.04 m yr^{-1}, between 1978–1986 (Zwally (1989), corresponding to a global sea-level depletion of 0.20 to 0.41 mm yr^{-1}. Systematic errors in the computed satellite orbits may however cast doubt on this result (Douglas *et al.*, 1990; van der Veen, 1993). Furthermore, the satellite coverage excluded large portions of Greeland, and moreover, the growth occurred during a decade of regional (although not global) cooling (Schneider, 1992). Continuous radar altimetry from 1978 to 1987 showed no significant change, or even a thinning (Braithwaite, 1993), so that the previously-noted thickening, if real, may represent only a very short-term fluctuation.

Glacier mass balance studies in Greenland suggests a sensitivity to warming of 0.3 ± 0.2 mm yr^{-1} per degree C (Warrick and Oerlemans, 1990). This corresponds to an estimated 23 ± 16 mm SLR between 1880 and 1980. An improved energy balance model for Greenland ice at higher spatial and temporal resolution, with a more detailed albedo parameterization produces a similar temperature sensitivity of 0.3 mm/yr SLR per degree K (van de Wal and Oerlemans, 1994). This is equivalent to a 25 mm SLR over the period 1890 to 1990 (Table V).

The extent of SLR contributions from the Antarctic ice sheet is even more uncertain. Field measurements of ice accumulation rates suggest a sea-level lowering of between 0.75 and 1.2 mm yr^{-1} (Meier, 1990; Morgan *et al.*, 1991; Bentley and Giovinetto, 1991). In contrast to the glaciological evidence, Jacobs (1992) found a negative mass balance equivalent to a SLR of 1.3 mm yr^{-1}. This conclusion was based on consideration of attrition processes, in particular basal melting of the ice shelves. Since, however, ice shelves are already afloat, not all of their mass loss will directly affect sea level. The contradiction between the glaciological and oceanographic findings could be resolved if there were sufficient shrinkage of the ice shelves (Bentley, 1993). While satellite imagery suggests that some of the ice shelves may in fact be retreating (e.g. the Wordie Ice Shelf, on the Antarctic Peninsula; Doake and Vaughan, 1991), it is not clear whether this is yet sufficient to affect sea level.

Anthropogenic activities such as changes in land cover and creation of artificial lakes could also influence sea level, as discussed above. However, estimates of these processes show such considerable spread that they may either be a source of *SLR* (Sahagian *et al.*, 1994) or a sink (Gornitz *et al.*, 1994).

Improvements in Sea Level Monitoring

The preceeding sections have pointed out shortcomings in the quality of sea level data and natural processes that tend to obfuscate the secular sea level trends of interest. Here we briefly review recent developments in sea level monitoring, which include improved tide-gauge instrumentation, implementation of global sea level data networks, and satellite geodetic methods for separating crustal motions from ocean height changes.

Many stations still employ traditional float gauges and record data on strip charts. These are being replaced by more automated technology. In the United States, the National Ocean Service (NOS, NOAA) has installed the Next Generation Water Level Management System (NGWLMS), which uses new sensors and microprocessor-based data collection and recording subsystems (Beaumariage and Scherer, 1987). A self-calibrating acoustic sensor takes 181 readings in three minutes. Data is stored every six minutes. The NGWLMS relays data every 3 hours from a data collection platform to a central computer facility in Rockville, MD, via NOAA's Geostationary Operational Satellite (GOES). In addition to sea level position, the data collection platform measures a suite of meteorological variables, including air and water temperatures, water density, wind speed and direction, and atmospheric pressure. The data can be linked directly to GPS and VLBI measurements (see below) for determination of absolute sea level change.

Another recent development is the creation of a global sea monitoring network, GLOSS (Global Sea Level Observing System) consisting of around 300 high-quality tide-gauges, of which ~ 200 are operational, spaced $\sim 1,000$ km apart along coastlines and on oceanic islands (Pugh, 1993; Woodworth, 1991, 1993). GLOSS, coordinated by the Intergovernmental Oceanographic Commission (IOC–UNESCO) through the PSMSL, provides hourly-resolution standardized sea level data useful to other global oceanographic programs, such as the World Ocean Circulation Experiment (WOCE, Clarke, 1993) and the Tropical Ocean Global Atmosphere (TOGA) experiment. The GLOSS gauges also serve as a source of 'ground truth' for validating radar altimetry measurements during the WOCE period (1990–1997).

A great advantage of satellite altimeter observations for sea level studies is their ocean-wide coverage over large areas where no conventional sea level data exist. Sea surface height relative to the earth's center is obtained by differencing the results from radar altimetry and precision tracking of the satellite's orbit. The former entails the exact measurement of the satellite's altitude above the ocean, while the latter accurately determines the satellite's distance with respect to the earth's center. An early comparison of sea surface topography, derived from GEOSAT altimetry, with sea level records from several Pacific island tide-gauge stations has demonstrated the potential utility of the satellite instrumentation (Wyrtki, 1987). More recently, Pacific tide-gauge data have been merged with ERS-1 and GEOSAT altimetry data (Miller *et al.*, 1993). The two data sets agree within a few centimeters at a near-equatorial site, but larger discrepancies occurred at higher latitudes.

Satellite altimetry has become a major tool for investigating ocean basin scale variability. For example, the propagation of oceanic waves over much of the North Pacific, originating in the 1982–83 El Niño event, was tracked over a decade, using combined GEOSAT and ERS-1 altimetry measurements of sea surface heights (Jacobs *et al.*, 1994). However, more precise orbit tracking and calibration against known standards will be necessary before altimetry can be used to measure global sea level rise (Wagner and Cheney, 1992). The TOPEX/POSEIDON space mission provides a significant improvement in orbit tracking and atmospheric corrections (Fu *et al.*, 1994). Yet, preliminary measurements of global sea level rise over a 1 1/2 year period (Nerem *et al.*, 1994) are inconsistent with longer-term sea level records and show a high variability stemming from the large amplitude of interannual and periodic sea level variations.

Satellite geodesy will enable to determination of absolute sea level change, by decoupling crustal movements from changes in mean ocean elevation, using Very Long Baseline Interferometry (VLBI), the Global Positioning System (GPS), and Satellite Laser Ranging, linked to a network of quality tide-gauge stations. The VLBI technique involves two or more radio telescopes spaced 1000's km apart that simultaneously observe the same extragalactic radio source (quasar). The difference in arrival time of the radio signal at the two stations is a measure of the distance between them. Variations over time of this distance provides a measure of the extent of land movements. The GPS system is analogous to VLBI, but uses portable ground-based radio receivers, stationed 100's km apart, to record radio transmissions from a network of around 20 geostationary satellites at altitudes of $\sim 20,000$ km, which take the place of the quasar.

In satellite laser ranging, focussed, coherent light beams are bounced off satellite retroreflector arrays (e.g. LAGEOS). The transit time between transmission and return of the light beam from the source represents $2\times$ the distance to the satellite. Simultaneous observations of the satellite from different locations and at different times provide an accurate measurement of the separation between stations. Most applications of satellite geodesy to data have involved study of tectonic plate motions, since at present accuracies in the horizontal direction are at least twice as good (5–10 mm) as in the vertical (10–20 mm; Bilham, 1991).

A network of around 30–40 stations are planned for the global VLBI geodedic reference system. Key tide-gauge stations will be tied to the VLBI network, through conventional geodetic levelling and/or GPS techniques. The VLBI/GPS network will be tied in which the satellite laser ranging system at key VLBI stations, in order to establish positions relative to the center of the earth. By subtracting the long-term land motion detected by the VLBI or GPS system from the apparent or

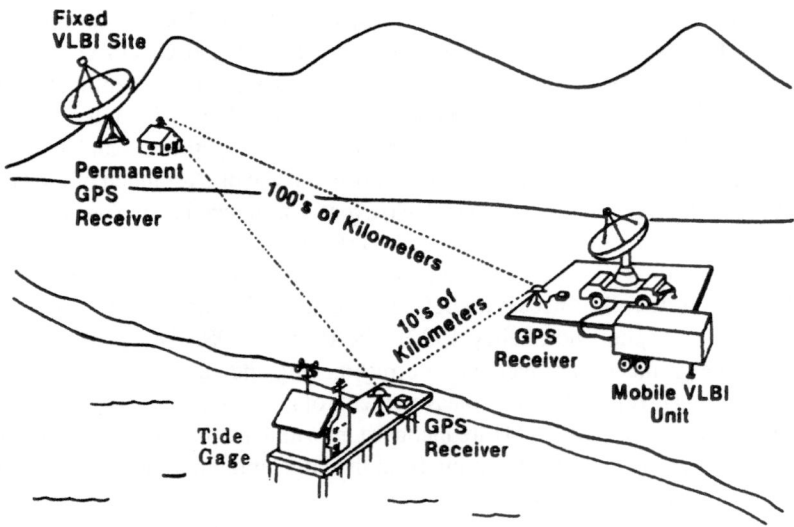

Fig. 6. Schematic diagram illustrating the measurement of absolute sea level. The tide-gauge station is tied geodetically to the GPS or VLBI system. The tide-gauge registers the relative change of the ocean surface over time, while the GPS or VLBI measurement gives the change in position of the earth's crust with respect to the earth's center. By difference the two measurements, the absolute change in sea level is determined (after Diamante *et al.*, 1987).

relative sea level change recorded by the tide-gauge, the absolute sea level signal can be established. Absolute gravity measurements taken both near the tide gauge and at the GPS/VLBI station will provide an independent measurement of vertical crustal motion (Figure 6).

Summary and Conclusions

Reported sea level trends range between 1–2 mm yr^{-1} during the last 100 years (Table I). However, longer records are more coherent and consistently yield values closer to 2 mm/yr. The reliability of sea level trends is adversely affected by the natural and anthropogenic processes that contaminate the tide-gauge data. These include vertical land movements (Emery and Aubrey, 1991), coupled ocean-atmospheric phenomena, anthropogenic effects (Tables II and IV), and data of variable quality (Pirazzoli, 1993; Gröger and Plag, 1993). Recent studies have attempted to filter out crustal motions by use of geological data (Gornitz and Lebedeff, 1987), glacial rebound models (Peltier and Tushingham, 1989, 1991; Trupin and Wahr, 1990; Douglas, 1991), or spherical harmonic analysis (Nakiboglu and Lambeck, 1991). The application of these techniques is restricted to passive plate margins. In convergent plate settings, additional geophysical and geodetic information is required to gain a better understanding of the earthquake cycle and to separate interseismic and long-term deformation from the eustatic signal.

Sea level trends do not show a statitically significant acceleration within the last 150 years (Douglas, 1992). Nevertheless present trends are much higher than those of the last few thousand years (Gornitz, 1995a; Thomas *et al.*, 1993; Varekamp *et al.*, 1992; Tanner, 1992; Shennan and Woodworth, 1992; Gornitz and Seeber, 1990). The average difference between paired tide-gauge and late Holocene trends from the same localities over widely separated regions, ranges between 1.0–1.8 mm yr^{-1} (Table II). The broadly similar increase in SLR from both glacial-isostatically uplifting and subsiding regimes, and from other diverse geologic environments, implies a comparatively recent onset of accelerated sea level rise, perhaps within the last few centuries. However, sea level data prior to 1850 are of limited extent and do not yet provide accurate dating for the beginning of this accelerated phase.

In spite of these difficulties, it is plausible to infer that the residual sea level trend of the last 150 years (or more) is related to the recent period of global warming (Hansen and Lebedeff, 1988; Folland *et al.*, 1990). This possibility is supported by the fairly high correlation between historic sea-level rise and global mean temperatures, lagging the latter by some 18 years (Gornitz *et al.*, 1982). However, the thermal expansion of the upper ocean layers and melting of mountain glaciers can account for only about 1/2 of the recent observed SLR of ~ 2 mm/yr, which is based on the longer, more reliable records. As discussed above, additional sources of SLR are more problematical. Melting of Greenland ice may have contributed up to 2.3 to 2.5 cm, but the role of the Antarctic remains much more uncertain, with estimates ranging from sink to source (Table V). The lack of thorough documentation of direct anthropogenic changes precludes any reliable assessment of their effects on SLR, although preliminary estimates have ranged from strongly positive (Sahagian *et al.*, 1994) to negative (Table IV; Gornitz *et al.*, 1994).

In the near future, a tie-in between high quality tide-gauges and global geodetic networks and satellite monitoring of ocean topography will enable delineation of absolute SL position (Bilham, 1991; Wagner and Cheney, 1992; Baker, 1993). This will permit separation of vertical land movements from changes in ocean height, thus providing a more accurate determination of true sea-level rise.

Acknowledgements

A portion of this research was supported by the NASA EOS Interdisciplinary Science Program and the NASA Columbia Cooperative Agreement NCC 5-61. The programming assistance of Mr. Z. Y. Zhang is also gratefully acknowledged. The manuscript has benefitted from the reviewer's comments.

References

Allen, J. R. L. and Rae, J. E.: 1988, 'Vertical Salt-Marsh Accretion since the Roman Period in the Severn Estuary, Southwest Britain', *Mar. Geol.* **83**, 225–235.

Bacon, S. and Carter, D. J. T.: 1993, 'A Connection between Mean Wave Height and Atmospheric Pressure Gradient in the North Atlantic', *Intl. J. Clim.* **13**, 423–436.

Baker, T. F.: 1993, 'Absolute Sea Level Measurements, Climate Change and Vertical Crustal Movements', *Glob. Planet. Change* **8**, 149–159.

Barnett, T. P.: 1983, 'Recent Changes in Sea Level and Their Possible Causes', *Clim. Change* **5**, 15–38.

Barnett, T. P.: 1984, 'The Estimates of "Global" Sea Level Change: A Problem of Uniqueness', *J. Geophys. Res.* **89**, 7980–7988.

Barnett, T. P.: 1988, 'Global Sea Level', in *NCPO, Climate Variations over the Past Century and the Greenhouse Effect*, A report based on the First Climate Trends Workshop, 7–9 Sept., 1988, Washington, D.D., Nat'l Clim. Program Office/NOAA, Rockville, MD.

Beaumariage, D. C. and Scherer, W. D.: 1987, 'New Technology Enhances Water Level Measurement', *Sea Technol.* May, 1987.

Beltrami, H. and Mareschal, J. C.: 1991, 'Recent Warming in Eastern Canada Inferred from Geothermal Measurements', *Geophys. Res. Lett.* **18**, 605–608.

Bentley, C. R.: 1993, 'Antarctic Mass Balance and Sea Level Change', *EOS* **74**, 585–586.

Bentley, C. R. and Giovinetto, M. B.: 1991, 'Mass Balance of Antarctica and Sea level Change', in Weller, G., Wilson, C. L., and Severin, B. A. B. (eds.), *Int'l Conf. on the Role of the Polar Regions in Global Change:* Proc. Conf. June 11–15, 1990, U. Alaska, Fairbanks, Geophysical Institute, U. Alaska, pp. 481–488.

Bilham, R.: 1991, 'Earthquakes and Sea Level: Space and Terrestrial Metrology on a Changing Planet', *Rev. Geophys.* **29**, 1–29.

Bird, E. C. F.: 1985, *Coastline Changes, A Global Review* J. Wiley and Sons, 219 pp.

Braithwaite, R. J.: 1993, 'Interpretation of Short-Term Ice-Sheet Elevation Changes Inferred from Satellite Altimetry', *Clim. Change* **23**, 383–405.

Camuffo, D.: 1993, 'Analysis of the Sea Surges at Venice from A.D. 782 to 1990', *Theor. Appl. Climatol.* **47**, 1–14.

Chao, B. F.: 1994, 'Man-Made Lakes and Sea-Level Rise', *Nature* **370**, 258.

Chen, Z., Warne, A. G., and Stanley, D. J.: 1992, 'Late Quaternary Evolution of the Northwestern Nile Delta between the Rosetta Promentory and Alexandria, Egypt', *J. Coast. Res.* **8**, 527–561.

Chi, S. C. and Reilinger, R. E.: 1984, 'Geodetic Evidence for Subsidence due to Groundwater Withdrawal in Many Parts of the USA', *J. Hydrol.* **67**, 155–182.

Church, J. A., Godfrey, J. S., Jackett, D. R., and MacDougall, T. J.: 1991, 'A Model of Sea Level Rise Caused by Ocean Thermal Expansion', *J. Clim.* **4**, 438–456.

Clarke, R. A.: 1993, 'The World Ocean Circulation Experiment (WOCE)', *W.M.O. Bull.* **42**, 28–33.

Colhoun, E. A., Mabin, M. C. G., Adamson, D. A., and Kirk, R. M.: 1992, 'Antarctic Ice Volume and Contribution to Sea-Level Fall at 20,000 yr BP from Raised Beaches', *Nature* **358**, 316–319.

Diamante, J. M., Pyle, T. E., Carter, W. E., and Scherer, W.: 1987, 'Global Change and the Measurement of Absolute Sea Level', *Prog. Oceanog.* **18**, 1–21.

Diaz, H. F. and Pulwarty, R.: 1993, 'A Comparison of Southern Oscillation and El Niño Signals in the Tropics', in Diaz, H. F. and Markgraf, V. (eds.), *El Niño: Historical and Paleoclimatic Aspects of the Southern Oscillation*, Cambridge University Press, pp. 175–192.

Dickson, R. R., Meincke, J., and Lee, A. J.: 1988, 'The "Great Salinity Anomaly" in the northern Atlantic, 1968–82', *Prog. Oceanogr.* **20**, 103–151.

Dixon, M. J. and Tawn, J. A.: 1992, 'Trends in UK Extreme Sea-Levels: A Spatial Approach', *Geophys. J. Int.* **111**, 607–616.

Doake, C. S. M. and Vaughan, D. G.: 1991, 'Rapid Disintegration of the Wordie Ice Shelf in Response to Atmospheric Warming', *Nature* **350**, 328–330.

Dolan, R. and Goodell, H. G.: 1986, 'Sinking Cities', *Am. Sci.* **74**, 38–47.

Dolan, R., Lins, H., and Hayden, B.: 1988, 'Mid-Atlantic Coastal Storms', *J. Coast. Res.* **4**, 417–433.

Douglas, B.: 1991, 'Global Sea Level Rise', *J. Geophys. Res.* **96**, 6981–6992.

Douglas, B.: 1992, 'Global Sea Level Acceleration', *J. Geophys. Res.* **97**, 12, 699–12, 706.

Douglas, B. C., Cheney, R. E., Miller, L., Agreen, R. W., Carter, W. E., and Robertson, D. S.: 1990, 'Greenland Ice Sheet: Is It Growing or Shrinking?', *Science* **248**, 288–289.

Downs, L. L., Nicholls, R., Leatherman, S. P., and Hautzenroder, J.: 1994, 'Historic Evolution of a Marsh Island: Bloodsworth Island, Maryland', *J. Coast Res.* **10**, 1031–1044.

Ebersole, B. A.: 1984, *Atlantic Coast Water-Level Climate*, WIS Report 7, U.S. Army Corps of Engineers, Vicksburg, MS.

Ekman, M.: 1988, 'The World's Longest Continued Series of Sea level Observations', *Pageoph.* **127**, 73–77.

Emery, K. O.: 1980, 'Relative Sea Levels from Tide-Gauge Records', *Proc. Natl. Acad. Sci., Washington, D.C.* **77**, 6968–6972.

Emery, K. O. and Aubrey, D. G.: 1991, *Sea Levels, Land Levels, and Tide Gauges*, Springer-Verlag Inc., New York, 237 pp.

Fairbanks, R.: 1989, 'A 17,000-Year Glacio-Eustatic Sea-Level Record: Influence of Glacial Melting Rates on the Younger Dryas Event and Deep Ocean Circulation', *Nature* **342**, 637–642.

Fairbridge, R. W. and Krebs, O. A.: 1962, 'Sea Level and the Southern Oscillation', *Geophys. J.* **6**, 532–545.

Fenster, M. and Dolan, R.: 1994, 'Large-Scale Reversals in Shoreline Trends along the U.S. Mid-Atlantic Coast', *Geology* **22**, 543–546.

Flemming, N. C. and Woodworth, P. L.: 1988, 'Monthly Mean Sea-Levels in Greece 1969–1983 Compared to Relative Vertical Land Movements Measured over Different Timescales', *Tectonophysics* **148**, 59–72.

Folland, C. K., Karl, T., and Vinnikov, K. Ya.: 1990, 'Observed Climate Variations and Change', in Houghton, J. T., Jenkins, G. J., and Ephraums, J. J. (eds.), *Climate Change: the IPCC Scientific Assessment*, Cambridge University Press, Cambridge, pp. 195–238.

Froomer, N. L.: 1980, 'Sea Level Changes in the Chesapeake Bay During Historic Times', *Mar. Geol.* **36**, 289–305.

Fu, L. L., Christensen, E. J., Yamarone, C. A., Jr., Lefebvre, M., Menard, Y., Dorrer, M., and Escudier, P.: 1994, 'TOPEX/POSEIDON Mission Overview', *J. Geophys. Res.* **99**, 24,369–24,381.

Führböter, A., Dette, H. H., and Töppe, A.: 1990, 'Recent Changes in Sea Level', *J. Coast. Res. Spec. Issue No.* **9**, 146–159.

Gleick, P. H.: 1992, 'Environmental Consequences of Hydroelectric Development: The Role of Facility Type and Size', *Energy* **17**, 735–747.

Gornitz, V.: 1995a, 'A Comparison of Differences between Recent and Late Holocene Sea Level Trends from Eastern North America and Other Selected Regions', *J. Coastal Res. Spec. Issue*, (in press).

Gornitz, V.: 1995b, 'Sea Level Rise: A Review of Recent Past and Near-Future Trends', *Earth Surface Proces. and Landforms* **20**, 7–20.

Gornitz, V. and Lebedeff, S.: 1987, 'Global Sea Level Changes During the Past Century', in Nummedal, D., Pilkey, O. H., and Howard, J. D. (eds.), *Sea Level Fluctuation and Coastal Evolution*, SEPM Spec. Publ. **41**, 3–16.

Gornitz, V., Lebedeff, S., and Hansen, J.: 1982, 'Global Sea Level Trend in the Past Century', *Science* **215**, 1611–1614.

Gornitz, V., Rosenzweig, C., and Hillel, D.: 1994, 'Is Sea Level Rising or Falling?', *Nature* **371**, 481.

Gornitz, V. and Seeber, L.: 1990, 'Vertical Crustal Movements along the East Coast, North America, from Historic and Late Holocene Sea Level Data', *Tectonophys.* **178**, 127–150.

Gornitz, V. and Solow, A.: 1991, 'Observations of Long-Term Tide-Gauge Records for Indicators of Accelerated Sea Level Rise', in Schlesinger, M. E. (ed.), *Greenhouse Gas-Induced Climatic Change: A Critical Appraisal of Simulations and Observations*, Elsevier, Amsterdam, pp. 347–367.

Grabherr, G., Gottfried, M., and Pauli, H.: 1994, 'Climate Effects on Mountain Plants', *Nature* **369**, 448.

Gregory, J. M.: 1993, 'Sea-Level Changes under Increasing Atmospheric CO_2 in a Transient Coupled Ocean-Atmosphere GCM Experiment', *J. Clim.* **6**, 2247–2262.

542 VIVIEN GORNITZ

Gröger, M. and Plag, H.-P.: 1993, 'Estimations of a Global Sea Level Trend: Limitations from the Structure of the PSMSL Global Sea Level Data Set', *Glob. and Planet. Change* **8**, 161–179.

Grove, J. M.: 1988, *The Little Ice Age*, Methuen, London, 498 pp.

Gutenberg, B.: 1941, 'Changes in Sea Level, Post-glacial Uplift and Mobility of the Earth's Interior', *Bull. Geol. Soc. Am.* **52**, 721–722.

Hansen, J., Lacis, A., Rind, D., Russell, G., Stone, P., Fung, I., Ruedy, R., and Lerner, J.: 1984, 'Climate Sensitivity: Analysis of Feedback Mechanisms', in Hansen, J. E. and Takahashi, T. (eds.), *Climate Processes and Climate Sensitivity*, Monogr. Ser., Vol. 29, Am. Geophys. Union, Washington, D.C., pp. 130–163.

Hansen, J. and Lebedeff, S.: 1988, 'Global Surface Air Temperatures: Update through 1987', *Geophys. Res. Lett.* **15**, 323–326.

Hayden, B. P.: 1981, 'Secular Variations in Atlantic Coast Extratropical Cyclones', *Mon. Wea. Rev.* **109**, 159–167.

Henderson-Sellers, A., Dickinson, R. E., Durbridge, T. B., Kennedy, P. J., McGuffie, K., and Pitman, A. J.: 1993, 'Tropical Deforestation: Modeling Local- to Regional-Scale Climate Change', *J. Geophys. Res.* **98**, 7289–7315.

Hofstede, J. L. A.: 1991, 'Sea Level Rise in the Inner German Bight (Germany) since AD 600 and Its Implications upon Tidal Flats Geomorphology', in Bruckner, H. and Radtke, U. (eds.), *Von der Nordsee bis zum Indischen Ozean*, Franz Steiner, Stuttgart, pp. 11–27.

Hoozemans, F. M. J. and Wiersma, J.: 1992, 'Is Mean Wave Height in the North Sea Increasing', *Hydrogr. J.* **63**, 13–15.

Horner, R. W.: 1972, 'Current Proposals for the Thames Barrier and the Organization of the Investigations', *Phil. Trans. Roy. Soc. London A* **272**, 179–185.

IPCC: 1992, *Climate Change 1992*, Houghton, J. T., Callander, B. A., and Varney, S. K. (eds.), *Climate Change 1992: The Supplementary Report to the IPCC Scientific Assessment*, Cambridge University Press, Cambridge.

Jacoby, G., Jr. and D'Arrigo, R.: 1989, 'Reconstructed Northern Hemisphere Annual Temperature since 1671 Based on High Latitude Tree-Ring Data from North America', *Clim. Change* **15**, 39–59.

Jacobs, G. A., Hurlburt, H. E., Kindle, J. C., Metzger, E. J., Mitchell, J. L., Teague, W. J., and Wallcraft, A. J.: 1994, 'Decade-Scale trans-Pacific Propagation and Warming Effects of an El Niño Anomaly', *Nature* **370**, 360–363.

Jacobs, S. S.: 1992, 'Is the Antarctic Ice Sheet Growing?', *Nature* **360**, 29–33.

Kelsey, H. M., Engbretson, D. C., Mitchell, C. E., and Ticknor, R. L.: 1994, 'Topographic Form of the Coast Ranges of the Cascadia Margin in Relation to Coastal Uplift Rates and Plate Subduction', *J. Geophys. Res.* **99**, 12, 245–12, 255.

Kearney, M. S. and Stevenson, J. C.: 1991, 'Island Land Loss and Marsh Vertical Accretion Rate Evidence for Historical Sea-Level Change in Chesapeake Bay', *J. Coastal Res.* **7**, 403–415.

Kearney, M. S., Stevenson, J. C., and Ward, L. G.: 1994, 'Spatial and Temporal Changes in Marsh Vertical Accretion Rates at Monie Bay: Implications for Sea-Level Rise', *J. Coast. Res.* **10**, 1010–1020.

Kerr, R. A.: 1994, 'How High was Ice Age Ice? A Rebounding Earth May Tell', *Science* **265**, 189.

Klige, R. K.: 1982, 'Oceanic Level Fluctuations in the History of the Earth', in *Sea and Oceanic Level Fluctuations for 15,000 Years*, Acad. Sc. U.S.S.R., Institute of Geography, Moscow, Nauka, pp. 11–22, (in Russian).

Komar, P. D. and Enfield, D. B.: 1987, 'Short-Term Sea Level Changes and Shore-Line Erosion', in Nummedal, D. *et al.* (eds.), *Sea Level Fluctuation and Coastal Evolution*, SEPM Spec. Publ. 41, pp. 17–27.

Lachenbruch, A. and Marshall, B. V.: 1986, 'Changing Climate: Geothermal Evidence from Permafrost in the Alaskan Arctic', *Science* **234**, 689–696.

Lambeck, K.: 1990, 'Late Pleistocene, Holocene and Present Sea-Levels: Constraints on Future Change', *Paleogeog. Paleoclim. Paleoecol.* **89**, 205–217.

Lewis, T. (ed.): 1992, 'Climatic Change Inferred from Underground Temperatures', *Global and Planet. Change Spec. Issue* **6**, 71–281.

Lisitzin, E.: 1974, *Sea Level Changes*, Elsevier, New York.

Matthews, R. K.: 1990, 'Quaternary Sea-Level Change', in *Sea-Level Change*, National Research Council, Washington, D.C., pp. 88–103.

Maul, G. A. and Hanson, K.: 1991, 'Interannual Coherence between North Atlantic Atmospheric Surface Pressure and Composite Southern U.S.A. Sea Level', *Geophys. Res. Lett.* **18**, 653–656.

Maul, G. A. and Martin, D. M.: 1993, 'Sea Level Rise at Key West, Florida, 1846–1992: America's Longest Instrument REcord?', *Geophys. Res. Lett.* **20**, 1955–1958.

Meade, R. H. and Emery, K. O.: 1971, 'Sea Level as Affected by River Runoff, Eastern United States', *Science* **173**, 425–427.

Meier, M. F.: 1984, 'The Contribution of Small Glaciers to Global Sea Level Rise', *Science* **226**, 1418–1421.

Meier, M. F.: 1990, 'Reduced Rise in Sea Level', *Nature* **343**, 115–116.

Miller, L., Cheney, R., and Lillibridge, J.: 1993, 'Blending ERS-1 Altimetry and Tide-Gauge Data', *EOS* **74**, 185, 197.

Milliman, J. D., Broadus, J. M., and Gable, F.: 1989, 'Environmental and Economic Implications of Rising Sea Level and Subsiding Deltas: The Nile and Bengal Examples', *Ambio* **18**, 340–345.

Mitchell, C. E., Vincent, P., Weldon, R. J. II, and Richards, M. A.: 1994, 'Present-Day Deformation of the Cascadia Margin, Pacific Northwest, U.S.A.', *J. Geophys. Res.* **99**, 12, 257–12, 277.

Morgan, V. I., Goodwin, I. D., Etheridge, D. M., and Wookey, C. W.: 1991, 'Evidence from Antarctic Ice cores for Recent Increases in Snow Accumulation', *Nature* **354**, 58–60.

Mörner, N. A.: 1973, 'Eustatic Changes During the Last 300 Years', *Paleogeog. Paleoclim. Paleoecol.* **13**, 1–14.

Moses, T., Kiladis, G. N., Diaz, H. F., and Barry, R. G.: 1987, 'Characteristics and Frequency of Reversals in Mean Sea Level Pressure in the North Atlantic Sector and Their Relationship to Long-Term Temperature Trends', *J. Clim.* **7**, 13–30.

Mysak, L. A. and Manak, D. K.: 1989, 'Arctic Sea-Icea Extent and Anomalies, 1953–1984', *Atmos.-Ocean* **27**, 376–405.

Mysak, L. A., Manak, D. K., and Marsden, R. F.: 1990, 'Sea-Ice Anomalies Observed in the Greenland and Labrador Seas During 1901–1984 and Their Relation to an Interdecadal Arctic Climate Cycle', *Clim. Dynam.* **5**, 111–133.

Nakiboglu, S. M. and Lambeck, K.: 1991, 'Secular Sea Level Change', in Sabadini, R., Lambeck, K., and Boschi, E. (eds.), *Glacial Isostasy, Sea Level and Mantle Rheology*, Kluwer Academic Publ., Dordrecht, pp. 237–258.

Nerem, R. S., Schrama, E. J., Koblinsky, C. J., and Beckley, B. D.: 1994, 'A Preliminary Evaluation of Ocean Topography from the TOPEX/POSEIDON Mission', *J. Geophys. Res.* **99**, 24, 565–24, 583.

Oerlemans, J.: 1994, 'Quantifying Global Warming from the Retreat of Glaciers', *Science* **264**, 243–245.

Oerlemans, J. and Fortuin, J. P. F.: 1992, 'Sensitivity of Glaciers and Small Ice Caps to Greenhouse Warming', *Science* **258**, 115–117.

Peltier, W. R.: 1994, 'Ice Age Paleotopography', *Science* **265**, 806–810.

Peltier, W. R. and Tushingham, A. M.: 1989, 'Global Sea Level Rise and the Greenhouse Effect: Might They Be Connected?', *Science* **244**, 806–810.

Peltier, W. R. and Tushingham, A. M.: 1991, 'Influence of Glacial Isostatic Adjustment on Tide-Gauge Measurements of Secular Sea Level Change', *J. Geophys. Res.* **96**, 6779–6796.

Penland, S. and Ramsey, K. E.: 1990, 'Relative Sea Level Rise in Louisiana and the Gulf of Mexico, 1908–1988', *J. Coast. Res.* **6**, 323–342.

Pirazzoli, P. A.: 1991, 'Possible Defenses against a Sea-Level Rise in the Venice Area, Italy', *J. Coast. Res.* **7**, 231–248.

Pirazzoli, P. A.: 1993, 'Global Sea-Level Changes and Their Measurement', *Glob. and Planet. Change* **8**, 135–148.

Pugh, D. T.: 1993, 'Improving Sea Level Data', in Warrick, R. A., Barrow, E. M., and Wigley, T. M. L. (eds.), *Climate and Sea Level Change: Observations, Projections and Implications*, Cambridge University Press, Cambridge, pp. 57–71.

Rodenburg, E.: 1994, 'Man-Made Lakes and Sea Level Rise', *Nature* **370**, 258.

Roemmich, D.: 1992, 'Ocean Warming and Sea Level Rise along the Southwest U.S. Coast', *Science* **257**, 373–375.

Rohde, H.: 1980, 'Changes in Sea Level in the German Bight', *Geophys. J. R. Astr. Soc.* **62**, 291–302.

Sahagian, D. L., Schwartz, F. W., and Jacobs, D. K.: 1994, 'Direct Anthropogenic Contributions to Sea Level Rise in the Twentieth Century', *Nature* **367**, 54–57.

Schinke, H.: 1992, 'Zum Auftreten von Zyklonen mit Niedrigen Kerndrucken im Atlantisch-Europäischen Raum von 1930 bis 1991', *Wiss. Zeitschrift der Humboldt-Universität zu Berlin, R. Math./Naturwiss.* **41**, 17–28.

Schmidt, H. and von Storch, H.: 1993, 'German Bight Storms Analyzed', *Nature* **365**, 791.

Schneider, S.: 1992, 'Will Sea Levels Rise or Fall?', *Nature* **356**, 11–12.

Shennan, I. and Woodworth, P. L.: 1992, 'A Comparison of Late Holocene and Twentieth-Century Sea-Level Trends from the UK and North Sea Region', *Geophys. R. Int.* **109**, 96–105.

Spencer, N. E. and Woodworth, P. L.: 1993, *Data Holdings of the Permanent Service for Mean Sea Level (November 1993)*, PSMSL, Bidston Observatory, Birkenhead, U.K., 81 pp.

Stapor, F. W., Jr., Mathews, T. D., and Lindfors-Kearns, F. E.: 1991, 'Barrier-Island Progradation and Holocene Sea-Level History in Southwest Florida', *J. Coast. Res.* **7**, 815–838.

Tanner, W. F.: 1992, '3000 Years of Sea Level Change', *Bull. Am. Met. Soc.* **73**, 297–303.

Thomas, E., Nydick, K., Scholand, S. J., Varekamp, J. C.: 1993, 'Accelerated Sea Level Rise over the Last 200 Years?', *Geol. Soc. Am. Abst. with Prog.* **X**, A61–A62.

Thorarinsson, S.: 1940, 'Present Glacier Shrinkage and Eustatic Changes in Sea Level', *Geogr. Ann.* **22**, 131–159.

Trupin, A. and Wahr, J.: 1990, 'Spectroscopic Analysis of Global Tide-Gauge Sea Level Data', *Geophys. J. Intl.* **100**, 441–453.

Tushingham, A. M. and Peltier, W. R.: 1991, 'ICE-3G: A New Model of Late Pleistocene Deglaciation Based upon Geophysical Predictions of Post-Glacial Sea Level Change', *J. Geophys. Res.* **96**, 4497–4523.

Van der Veen, C. J.: 1993, 'Interpretation of Short-Term Ice-Sheet Elevation Changes Inferred from Satellite Altimetry', *Clim. Change* **23**, 383–405.

van de Wal, R. S. W. and Oerlemans, J.: 1994, 'An Energy Balance Model for the Greenland Ice Sheet', *Glob. Plan. Change* **9**, 115–131.

Varekamp, J. C., Thomas, E., and Van de Plassche, O.: 1992, 'Relative Sea-Level Rise and Climate Change over the Last 1500 Years', *Terra Nova* **4**, 293–304.

Wagner, C. A. and Cheney, R. E.: 1992, 'Global Sea Level Change from Satellite Altimetry', *J. Geophys. Res.* **97**, 15, 607–15, 615.

Wahr, J. M. and Trupin, A. S.: 1990, 'New Computation of Global Sea Level Rise from 1990 Tide-Gauge Data', *EOS* **71**, 1267.

Warrick, R. A. and Oerlemans, J.: 1990, 'Sea Level Rise', in Houghton, J. T., Jenkins, G. J., and Ephraums, J. J. (eds.), *Climate Change: The IPCC Scientific Assessment*, Cambridge University Press, Cambridge, pp. 257–281.

Wigley, T. M. L. and Raper, S. C. B.: 1987, 'Thermal Expansion of Sea Water Associated with Global Warming', *Nature* **330**, 127–131.

Wigley, T. M. L. and Raper, S. C. B.: 1992, 'Implications for Climate and Sea Level of Revised IPCC Emission Scenarios', *Nature* **357**, 293–300.

Woodworth, P. L.: 1990, 'A Search for Acceleration in Records of European Mean Sea Level', *Int'l. J. Clim.* **10**, 129–143.

Woodworth, P. L.: 1991, 'The Permanent Service for Mean Sea Level and the Global Sea Level Observing System', *J. Coast. Res.* **7**, 699–710.

Woodworth, P. L. (ed.): 1993, *Sea Level Changes: Measurements and Analysis. Sixtienth Anniversary Meeting of the Permanent Service for Mean Sea Level, 9–10 Dec., 1993*, 37 pp.

Wyrtki, K.: 1987, 'Comparing GEOSAT Altimetry and Sea Level', *EOS* **35**, 731.

Zwally, H. J.: 1989, 'Growth of Greenland Ice Sheet: Interpretation', *Science* **246**, 1589–1591.

(Received 23 January, 1995; in revised form 18 May, 1995)

LAND SURFACE TEMPERATURES – IS THE NETWORK GOOD ENOUGH?

P. D. JONES

Climatic Research Unit, University of East Anglia, Norwich NR4 7TJ, U.K.

Abstract. The land surface air temperature record is shown to be adequate for monitoring hemispheric and global scale temperatures. This is only one of a myriad of uses to which station temperatures can be put. For many of the other uses the record is inadequate, but could easily be improved by gaining access to much more data that is collected by all the countries of the world, but not passed on internationally. The situation affects other fields measured at the surface, notably precipitation and mean-sea-level pressure.

1. Introduction

The average temperature of the Earth's surface is probably the most important single measure of the state of the climate system. However, if we want to understand the system better and to eventually detect anthropogenic influences on the system, by itself it is practically useless. For detection and understanding we require temperature and many other variables measured spatially across the Earth's surface and at several levels in the atmosphere and in the ocean.

The purpose of this article is to provide some 'personal' insights into the surface air temperature record, based on my 15 years of experience of working with such data. How well are we monitoring the variable and how accurate are resulting series?

2. Monitoring

2.1. HOW WIDELY IS THE LAND SURFACE AIR TEMPERATURE MONITORED?

The short answer to this question is that the potential nework af observing sites listed in World Meteorological Organisation (WMO) reports (for example, WMO, 1978) is impressive. Over 9000 stations are listed in a 1992 digital version of this WMO report. Monitoring the network month by month, however, illustrates many of the problems. The number of potential stations may be high in WMO publications, but the number reporting in real time over land regions during 1994 has only been about 1250. Furthermore, 1250 may report but, in order for the data to be used, reference data for a common period for each station are required. Additionally, some of the sites cannot be used because they are affected by urbanization effects while others have had, over the years, a number of site moves and/or procedural changes. Consequently, the number of stations available for use is only about 900 per month. Figure 1 shows a typical month's reports during 1994.

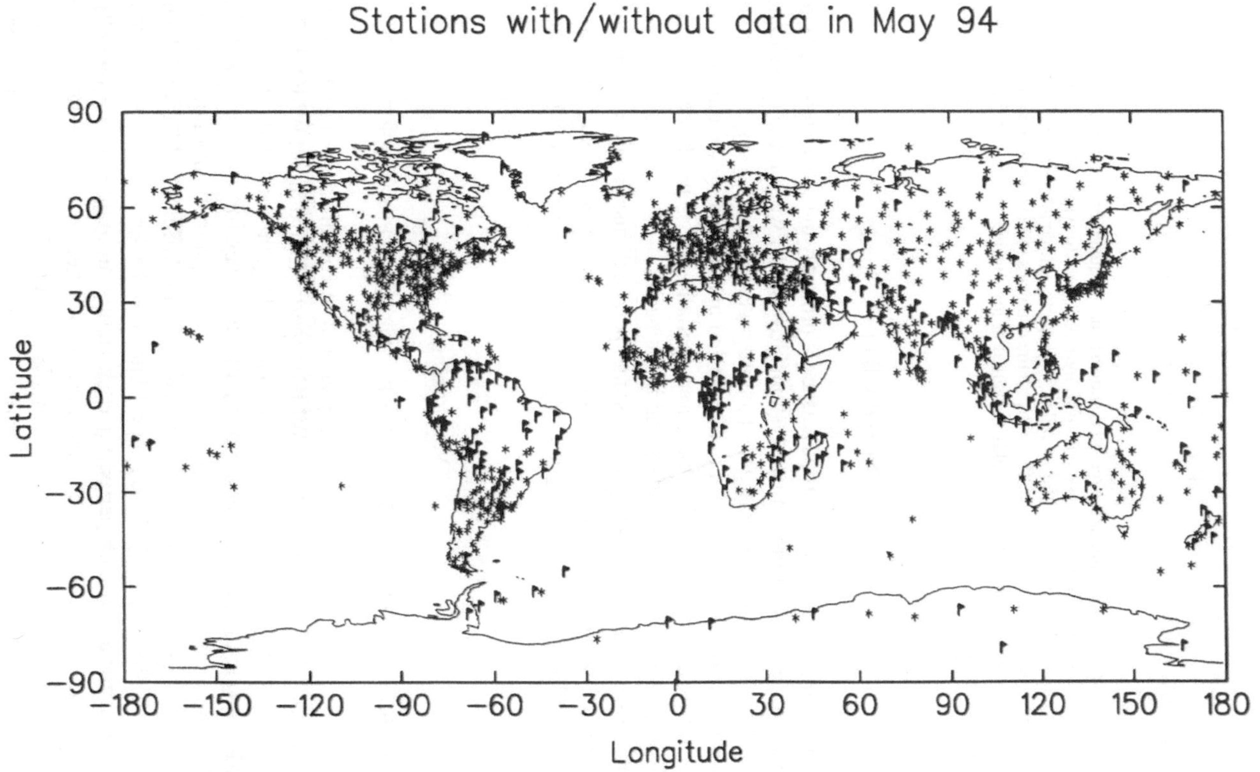

Fig. 1. Climatological stations reporting over the GTS in May 1994. 1264 stations reported this month. Of these, 913 had 1961–90 reference period data and had passed the homogeneity tests detailed in Jones *et al.* (1985, 1986a). These are coded as *. 312 stations, with reference period data, have at least one report during the last 10 years, but did not report this month. They are marked with a ▲. 351 (1264–913) stations were not useable. They are not marked, some of their locations not being known.

So one simple conclusion is that one quarter of the data reported through WMO are of little use in the construction of grid box temperature datasets. A larger fraction of the data could be used, however, if reference period data were available. My latest gridded temperature data (Jones, 1994) uses 1961–90 as the reference period, and incorporates about 1000 more stations than earlier work (Jones, 1988).

Publication of the 1961–90 averages received by WMO should be accomplished later in 1995. In the Climatic Research Unit we are attempting to collect such normals from the individual countries. For many countries we have received more stations and more variables than have been sent to the National Climatic Data Center NCDC) in Asheville for publication by WMO (Heim, pers. comm.). Figure 2 shows our current status, while Hulme (1994) and Hulme et al. (1995) highlight many of the problems found in gaining access to such data for Europe. Similar problems have occurred in many countries outside Europe. The obvious conclusion from Hulme (1994) is that, if more resources were available nationally and meteorological services did not charge, sometimes exorbitant rates for such data, we would to be in a better position.

2.2. IS THIS SITUATION ADEQUATE AND IS IT LIKELY TO IMPROVE?

If the construction of the global or hermispheric average land-based air temperature series is all that is required, it is easily adequate. Briffa and Jones (1993) and Jones (1994) have both shown that about 170 stations, well spaced around the world, would adequately capture 95% ofthe variations in the global and hemispheric land-based air temperature series. We are only in a position to be able to know this answer, however, because of the amount of station data available during the twentieth century.

For most other uses to which the station data might be put (e.g., regional monitoring, climate impact studies etc.), the present network of received data can and should be improved. Maps such as Figure 1 clearly show where improvements are most needed. No additional resources are required, just a commitment by a few countries that currently release no data internationally (e.g. over the Global Telecommunication System (GTS) or to *Monthly Climatic Data for the World* (MCDW)) to start, and by the majority of countries to release data for a greater number of stations. The additional station data are being monitored within certain countries but not released over the GTS or to MCDW. National datasets have been incorporated into global data sets such as those held at the Climatic Research Unit and NCDC, Asheville as a result of specific efforts or projects (e.g. Karl et al., 1993; and Razuvayev et al., 1993). These data are never real time and exercises like this are constantly being required to improve the network of available data.

Figure 3 shows the number of statlons with near-complete records (at least 10 of 12 months in a given year), and the number of 5° boxes in which these stations are located and the percentage of the Earth's surface represented through time. The number of available stations has been falling since the 1960s. The effect on the

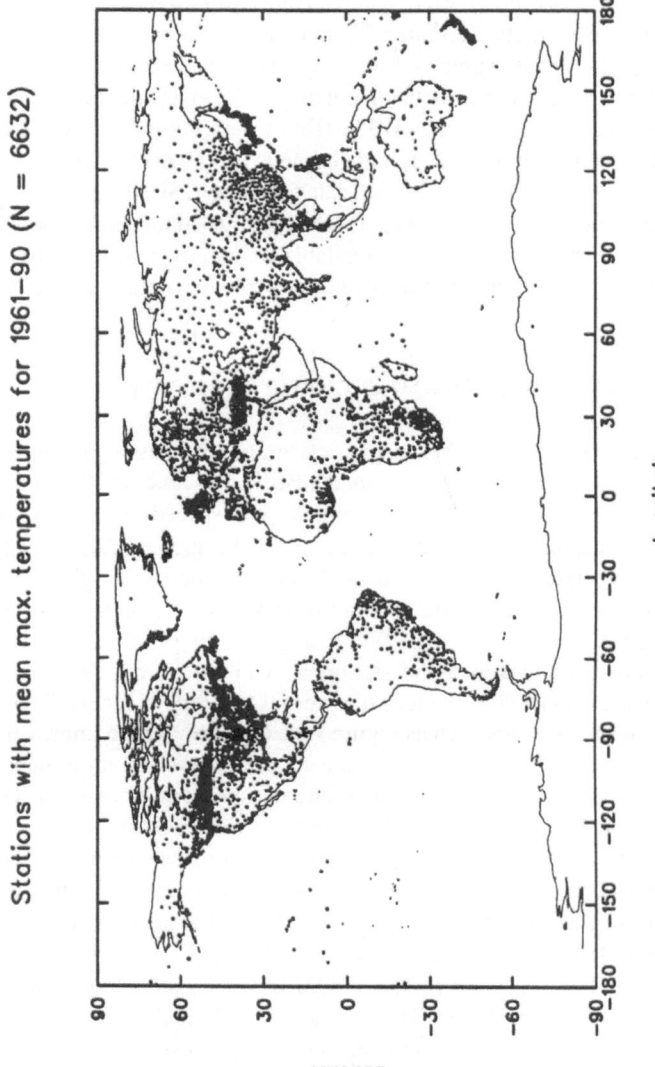

Fig. 2. Climatological stations with 1961–90 normals data for mean maximum temperature. All station data has been received by CRU from the National Meteorological Agencies. 6632 locations are shown.

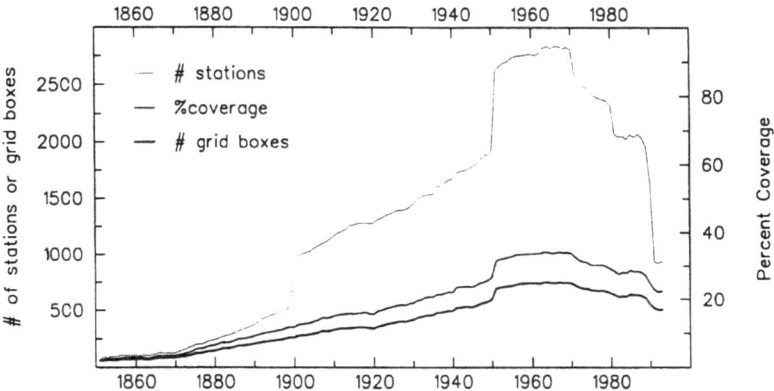

Fig. 3.   Time series for 1851 to 1993 of the number of stations with near complete records, the number of 5° boxes in which these stations are located and the percentage of the Earth's surface with data.

area represented is less pronounced but still clearly apparent. The decline in station numbers mainly results from the transition from GTS/MCDW data, boosted by the specific efforts outlined above, to solely GTS data. Some volumes of World Weather Records for the 1970s and 1980s have not been published. Some of the decline during the late 1980s and early 1990s is due to the break-up of the former Soviet Union and Yugoslavia, poor reporting from parts of Africa and the closing of some remote stations in some countries.

The only sure way to improve the accuracy of the *grid box estimates* is to increase the numbers of contributing stations, particularly in regions which never have had data or have only a sparse network (see Figure 1). The number of stations on the GTS should be increased and the system upgraded to take advantage of the information superhighway. Improvements to the network have been achieved by specific efforts (see above) and throught requests to National Meteorological Agencies (NMAs). If national governments see the Framework Convention on Climate Change (FCCC) as important, they should instruct their meteorological agencies to release as many data as possible, through publications/reports and/or to designated data centres around the world. This may be difficult for a number of countries whose NMAs are not state-owned. At present a number of the signatories release few, if any, data each month.

The Climatic Research Unit attempts to collect 1961–90 reference period data from European NMAs illustrates many of the issues with regard to data exchange. The charges requested by some countries mean that, for all countries that have charged us for the data, there is an inverse linear relationship between the charges and the number of stations we have. Hulme (1994) shows no relationship between the costs of the data and national wealth and some charges were reduced by large actions after an appeal.

A two-tier system for data exchange is currently being addressed by WMO, partially as an attempt by some nations to recoup costs by charging for the second tier. Tier 1 would have no restrictions on use and would include data from the basic synoptic networks, in-situ marine observations and data needed for a good representation of climate (for a full list including satellite data and other products see World Climate Research Programme, WCRP, 1994). All data and products not included in Tier 1 would be included in Tier 2, and restrictions on and charges for their use might be made. The steering committee (JSC) of WCRP stressed the importance of maintaining free access and exchange of delayed-mode data sets, which are essential requirements of most climate studies. The vagueness of the wording has lead to much discussion in the popular scientific (EOS, 1995) and the learned society (White, 1994) literature. Any new framework has to be both explicit and unambiguous to avoid the need for organizations like WCRP to stress certain datasets.

Whatever the outcome, I believe active scientists (who are rarely consulted on the data exchange/charging issue) will do what they have always done – make do with what they can get and afford, and exchange, where possible, with like-minded individuals around the world. Without adequate access to the data it will be difficult for scientists to address important pressing global environmental issues. If the Global Climate Observing System (GCOS) is to succeed for surface climate data (i.e., improve on the present situation), the data exchange policy needs to be addressed directly, with nominated international centres releasing and standardizing NMA data. The cost of getting access to NMA data is probably very small when compared to the cost of getting data from satellites.

3. Accuracy

3.1. HOMOGENEITY

Before any assessment of accuracy of the grid box, regional, hemispheric or global time series can begin, the homogeneity of the individual station time series must be considered. The various reasons for heterogeneity have been addressed before (see e.g. Bradley and Jones, 1985; Folland et al., 1990, 1992; and Jones, 1995). Some are site specific relating to moves, instruments and exposure changes, some are nationally specific relating to observation times and the methods used to calculate monthly means, while the third major source of error relates to the growth of cities around the sites (the urbanization effect). The latter almost always leads to warming in affected sites while the former two reasons can lead to changes of either sign and will thus have a tendency to cancel, especially in larger-scale regional time series and analyses. This fact should not be considered as a saving grace, however. Any compilation of all station and grid box temperature analyses should actively assess all stations for homogeneity, preferably using a combination of subjective (Jones *et al.*, 1985, 1986a) and automated checking procedures Easterling and Peterson, 1995). But, however automated the procedure, any homogeneity exercise will consume a considerable amount of resources.

The 'ideal' situation would be for each country to check and correct all its data and then release them to an international data centre with documentation. GCOS, through WMO, should actively promote workshops to pass on expertise in this area, remembering all the time the various climatic uses to which the data will be put. The reason this is 'ideal' is that each nation should know the most about its network and site histories and, therefore, should be best equipped to apply standard procedures. The 'ideal', however, will not produce a consistent homogenized station dataset. In reality, despite workshops, the whole exercise would probably be best achieved at a few institutes around the world, who have the resources and expertise to achieve the task.

3.2. ACCURACY OF SURFACE TEMPERATURE DATA SETS

3.2.1. *Stations*
Currently, stations that issue monthly mean temperature* over the GTS do so with a precision of 0.1 °C. Common mistakes still regularly found on 'real-time' basis are the omission/addition of the sign, and values with 1 or 10 °C added/subtracted. Scaling by 10 or 0. 1 occurs commonly for precipitation. Such common mistakes are one reason why users would have little faith in homogeneity adjustments made in some regions if the 'ideal' situation in 3.1. was followed. Even when the data are free of mistakes, without a long reference period of 20 to 30 years the data issued are practically useless for climate monitoring. Except for remote regions, where it may not be possible, countries should be encouraged to issue reference period data for each station (see section 2.1). If this is not possible they should be encouraged to release (to designated centres) or publish all historical monthly data for their CLIMAT network stations, with metadata (information about site changes etc.). This would enable station homogeneity to be assessed and reference period averages to be calculated.

3.2.2. *Grid Boxes*
Any attempt to produce regional and larger-scale temperature time series is beset by variations in the spatial coverage of data across a region and in changes in that coverage through time (Jones and Briffa, 1992). Changes in station density are partially overcome by interpolating the station anomaly values to a grid (e.g. Jones *et al.*, 1986b) or by averaging, possibly with some form of weighting, to grid boxes (Jones, 1994; and Hansen and Lebedeff, 1987).

In any grid box dataset the accuracy of the temperature anomaly estimates will depend on the number of stations within the box and on the size of the box relative

---

* From November 1994, many stations also began issuing monthly mean maximum and minimum temperatures. Although these data will be particularly important in improving understanding of the climate system, they will only be useful where appropriate reference period data exist.

to the spatial scales of the anomaly patterns. If the box is small it is likely that any temperature anomaly gradient across the box will be small. The selection of optimum box size, therefore, depends upon this gradient, data availability and uses of the resulting data set. The $5 \times 5°$ grid-box dataset developed by Jones (1994) is a compromise for temperature. For precipitation, for example, a smaller box size would be preferable but the data availability, over the GTS and MCDW, is approximately the same.

Grid-box accuracy will vary with time as the number of stations decreases in the earlier (and in some boxes also, the most recent) years. Since the temperature anomaly gradients across the box are relatively small, and all the station data homogeneous, this will not result in biased estimates of the mean anomalies. The variance of individual grid-box time series will be inflated, however, when fewer stations are available. The number of stations required for a given accuracy in a grid box will depend on factors such as the climatic type, the orography, distance firom coast and the season. It would be possible to assess this by calculating the mean inter-station correlation coefficients (Wigley et al., 1984) between all stations within each box. Whilst such an exercise may be worthwhile, the basic conclusion that only more stations in each box will lead to increases in accuracy is beyond doubt. Figure 4 shows the grid boxes with. (a) at least one; and (b) at least three usable stations per box throughout the 1961–90 period.

3.2.3. *Regions and Hemispheres*
Larger scale averages can be made by areally weighting a set of grid-box temperature anomalies using the following equation:

$$R_i = \sum_{j=1}^{N} w_j t_{ij} \bigg/ \sum_{j=1}^{N} w_j$$

where
R_i is the resulting time series ($i = 1,$M months or years);
t_{ij} the temperature anomaly in grid box j in month or year;
w_j the weight of the grid box (cosine of the latitude of the central point in the $5 \times 5°$ box formulation);
and N the number of boxes.

The accuracy of resulting time series depends upon the number of available boxes and their spread across the study region or the hemisphere.

Most attempts to assess the problem have been concerned with relative rather than absolute accuracy. The region or hemisphere may have most of the boxes available during the recent 40 years but the issue is: 'How well do the sparser networks of the late 19th and early 20th century perform compared with the period of best data availability'. This issue has been addressed by three basic methods. The first has been termed 'frozen grids' (Jones et al. 1986b,c, 1991; Folland et al., 1990, and Parker et al., 1994). In this technique, a fixed grid-box network

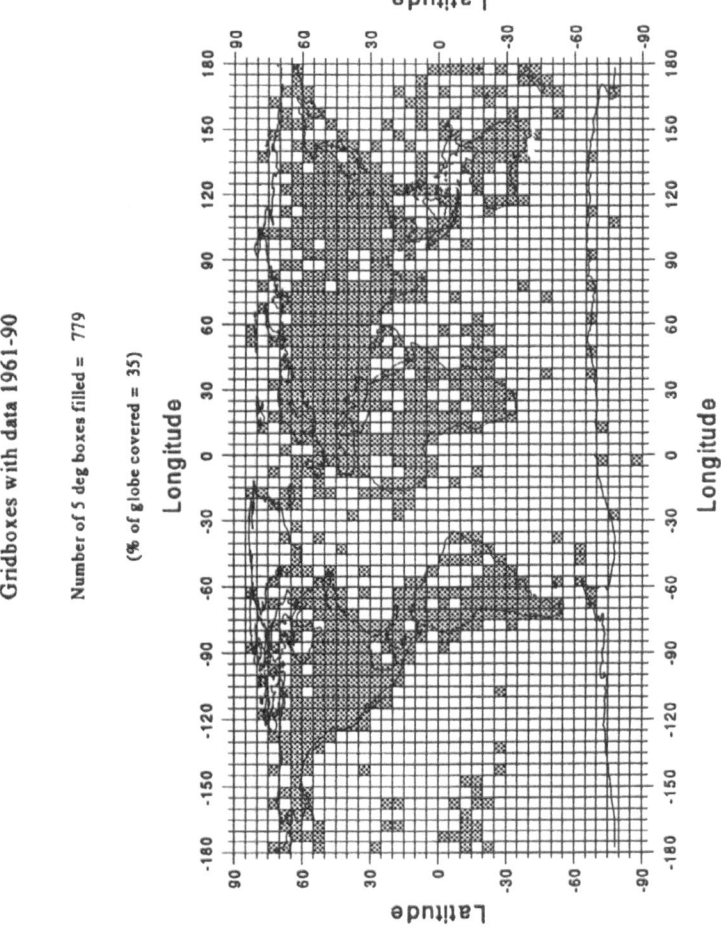

Gridboxes with data 1961-90

Number of 5 deg boxes filled = 779

(% of globe covered = 35)

Fig. 4a.

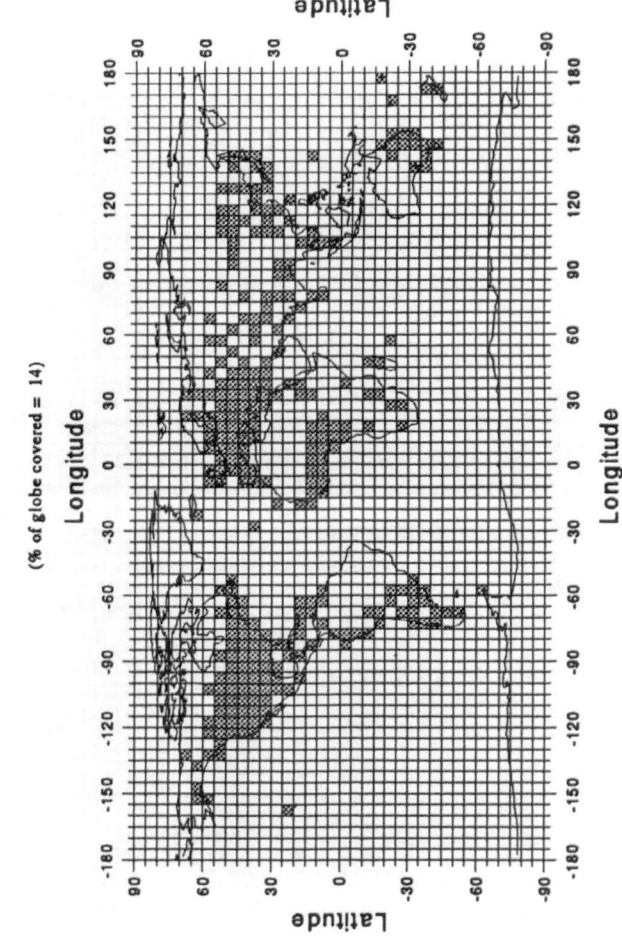

Gridboxes with 3 stations 1961-90

Number of 5 deg boxes filled = 295

(% of globe covered = 14)

Fig. 4b.

Figs. 4(a)–(b) The two maps show: (a) the location of the grid boxes with at least one station; and (b) grid boxes with at least three stations per box during 1961–90.

representing the coverage in some sparser period (e.g., the 1880s) is used to create a time series for the entire period. Comparison with a time series based on the full dataset then yields an assessment of how well the sparser network would have fared over more recent periods. The results generally indicate very small errors on multiannual timescales, particularly when surface temperature data from land and marine regions are combined.

Robeson (1995) used a related sampling method to estimate confidence intervals for annual hemispheric and global land surface air temperatures and their trends. Instead of using the network of grid boxes, however, Robeson selected subsets of stations. The subset number was chosen to be that available during a particular decade in the nineteenth century. By chance, the random selection of stations sometimes omitted data from whole continents. Robeson's errors were, therefore, larger than those implied by the 'frozen grids'.

The second technique is a variant. Incomplete networks are used to sample globally complete fields of tropospheric average temperatures, based on satellite derived Microwave Sounding Unit (MSU) data (Karl *et al.*, 1994) or General Circulation Model (GCM) fields of surface temperature (Hansen and Lebedeff, 1987; and Madden *et al.* 1993). The results are similar to the first technique. This approach has the advantage of assessing absolute as well as relative error (How important are areas always missing?), but do the GCMs and the short satellite records contain the full scope of temperature variability on the decadal-to-century timescales?

The third technique is optimum averaging (Kagan, 1979) and this has recently been applied to oceanic (Smith *et al.*, 1994) as well as land surface air temperatures (Shen *et al.*, 1994). To use the technique, the spatial structure of the variability must be known and expressed through covariance matrices, eigenfunctions or structure functions. The technique gives similar results to the other two methods.

Estimates of the relative accuracy of specific networks can be made using data on different timescales. If, for example, the errors of successive annual hemispheric averages are independent, the relative accuracy on longer timescales will be greater. Thus, although we may not be able to distinguish between two years unless one is warmer/colder than the other by a certain amount, we can say that one decade is warmer than another given a smaller temperature difference. Analogous to this argument, we expect to need markedly fewer paleoclimatic data reconstructions than temperature stations, because we are more interested in the longer timescales (Bradley and Jones, 1993; and Jones and Briffa, 1995). Taken to the extreme, this is why we tentatively accept the global temperature inferences from ice cores in Greenland and Antarctica on timescales greater than a thousand years over the last 200,000 years.

4. Possible Improvements

Improvements may be made to the land surface data base by incorporating satellite estimates of land surface temperature in a similar manner to that currently used for sea surface temperature data (Reynolds and Marsico, 1993). Such improvements, however, will have no effect on the detection issue in the short term. What are required here are long records for as many regions of the world as possible (Santer *et al.*, 1993). The best improvements to the network for this purpose, and most others, will come from gaining access to many more of the temperature data (and also precipitation and pressure data) currently being collected in real time at the Earth's surface, and historic data for more sites. No major resource increases are required for this. Many of the required data may have been digitized (or at least saved by microfilming) through WMO projects such as DARE (DAta REscue) and CLICOM (CLImate COMputing), but have not been made available to the research community. These projects improve the data archiving in developing countries, but there has been little resulting impact on many Global Change issues.

5. Conclusions

The station surface air temperature data set is shown to be adequate for monitoring hemispheric- and global-scale temperatures. At the grid-box and regional level, the network can be improved at relatively minimal cost by requesting member nations of WMO to issue more station data in real time and to make more historic time series available. In the short term, the issuing of more data in real time will make the most significant improvement. In terms of providing information for a number of issues in the Global Change area such as the detection of ciimate change in the observations, new datasets from, for example, satellites are going to make little immediate impact. What are needed are better quality and longer datasets to help improve understanding of climatic variability on the decadal-to-century timescales. This will be facilitated by improvements to the availability of surface and possibly upper air data. Most improvements have come, during the last decade, through the work of a number of individuals, rather than through the efforts of NMAs.

Acknowledgements

The author thanks Mike Hulme for comments on an earlier draft of this manuscript. David Parker and an anonymous reviewer provided comments on an earlier draft of the manuscript that significantly improved the present version. This work was supported by the U.S. Department of Energy, Atmospheric and Climate Research Division under Grant DE-FG02-86ER60397.

References

Bradley, R. S. and Jones, P. D.: 1985, 'Data Bases for Isolating the Effects of Increasing Carbon Dioxide Concentrations', in: MacCracken, M. C. and Luther, F. M. (eds.), *Detecting the Climatic Effects of Increasing Carbon Dioxide*, U.S. Dept. of Energy, Washington D.C., pp. 29–53.

Bradley, R. S. and Jones, P. D.: 1993, 'Little Ice Age' Summer Temperature Variations: Their Nature and Relevance to Recent Warming Trends', *The Holocene* **3**, 367–376.

Briffa, K. R. and Jones, P. D.: 1993, 'Global Surface Air Temperature Variations over the Twentieth Century, Part 2: Implications for Large-Scale Paleoclimatic Studies of the Holocene', *The Holocene* **3**, 77–88.

Easterling, D. R. and Peterson, T. C.: 1995, 'A New Method for Detecting Undocumented Discontinuities in Climatological Time Series', *Int. J. Climatol.* **15**, 369–377.

EOS: 1995, 'Debate over Free Exchange of Data Roils Geophysical World', *EOS* **76**, Feb. 14, AGU, Washington D.C.

Folland, C. K., Karl, T. R., Nicholls, N., Nyenzi, B.S., Parker, D. E., and Vinnikov, K. Ya.: 1992, 'Observed Climate Variability and Change', in: Houghton, J. T., Callander, B. A., and Varney, S. K. (eds.), *Climate Change 1992, The Supplementary Report to the IPCC Scientific Assessment*, Cambridge University Press, pp. 135–170.

Folland, C. K., Karl, T. R., and Vinnikov, K. Ya.: 1990, 'Observed Climate Variations and Change', in: Houghton, J. T., Jenkins, G. J., and Ephraums, J. J. (eds.): *Climate Change: The IPCC Scientific Assessment*, Cambridge University Press, pp. 195–238.

Hansen, J. E. and Lebedeff, S.: 1987, 'Global Trends of Measured Surface Air Temperature', *J. Geophys. Res.* **92**, 13345–13372.

Hulme, M.: 1994, 'The Cost of Climate Data – a European Experience', *Weather* **49**, 168–175.

Hulme, M., Conway, D., Jones, P. D., Jiang, T., Barrow, E. M., and Turney, C.: 1995, 'Construction of a 1961–90 European Climatology for Climate Change Modelling and Impact Applications', *Int. J. Climatol.* **15** (in press).

Jones, P. D.: 1988, 'Hemispheric Surface Air Temperature Variations: Recent Trends and an Update to 1987', *J. Clim.* **1**, 654–660.

Jones, P. D.: 1994, 'Hemispheric Surface Air Temperature Variations: A Re-Analysis and an Update to 1993', *J. Clim.* **7**, 1794–1802.

Jones, P. D.: 1995, 'Observations from the Surface–Projections from Traditional Meteorological Observations', in Henderson-Sellers, A. (ed.), *Future Climates of the World*, Elsevier, (in press).

Jones, P. D. and Briffa, K. R.: 1992, 'Global Surface Air Temperature Variations over the Tentieth Century, Part 1: Spatial, Temporal and Seasonal Details', *The Holocene* **2**, 174–188.

Jones, P. D. and Briffa, K. R.: 1995, 'What the Instrumental Record Can Tell Us About Longer Timescale Paleoclimatic Reconstructions', in Jones, P. D., Bradley, R. S., and Jouzel, J. (eds.), *Climatic Variations and Forcing Mechanisms of the Last 2000 Years*, Springer-Verlag, Berlin, (in press).

Jones, P. D., Raper, S. C. B., Santer, B. D., Cherry, B. S. G., Goodess, C. M., Kelly, P. M., Wigley, T. M. L., Bradley, R. S., and Diaz, H. F.: 1985, 'A Grid Point Surface Air Temperature Data Set for the Northern Hemisphere', *U.S. Dept. of Energy, Carbon Dioxide Research Division, Tech. Rep. TR022*, 251 pp.

Jones, P. D., Raper, S. C. B., Cherry, B. S. G., Goodess, C. M., Wigley, T. M. L.: 1986a, 'A Grid Point Surface Air Temperature Data Set for the Southern Hemisphere', *U.S. Dept. of Energy, Carbon Dioxide Research Division, Tech. Rep. TR027*, 73 pp.

Jones, P. D., Raper, S. C. B., Bradley, R. S., Diaz, H. F., Kelly, P. M., Wigley, T. M. L.: 1986b, 'Northern Hemisphere Surface Air Temperature Variations, 1851–1984', *J. Clim. Appl. Meteorol.* **25**, 161–179.

Jones, P. D., Raper, S. C. B., and Wigley, T. M. L.: 1986c, 'Southern Hemisphere Surface Air Temperature Variations, 1851–1984', *J. Clim. Appl. Meteorol.* **25**, 1213–1230.

Jones, P. D., Wigley, T. M. L., and Farmer, G.: 1991, 'Marine and Land Temperature Data Sets: A Comparison and A Look at Recent Trends', in: Schlesinger, M. E. (ed.), *Greenhouse-Gas-Induced Climatic Change: A Critical Appraisal of Simulations and Observations*, Elsevier, pp. 153–172.

Kagan, R. L.: 1979, *Averaging Meteorological Fields*, Gidrometeoizdat, Leningrad, 212 pp. (in Russian).

Karl, T. R., Jones, P. D., Knight, R. W., Kukla, G., Plummer, N., Razuvayev, V., Gallo, K. P., Lindesay, J., Charlson, R. J., and Peterson, T. C.: 1993, 'Asymmetric Trends of Daily Maximum and Minimum Temperature', *Bull. Amer. Meteor. Soc.* **74**, 1007–1023.

Karl, T. R., Knight, R. W., and Christy, J. R.: 1994, 'Global and Hemispheric Temperature Trends: Uncertainties Related to Inadequate Spatial Sampling', *J. Clim.* **7**, 1144–1163.

Madden, R. A., Shea, D. J., Branstator, G. W., Tribbia, J. J., and Weber, R. O.: 1993, 'The Effects of Imperfect Spatial and Temporal Sampling on Estimates of the Global Mean Temperature: Experiments with Model Data', *J. Clim.* **6**, 1057–1066.

Parker, D. E., Jones, P. D., Folland, C. K., and Bevan, A.: 1994, 'Interdecadal Changes of Surface Temperature since the Late Nineteenth Century', *J. Geophys. Res.* **99**, 14373–14399.

Razuvayev, V. N., Apasova, E. G., Martuganov, R. A., Vose, R. S., and Steurer, P. M.: 1993, 'Daily Temperature and Precipitation Data for 223 USSR Stations', *Numerical Data Package NDP-040*, Carbon Dioxide Information and Analysis Center, Oak Ridge, Tennessee, 47 pp. plus two Appendices.

Reynolds, R. W. and Marsico, D. C.: 1993, 'An Improved Real-Time Global Sea Surface Temperatue Analysis', *J. Clim.* **6**, 114–119.

Robeson, S. M.: 1995, 'Resampling of Network-Induced Variability in Estimates of Terrestrial Air Temperature Change', *Clim. Change* **29**, 213–229.

Santer, B. D., Wigley, T. M., and Jones, P. D.: 1993, 'Correlation Methods in Fingerprint Detection Studies', *Clim. Dynam.* **8**, 265–276.

Shen, S. S. P., North, G. R., and Kim, K.-Y.: 1994, 'Spectral Approach to Optimal Estimation of the Global Average Temperature', *J. Clim.* **7**, 1999–2007.

Smith, T. M., Reynolds, R. W., and Ropelewski, C. F.: 1994, 'Optimal Averaging of Seasonal Sea Surface Temperatures and Associated Confidence Intervals (1860–1989)', *J. Clim.* **7**, 949–964.

White, R. M.: 1994, 'Report of the Ad Hoc Committee on International Data Exchange', *Bull. Amer. Meteorol. Soc.* **75**, 549–551.

Wigley, T. M. L., Briffa, K. R., and Jones, P. D.: 1984, 'On the Average Value of Correlated Time Series with Applications in Dendroclimatology and Hydrometeorology', *J. Clim. Appl. Meorol.* **23**, 201–213.

WCRP: 1994, 'Report of the Fifteenth Session of the Joint Scientific Committee', *WMO/TD – No. 632*, WMO, Geneva, Switzerland.

WMO: 1978, 'Weather Reporting Observing Stations', *WMO No. 9*, WMO, Geneva, Switzerland.

(Received 23 January, 1995; in revised form 10 July, 1995)

MARINE SURFACE TEMPERATURE: OBSERVED VARIATIONS AND DATA REQUIREMENTS *

D. E. PARKER, C. K. FOLLAND and M. JACKSON

Hadley Centre, Meteorological Office, London Rd., Bracknell, Berks, RG12 2SY, U.K.

Abstract. Measurements of temperature at the ocean surface are an indispensible part of the Global Climate Observing System (GCOS). We describe the varying coverage of these measurements from the mid-nineteenth century through to the present era of satellite data, along with ongoing attempts to augment the available digitized data base. We next survey attempts to remove systematic biases from both sea surface temperature (SST) and marine air temperature (MAT) data and to combine *in situ* and satellite SSTs in a consistent manner. We also describe new or planned geographically complete climatologies of SST and night MAT for 1961–90. These are expected to be more reliable than existing climatologies in the Southern Ocean and other sparsely-observed areas. The new SST climatology has been used in the construction of an improved geographically-complete data set of sea ice and SST: the techniques used are briefly reviewed, as are other methods of analysis and assessment of worldwide SST.

We present global and regional time series of anomalies (i.e. deviations from reference climatology) of SST and night MAT for 1856 to 1994 constructed using the most complete data and best-estimate bias-corrections hitherto available. These series are compared with earlier published series, and are validated by means of comparisons with anomalies of air temperature from coastal and island stations. The sensitivity of the time series to imperfect coverage is assessed by means of frozen grid experiments. The results underscore the need for ongoing development of SST and MAT data bases within GCOS for the detection of climatic change, and for improved methods of analysis to optimally isolate the signals from incomplete data.

1. Introduction

To monitor climate and detect its changes, long, complete, homogeneous data sets are required. We will concentrate on sea surface temperature (SST) and marine air temperature (MAT).

In this paper, we define SST as the bulk temperature between 1 m and 10 m depth in the ocean, which is the depth range sampled by engine-intake (inlet, injection) and hull sensors (e.g. Kent *et al.*, 1993). This layer is usually well-mixed, without major vertical gradients of temperature (Kent *et al.*, 1993). Although buckets, which predominated before the 1940s, sample the upper 0.5 m, our corrections to the bucket data (Section 3) are designed to make them consistent with the mix of data in 1951–80 when engine-intake measurements dominated (Folland and Parker, 1995). We also adjust any satellite data to be consistent with our own modern bulk data, even if they were originally calibrated using bulk SSTs from buoys, and we make no direct use of skin temperatures (Section 3). Our bulk data avoid the strong diurnal cycles characteristic of the upper few centimetres, where daytime warming

---

* The British Crown right to retain a non-exclusive royalty-free license in and to any copyright is acknowledged.

can exceed 1 °C so that even the relatively cool (by several tenths °C) skin is then warmer than the bulk temperature as we have defined it (see Schluessel *et al.*, 1990.)

Our MAT data are adjusted to an effective elevation of 15 m above sea level using profiles based on surface-layer similarity theory (Bottomley *et al.*, 1990).

Digitized archives begin in the 1850s, a long-term fruit of the initiatives of Maury (1858), but coverage has varied for technical, political and economic reasons, and many data exist only in manuscript form. So, improvements are needed. We describe progress in this area in Section 2.

Sections 3 and 4 review the adjustments which need to be made to the SST and MAT data to compensate for historical variations in instrumentation and observing practices. Section 3 also includes a description of how we compensate for systematic differences between *in situ* and satellite-based observations of SST.

An implicit requirement is for reliable base climatologies to use as a reference when assessing climatic variations: ongoing improvements in this field are addressed in Section 5.

Given imperfect coverage, how can best use be made of the data? Developments in analysis methods are considered in Section 6.

Finally, in Section 7, we present time series of SST and MAT from the 1850s to the present, compare anomalies of marine with those of nearby land data, and illustrate the effects of improved bias-adjustments and incomplete coverage on the marine results.

2. Historical Variations of Data Coverage

The geographical distribution and density of *in situ* marine surface temperature observations (and other marine meteorological data) have varied markedly over the past 140 years. The main features have been illustrated by Woodruff *et al.* (1987), Bottomley *et al.* (1990), Trenberth *et al.* (1992), Parker *et al.* (1994) and Folland and Parker (1995). Coverage became widespread in much of the Atlantic and Indian Oceans during the late nineteenth century, but much of the Pacific Ocean has very sparse coverage in digitized data bases until the 1930s, and many areas suffered marked decreases in availability of data for the years of and around the two World Wars. The opening of the Suez (1869) and Panama (1914) canals caused major changes in predominant shipping routes. In particular, after about 1914 far fewer ships circumnavigated Cape Horn, so that coverage decreased dramatically in the midlatitude southeast Pacific (compare Plates 2–4 of Parker *et al.* (1994) with their subsequent Plates).

Here, we do not reproduce the coverage maps given in the above mentioned references, but supplement their results with some maps of numbers of available digitized data. Figure 1 shows examples of the numbers of SST data per 5° latitude × longitude area from the U.K. Meteorological Office Marine Data Bank (MDB) and the Global Telecommunication system (GTS) contributing to the new Meteorological Office Historical Sea Surface Temperature (MOHSST) data set version 6 (MOHSST6), before quality control, in January and July of 1878 (a major El Niño year), 1918 (a war year), 1958 (International Geophysical Year), and 1994. Because we then included in MOHSST6 some analysed data (as opposed to raw observations) from the Comprehensive Ocean-Atmosphere Data Set (COADS) (Woodruff *et al.*, 1987) in some data-sparse areas (as described for MOHSST5 by Parker *et al.*, 1994), the numbers in Figure 1 are under-estimates of MOHSST6 coverage in such areas, e.g. the eastern Pacific in the 1878 fields. At present we have no method of estimating the number of original COADS observations we have effectively used. Precise estimates await the planned observation – by observation blend of the Meteorological Office Marine Data Bank with COADS. Figure 2 is a similar summary for the entire years 1878, 1918 (almost the worst coverage since 1860), 1958 and 1994. Figure 3(a) repeats the analysis but for the decades 1871–80, 1911–20, 1951–60 and 1985–94 and Figure 3(b) is as Figure 3(a) but for night MAT, Meteorological Office Historical Marine Air Temperature (MOHMAT) data set version 3.1 (MOHMAT31). We use night MAT (NMAT) to avoid daytime heating of ships' decks (Section 4). The NMAT data base does not include any contributions from COADS, so Figure 3(b) exactly represents the contents of MOHMAT31. Figure 4 gives an annual global count of SST and NMAT observations in our data bases, with approximate allowance for the incorporation of COADS into MOHSST6: the largest proportionate effect of COADS is in the late nineteenth century.

The results in Figures 1 to 3 are in good accord with the cited literature and can be used, along with considerations discussed by Trenberth *et al.* (1992), in the estimation of error statistics for 5° area monthly, seasonal, annual and decadal anomalies of SST and NMAT. Parker *et al.* (1994) attempted to estimate error statistics for seasonal and decadal fields of SST, through they used a technique based on the spatial and temporal coherence of seasonal 5° area SST anomalies, rather than on numbers of observations.

The data shown are not all that exist. Over 20 million sets of historical marine observations (which include pressure, wind and other elements as well as SST and MAT), are either undigitized or have not yet been incorporated into COADS (Elms *et al.*, 1993) or into our MDB. About 3 million records from U.S.A. ships between 1912 and 1946 have been digitized but await quality-control. The Maury collection of 1.8 million observations for the early and mid-nineteenth century is being digitized, as is a Norwegian collection of half a million observations for 1867–90. But a major manuscript archive, from Kobe, Japan, and kept in microfiche form in the U.S.A., containing over 5 million observations for 1890–1941, remains undigitized. These data are mainly from the western North Pacific and would fill major gaps in SST and MAT analyses (see, for example, the Plates in Parker *et al.*, 1994), as well as reducing uncertainties in time series of regional and global surface temperature. They are critical to a full analysis of interdecadal climatic variations, and therefore to climate change detection, not only for SST and MAT but also

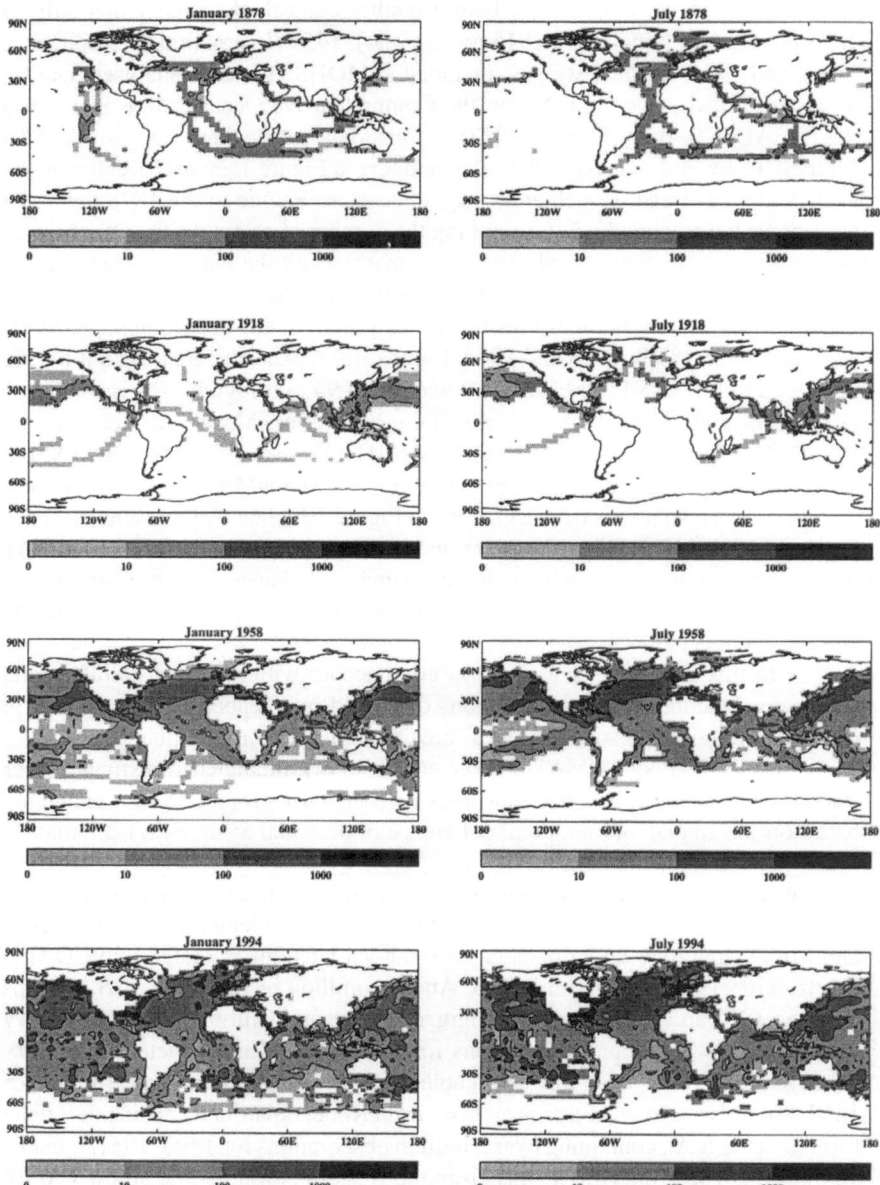

Fig. 1. Numbers of SST data per 5° latitude × longitude area contributing to MOHSST6: January (left) and July (right), 1878 (top), 1918 (upper middle), 1958 (lower middle), 1994 (bottom). No account is taken of analysed data from COADS used in MOHSST6.

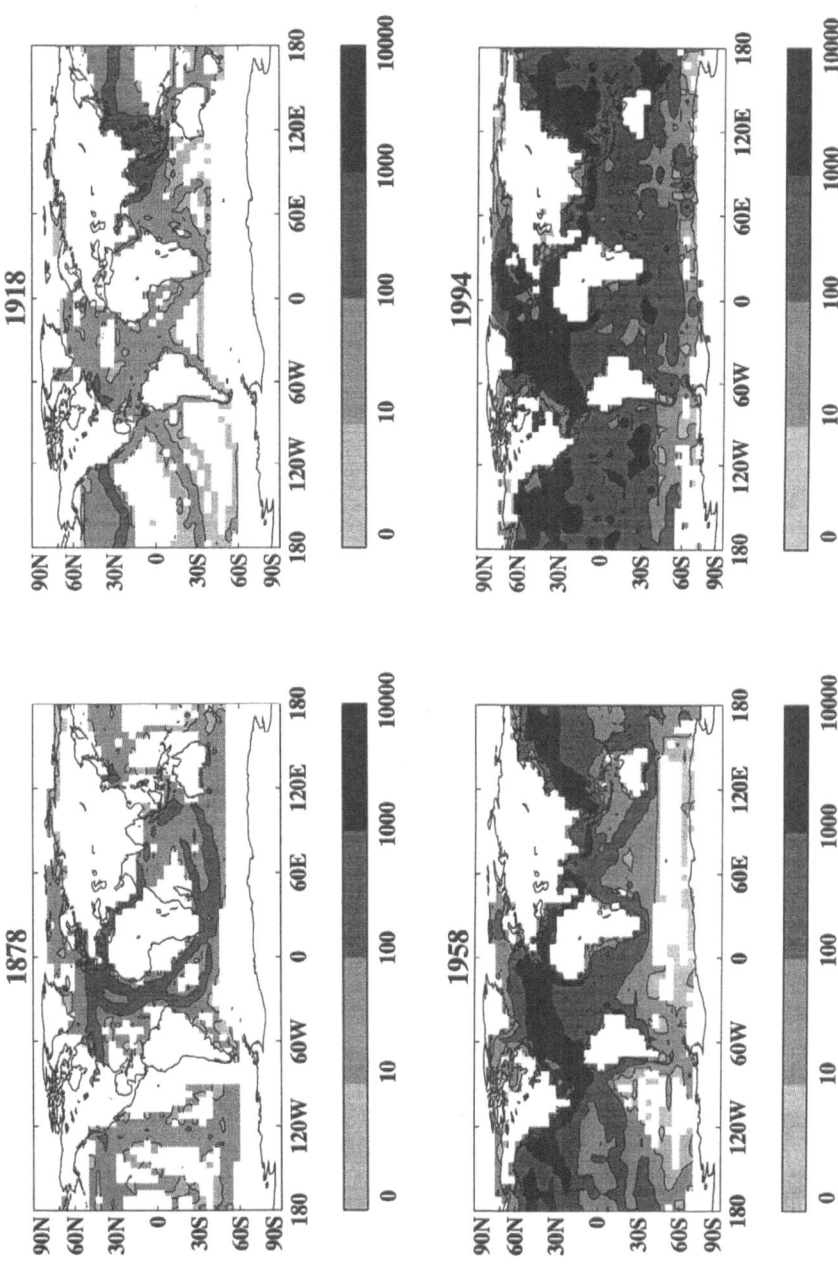

Fig. 2. Numbers of SST data per 5° latitude × longitude area contributing to MOHSST6 in 1878, 1918, 1958, and 1994. No account is taken of analysed data from COADS used in MOHSST6.

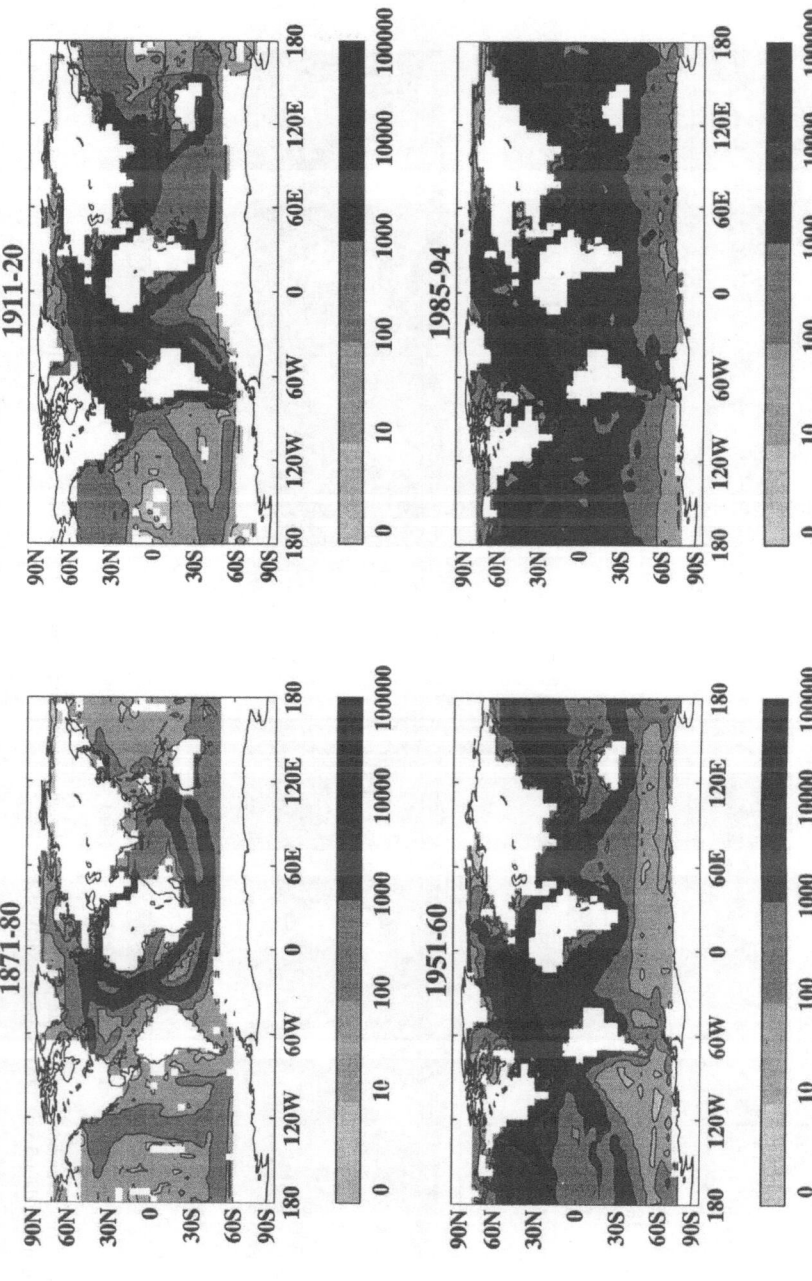

Fig. 3(a).   Numbers of SST data per 5° latitude × longitude area contributing to MOHSST6 in 1871–80, 1911–20, 1951–60 and 1985–94. No account
is taken of analysed data from COADS used in MOHSST6.

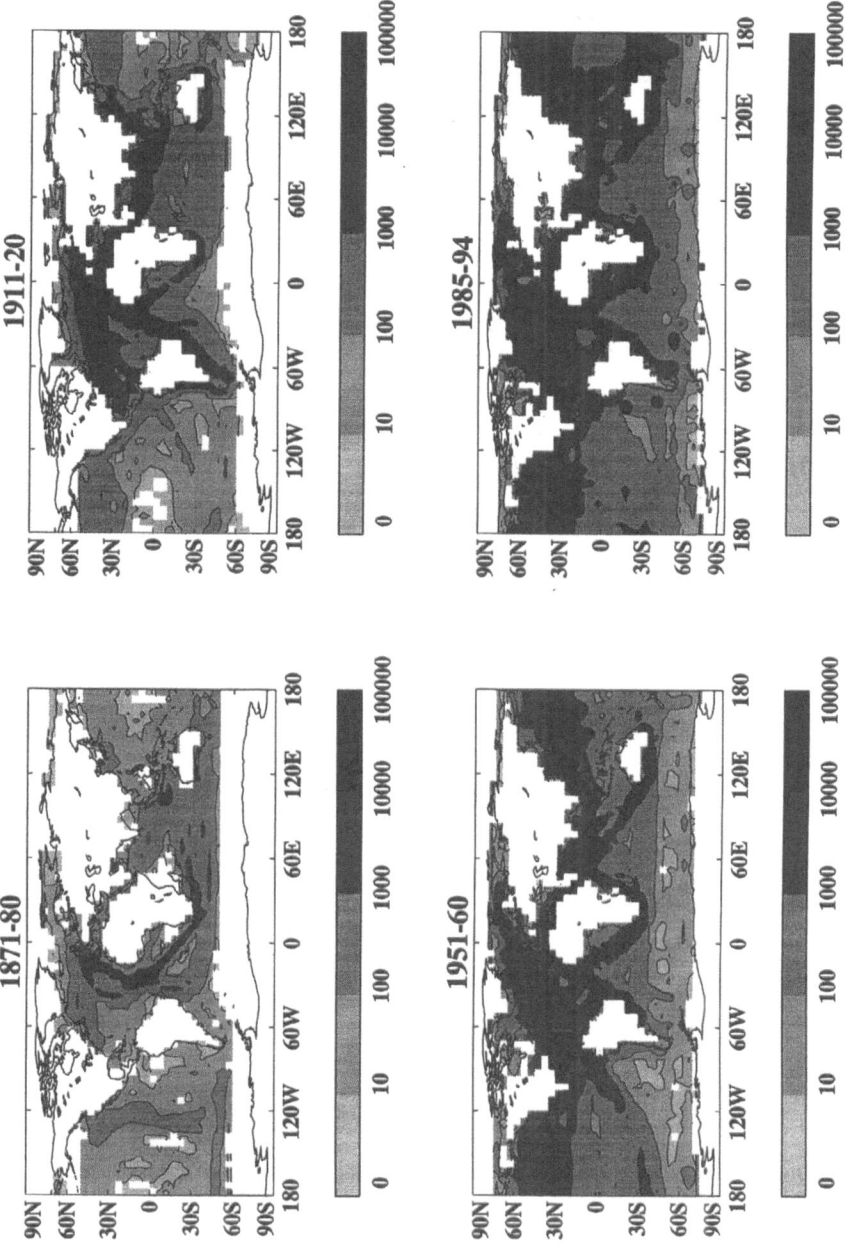

Fig. 3(b). Numbers of NMAT data per 5° latitude × longitude area contributing to MOHMAT31 in 1871–80, 1911–20, 1951–60, and 1985–94.

Annual numbers of observations for the globe

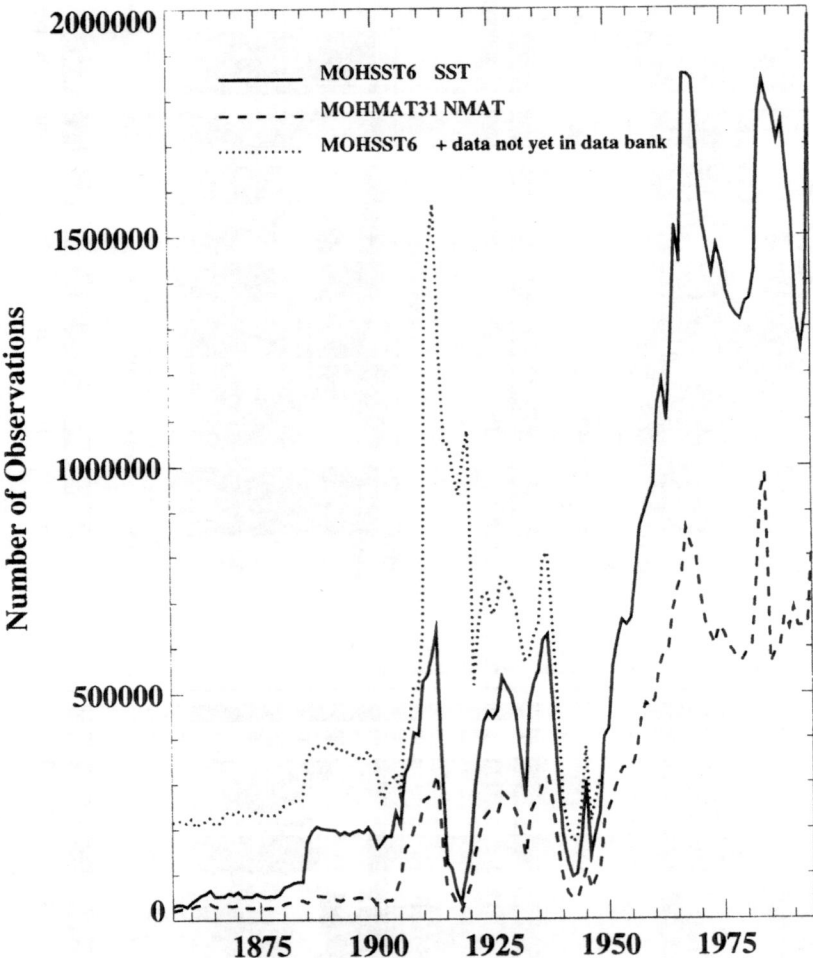

Fig. 4. Annual numbers of MOHSST6 SST (solid) and MOHMAT31 NMAT observations (dashed) for the globe, 1856–1994. Approximate allowance is made for the incorporation of COADS data into MOHSST6. The dotted line estimates numbers of SST including all known additional data sources (see text).

for surface atmospheric variation. They may reveal or elucidate past interdecadal changes in the North Pacific similar to that in the 1970s–1980s (Graham, 1994; Trenberth and Hurrell, 1994). Their digitization is vital to GCOS. Recently, about a million undigitized records from U.K. ships for 1936–48, 7 million for 1911–20,

[436]

and 8 million for the mid- and late nineteenth century, have been located. These data are mainly, but not entirely for the Atlantic: in view of the worldwide sparsity of available digitized data for these periods (Figures 1 to 4), their digitization is needed also. Some smaller sources also need to be digitized, e.g. nearly 20,000 recently released observations from Japanese whaling ships for 1946–1983 for the Antarctic (R. E. Newell, pers. comm.) where data have never been plentiful. The dotted line in Figure 4 is an estimate of the annual numbers of SST data which would be in MOHSST6 if all these extra data were included.

From 1982 onwards, logbook data were combined with buoy data and with real time data received via the GTS in the construction of MOHSST6 and its predecessor, MOHSST5E. A still earlier version, MOHSST55 (Parker *et al.*, 1994, 1995), used only GTS data from 1982 ownards, yielding slightly different results especially in the Southern Hemisphere. Figure 5(a) shows an SST difference map and Figure 5(b) compares series of seasonal hemispheric anomalies. The differences, which are minor on a hemispheric scale (Figure 5(b)), appear to result not only from the extra data but also from improved quality-control routines in MOHSST5E and MOHSST6. In these routines, individual ships are tracked, blacklisted ships are deleted, and observations are assessed with respect to neighbouring and recent anomalies (Appendix 1). The latter process appears to exclude fewer good observations than were excluded by Bottomley *et al.* (1990), whose criterion for exclusion was a deviation from climatology of more than 6 °C. This is because it is more realistic to assess scatter relative to recent, persistent, coherent real anomalies than relative to climatology. The differences between MOHSST55 and MOHSST5E are consistent with expectations based on the analyses of Folland *et al.* (1993) and Trenberth *et al.* (1992). MOHSST6 shows additional differences because it is referred to a new 1961–90 climatology (Section 5).

Much more complete coverage of SST is available from 1982 onwards using the Advanced Very High Resolution Radiometer (AVHRR) satellite data (Reynolds, 1988). Additional SST data have also become available from the Along Track Scanning Radiometer (ATSR) which was launched in 1992 (Saunders *et al.*, 1993). However all satellite data are likely to differ systematically from our SST data which are mainly based on bulk measurements from ships. The AVHRR satellite data have in general been converted from surface-skin to bulk temperatures (see Section 1) by calibration against buoys, but the calibrations may not have taken full account of satellite sensor drift, atmospheric aerosols, and temporal and geographical variations of skin-bulk temperature differences. The ATSR data are available as skin temperatures (Barton *et al.*, 1995) and so differ intrinsically from our bulk data. Therefore, all satellite SST data have to be adjusted to be consistent with *in situ* data if one of the main aims of GCOS, climate change detection, is to be fulfilled. These adjustments are discussed further, in the context of systematic adjustments to SST in general, in Section 3. ATSR data have not, so far, been used in our analyses.

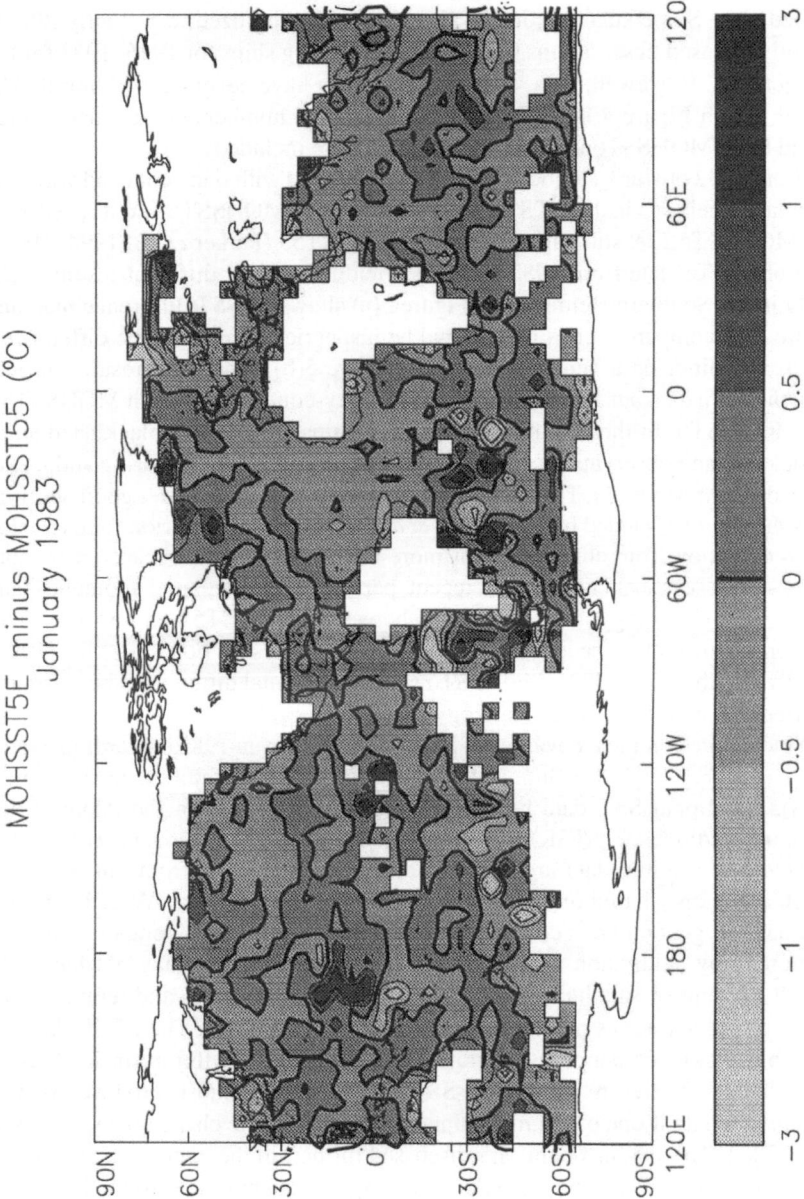

Fig. 5(a).   MOHSST5E minus MOHSST55, January 1983.

(1)

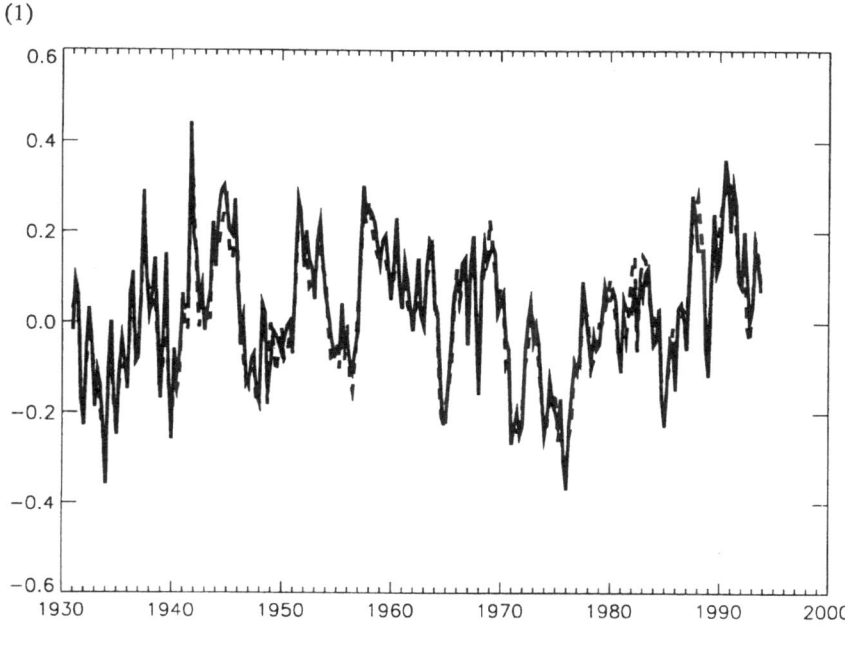

(2)

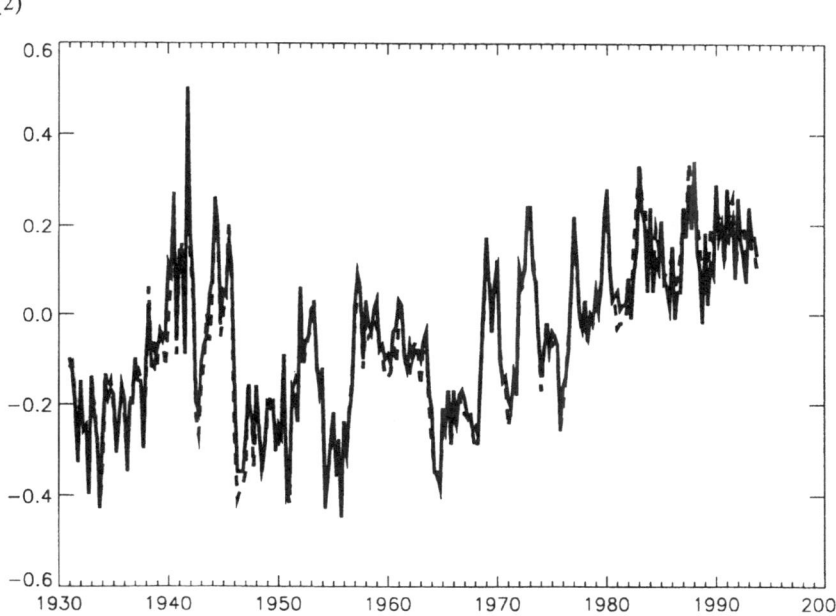

Fig. 5(b). Seasonal anomalies, relative to 1951–80, MOHSST5E (solid) and MOHSST55 (dashed).
(1) Northern Hemisphere; (2) Southern Hemisphere.

3. Systematic Adjustments to SST

The realisation that SST data require systematic adjustments stemmed from studies such as James and Fox (1972). These authors found that engine intake (inlet, injection) SSTs, which have been common since the 1940s, were systematically warmer, by several tenths °C, than SSTs measured using uninsulated buckets, which were in common use in the early 20th century and before. On this basis, Folland and Kates (1984) improved on the uncorrected SST analysis of Paltridge and Woodruff (1981) by applying an adjustment which was 0.15 °C up till 1930 and decreased linearly to –0.1 °C in the 1970s (their reference period was 1951–60). However, comparison with NMAT suggested that the change in instrumentation took place rather suddenly around the Second World War, so Folland *et al.* (1984) added 0.3 °C until early 1940, 0.25 °C thereafter through 1941, and nothing subsequently. Using an alternative approach, Jones *et al.* (1986) adjusted MAT anomalies (i.e. deviations from a reference climatology) to ensure consistency in trends with nearby (though often including inland) regional land surface air temperature anomalies. They then adjusted SST anomalies to give consistency with MAT anomalies. Their corrections to (COADS) SST were less than 0.1 °C before 1890 but nearly 0.5 °C in the early twentieth century. All the above mentioned corrections were large enough and uncertain enough that the magnitude and nature of climate change over the oceans remained uncertain.

A more rigorous and complex physically-based method of correcting the early SSTs for systematic biases has since been developed, culminating in the work of Folland and Parker (1995) on which the corrections now applied to MOHSST6 are based. These corrections are important because misconceptions about climatic change are much more likely to arise from persistent systematic errors than from often larger, almost random, errors that affect individual observations. The correction method is based on:

(i) The observation that the earlier SSTs, expressed as anomalies from recent averages, are not only too cold relative to NMATs similarly expressed (Barnett, 1984), but also, outside the tropics, show enhanced annual cycles, presumably because more heat is lost from uninsulated buckets in winter when stronger, colder winds blow over relatively warm water (Wright, 1986; Bottomley *et al.*, 1990);

(ii) A knowledge of past observing techniques and instruments. Folland and Parker (1995) compiled, from a wide range of historical instructions to observers and reports of voyages, information regarding:
  – Instrumentation including types of buckets and designs of thermometers;
  – Location of sampling, e.g. it was to be ahead of the engine outlet;
  – Placing of the bucket on deck (e.g. in the shade) after withdrawal from the sea;
  – Time-lapse before the thermometer was read;
  – Stirring the water in the bucket;

- Resampling and other techniques to reduce the influence of the initial
  temperature of the bucket.

In addition, Folland and Parker (1995) compiled a history of mean ships'
speeds, which affect ventilation on deck. The mean changes in voluntary
observing ships' speeds were assessed from logbook records.

(iii) Modelling of the heat and moisture transfers affecting uninsulated (canvas or
metal) or partially insulated (wooden) buckets. Sensible, latent and radiative
(solar and longwave) heat fluxes are included. A seasonally and geographically varying climtological average environment is assumed, modified by the
influence of the ship. As the duration of exposure of the bucket is imperfectly
known, the model is integrated until the implied corrections make the annual
cycles of SST (see (i)) as homogeneous as possible through time. This is
probably the key aspect of the method.

Recent shipboard measurements analysed by Downey (pers. comm.), along with
earlier measurements by the American Sea Education Association published in
Folland and Parker (1995), have confirmed the reliability of the uninsulated-bucket
models used by Folland and Parker (1995).

Differences in published bucket-corrections have resulted from different assumptions or simplifications under (ii) and (iii).

(a) For example, Jones et al. (1991) assumed that all buckets were wooden in the
mid nineteenth century, and that wooden buckets were perfectly insulated. By
contrast, Folland and Parker (1990) and Parker and Folland (1991)), assumed
only canvas buckets, and Bottomley et al. (1990), who also reviewed earlier
works, assumed that only 25% of buckets were wooden in 1856 (the start
of the data base), linearly decreasing to zero by 1905. Folland and Parker
(1995), weighing up the documented and empirical evidence, assumed that
80% of buckets were wooden in 1856, linearly decreasing to zero in 1920.
The proportion of wooden buckets in 1856 was chosen to yield optimum
agreement between anomalies (not actual values) of SST and of night marine
air temperature (NMAT) in the 19th century in the tropical Indian Ocean and
tropical Pacific. Other areas could not be used in this way because data were
too sparse (North Pacific) or because NMAT anomalies had already been
adjusted using provisionally corrected SST anomalies (Atlantic; see Section 4
and Bottomley et al., 1990).

(b) Bottomley et al. (1990) used an over-simplified formulation for wooden
buckets, whereas Folland and Parker (1995) used the full non-linear heat-
conduction equations.

(c) Folland and Parker (1995) found that the implied exposure durations (see (iii))
for wooden buckets were unrealistically long, probably owing to the sparsity
of data in the nineteenth century when these were used; so the exposure time
was chosen in accord with published instructions to observers (e.g. Maury,
1858).

(d) The corrections presented by Folland and Parker (1995) incorporate an as-
    sumed linear changeover from slow (4 ms^{-1}) to fast (7 ms^{-1}) moving ships
    between 1870 and 1940.

The geographical and seasonal variations of the systematic bucket-corrections to
SST derived by Folland and Parker (1995) (Figure 6(a)) are qualitatively similar
to those computed by Bottomley *et al.* (1990) and by the other cited authors. The
largest, positive corrections are in early winter (December) over the Gulf Stream
and the Kuroshio, where warm water, cold dry air, and strong winds cause rapid
evaporative heat loss from the buckets. The corrections approach 1 °C by 1940 in
these regions in early winter. Corrections are also large (around 0.4 °C to 0.5 °C by
1940) in all seasons in the tropics because of the high rate of evaporation when SST
is high. Some negative corrections are made in mid latitudes in summer, mainly
where the mean air temperature around the bucket exceeds the mean SST.

   Bottomley *et al.* (1990) compare (in their Figure 9c) their globally-averaged
SST corrections with those reported in some of the earlier papers. Here we compare
the global average corrections of Bottomley *et al.* (1990) with those of Folland
and Parker (1995) (Figure 6(b)), and the hemispheric average corrections of Jones
et al. (1991) with those of Folland and Parker (1995) (Figure 6(c)). Overall, the
corrections of Folland and Parker (1995) give a little less climatic change since
the 1860s than those of Jones *et al.* (1991), and markedly less from 1890, but a
similar amount since 1900. The Bottomley *et al.* (1990) corrections give somewhat
less climatic change since the 1860s than those of Folland and Parker (1995).
Comparison of globally-averaged uncorrected and corrected (Folland and Parker,
1995) SST (Figure 7) proves that good corrections for systematic biases in SST are
crucial to climate change analysis.

   How do we know that these corrections are trustworthy? The agreement of SST
anomalies with largely independently corrected NMAT anomalies (Section 7 and
Figure 7) is the strongest support to the results, and suggests that the impacts of
future refinements and reduction of uncertainties in this area will be small. On a
global decadal average, error bars of the systematic corrections do not appear to
exceed 0.1 °C (Figure 7), and are much smaller than the 0.5 °C climate signal.
Note that we expect SST anomalies and NMAT anomalies to be similar over large
ocean areas over a season or more, even though there is a variable relationship
geographically and seasonally between their absolute values, and NMAT is mostly colder than SST (see Bottomley *et al.*, 1990). See Barnett (1984) for further
discussion.

   The adjustments to the earlier SSTs have been made relative to over 30 million observations made during 1951–80 using a mix of engine-intakes (the largest
contribution), hull-sensors, insulated buckets, and some uninsulated buckets. Very
few buoy data were involved. As the mix of observation techniques is likely to be
changing, however, modern in situ SSTs cannot be regarded as free from changing
biases. Folland *et al.* (1993) used about 5 million observations to document zonal, seasonal mean differences averaging 0.08 °C between bucket and non-bucket

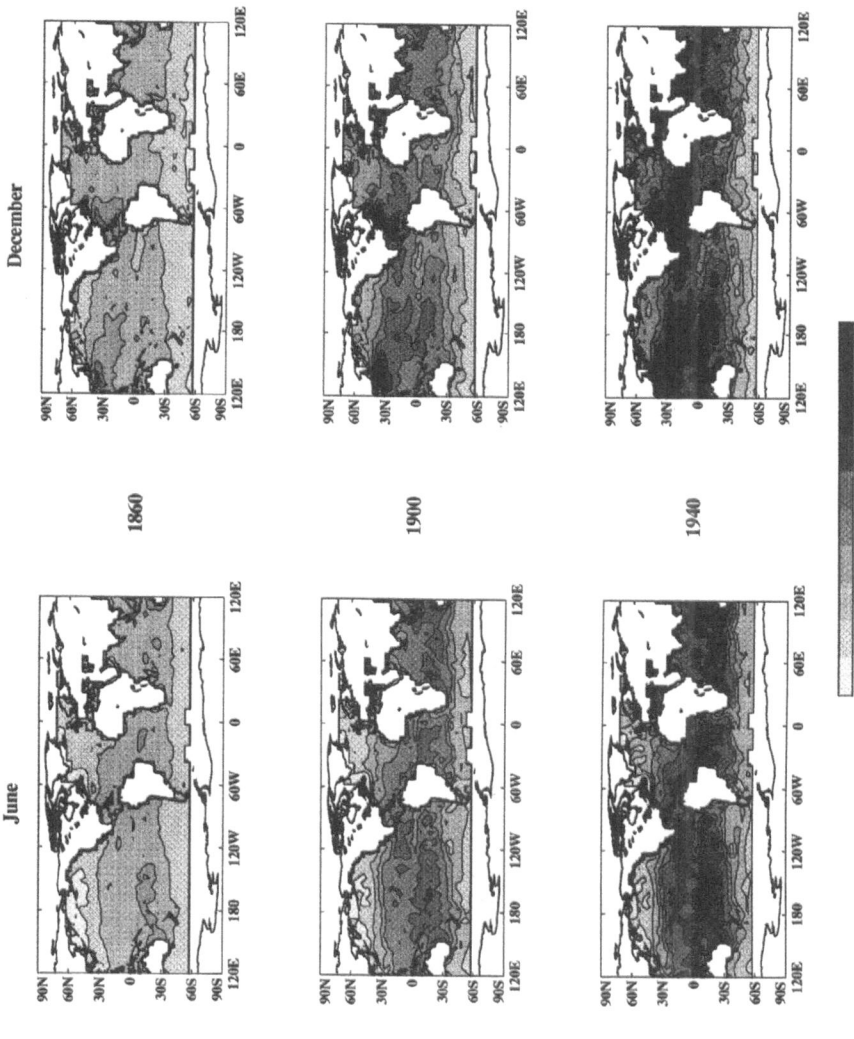

Fig. 6(a). Bucket-corrections to SST (°C) (from Folland and Parker, 1995).

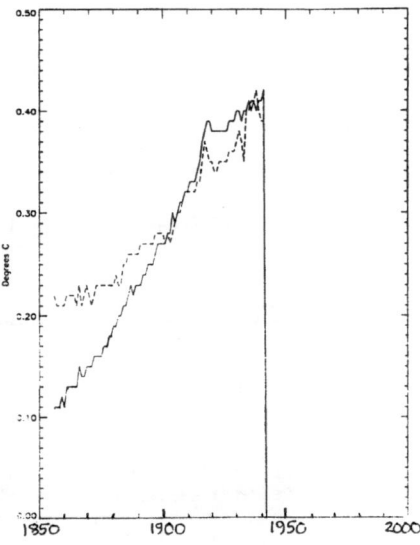

Fig. 6(b).   Global average bucket corrections applied by Bottomley *et al.* (1990) (dashed), and Folland and Parker (1995) (solid).

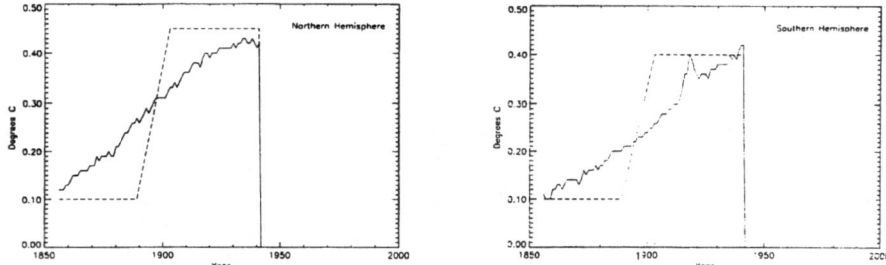

Fig. 6(c).   Hemispheric average bucket corrections applied by Jones *et al.* (1991) (dashed) and Folland and Parker (1995) (solid).

SST, the bucket SSTs being generally the colder. However, readily available documentation of ships' instrumentation (World Meteorological Organization (WMO) No. 47) does not specify the type of bucket used. Kent *et al.* (1993) have found larger differences, up to 0.3 °C between buckets and engine intakes, and have assembled metadata, but only for a much more limited sample of ships. More detailed and complete metadata are required before differences in the behaviour of recent instruments can be assessed and any time-varying biases in the overall data base quantified. This is an urgent matter in view of the key part that SST plays

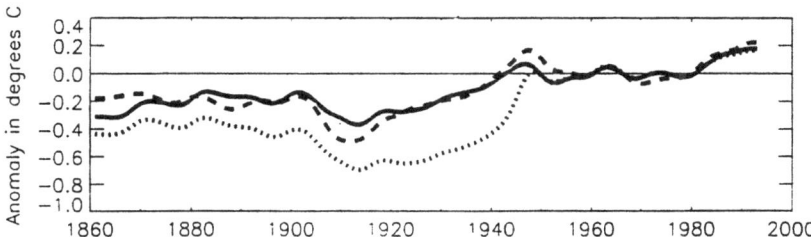

Fig. 7. Globally-averaged anomalies of uncorrected (dotted) and corrected (solid) SST. Corrected NMAT (dashed) is shown for comparison. From Folland and Parker (1995). Reference period is 1951–80.

in ongoing temperature change assessments (Intergovernmental Panel on Climate Change, 1992). However, any systematic warm or cold biases in recent data would be *relative to those in 1951–80 when many SSTs were measured using techniques similar to those still in use.* So a mean bias, globally, of as much as 0.1 °C in the most recent decade seems unlikely.

Satellite-based SSTs yield more complete coverage but require adjustment to make them consistent with *in situ* data (Section 2). This process is particularly necessary for AVHRR because of the systematic effects, sometimes exceeding 1 °C, of intervening stratospheric (volcanic) and tropospheric aerosols on the retrievals (Reynolds *et al.*, 1989; Reynolds, 1993; Folland *et al.*, 1993). Also, varying atmospheric moisture may not be fully accounted for by AVHRR calibrations, which use global statistical relationships between the radiances and the SSTs measured by a network of drifting buoys (Reynolds and Smith, 1994). In addition, these calibrations cannot take account of any temporal and geographical variations of the difference between oceanic skin temperature which satellites estimate, and bulk temperature. These latter problems may bias AVHRR by several tenths °C (see Schluessel *et al.* (1990) and e.g. Figure 10 of Reynolds and Smith (1994)). We have therefore used the Laplacian blending method of Reynolds (1988) to combine *in situ* and AVHRR SSTs in version 1.1 of our Global sea-Ice and Sea Surface Temperature data set (GISST1.1) (see Section 6). In Reynolds' technique, the satellite SSTs are adjusted to remove mean biases, and linear gradients in these biases, relative to *in situ* SSTs, while largely retaining as far as possible the geographical patterns of the rate of change of temperature gradient in the satellite data. Recently, Reynolds and Smith (1994) have created a new blend based on optimum interpolation, but systematic biases must still be removed from satellite-based SSTs before their use in this technique (Section 6.2).

The proper adjustment of all satellite SST data is essential to GCOS. However, the ATSR's combination of nadir and slantwise views (Saunders *et al.*, 1993) allows skin temperature to be calculated within about 0.2 °C even when there are aerosols or anomalous water vapour in the atmosphere (Barton *et al.*, 1995), so that remaining biases (i.e. skin-bulk differences) may be smaller and more easily removed. So ATSR2 and the Advanced ATSR are in principle key future instruments for GCOS, though experience of AVHRR satellite data shows the vital need for careful monitoring of their performance, in a climate context, using ground truth data and advanced analytical techniques.

4. Systematic Adjustments to NMAT

Folland *et al.* (1984) applied corrections to NMAT to compensate for the historical increases of the average height of ship's decks. These rose from about 6 m before 1890 to 15 m by the 1930s and 17 m by the 1980s. The corrections, based on surface layer similarity theory, removed a spurious cooling of about 0.2 °C between the late nineteenth century and 1980. On the other hand, Jones *et al.* (1986) used anomalies (not actual values) of regional, mainly coastal, land surface air temperature to adjust anomalies of nearby MAT. This was possible because anomalies of MAT and nearby 'coastal' land surface air temperature are found to be similar in recent data over periods as long as a decade, even though the absolute values differ considerably. However, because Jones *et al.* (1986) used COADS summaries, they were unable to separate NMAT from day MAT which are affected by historically-varying, on-deck solar heating: their corrections therefore differed from those of Folland *et al.* (1984). In both these early studies, about 0.5 °C was subtracted from MAT for 1942–5, a period of non-standard measurement practices owing to war.

Bottomley *et al.* (1990) applied similar height-related corrections to NMAT as Folland *et al.* (1984) but added two further adjustments:

(i) 5° area monthly NMAT anomalies (not actual values) in the Mediterranean and Northern Indian Ocean in 1876–93 were found to be consistently much too warm relative to SST anomalies. Plausible reasons were found for this problem. So the NMAT anomalies were replaced by corresponding SST anomalies. This was regarded as a valid procedure because of good similarity between anomalies of NMAT and SST in subsequent years, and the unlikelihood of substantial systematic changes in air-sea heat fluxes over decades (see Barnett, 1984).

(ii) 5° area monthly NMAT anomalies in the Atlantic in 1856–85 were also found to be consistently too warm relative to SST anomalies. The problem was considerably worse at high wind speeds, and suggested the use of inappropriate observational practices. So the NMAT anomalies were adjusted so that, at any location in any given calendar month, the average NMAT anomaly for this 30-year period equalled the average anomaly of SST.

We are now improving the NMAT archive in several ways.

(iii) We have enhanced the recent data base and developed better quality-control routines in parallel with similar improvements for SST (Section 2).

(iv) The SST anomalies used to adjust the nineteenth century NMAT as in (i) and (ii) above, have been improved as outlined in Section 3.

(v) A more reliable and complete reference climatology is being created (Section 5) and is to be used in the quality-control of the observations.

(vi) A study of MAT anomalies as a function of latitude, calendar month and observing time has shown that NMAT shortly after sunset are affected by residual solar heating of decks, while MAT within an hour after sunrise are relatively unaffected (e.g. Figure 8). As a result, we plan to recompile NMAT, using data from an hour after sunset through to an hour after sunrise. Until about 1930, observations were at 00, 04, 08, 12, 16, 20 local time; but then there was a change to 00, 06, 12, 18 GMT (Bottomley *et al.*, 1990). There were thus fewer NMAT data just after sunset before 1930, creating a **negative** bias relative to modern data especially at longitudes where sunset now shortly precedes an observing hour, e.g. 0°–15° E for 18 GMT particularly in the tropics where sunset remains close to 18 local time. The recompilation should remove most of this bias.

Figure 9 compares anomalies of hemispheric NMAT, incorporating all developments to date (i.e. (i) to (iv)), with anomalies of SST from MOHSST6 and with anomalies of NMAT from Bottomley *et al.* (1990).

The recompilation of NMAT to avoid residual heating of decks is expected to improve the agreement in the Southern Hemisphere by raising the NMAT anomalies by about 0.05 °C in the late nineteenth and early twentieth century. There is likely to be a lesser effect in the Northern Hemisphere because more of the longitudes where biases are greatest are over land.

5. New Climatologies

Reliable climatologies are essential for the assessment of climatic anomalies and the detection of climatic changes, on both global and regional scales. They are also needed as a reference in the quality-control of observations. We have constructed a new climatology of SST and are constructing one for NMAT. The reference period is 1961–90, the period of most plentiful data, including satellite-based observations of SST, and we anticipate very worthwhile improvements to the assessment of climatological gradients in data-sparse regions. 1961–90 is also the new World Meteorological Organization standard reference period, for which new land station climatologies have been constructed (Jones, 1994).

5.1. SST

In parts of the midlatitude Southern Ocean, especially in the eastern Pacific sector, our existing climatologies for 1951–80 are substantially biased, by up to 1 °C or more (Parker *et al.*, 1994). The evidence for this is the persistence of anomalies of

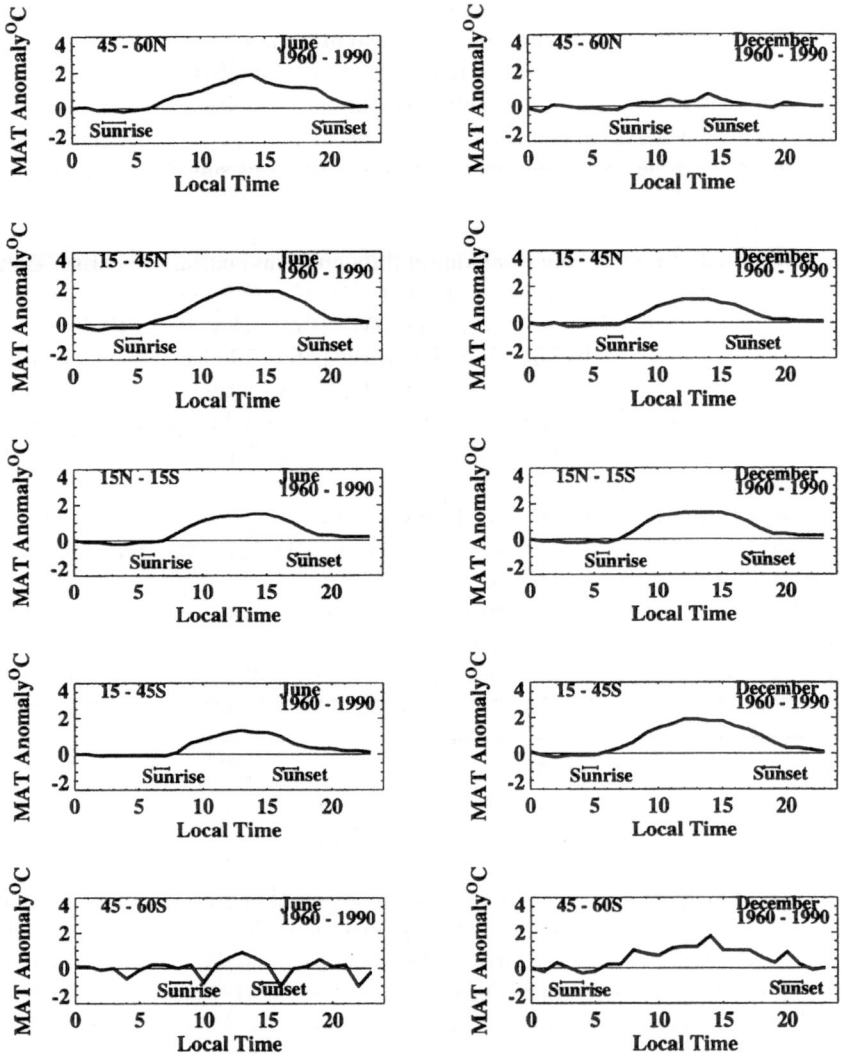

Fig. 8. Composite diurnal cycles of observed MAT. Each panel indicates the latitude belt and calendar month to which the cycles apply.

one sign and of this size in recent decades in data which are referenced to these climatologies. The biases arose because 1951–80 *in situ* data were too sparse to permit Bottomley *et al.* (1990) to base their climatology on them: instead they made substantial use of the climatology of Alexander and Mobley (1976) who had used earlier data and, in the Southern Ocean, a bilinear interpolation. This interpolation

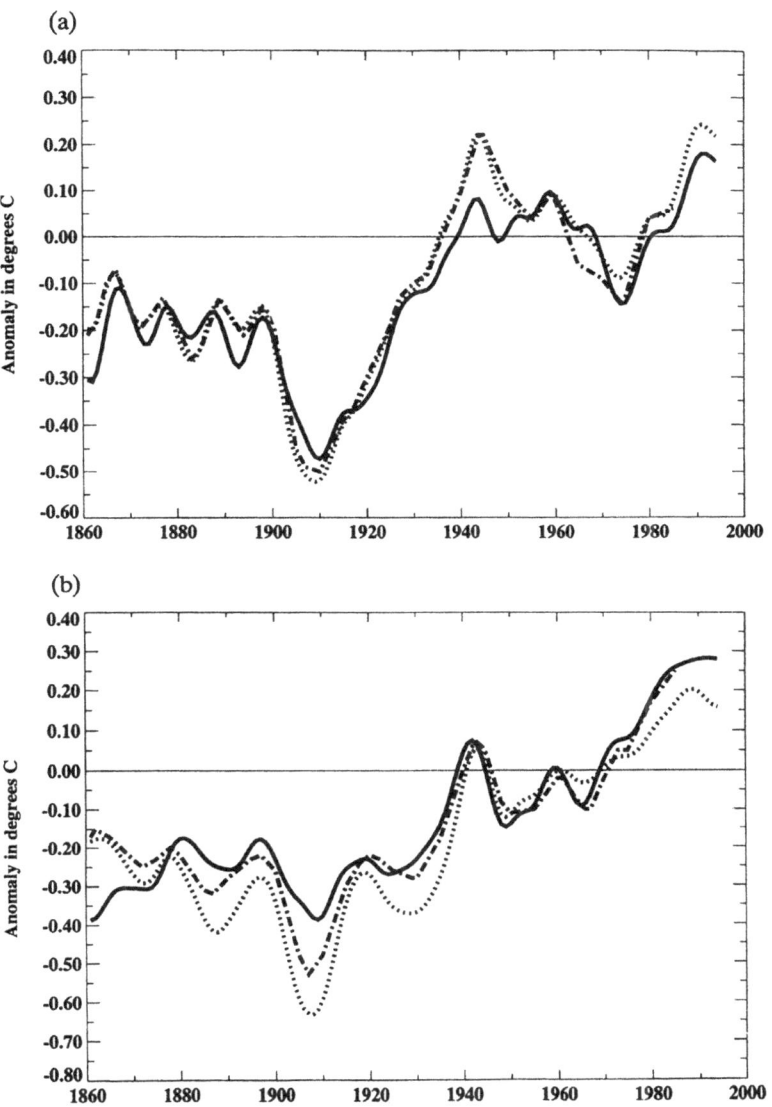

Fig. 9. Corrected NMAT, this analysis (dotted); corrected NMAT (Bottomley *et al.*, 1990) (dash-dot); corrected SST, MOHSST6 (solid). (a) Northern Hemisphere; (b) Southern Hemisphere. The smoothed curves except that from Bottomley *et al.* (1990) are the result of application of a 21-point binomial filter to annual values. The Bottomley *et al.* smoothing was a 41-point triangular filter applied to seasonal values. Reference period is 1951–80.

was especially prone to error in regions of strong climatological SST gradient near the oceanic convergence zones. The GISST1.1 climatology for 1951–80 had smaller biases than that of Bottomley *et al.* (1990), because of the interpolation of MOHSST55 anomalies into some missing areas, and because SSTs near ice-edges were modified in a manner consistent with anomalies of ice coverage (see Parker *et al.*, 1995).

Using the 1982–94 average of GISST1.1 as a background field, we have now created new normals. We chose this background field because it includes satellite data, with their biases relative to *in situ* data reduced by Reynolds' (1988) Laplacian blending technique (Parker *et al.*, 1995). Because of the satellite data, there are few areas in which the Bottomley *et al.* (1990) climatology dominates the background field. We Laplacian-blended the background field with *in situ* data, along with marginal ice zone SSTs specified statistically from ice concentrations (Section 6.1.), to yield $1° \times 1°$ resolution fields for each month from January 1961 to December 1990. After light smoothing, these fields were averaged to form a new monthly 1961–90 climatology. This climatology was then interpolated to pentad resolution using harmonic synthesis, and further interpolated linearly to give a daily $1° \times 1°$ resolution climatology.

Figures 10(a), (b) show the new SST climatology for January and July; differences from the Bottomley *et al.* climatology are shown in Figures 10(c), (d) and display changes exceeding 1 °C over parts of the Southern Ocean, as well as smaller but real worldwide differences owing to the change of reference period, e.g. a cooling in the northwest Atlantic (see Parker *et al.*, 1994) and the midlatitude north Pacific (Trenberth and Hurrell, 1994). A new optimum-interpolation-based climatology has also been created by Reynolds and Smith (1995), but this is for 1950–79 and therefore not exactly comparable with ours.

The daily climatology for 1961–90 has been used to convert individual SST observations to anomalies for quality-control and compositing during the creation of MOHSST6. When normals for the coarser pentad periods (Jan 1–5, 6–10 etc.) are used to calculate SST anomalies, as was done by Bottomley *et al.* (1990), some additional scatter occurs in the anomalies (Trenberth *et al.* (1992)). Our new precedure avoids this problem.

5.2. NMAT

The NMAT climatology for 1951–80 created by Bottomley *et al.* (1990) was geographically incomplete. This is not only because of data sparsity, but also because there is no constraint on air temperature to approach -1.8 °C, the freezing point of sea water, at the ice-edges. We have computed a provisional climatology for 1961–90 based on the improved NMAT data (Section 4(i) to (iv)), but this is also geographically incomplete for the same reasons. It will also need to be re-created using the recompiled NMAT to avoid solar heat which continues to affect decks just after sunset (Section 4).

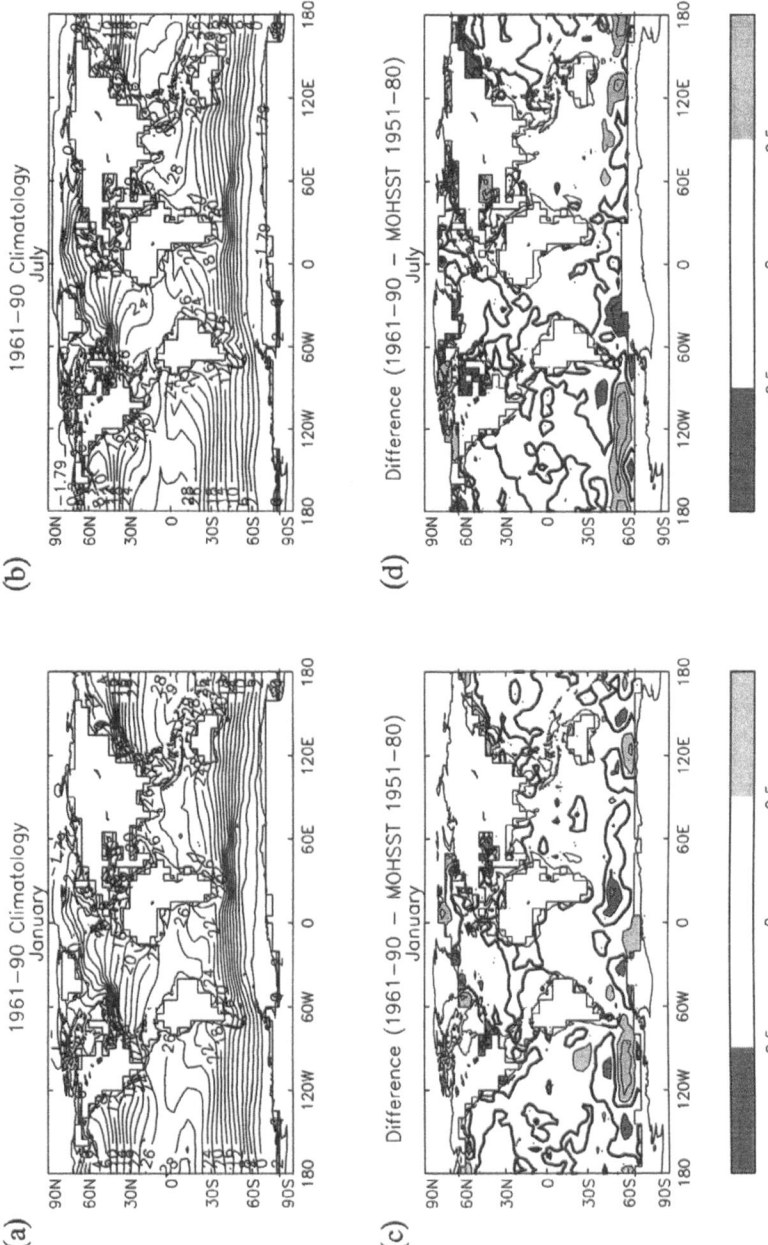

Fig. 10. New SST climatology for 1961–90: (a) January; (b) July; (c) January difference from Bottomley *et al.* (1990); (d) as (c) but for July. Values in °C. The zero contour is heavy in (c) and (d).

6. Analysis Techniques

6.1. GISST1.1 AND 2.1

Bottomley *et al.* (1990) created monthly 5° area SST anomalies, via pentad 1° area SST anomalies, from observations made within the confines of a given 5° area and historical month, e.g. 35°–40° S, 145°–150° E, August 1914. They did not interpolate these monthly 5° area anomalies geographically or temporally, and therefore large gaps remained. Also, the addition of 5° area anomalies to their 1° area climatology sometimes produced slight discontinuities at the borders of 5° areas in their resulting 1° area values. We now discuss the GISST data sets where, amongst other developments, attempts have been made to create a 'globally complete' data set without these problems, suitable for use with climate models. Full details of the first version (GISST1.1) are given by Parker *et al.* (1995): here we show an outline (Figure 11) and give a brief summary of the creation of GISST1.1 and the improvements made in GISST2.1.

To create MOHSST55, the Bottomley *et al.* (1990) monthly 5° area SST data were augmented with monthly COADS SST data (Woodruff *et al.*, 1987) which had been interpolated to 5° × 5° resolution, and the Folland and Parker (1995) bucket-corrections (Section 3) were applied. To create GISST1.1, missing and extreme 5° area monthly MOHSST55 anomalies were replaced by averages of spatially or temporally adjacent 5° area monthly anomalies if sufficient of these were available. 'Extreme' was defined as a deviation of more than 2.25 °C from the latter averages, which had to be based on at least 4 (spatial average) or 2 (temporal average) 5° area monthly anomalies. These interpolations, which had been optimised by trial and error, were applied in 3 iterations. As a result, the interpolation of missing values augmented the original data coverage substantially. This nominal coverage was further enhanced from 1982 onwards by using satellite data, adjusted to be consistent with the *in situ* fields. Before that, weighted anomalies from up to 5 months before or after a missing value were used. This interpolation assumed a Markov-type persistence of anomalies with a one-month lag autocorrelation of 0.6. It tended to generate rather weak anomalies in data-sparse areas. The value of 0.6 was chosen as a globally representative value on the basis of monthly SST autocorrelations presented by Bottomley *et al.* (1990). Sea-ice extent was taken from a variety of sources as shown in Figure 11. SSTs were assinged along ice-edges with the aid of climatological SST statistics and ice-extent anomalies, and the SSTs were extended into remaining missing areas using the Laplacian of the Bottomley *et al.* (1990) 1951–80 climatology. A final smoothing retained anomaly variations on a spatial resolution rather coarser than 5° × 5°.

The more recent GISST2.1 incorporates the following improvements:

(a) The basic *in situ* SSTs are from MOHSST6 instead of MOHSST55. Thus, GISST2.1 benefits from the extra recent logbook and buoy data, new accessing and quality-control algorithms, and the use of the new daily normals (Sections

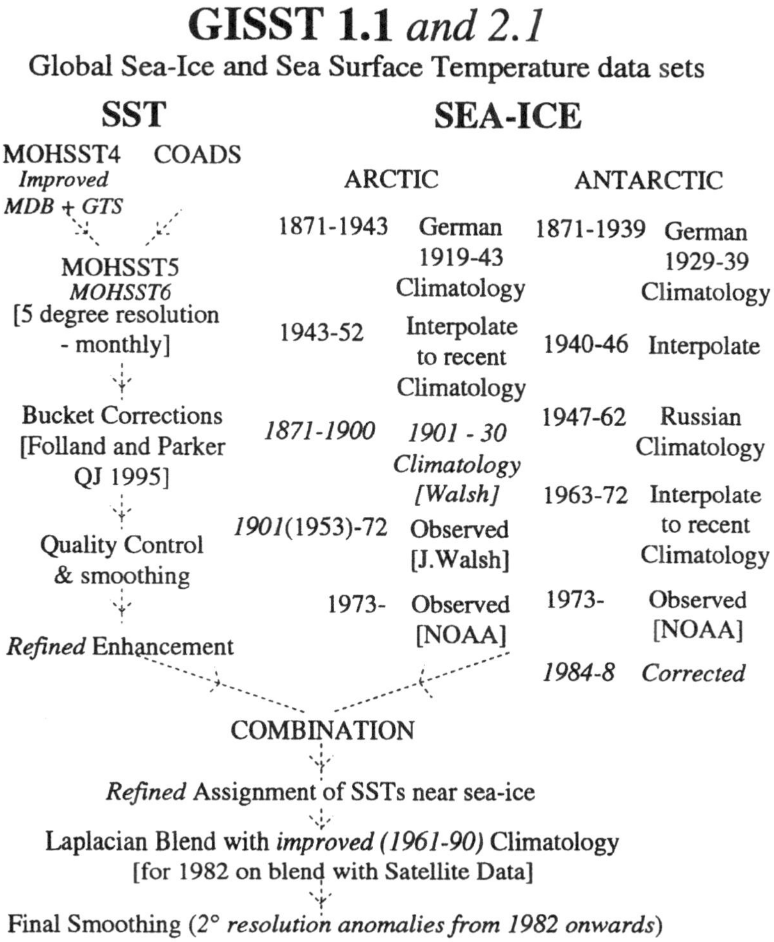

GISST **1.1** *and 2.1*

Global Sea-Ice and Sea Surface Temperature data sets

SST	SEA-ICE

SST	ARCTIC		ANTARCTIC	
MOHSST4 COADS *Improved* *MDB + GTS*				
	1871-1943	German 1919-43 Climatology	1871-1939	German 1929-39 Climatology
MOHSST5 *MOHSST6* [5 degree resolution - monthly]	1943-52	Interpolate to recent Climatology	1940-46	Interpolate
Bucket Corrections [Folland and Parker QJ 1995]	*1871-1900*	*1901 - 30 Climatology [Walsh]*	1947-62	Russian Climatology
Quality Control & smoothing	*1901(1953)-72*	Observed [J.Walsh]	1963-72	Interpolate to recent Climatology
	1973-	Observed [NOAA]	1973-	Observed [NOAA]
Refined Enhancement			*1984-8*	*Corrected*

COMBINATION

Refined Assignment of SSTs near sea-ice

Laplacian Blend with *improved (1961-90)* Climatology
[for 1982 on blend with Satellite Data]

Final Smoothing (*2° resolution anomalies from 1982 onwards*)

Fig. 11. Creation of GISST1.1 and 2.1. Features of GISST2.1 are italicised.

2 and 5.1. and Appendix 1). So the initial fields of *in situ* data have typically 2% more 5° area-months with data than did MOHSST55.

(b) The temporal interpolation scheme has been refined by allowing for spatial variations of the one-month lag-autocorrelation of SST anomalies, using a geographical cluster analysis of the autocorrelation functions.

(c) Erroneous Antarctic sea-ice extent data for 1984–8 have been replaced, and we have used Arctic sea-ice concentration data for 1901–72 provided by J. Walsh (Univ. Illinois) (Figure 11). Marginal ice zone SSTs have been estimated on the basis of quadratic regressions between observed *in situ* SST in 1° areas and sea-ice concentration averaged over the surrounding 5° area, using data for

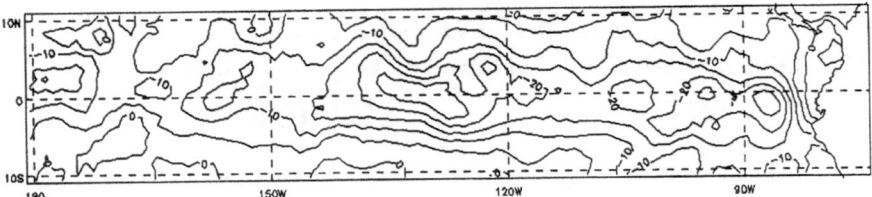

Fig. 12.   Provisional GISST2.1 anomalies (tenths °C, with respect to 1961–90) in the eastern tropical Pacific, in La Niña conditions, July 1988. Note the narrow equatorial tongue of cold anomalies.

1961–90. SST was constrained to equal −1.8 °C for 100% ice cover. For the Arctic, the regressions were computed separately for each season (December to February etc.) for separate 30° longitude sectors, which were overlapped to avoid discontinuities in the results. For the data-sparse Antarctic, we used a single regression based on the entire *Arctic* for the corresponding season, e.g. the December to February Arctic regression was used to specify June to August Antarctic SSTs from observed Antarctic sea-ice.
(d) The background field for the Laplacian interpolation was the improved SST climatology described in Section 5.1.
(e) GISST2.1 anomalies retain 2° resolution from 1982 onwards, before any smoothing, so that narrow equatorial cold tongues in La Niña events are considerably better monitored (Figure 12).

6.2.   OPTIMUM INTERPOLATION AND AVERAGING SCHEMES

Globally-complete SST fields have also been created from a combination of *in situ* and satellite data by optimum interpolation (Reynolds and Smith, 1994). This technique is computationally intensive, requiring the repeated inversion of large matrices relating to the spatial correlation structure of the SST anomalies (Gandin, 1963), if error estimates are to be made. Its major advantage is that Gaussian standard error bars are assigned to the resulting analysis, so that the statistical significance of local changes can in principle be assessed. Another advantage of the optimum interpolation technique is that different data-types (e.g. ship, buoy, satellite) can be given different weightings, which determine their influence on the final analysis. The weightings depend on the estimated standard errors of the respective data-types: ships' SSTs were estimated by Reynolds and Smith (1994) to have typical standard errors about three times greater than satellite SSTs, so the ships' SSTs were given lower weightings. However, optimum interpolation cannot take account of *systematic biases*, and Reynolds and Smith (1994) had to apply prior systematic adjustments to the satellite data, using *in situ* data and Reynolds' (1988) Laplacian blending technique as described in Section 3. Reynolds and Smith (1994) found that their optimum interpolation technique, with the adjusted satellite data, had better spatial and temporal resolution than the original blended analysis

(a)

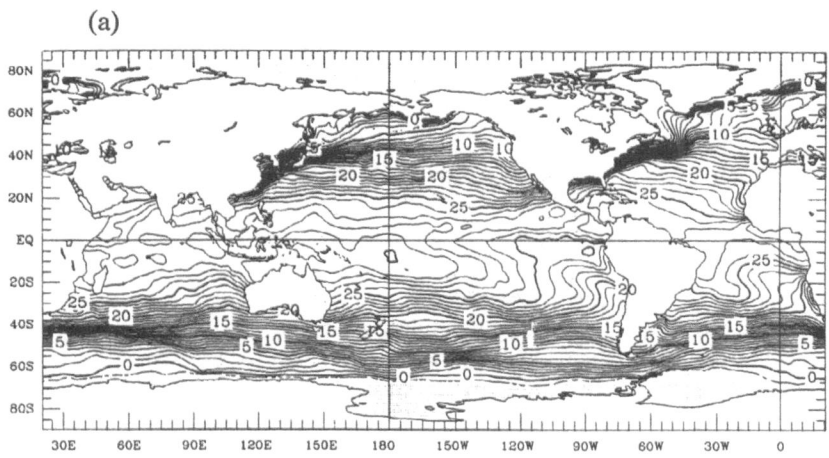

(b)

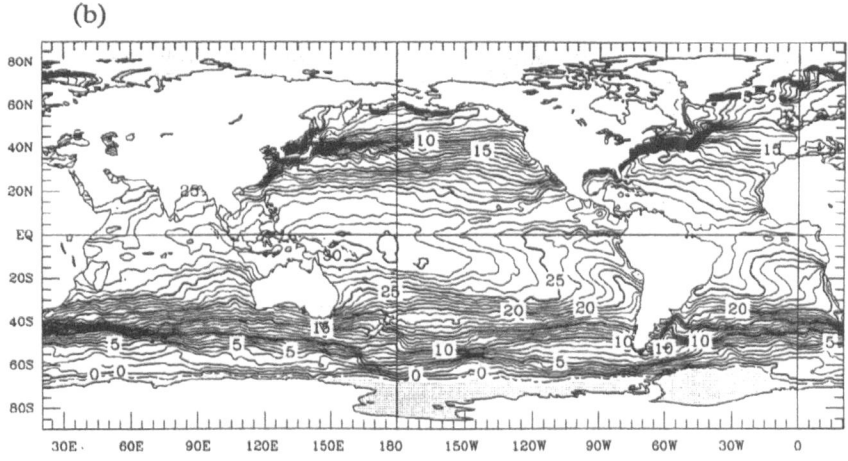

Fig. 13.   Mean SST (a) from the Reynolds (1988) blended analysis for the 15-day period 9–23 January 1991; and (b) from the optimum interpolation with refined statistics for the week 13–19 January 1991. From Reynolds and Smith (1994).

created by Reynolds (1988). For example, areas with SST exceeding 30 °C in the equatorial western Pacific were better resolved, as was the equatorial eastern Pacific cold tongue (Figure 13). Reynolds and Smith (1994) used greater zonal than meridional correlation scales in their analysis, i.e. they took account of anisotropy in their data, but their results were insensitive to this aspect.

Regionally-averaged SST anomalies have also been estimated by a related technique, optimum averaging (Smith *et al.*, 1994; Kagan, 1979). This technique also assigns standard errors to the results, so that the statistical significance of regional changes can be gauged. However, because of the large matrix inversions that would have been involved, Smith *et al.* (1994) did not estimate hemispheric or global anomalies directly by this method; instead, they composited the regional (20° latitude × longitude) anomalies and their errors. So estimates of the resulting global anomalies and their associated errors were close to, but not absolutely, optimal. Because of interdecadal variations of SST, results were greatly improved by using persistence, rather than climatology, as the first guess in their season-by-season optimum averaging procedure.

Given further increases in computing power, the 'optimum' techniques described in this section may, because of their potential for assessment of intrinsic errors, be among the best tools for analysing marine surface temperatures and their climatic changes. In their present form the question must be asked, however, whether they make the maximum possible use of available information about the data fields. A simpler but effective new interpolation technique, which attempts this but does not estimate error bars, is therefore reviewed in Section 8. Some form of optimum interpolation *is* needed to provide high-quality random error estimates currently lacking in accepted analyses of ocean surface temperature changes.

7. Global and Regional Time Series

7.1. RECENT IMPROVEMENTS

We compare, in Figure 14, smoothed global SST anomalies from Folland *et al.* (1984) and from Bottomley *et al.* (1990) with smoothed global SST anomalies and annual bars using MOHSST6 with the Folland and Parker (1995) corrections. Also shown for reference is smoothed global NMAT, also using the latest data and corrections (Section 4(i) to (iv)). The series are all expressed as anomalies relative to 1951–80, for ease of comparison. Overall global warming in SST between the 1860s and the 1970s is about 0.3 °C greater in the present analysis than in Folland *et al.* (1984), mainly owing to reduced early corrections applicable under the assumption of the predominant use of wooden buckets (Section 3). There is little difference, however, since 1900. The overall warming in the Bottomley *et al.* (1990) analysis is intermediate between Folland *et al.* (1984) and the present analysis. Folland and Parker (1990) and Parker and Folland (1991) obtained marginally less overall global warming in SST than Bottomley *et al.* (1990), but more than Folland *et al.* (1984) (see Figure 9c of Bottomley *et al.* (1990)). NMAT in Figure 14 generally agrees well with the present SST analysis, but is too cold around 1910, as expected from Figure 9 and Section 4.

In Figure 15, similar comparisons are made with NMAT, but only using Bottomley *et al.* (1990) and the MOHSST5 analysis for SST, for six ocean basins,

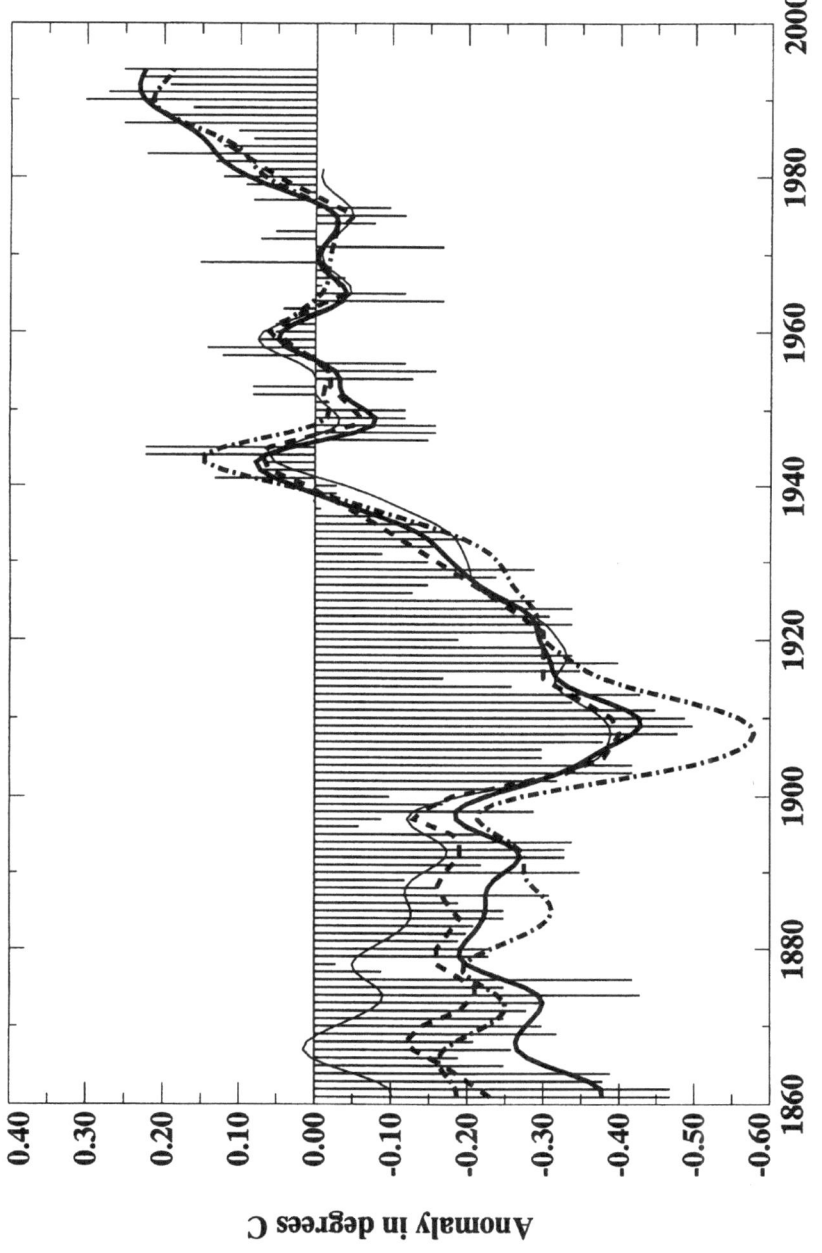

Fig. 14. Global anomalies of SST, after Folland *et al.* (1984) (light continuous smoothed curve) and using MOHSST6 with Folland and Parker (1995) corrections (heavy continuous smoothed curve and annual bars). The dash-dot smoothed curve is NMAT, from the present analysis. Smoothing as Figure 9. Reference period is 1951–80.

the two hemispheres, the tropics (20° N–20° S), and the Gulf Stream (35°–45° N, 50°–70° W) and Kuroshio (30°–40° N, 125°–160° E) areas. Note the expanded scale for the latter two areas. In most cases the latest data and corrections give fair agreement of SST anomalies with the largely independent NMAT anomalies, except in the southern Indian Ocean and South Pacific (and therefore the Southern Hemisphere) in 1880–1930. However SST and NMAT corrections are not independent everywhere before 1893 (Section 4).

Figure 16 shows the overall impact of the current bucket corrections on SST for the same regions as in Figure 15. Smoothed annual corrected and uncorrected MOHSST6 SST anomalies are compared with smoothed annual corrected NMAT anomalies. In all regions, the effectiveness of the bucket-corrections is demonstrated. Of the large ocean basins, only the Southern Indian Ocean shows persistent differences in the corrected anomalies exceeding 0.1 °C. Possible over-correction of SST, relative to NMAT, over the Gulf Stream, where corrections to SST are large, is not evident over the Kuroshio where bucket-corrections are also large.

We have not shown series based on GISST1.1 or 2.1. These data sets are at present less suitable for climate change estimation, because the interpolated anomalies, necessary for climate modelling, tend to be underestimated in magnitude. So GISST1.1 and 2.1 are slightly biased globally, and significantly so locally. However, GISST1.1 and 2.1 have an advantage over MOHSST5E and MOHSST6 for climate assessment since 1982 in that bias-adjusted satellite SSTs (Section 3) are used to interpolate, and little bias is expected in these analyses from that procedure (Section 6.1). The impact of the satellite SSTs on global and hemispheric anomalies is only of the order of 0.01 °C, according to Parker *et al.* (1994) who included satellite data in their final estimates of global SST changes, but the impact is considerably stronger locally.

7.2. COMPARISONS WITH LAND

Corrected MOHSST6 5° area anomalies (not actual values) and corrected NMAT 5° area anomalies have been compared with 'colocated' coastal land and island surface air temperature anomalies, i.e. the Jones (1994) anomalies for coincident 5° areas, and the results composited for the two hemispheres, the tropics, and the globe (Figures 17 and 18). The comparison is done on decadal time scales, for which a close match of anomalies is expected (Sections 3 and 4). Global agreement is similar to that obtained by Folland and Parker (1992). Relative warmth of about 0.2 °C in land air temperature anomalies relative to anomalies of both SST and NMAT in the tropics in the early twentieth century probably reflects biases in land air temperature. These biases are thought to have resulted from the widespread use of thermometers in cages under thatched or felted sheds at tropical stations (Parker, 1994). This bias has affected the hemispheric comparisons also, but part of the relative warmth in land air temperature in the Northern Hemisphere at that time is likely to have resulted from enhanced westerly circulation over the North Atlantic

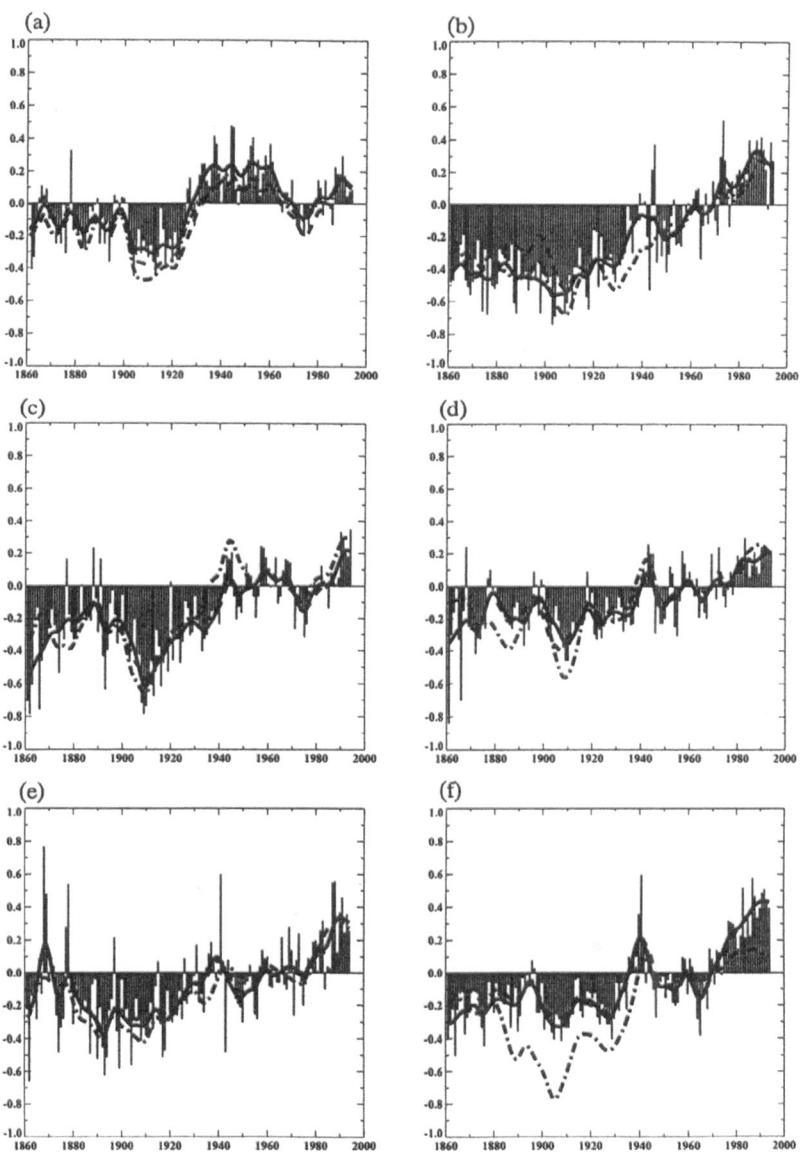

Fig. 15(a)–(f).

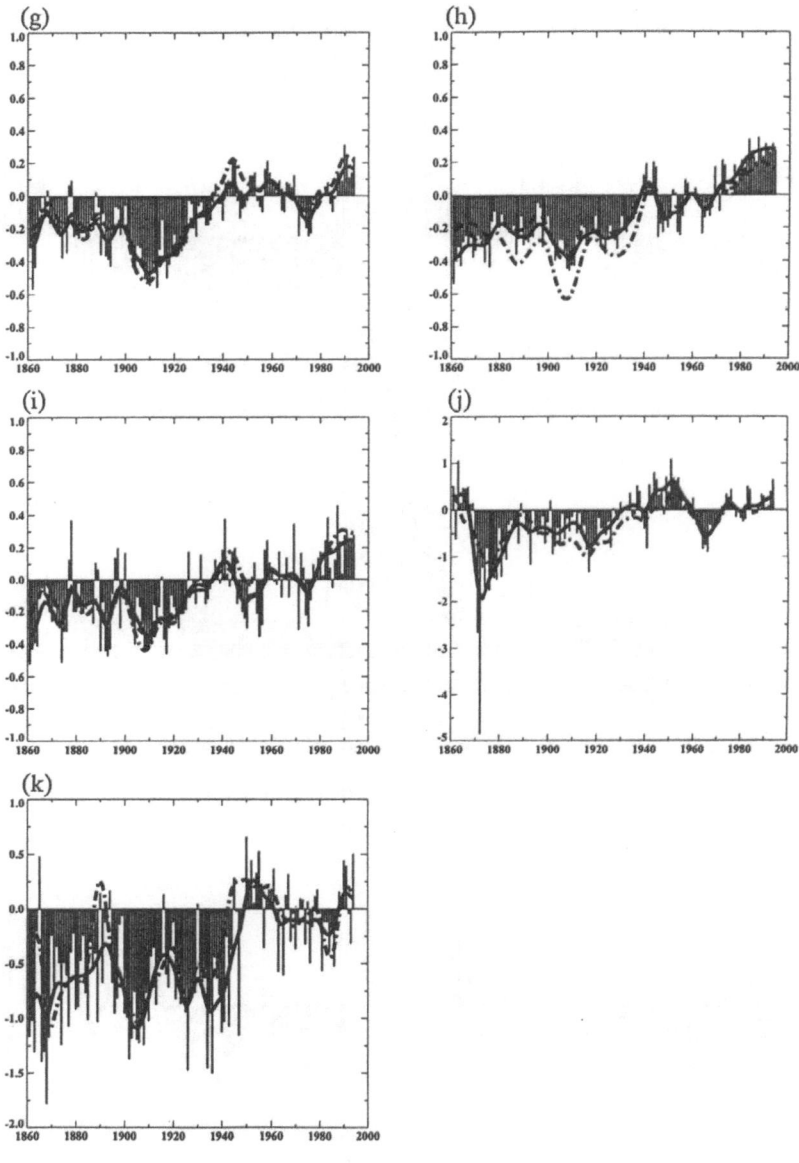

Fig. 15(g)–(k).

Fig. 15(a)–(k).   As Figure 14 but only showing Bottomley *et al.* (1990) SST (where available, dashed smoothed curves), the MOHSST6 SST analysis (solid smoothed curves and annual bars), and the latest NMAT analysis (dash-dot smoothed curves). Smoothing as Figure 9. (a) N Atlantic; (b) S Atlantic; (c) N Pacific; (d) S Pacific; (e) N Indian Ocean; (f) S Indian Ocean; (g) N Hemisphere; (h) S Hemisphere; (i) Tropics (20° N–20° S); (j) Gulf Stream area (35°–45° N, 50°–70° W); (k) Kuroshio area (30°–40° N, 125°–160° E). Reference period is 1951–80.

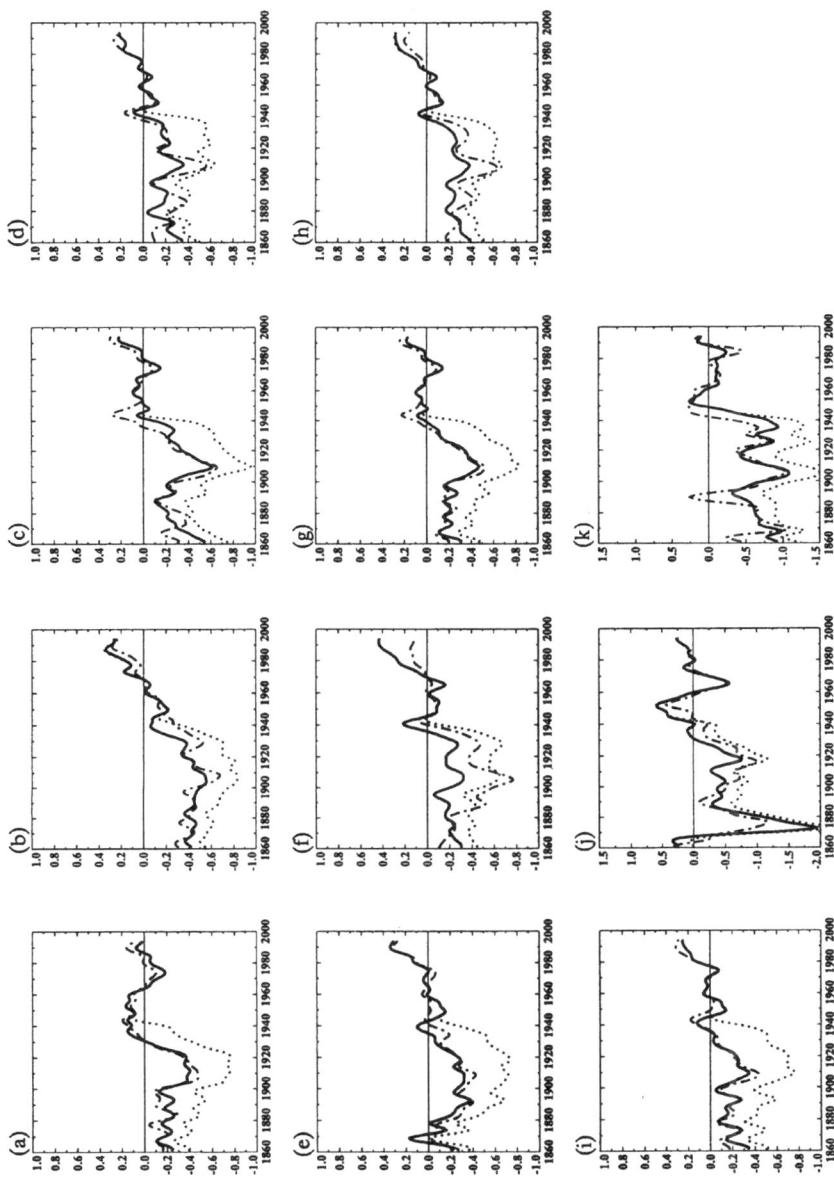

Fig. 16(a)–(k). Annual corrected (solid curve) and uncorrected (dotted curve) SST, based on MOHSST6 data, and smoothed corrected NMAT (dash-dot curve) as developed in this paper. Smoothing as Figure 9. (a) N Atlantic; (b) S Atlantic; (c) N Pacific; (d) S Pacific; (e) N Indian Ocean; (f) S Indian Ocean; (g) N Hemisphere; (h) S Hemisphere; (i) Tropics (20° N–20° S); (j) Gulf Stream area (35°–45° N, 50°–70° W); (k) Kuroshio area (30°–40° N, 125°–160° E). Reference period is 1951–80.

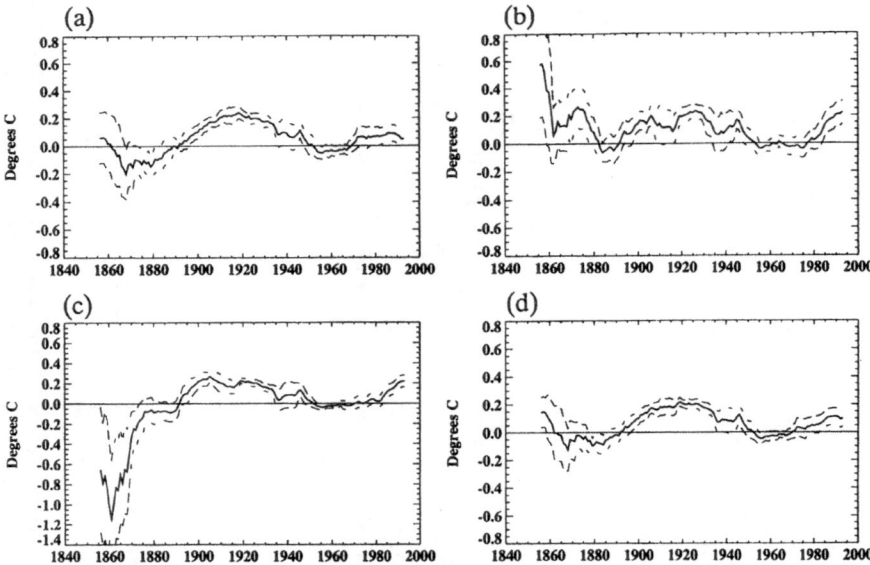

Fig. 17. 10-year running mean differences, coastal land and island (Jones, 1994) surface air temperature anomalies minus colocated corrected SST anomalies (MOHSST6), relative to 1951–80. Dashed
lines are twice the standard error. (a) N Hemisphere; (b) S Hemisphere; (c) Tropics (20° N–20° S);
(d) Globe.

in winter (Parker *et al.*, 1994), yielding large positive air temperature anomalies
over western Europe colocated with smaller coastal SST and NMAT anomalies.
There were generally fewer than 10 tropical locations for comparisons before the
mid 1880s.

7.3. FROZEN GRID EXPERIMENTS

The comparisons between SST, NMAT and land air temperature in Sections 7.1
and 7.2. have validated the corrections applied to SST and NMAT. Here, we assess
(rather imperfectly, see below) the influence of incomplete geographical coverage
on hemispheric anomalies of SST, by carrying out 'frozen grid' experiments.
Figures 19 and 20 show annual and winter hemispheric corrected SST anomalies
using MOHSST5E but with geographical coverage limited to that available in 90%
of all seasons in 1871–80 (dotted line) and 1911–1920 (dashed line). The solid
lines show the anomalies using all available data. Agreement is good on decadal
time-scales with biases generally less than 0.1 °C: long-term climatic trends are not
in doubt. Interannual agreement is poorer especially using the 1871–80 grid, and
especially for the seasonal series for which differences occasionally reach 0.5 °C.
These results are similar to those shown (for land surface air temperature and
SST combined and for different decades' coverage grids) by Folland and Parker

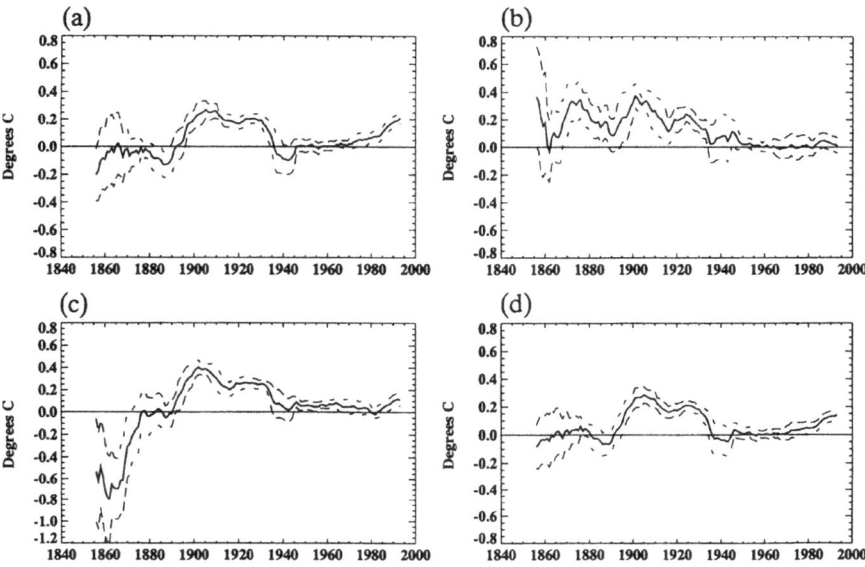

Fig. 18. As Figure 17 but for corrected NMAT (MOHMAT31N).

(1992). Note, however, that frozen grid experiments, while showing the sensitivity of the analysis to varying coverage, cannot assess the impact of regions, such as the Southern Ocean, which were never covered with *in situ* data.

The effects of inadequate coverage can also be assessed by varying the geographical averaging technique used to composite 5° area anomalies into hemispheric or global anomalies (Karl *et al.*, 1994; Parker *et al.*, 1994). Averaging 5° area anomalies into, for example, 30° latitude × 60° longitude area anomalies before averaging these into hemispheric or global anomalies can give more weight to isolated data, though the influence on hemispheric anomalies is quite small (Parker *et al.*, 1994). However, this method assumes the validity of extrapolation of anomalies into data voids, and some regions, e.g. the Southern Ocean and Antarctica, may still be completely omitted before the 1950s. Real, reliable, data are preferable to the most skilful extrapolation.

8. Conclusions

On decadal time scales and global, hemispheric and ocean-basin scales, the existing SST and NMAT data bases yield random errors of about 0.1 °C, and systematic errors of up to a similar magnitude. These result mainly from broken coverage and instrumental uncertainties respectively. So the overall uncertainty in the late nineteenth century estimates of global marine surface temperature anomalies is thought to be about 0.15 °C. This is to be compared with a climatic change signal

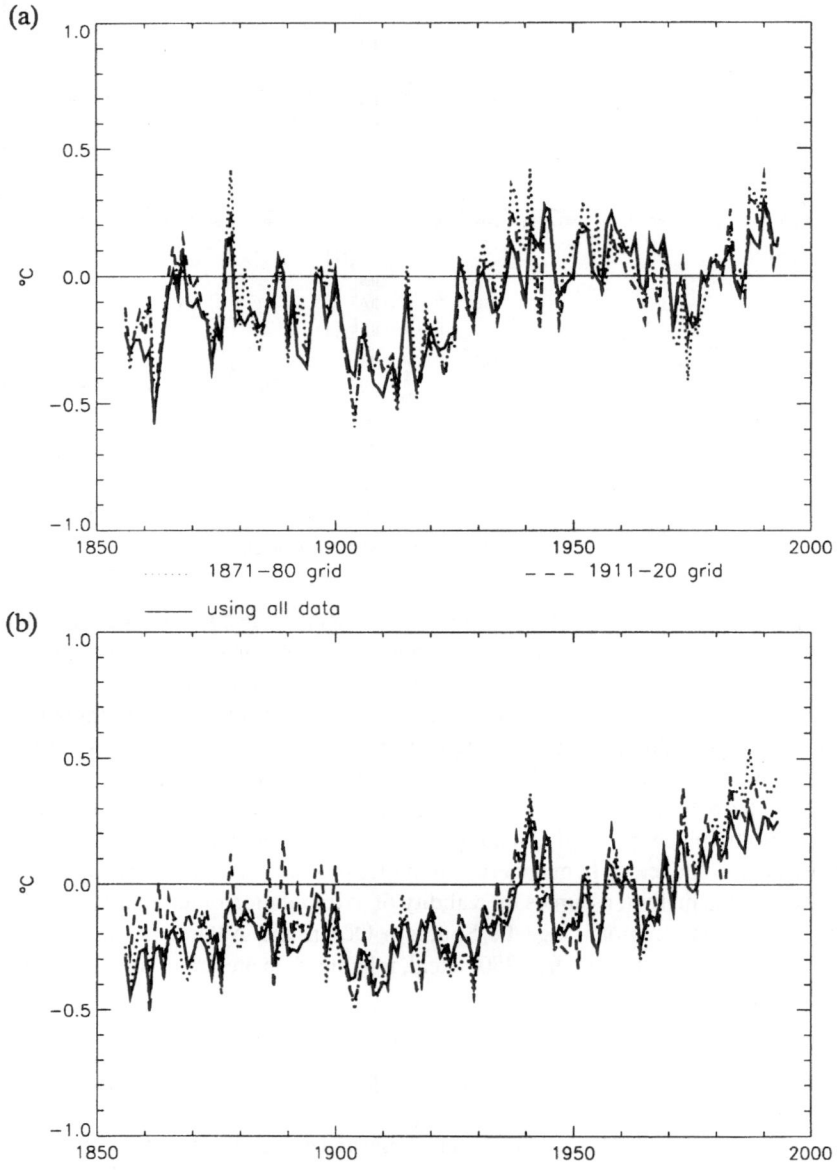

Fig. 19. Frozen grid tests of unsmoothed annual corrected SST (MOHSST5E). (a) N Hemisphere; (b) S Hemisphere. Reference period is 1951–80.

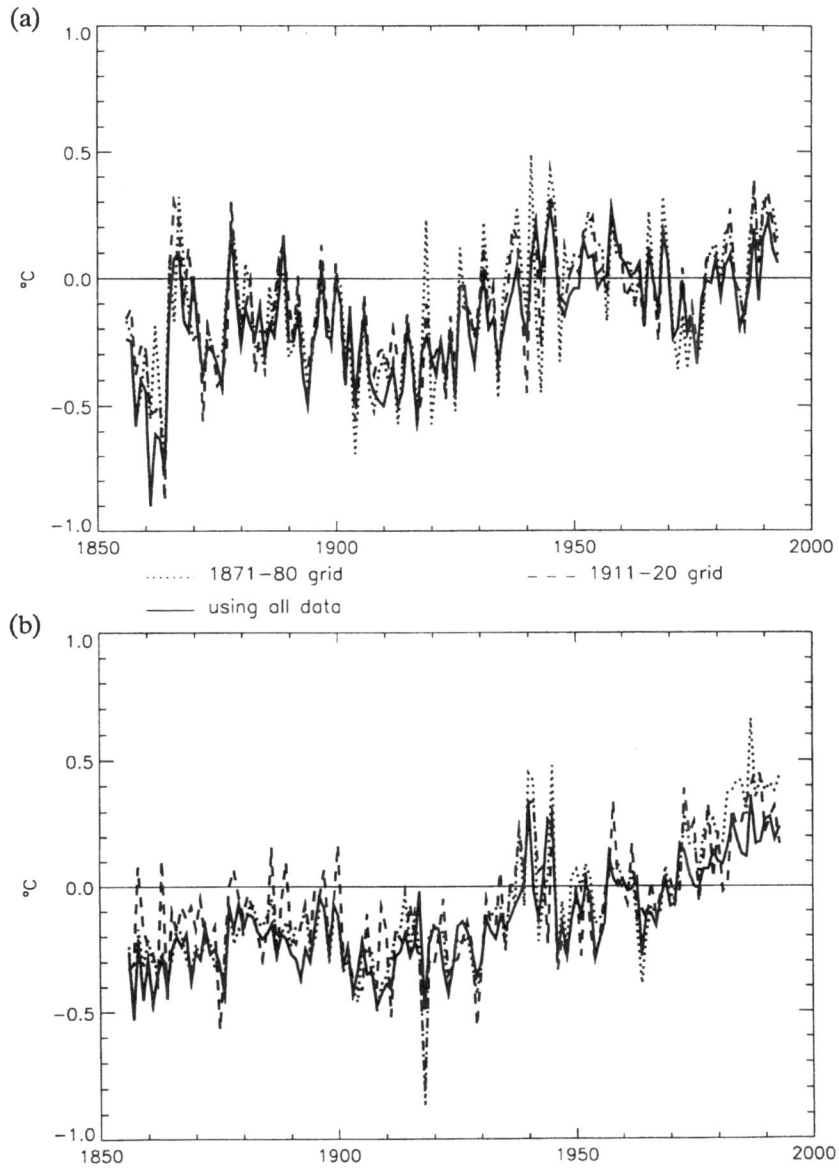

Fig. 20. Frozen grid tests of unsmoothed seasonal corrected SST (MOHSST5E). (a) N Hemisphere, winter (January to March); (b) S Hemisphere, winter (July to September). Reference period is 1951–80.

of about 0.5 °C. The likely errors are larger than 0.15 °C for smaller regions, and much larger in data-sparse areas in the Southern Hemisphere. So, the following improvements in marine surface temperature data are needed:

(i) The digitization of remaining manuscript observations (Section 2), and their quality-control and incorporation into the data base, is a priority task. Other analyses, e.g. of mean sea level pressure, will also benefit from this.

(ii) In parallel with (i), a blend, observation by observation, of COADS and the Meteorological Office Marine Data Bank, is needed, and is planned. Each of these data banks holds unique observations, so the blended archive will be greater than either (Parker, 1992). It is important, however, that duplicate observations be carefully identified and eliminated during the blend. Although our SST analyses already include COADS data in a simplified form, our NMAT analyses do not, because COADS statistics for MAT combine day and night.

(iii) NMAT should be recompiled to avoid biases from residual evening warmth of ships' decks, as explained in Section 4.

(iv) Improved adjustments are needed to nineteenth century MAT anomalies. Many of these are, at present, adjusted with the aid of SST anomalies (Section 4), so that we have lost the advantage of having two independent marine surface temperature records (Section 7.1.).

On the other hand, refinement of the existing techniques for deriving corrections to SST may not yield significant benefits. When the data base is augmented, the corrections will need to be recalculated using the Folland and Parker (1995) techniques and assumptions, because the augmented data may show spurious annual cycles which differ slightly from those in existing uncorrected data. However the use of individual observations of wind etc., rather than climatological environments, in the derivation of the corrections may be unprofitable owing to errors or missing values of the required parameters. Even the use of appropriate individual monthly values, while possible for some parameters, may not greatly reduce uncertainties in the corrections, because these uncertainties mainly result from complex factors such as the influence of the ship on wind strength on deck (Folland and Parker, 1995). However, this question is not settled.

A new analysis technique under development is the use of eigenvector patterns for the geographical interpolation of marine surface temperature fields (Rayner et al., 1995). The patterns are defined using recent, geographically extensive data, and can be finely resolved where appropriate, e.g. in the equatorial Pacific. Time coefficients of these patterns are then, in principle, estimated for earlier fields using the more limited data available, by an iterative procedure. The patterns are then used to fill the gaps. Experiments using recent data with coverage limited to that of the late nineteenth century demonstrate that the method is skilful, and has the capability of removing much of the bias towards small anomalies inherent in GISST1.1 and 2.1. However there is an underlying assumption that the eigenvector patterns derived from recent data are applicable to earlier years. The planned application is to create new analyses such as GISST2.2 for forcing simulations with atmospheric general circulation models. However, better analyses of, for example, SST in El Niño events, may, *subject to the above assumption*, also assist assessments of whether such aspects of climate are becoming more or less variable. A disadvantage of the eigenvector method, compared to optimum interpolation, is that no intrinsic error-estimate is made. So a desirable goal would be to combine the optimum interpolation method as currently practiced (Reynolds and Smith, 1994), with the eigenvector interpolation technique.

Historical SST and NMAT data are crucial, not only to analysis of climatic variations and detection of climate changes, but also to climate prediction. This arises firstly because the SSTs are used to force atmospheric general circulation models in the simulation of recent climate, thereby enabling validation of the atmospheric models: observed historical variations in NMAT can contribute to the validation. In addition, coupled ocean-atmosphere model simulations of SST and air temperature over the oceans can be compared with the observed SST and NMAT, supporting efforts to validate the oceanic heat fluxes and the air-sea interactions occurring in the models. Both improved SST and NMAT data bases are important for estimating heat fluxes. In future, coupled model integrations will need to be started from a realistic SST climatology, such as that for the mid-nineteenth century, before greenhouse forcing was significant. This puts extra emphasis on the SST data for that period. A good SST data base is also essential for studies leading to understanding and prediction of regional seasonal or multi-seasonal climatic variations, for example of precipitation over sub-Saharan Africa (Rowell *et al.*, 1995). Finally, SST is an essential input to models for the planned atmospheric reanalyses covering the past 40 years or more (World Meteorological Organization (WMO), 1992).

Acknowledgements

R. Hackett and M. O'Donnell contributed most of the computer graphics for this paper.

References

Alexander, R. C. and Mobley, R. L.: 1976, 'Monthly Average Sea-Surface Temperatures and Ice-Pack Limits on a 1° Global Grid', *Mon. Wea. Rev.* **104**, 143–148.

Barnett, T. P.: 1984, 'Long-Term Trends in Surface Temperature over the Oceans', *Mon. Wea. Rev.* **112**, 303–312.

Barton, I. J., Prata, A. J., and Cechet, R. P.: 1995, 'Validation of the ATSR in Australian Waters', *J. Atmos. Oceanic Tech.* **12**, 290–300.

Bottomley, M., Folland, C. K., Hsiung, J., Newell, R. E., and Parker, D. E.: 1990, *Global Ocean Surface Temperature Atlas (GOSTA)*, Joint Meteorological Office and Massachusetts Institute of

Technology Project, Project supported by U.S. Dept. of Energy, U.S. National Science Foundation and U.S. Office of Naval Research, Publication funded by U.K. Depts. of the Environment and Energy, HMSO, London, 20 + iv pp. and 313 Plates.

Elms, J. D., Woodruff, S. D., Worley, S. J., and Hanson, C. S.: 1993, 'Digitizing Historical Records for the Comprehensive Ocean-Atmosphere Data Set (COADS)', *Earth System Monitor* 4, No. 2, 4–10.

Folland, C. K. and Kates, F. E.: 1984, 'Changes in Decadally Averaged Sea-Surface Temperature over the World 1861–1980', in Berger, A. *et al.* (eds.), *Milankovitch & Climate, pt 2*, D. Reidel, pp. 721–727.

Folland, C. K. and Parker, D. E.: 1990, 'Observed Variations of Sea Surface Temperature', in Schlesinger, M. E. (ed.), *Climate-Ocean Interaction*, Kluwer Academic Press, Dordrecht, pp. 21–52.

Folland, C. K. and Parker, D. E.: 1992, 'The Instrumental Record of Surface Temperature: How Good Is It and What Can It Tell Us About Climate Change and Variability', *5th Internat. Mtg. Statist. Clim. and 12th Amer. Meteor. Assoc. Conf. Probability and Statistics in the Atmospheric Sciences*, Toronto, 22–26 June 1992, pp. J1–J6.

Folland, C. K. and Parker, D. E.: 1995, 'Correction of Instrumental Biases in Historical Sea Surface Temperature Data', *Quart. J. Roy. Meteor. Soc.* 121, 319–367.

Folland, C. K., Parker, D. E., and Kates, F. E.: 1984, 'Worldwide Marine Surface Temperature Fluctuations 1856–1981', *Nature* 310, 670–673.

Folland, C. K., Reynolds, R. W., Gordon, M., and Parker, D. E.: 1993, 'A Study of Six Operational Sea Surface Temperature Analyses', *J. Clim.* 6, 96–113.

Gandin, L. S.: 1963, *Objective Analysis of Meteorological Fields*, Gidrometeoizdat, St. Petersburg, in Russian, (English Translation, 1966, Israeli Program for Scientific Translations, Jerusalem, 242 pp.).

Graham, N. E.: 1994, 'Decadal-Scale Climate Variability in the Tropical and North Pacific during the 1970s and 1980s: Observations and Model Results', *Clim. Dyn.* 10, 135–162.

Intergovernmental Panel on Climate Change: 1992, 'Climate Change 1992, The Supplementary Report to the IPCC Scientific Assessment', Houghton, J. T., Callander, B. A., and Varney, S. K. (eds.), WMO/UNEP, Cambridge University Press, 200 pp.

James, R. W. and Fox, P. T.: 1972, 'Comparative Sea-Surface Temperature Measurements', *Marine Sci. Affairs Reprot* No. 5, WMO No. 336, Geneva, 27 pp.

Jones, P. D.: 1994, 'Hemispheric Surface Air Temperature Variations: A Reanalysis and an Update to 1993', *J. Clim.* 7, 1794–1802.

Jones, P. D., Wigley, T. M. L., and Farmer, G.: 1991, 'Marine and Land Temperature Data Sets: A comparison and a Look at Recent Trends', in Schlesinger, M. E. (ed.), *Greenhouse-Gas-Induced Climatic Change: A Critical Appraisal of Simulations and Observations*, Elsevier, Amsterdam, pp. 153–172.

Jones, P. D., Wigley, T. M. L., and Wright, P. B.: 1986, 'Global Temperature Variations between 1861 and 1984', *Nature* 322, 430–434.

Kagan, R. L.: 1979, *Averaging Meteorological Fields*, Gidrometeoizdat, St. Petersburg, 212 pp., in Russian, (Draft English translation available at the U.K. Meteorological Office and in the NOAA Climate Analysis Center).

Karl, T. R., Knight, R. W., and Christy, J. R.: 1994, 'Global and Hemispheric Temperature Trends: Uncertainties Related to Inadequate Spatial Sampling', *J. Clim.* 7, 1144–1163.

Kent, E. C., Taylor, P. K., Truscott, B. S., and Hopkins, J. S.: 1993, 'The Accuracy of Voluntary Observing Ships' Meteorological Observations – Results of the VSOP-NA', *J. Atmos. Oceanic Tech.* 10, 591–608.

Maury, M. F.: 1958, *Explanations and Sailing Directions to Accompany the Wind and Current Charts*, Vol. 1, Printed by W. A. Harris, Washington DC, 383 + xxxvi pp and 51 Plates.

Paltridge, G. and Woodruff, S.: 1981, 'Changes in Global Surface Temperature from 1880 to 1977 Derived from Historical Records of Sea Surface Temperature', *Mon. Wea. Rev.* 109, 2427–2434.

Parker, D. E.: 1992, 'Blending of COADS and UK Meteorological Office Marine Data Sets', Proc. International COADS Workshop, Jan 1992, Boulder, Co., U.S.A., Diaz, H. F., Wolter, K., and Woodruff, S. D. (eds), NOAA, pp. 61–72.

Parker, D. E.: 1994, 'Effects of Changing Exposure of Thermometers at Land Stations', *Int. J. Climatol.* **14**, 1–31.

Parker, D. E. and Folland, C. K.: 1991, 'Worldwide Surface Temperature Trends since the Mid-19th Century', in Schlesinger, M. E. (ed.), *Greenhouse-Gas-Induced Climatic Change: A Critical Appraisal of Simulations and Observations*, Elsevier, Amsterdam, pp. 173–193.

Parker, D. E., Jones, P. D., Folland, C. K., and Bevan, A.: 1994, 'Interdecadal Changes of Surface Temperature since the Late Nineteenth Century', *J. Geophys. Res.* **99**, 14373–14399.

Parker, D. E., Folland, C. K., Bevan, A., Ward, M. N., Jackson, M., and Maskell, K.: 1995, 'Marine Surface Data for Analysis of Climatic Fluctuations on Interannual to Century Time Scales', in Martinson, D. G. *et al.* (eds.), *Natural Climate Variability on Decade-to-Century Time Scales*, National Acad. Press, Washington, DC, (in press).

Rayner, N. A., Ward, M. N., Parker, D. E., and Folland, C. K.: 1995, 'Using EOF Analysis to Create a New GISST Data Set for Forcing GCMs', in Folland, C. K. and Rowell, D. P. (eds.), *Workshop on Simulations of the Climate of the Twentieth Century Using GISST, 28–30 November 1994, Hadley Centre, Bracknell, UK*, Climate Research Technical Note CRTN56, available from the Hadley Centre, pp. 52–53.

Reynolds, R. W.: 1988, 'A Real-Time Global Sea Surface Temperature Analysis', *J. Clim.* **1**, 75–86.

Reynolds, R. W.: 1993, 'Impact of Mount Pinatubo Aerosols on Satellite-Derived Sea Surface Temperatures', *J. Clim.* **6**, 768–774.

Reynolds, R. W., Folland, C. K., and Parker, D. E.: 1989, 'Biases in Satellite-Derived Sea-Surface-Temperature Data', *Nature* **341**, 728–731.

Reynolds, R. W. and Smith, T. M.: 1994, 'Improved Global Sea Surface Temperature Analyses Using Optimum Interpolation', *J. Clim.* **7**, 929–948.

Reynolds, R. W. and Smith, T. M.: 1995, 'A High Resolution Global Sea Surface Temperature Climatology', *J. Clim.* **8**, 1571–1583.

Rowell, D. P., Folland, C. K., Maskell, K., and Ward, M. N.: 1995, 'Variability of Summer Rainfall over Tropical North Africa (1906–92): Observations and Modelling', *Quart. J. Roy. Meteor. Soc.* **121**, 669–704.

Saunders, R. W., Smith, A. H., and Harrison, D. L.: 1993, 'Sea-Surface Temperature Measurements by the ATSR', *Meteor. Mag.* **122**, 105–113.

Schluessel, P., Emery, W. J., Grassl, H., and Mammen, T.: 1990, 'On the Bulk-Skin Temperature Difference and Its Impact on Satellite Remote Sensing of Sea Surface Temperatures', *J. Geophys. Res.* **95**, 13341–13356.

Smith, T. M., Reynolds, R. W., and Ropelewski, C. F.: 1994, 'Optimal Averaging of Seasonal Sea Surface Temperatures and Associated Confidence Intervals (1860–1989)', *J. Clim.* **7**, 949–964.

Trenberth, K. E., Christy, J. R., and Hurrell, J. W.: 1992, 'Monitoring Global Monthly Mean Surface Temperatures', *J. Clim.* **5**, 1405–1423.

Trenberth, K. E. and Hurrell, J. W.: 1994, 'Decadal Atmosphere-Ocean Variations in the Pacific', *Clim. Dyn.* **9**, 303–319.

WMO: 1956 *et seq.*, *International List of Selected, Supplementary and Auxiliary Ships*, WMO No. 47, Geneva, published annually.

WMO: 1992, *Report of the Seventh Session of the CAS/JSC Working Group on Numerical Experimentation*, WMO/ICSU, WCRP-70, WMO/TD No. 477, Geneva, pp. 27–28.

Woodruff, S. D., Slutz, R. J., Jenne, R. L., and Steurer, P. M.: 1987, 'A Comprehensive Ocean-Atmosphere Data Set', *Bull. Amer. Meteor. Soc.* **68**, 1239–1250.

Wright, P. B.: 1986, 'Problems in the Use of Ship Observations for the Study of Interdecadal Climate Changes', *Mon. Wea. Rev.* **114**, 1028–1034.

(Received 23 January 1995; in revised form 21 June, 1995)

Appendix I

Quality Control Limits for real time and historical SST and NMAT data.
Formation of 1-degree-area pentad mean anomalies

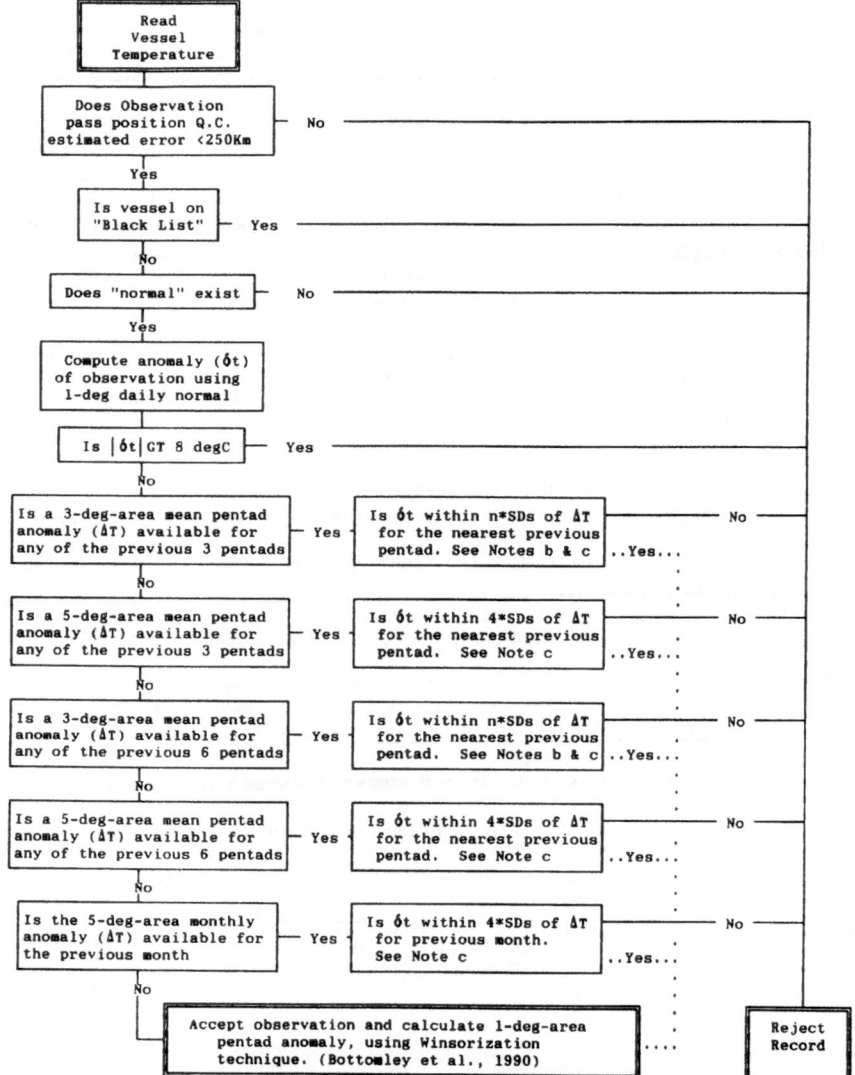

Notes:
a. The longitude of the area is increased by $[\cos(\text{lat})]^{-1}$ to allow equal areas to be compared.
b. If the number of observations in the pentad 1-deg-area is   1 -   5:   n = 4.0
                                                              6 -   15:   n = 3.5
                                                          16 - 100:   n = 3.0
                                                           GE 101:   n = 2.5
c. Standard Deviation SD previously calculated for 1961-90 for each 1-deg-area and pentad.

[470]

DOCUMENTING AND DETECTING LONG-TERM PRECIPITATION TRENDS: WHERE WE ARE AND WHAT SHOULD BE DONE

PAVEL YA. GROISMAN

Dept. of Geosciences, University of Massachusetts, Amherst, MA, 01002, U.S.A.

and

DAVID R. LEGATES

College of Geosciences, University of Oklahoma, Norman, OK, 73019, U.S.A.

Abstract. A brief review of problems and achievements in documenting precipitation changes during the period of instrumental measurements is presented. Concern is expressed that without appropriate studies in the coming period of a new generation of precipitation measurements, technological progress in instrumentation may adversely and inadvertently affect our capability for monitoring and detecting future changes in terrestrial precipitation. At the same time, only a new generation of instrumentation will be capable of resolving the problems of monitoring precipitation over oceans.

Special attention is paid to validation of the increasing trend in terrestrial precipitation observed during the past hundred years at high latitudes of the northern hemisphere – a feature of global warming predicted by most climate models.

Introduction

What will be the impact of climate change on the hydrological cycle? The answer to this question probably has more importance to our society than an increase of global air temperature. While these changes in the hydrological cycle and global air temperature are certainly related, the changes in water supply directly affect many human activities and may have a more substantial economic impact.

Precipitation is a key element of this cycle and the need to monitor it and document its variability has been realized since ancient times. For limited periods of time, rain and snow gages existed in India, China, and Palestine more than two thousand years ago (Biswas, 1970; Khrgian, 1948). The earliest moderately long, regular, precipitation record for period 1677–1704 at Burnley, England is still largely accessible (Folland and Wales-Smith, 1977). Precipitation measurements that have been taken at the same location for more than two hundred years are not rare in Europe (Tabony, 1981) and at present, there are about 10^5 rain gages worldwide. Biases of precipitation measurements and an absence of a reliable global archive of long-term homogeneous precipitation time series, however, create problems for monitoring large-scale precipitation variability and the detection of changes in precipitation. In this paper, we describe the present status of precipitation measurements and try to outline the major problems that still prevent or will prevent detection of climate change in precipitation.

Climatic Change **31**: 601–622, 1995.

The Role of Precipitation in Climate Change Detection

Terrestrial precipitation plays an important role in affecting and/or restricting a variety of human activities. Changes in precipitation (if they do occur) will be, therefore, of fundamental importance. *They require special consideration regardless of their role in detecting climatic change and attributing specific causes to it.*

Many scientists have reasoned that any global warming should be accompanied by the general intensification of hydrological cycle (cf., Budyko and Drozdov, 1976; IPCC, 1990, 1992, 1995). Most general circulation models (GCM) of the atmosphere and ocean foresee the prominent changes in precipitation occurring in the high latitudes (cf., Manabe *et al.*, 1991, 1992; Murphy and Mitchell, 1995). This also is supported by paleoclimatic reconstructions (Budyko and Izrael, 1987; Borzenkova, 1994) and by the analyses of precipitation trends over the past century (Bradley *et al.*, 1987; Diaz *et al.*, 1989; Groisman, 1991a, b; Groisman and Easterling, 1994). Thus, our knowledge of precipitation patterns based on observational data has already contributed a little towards the detection of climate changes. Forecasting these changes in the future, however, is extremely difficult.

On a regional scale, GCMs still are uncertain about the range and even the sign of the change in precipitation (cf., Mitchell *et al.*, 1990). When a regional change is considered, therefore, GCM prognostications are not reliable and existing precipitation data are adversely affected by problems that prevent their reliable use in climate change studies. The key point is, however, that unless we have the theoretical patterns of the changes we should expect, any drastic precipitation change that affect our society and even life itself (e.g., the Sahel drought) will be difficult to ascribe to the enhanced greenhouse effect.

In this paper, we will not recount the problems associated with GCM simulations of precipitation, but rather will discuss the current status of data measurement problems associated with precipitation and possible ways to resolve them. It should be noted, however that, *we do not identify precipitation as the highest priority for the detection of anthropogenic induced climate change, but a relatively important one.*

Traditional Precipitation Measurements

Precipitation is one of the key components of hydrological cycle although measuring it can be quite difficult due to a lack of a sufficient spatial coverage, biases associated with the gage measurement process, and homogeneity of precipitation time series. All three of these problems are critical to obtaining accurate estimates of precipitation variability and temporal trends.

The spatial distribution of terrestrial precipitation measurements is not uniform since far more gages are located along the coastlines and near urban areas. Even in

Fig. 1. Distribution of terrestrial precipitation stations in the Legates (1987) and Legates and Willmott (1990) climatologies (adapted from Legates, 1987).

the United States where gage densities are often greatest, the distribution of stations is inadequate to resolve the spatial variability over the Rocky Mountains and the desert southwest (cf., Figure 1). This apparent availability of data is misleading, however, in that while long-term means generally are available for the station distributions given in Figure 1 (although some additional sources could be used to greatly enhance the networks of the United States, Australia, and many European countries), long-term station time series of precipitation are available only for a small subset of these stations (Figure 2). Thus for climate change studies, much fewer stations are available for the analysis of possible trends.

Over the oceans, precipitation observations are very sparse and exist only for selected island and atoll stations. Ship-borne precipitation measurements do not provide for a continuous record (except for a handful of fixed-position ships) and are subject to biases resulting from the roll and pitch of the ship, spray from the surf, and a fair-weather bias (Quayle, 1974). Indirect techniques have been developed to estimate precipitation from surface synoptic reports recorded on board ships (e.g., Tucker, 1961; Reed and Elliott, 1977; Dorman and Bourke, 1978, 1979, 1981), from nearby land-based gages (e.g., Eliott *et al.*, 1971; Elliott and Reed, 1973; Reed and Elliott, 1973; Reed, 1980), or through generalized precipitation frequency approaches (e.g., Jacobs, 1968; Reed, 1979, Reed and Elliott, 1979; Bogdanova, 1986). Such techniques, however, often have considerable limitations owing largely to the fact that surface synoptic reports are affected by the fair-weather bias of ships and offshore precipitation can differ considerably from precipitation over adjacent land areas (Dorman, 1982). Satellite estimates, using both the microwave and infrared portions of the electromagnetic spectrum, will possibly remedy this situation in the future (cf., Rao and Theon, 1977; Adler and Mack, 1984; Simpson

Fig. 2.   Meteorological network of precipitation measurements provided by the first version of the Global Historical Climatology Network (Vose *et al.*, 1992).

et al., 1988; Arkin and Ardanuy, 1989; Wilheit *et al.*, 1991; Janowiak, 1992; Spencer, 1993; Janowiak *et al.*, 1995) but satellite estimates still rely on an adequate correlation between the measured gage catch (and its associated bias both in space and in measurement) and the received energy. Even if this correlation is found, biases in the gage catch and doubtful representativeness of the islands for the surrounding ocean will still adversely affect our estimates of oceanic precipitation. We are not aware of any large-scale attempts to analyze ship rain gage data except for some internal seas (cf., Golitsyn *et al.*, 1990).

Biases in precipitation gage measurements are substantial as has been documented by many researchers over the past century (see Kurtyka (1953); Larson (1971); and WMO (1973) for a bibliography of many early researchers). These systematic biases can be attributed to seven sources: the effect of the wind, wetting losses, evaporation from the gage, splashing effects, blowing and drifting snow, the treatment of trace precipitation events, and the impact of automatic recording techniques (Bogdanova, 1966; Rodda, 1971; Larson and Peck, 1974; Sevruk, 1982; Legates, 1987; Folland, 1988). Random biases also exist and add an unsystematic component. The effect of the wind is the largest source of gage undercatch bias (Golubev, 1965, 1969; Allerup and Madsen, 1980; Sevruk, 1982) and accounts for a decrease in the annually-averaged, global precipitation of about eight percent (Legates, 1987). This error tends to increase as latitude and elevation increase due to the greater impact of wind on snowfall. The reason for this wind-induced bias arises from the fact that a precipitation gage acts as an obstruction to the wind which forces air to flow around the sides and top of the gage. Near the gage orifice, air flows over the top of the gage producing a small updraft and the convergence of the horizontal wind immediately adjacent to the gage orifice caused by the flow

TABLE I

Biases in the measurements of the standard national gages of the United States (8-inch non-recording U.S. raingage with and without Alter wind shield), Canada (Nipher shielded snow gage), and Russia (Tretiyakov shielded gage), in % of the 'ground truth' precipitation as revealed at the Valdai experimental site, Russia (adapted from Golubev et al., 1995). These biases vary from site to site depending mostly from the gage exposure and wind speed and are extremely large for solid precipitation measured by unshielded gages

Country	Shield on the gage	Precipitation type		
		Snow	Mixed	Rain
Canada	Yes	16	13	7
Russia	Yes	42	23	8
United	Yes	34	23	10
States	No	61	36	14

of air over the top of the gage both lead to an increase in the wind speed across the gage orifice. Wind tunnel experiments have indicated that the wind speed increases by as much as twenty percent across the entire orifice (Sevruk, 1988). Thus, the increase in wind speed coupled with the small updraft of wind lead to a decrease in the measured precipitation. Since national standard precipitation gages vary in size, shape, and designs as well as in the elevation of their orifice above ground level, the effect of the wind is gage-dependent. Consider, for example, Table I where the biases for rain, snow, and mixed precipitation events are given for four gage and shield designs. While the biases for snowfall events are much greater than those for rainfall owing to the fact that snowflakes are much more affected by the wind than are raindrops (and similarly, light drizzle is more affected by wind than heavy rain showers, cf., Folland, 1988), note that they differ considerably among the different gages. This has significant ramifications for climate change studies since gage designs have been changed in many countries throughout the last century.

Inhomogeneities in the precipitation time series caused by changes in instrumentation and recording practices, siting characteristics, and station location (Eischeid et al., 1991; Karl et al., 1993b) also may adversely affect precipitation time series, particularly for climate change studies. As improvements in gage designs have been made, the adoption of new gages have introduced a marked discontinuity into the precipitation time series of various countries. For example, the Soviet Union changed from the Nipher-shielded to the Tretyakov-shielded gage between 1948 and 1953 while the United States adopted the use of Alter wind shields for some stations, particularly in the northwest, in the 1940s (Groisman, 1991b;

Groisman *et al.*, 1991a, b). Canada, Japan, Finland, Norway, Poland, Iceland, Sweden, Czechoslovakia, and Switzerland also have changed gage designs or shields in the past century (Sevruk and Klemm, 1989; Karl *et al.*, 1993b). These changes affect the magnitude of the bias in the gage measurement and, as a result, make climate change detection difficult.

Changes in the standard height of the gage orifice also is one of the most important factors that affect the homogeneity of a precipitation time series. Since the 1940s, the elevation of the gage for the national precipitation gage network has been decreased in both The Netherlands and China. This leads to an apparent increase in the actual precipitation since the wind-induced error increases as the wind speed increases which, in turn, increases with height. These inhomogeneities are particularly significant when the gage is moved from or to the roofs of buildings (Groisman *et al.*, 1991b). Changes in standard recording practices, particularly with respect to the measurement of snowfall, also adversely affect the accuracy of precipitation measurements (Goodison, 1981; Metcalfe and Goodison, 1992).

The environment surrounding the gage may change considerably over time. Since each of these changes will affect the wind speed across the gage orifice, the gage catch will be adversely affected (cf. Eischeid *et al.*, 1991). Immediate changes to the gage environment (caused, for example, but the cutting down of trees) can produce a discontinuity in the station record which may be detectable through visual inspection or double-mass analysis with a nearby station. More gradual changes (such as the growth of trees over time or urban development) may introduce a time-varying bias which can be difficult to detect by looking for discontinuities in the record.

Some precipitation stations will be relocated during the time period of record. Movement of downtown (urban) stations to airports and other rural areas has been particularly important (Eischeid *et al.*, 1991; Groisman, 1991b; Groisman *et al.*, 1991b). Such relocations can introduce a discontinuity into the station record resulting from changes in the local environment. The combined effects of these potential discontinuities in the station record instrumental and observational changes, variations in siting characteristics, and station relocation can yield a misleading picture of temporal climate changes in precipitation.

Impact of Biases and Inhomogeneities on the Precipitation Change Studies

A significant portion of precipitation variability occurs on the micro-scale and spatial coherence is quite small (Gandin *et al.*, 1976). For example, two identical gages located 100 meters apart may differ in total precipitation by a factor of two for a single storm (Golubev, 1975). Correlation radii (the e-folding distance for the correlation between two time series) for monthly precipitation can be as large as several hundred kilometers over level terrain, but can decrease to as little as 30 km in mountainous terrain (Huff and Shipp, 1969; Golubev, 1965, 1975; Nørdø and

Hjortnaes, 1967; Gandin *et al.*, 1976; Groisman and Easterling, 1994; Groisman and Legates, 1994). As a result, if we intend to draw conclusions that indeed are representative of large areas (country, continents, or the globe) a much more dense network is required than with surface air temperature or sea-level pressure.

The precipitation network is subject to considerable changes in station distributions over time. Willmott and Legates (1991) examined the impact of these temporally-varying network densities on estimates of terrestrial precipitation by randomly sampling from the 24,635 stations in the Legates (1987) and Legates and Willmott (1990) climatology to produce sub-networks consisting of sample sizes ranging from 200 to 2000 stations. Their approach specifically applied a *formal* averaging procedure to show the effect of station distribution on spatially-averaged precipitation totals. In practice, climatologists may use their knowledge of spatial distribution of precipitation and/or averaging of anomalies to avoid these biases (United States Dept. of Commerce, 1968; WMO, 1979; WWB, 1974; Bradley *et al.*, 1987; Groisman, 1991a, b, etc.) but quite often such is not the case. For smaller network densities, Willmott and Legates (1991) found that precipitation is generally overestimated, while as sample size increases, the network bias decreases. This overestimation of precipitation with smaller networks occurs because gages tend to be co-located with urban development which is biased toward wetter regions. Willmott *et al.* (1994) further concluded that these results were valid for all continents except South America where underestimates occur with sparse networks due to an inadequate sampling of the Amazon Basin in most earlier gage networks. Willmott and Legates (1991) also computed long-term mean terrestrial precipitation from the Legates and Willmott climatology using the yearly station distribution given by NCAR World Monthly Surface Station Climatology for 1900 through 1984 (Spangler and Jenne, 1990). As the number of stations increased from 1900 to about 1945, the network bias in the estimate of the area-averaged long-term precipitation decreased. At the same time, sparse gage networks may 'overlook' individual storm events and, thus, produce lower values of short-term (day to month) terrestrial precipitation in dry climates (Kay and Kutiel, 1994). Therefore, region-scale estimates of long-term (climatological) terrestrial precipitation made from the gage record using objective interpolation procedures, may be considerably overestimated prior to the 1930s due to the sparseness of the gage network although this overestimation may be mitigated in arid climates owing to the small spatial coherence of infrequent storms in these regions.

As previously discussed, conventional precipitation measurements are quite sensitive to the changes in the gage surroundings, gage design, and its installation, which can result in a substantial bias (Golubev, 1969, 1993; Sevruk, 1982; World Water Balance, 1974; Legates, 1987; Legates and Willmott, 1990; WMO, 1991; Groisman and Legates, 1994). This bias would not be important for the detecting of the climate change if this bias were constant or even if it were to randomly fluctuate. It is not a constant, however, and thus a major problem with documenting changes in precipitation is the temporary variability of the bias which introduces artificial trends in the time series.

No accepted standards in precipitation instrumentation are used on an operational basis. Pit gauges and double fence shielded gauges that have been recommended by the World Meteorological Organization for international intercomparisons of liquid (Sevruk and Hamon, 1984) and solid (WMO, 1991; Goodison *et al.*, 1992) precipitation measurements, respectively, have never been introduced into the national network of any countries because of difficulties associated with their maintenance. Instead, every country has chosen to develop and modernize its own gage design. A 'more accurate' gage that replaces the old national gage may record systematically different precipitation totals. When such changes in instrumentation are introduced nationwide, large-scale precipitation changes may be documented which are not associated with climate change (see Figures 3 and 4). This has occurred often over the past century in many countries of the world (cf., Sevruk and Klemm, 1989; Karl *et al.*, 1993b) and adversely affect our ability to filter climate-related signals from the natural variability of precipitation over the last century.

Moreover, gage-measured precipitation is affected by other meteorological variables including the wind speed, type of precipitation (solid, liquid, or mixed), and air temprature (which affects the structure of solid precipitation), in addition to gage exposure and gage type. Climate changes also certainly will affect some (or all) of these meteorological variables. These changes may mask or enhance the change in measured precipitation by affecting the bias in the gage measurement, even when the true precipitation does not change. Building on the hypothetical analysis of Legates (1992), Legates (1995b) presented an example of the consequences of moderate warming and urbanization on gage measurements of precipitation which illustrates how they may adversely affect our judgment of the precipitation change. He used the standard United States rain gage installed on an open site in a climate comensurate with the upper Great Plains (Madison, Wisconsin) and estimated mean air temperature, precipitation, and wind speed and their variability for the period of instrumental observations. February climate conditions (when approximately half of the precipitation falls as snow) were simulated for two scenarios; one of an unchanging climate (the control) and the second exhibiting an increase of air temperature by 1 °C per 100 years (a moderate warming trend) and wind speed at anemometer height (33 feet or 10.0584 m) decreasing by 1.0 knot (0.514 m s^{-1}) per 100 years (to simulate a possible urban effect due to city growth or weakening of atmospheric circulation due to warming in the mid-latitudes). Although the 'true' distribution of precipitation was unchanged in this 'climate change scenario' simulation, the bias in the measured precipitation decreased by 8% in 100 years and, during the last 20 years of simulation, the measured precipitation was 13% higher than the control scenario (cf., Figure 5). While these changes occurred on a background of high natural variability, the 'perceived' increase in measured pre-

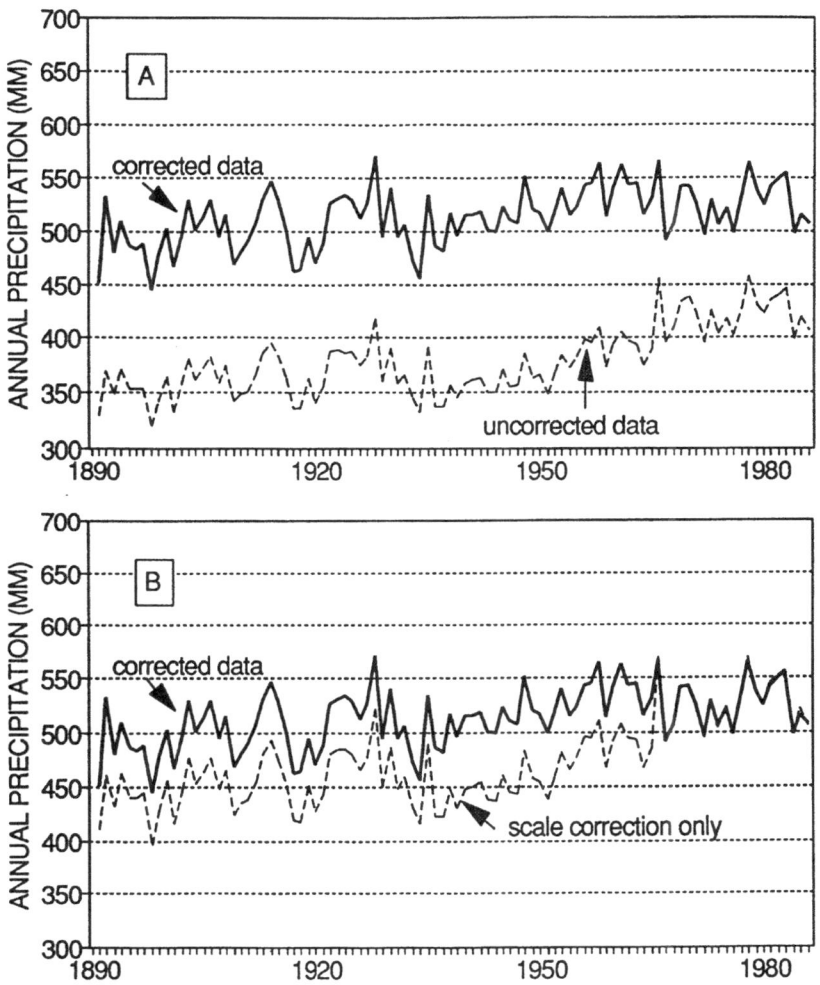

Fig. 3.   Annual precipitation estimates for the former U.S.S.R. based on corrected and uncorrected data of the international exchange. The corrections applied are those used by Groisman *et al.* (1991a). On the b-graph the scale correction used in corrected data has been applied to uncorrected data also (adapted from Groisman, 1991b). *Enormous biases and inhomogeneities of unadjusted data may mislead any unaware user who tries to study precipitation change over north Eurasia.*

cipitation for the climate change scenario nevertheless could be easily mistaken for an increase in precipitation.

Studies that analyze changes in precipitation require (a) relatively dense networks to reduce the effects of the microscale and weather 'noise' on the results; and (b) careful preprocessing of the data to avoid the effect of large-scale inho-

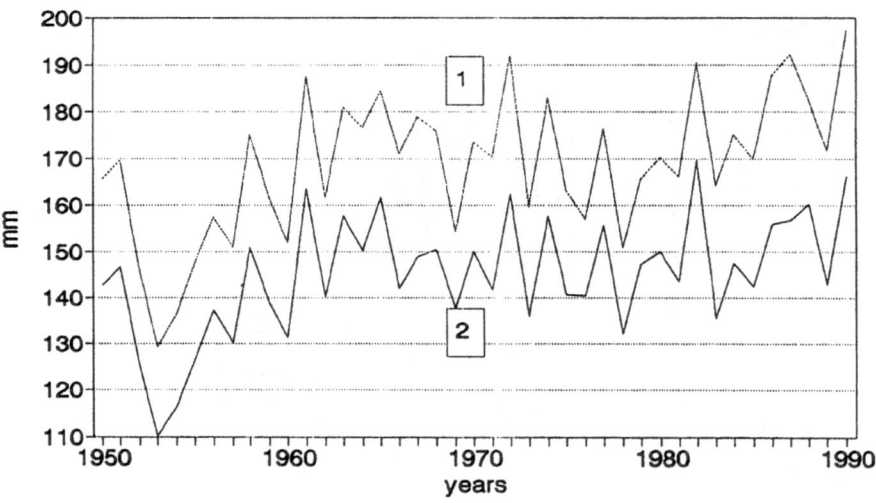

Fig. 4.   Annual solid precipitation over northern Canada (from 55°–70° N). These estimates are based on: (1) the snow-stick measurements; and (2) on the elevated gage measurements at the same primary Canadian network (adapted from Groisman, 1993). Without wide announcement, the primary Canadian network started to report solid precipitation using elevated snow gages since the early 1960s (previously only snow stick measurements were in use). *Although the old and new methods are quite closely correlated, the systematic difference may mislead unaware users of Canadian precipitation data.* The situation is complicated by the fact that the residual Canadian network (except about 300 primary stations) still use snow stick measurements to report solid precipitation.

mogeneities and the contribution of changes in other meteorological variables on conclusions about climate changes in precipitation.

Attempts to Overcome the Problems with Precipitation Data

The influence of the micro-scale variability of storm total precipitation can be substantially reduced when a large number of stations are used to analyze climate change (Bradley *et al.*, 1987; Diaz *et al.*, 1989; Groisman and Easterling, 1994). A knowledge of the spatial structure of the precipitation field (Gandin *et al.*, 1976; Huff, 1979; Huff and Shipp, 1969) and the theory of optimal spatial averaging (Kagan, 1979; Gandin, 1993) provide a reliable guidance of the necessary density of a precipitation network to be used for different applications (cf., Czelnai *et al.*, 1963; Hendrick and Comer, 1970; Kagan, 1972; Peck and Schaake, 1990). Experience with climate change studies (Bradley and Groisman, 1989; Groisman and Easterling, 1994; Groisman and Legates, 1994; Hulme, 1995) shows that the large scale changes in annual precipitation over land areas can be documented when

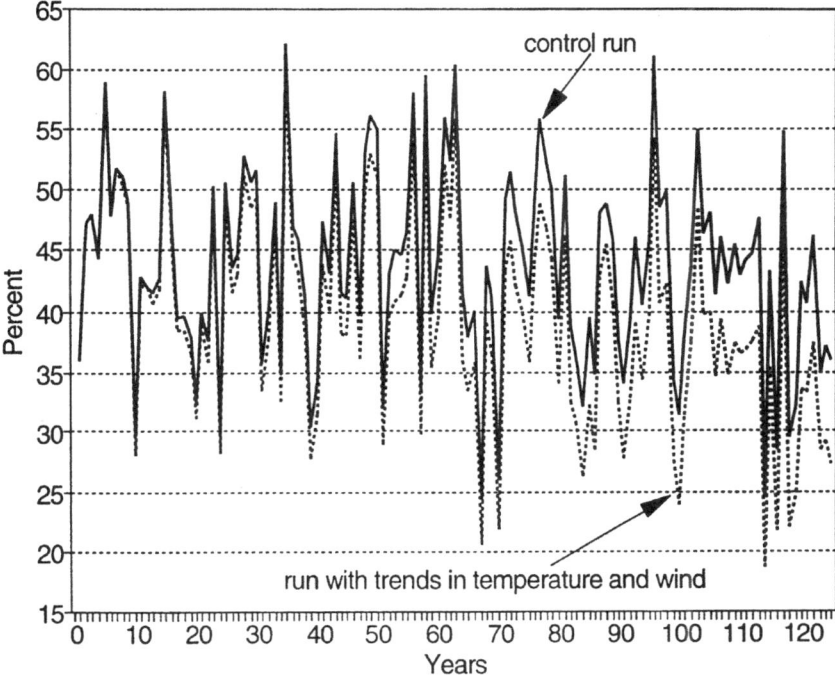

Fig. 5.   Variability of bias in the gage-measured February precipitation at Madison, Wisconsin for: (a) an unchanged climate; and (b) for a climate change scenario: the increase of temperature by 1 °C and the decrease of wind speed at the anemometer height by 1.0 knot (0.5144 m s^{-1}) per century. The 'true' precipitation remains *unchanged* in both scenarios. A systematic decrease in the bias of the measured precipitation (and corresponding increase in the measured precipitation) occurred due to climate change that is not associated with the change of the 'ground truth' precipitation (adapted from Legates, 1995b).

data of several thousand stations are used – that is, five to ten times more stations than are required for documenting global changes in surface air temperature (cf., Vinnikov *et al.*, 1990).

Problems in precipitation measurement, specifically with respect to the measurement of solid precipitation, have been carefully considered by several countries. For example, the national weather services of Norway (Hanssen-Bauer and Førland, 1994), Sweden (Dahlström, 1986; Alexandersson, 1994), and Finland (Heino, 1994) have created subsets of long-term homogenous precipitation time series for their countries that are appropriate for climate change studies. Similarly, the United Kingdom Meteorological Office (Tabony, 1980, 1981) has made a concerted effort to restore the homogeneity of precipitation time series over the past two hundred years for much of Europe. These initial efforts are being continued by Deutscher Wetterdienst on a global scale (Rudolf *et al.*, 1992) and by the joint efforts of eleven European countries for the Northern Europe (Frich *et al.*, 1994). In Switzerland, a Hydrological State Atlas was also prepared (Spreafico *et al.*, 1992). For its production, the entire data set of precipitation measurements was reanalyzed and all sources of bias were removed (Sevruk *et al.*, 1993). The experience of decadal-long studies in Russia were summarized in a series of publications (Shver, 1976, 1984; World Water Balance, 1974) and a subset of century-long monthly precipitation time series that have incorporated adjustments necessary for climate change analyses (Groisman *et al.*, 1991a).

In North America, the reanalysis of Canadian precipitation data made available recently by Karl *et al.* (1993a) and Groisman and Easterling (1994) revealed that significant precipitation changes were evident after the instrumental inhomogeneities were taken into account and the representativeness of the network was tested. Most of these results have been confirmed by an independent analysis by Canadian scientists (IPCC, 1995). The United States National Climatic Data Center has created and maintained a baseline subset of long-term precipitation time series (Historical Climatology Network, HCN) that has been appended by an extensive metadata archive (Karl *et al.*, 1990; Groisman *et al.*, 1995, 1994). The use of this metadata can provide a proper assessment of data homogeneity problems for more than 1500 stations in the United States. The use of the HCN itself provides reliable first-cut estimates of precipitation trends over the United States (Groisman and Easterling, 1994; Lettenmaier *et al.*, 1994).

More than 300 instrumentally homogeneous precipitation time series (since 1950) have been prepared by the National Meteorological Service of China and incorporated into the Global Historical Climatology Network (GHCN – Vose *et al.*, 1992). Reliable sources of information on precipitation during the period of instrumental observations also are available for Africa (Nicholson, 1995), Australia (Lavery *et al.*, 1992), and India (Parthasarathy *et al.*, 1992, 1994). These databases together with efforts to create global precipitation data sets (Diaz *et al.*, 1989; Vose *et al.*, 1992; Hulme, 1992; Eischeid *et al.*, 1991) have begun to yield significant dividends. *From the reports of the Intergovernmental Panel on Climate Change, we point out there is more and more confidence in answering questions such as "What large-scale changes of precipitation happened during the period of instrumental observations over the most of the land areas of the globe?"* (cf., IPCC, 1990, 1992, 1995).

An Increasing Danger to Our Ability to Monitor and Detect Precipitation Changes

Changes in observational methods pose a substantial problem for our ability to monitor and detect precipitation trends. In the past, each such change has introduced significant inhomogeneities into the precipitation record and the national meteorological services of many countries have developed different solutions to this problem. These solutions vary from the development of costly systems of parallel observations of both the new and old techniques (e.g., Russia and Norway) to a nearly complete neglect of the problem.

At present, the requirements of a better spatial and temporal coverage of precipitation have resulted in a generation of new precipitation measuring techniques which include different types of recording gages, optical rain gages, and remotely-sensed estimates of precipitation (satellites and radars). Unfortunately, the principles that underlie these new techniques are quite diverse and, as a consequence, the precipitation measurements and estimates are themselves sometimes considerably different. Traditional gage measurements are based on the volume or weight of the precipitation caught while optical gages measure the precipitation rate (determined by the rate of raindrops or snowflakes passing through the laser beam). By contrast, satellite estimates are based on correlations between the amount of energy emitted or reflected in either the microwave or infrared portions of the electromagnetic spectrum while radar-based estimates are based on a Z–R (reflectivity to rainfall) relationship of microwave reflectivities (e.g., 10 cm wavelength for the U.S. National Weather Service's WSR-88D radars). Note that precipitation gage measurements are made at a single point whereas satellite and radar estimates purport to be areal averages. This diversity of principles and techniques poses a serious problem in that the result is a new set of precipitation measurements and estimates that are incompatible with past observations. Intercomparison of old and new instrumentation has been conducted in many countries but sometimes these comparisons reveal more problems than they resolve. For example, radars developed in Russia have never been able to quantitatively estimate the true precipitation during mixed precipitation events (snow and rain concurrently) while calibration of the modern WSR-88D (NEXRAD) radars in the United States are made with recording gages which themselves represent a sparse spatial coverage and are prone to considerable measurement biases.

The need to automate the measurement of precipitation and to estimate spatial averages (e.g., satellite and radar estimates) rather than obtain point measurements (i.e., rain gages) raises a serious question of yet another significant inhomogeneity being introduced into the precipitation record. If the traditional gage measurement of precipitation is not maintained or if the issue of compatibility of measurements and estimates is not properly addressed, our ability to detect and monitor subtle, but significant, trends and changes in precipitation from the past century and extending into the future will be lost. This will have a considerable and detrimental impact on future analyses of climatic change.

Final Notes and Recommendations

In summary, our main conclusions regarding the present status of precipitation measurements are:

- Precipitation changes require special consideration regardless of their role in detecting climatic change and attributing specific causes to it; we, however, do not identify precipitation as the highest priority for the detection of anthropogenic induced climate change.

- Studies that analyze changes in precipitation require: (a) relatively dense networks to reduce the effects of the microscale and weather 'noise' on the results; and (b) careful preprocessing of the data to avoid the effect of large-scale inhomogeneities and the contribution of changes in other meteorological variables on conclusions about climate changes in precipitation.

- During the past five years, we have acquired more confidence in answering questions such as "What large-scale changes of precipitation happend during the period of instrumental observations over the most of the land areas of the globe?"

- If the traditional gage measurement of precipitation is not maintained or if the issue of compatibility of new and old technologies of precipitation measurements is not properly addressed, our ability to detect changes in precipitation will be lost.

Although global many climatologies have been developed (e.g., Schutz and Gates, 1971: Jaeger, 1976), no reliable precipitation climatology adjusted for gage measurement biases were digitally available before 1987 (results of the World Water Balance (1974) studies are in paper form and are practically non-reproducible). Legates (Legates, 1987; Legates and Willmott, 1990) provided the first such climatology on a global scale (see Legates, 1995a, for an intercomparison of several digitally-based global and terrestrial precipitation archives) while, about the same time, the first clues about continental-scale precipitation changes during the period of instrumental observations were published (Bradley *et al.*, 1987; Diaz *et al.*, 1989). Since then, more insight into the problem of documenting terrestrial precipitation changes have evolved (Table II). New studies start with an assessment of the instrumental homogeneity of precipitation time series (Karl *et al.*, 1993a; Hanssen-Bauer and Førland, 1994; Groisman and Easterling, 1994; Georgievsky *et al.*, 1995), introduce necessary corrections/adjustments into the data, and discuss relationship between trends (if any are found) with analyses of other meteorological characteristics of the region under consideration (Karl *et al.*, 1993a; Georgievsky *et al.*, 1995). This explains our optimism but at the same time, we wish to note the key problems that remain unresolved and required further analysis:

1. Data availability. Only about 5% of precipitation measurements can be obtained from the largest international archives (Eischeid *et al.*, 1991; Vose *et al.*, 1992). To study the problems of regional-scale precipitation change detection, we probably do not need much more (several thousand stations

TABLE II

Statements about the increase of terrestrial precipitation in the extratropical regions during the past period of instrumental observations. The unusually warm (globally) 1980s gave a record high decadal precipitation totals in several regions of North America (Karl *et al.*, 1993a) and Eurasia (Georgievsky *et al.*, 1995), thus somewhat exagerrated the trend estimates making them 'visible' above the level of natural variability

Authors	Region	Type of precipitation	Period	Assessment of inhomogeneities
Bradley *et al.* (1987)	35–70° N	Annual indices	120 yrs	No
Diaz *et al.* (1989)	25–60° N 25–60° S	Seasonal indices	97 yrs	No
Groisman (1991a, b)	35–70° N except N. Canada & China	Annual totals	98 yrs	only for the U.S.S.R.
Karl *et al.* (1993a)	Canada	Annual totals	55–70° N, 41 yrs < 55° N, 100 yrs	Yes
Hanssen-Bauer and Førland (1994)	Norway	Annual totals	90 yrs	Yes
Dahlström, (1994)	Northern Europe	Annual totals	100 yrs	No
Groisman and Easterling (1994)	North America	Annual and cold season totals	45–55° N, 100 yrs 55–70° N, 43 yrs	Yes
Georgievsky *et al.* (1995)	European part of the U.S.S.R.	Annual and warm season totals	41 yrs	Yes

evenly distributed over land areas would be enough). But for an assessment of the requirements necessary for a reasonable spatial coverage through time, the presence of long-term time series and their homogeneity are a must. Metadata (information about the data) are also required, particularly for regions where a significant portion of precipitation falls in frozen form. Existing global data sets do not completely match these requirements. Access to additional data, however, is often restricted by national meteorological services or the data themselves become an object of commercial sale (cf. Hulme, 1994). This is why we expect that international cooperation, if it occurs, may substantially rectify the existing gaps in spatial coverage of terrestrial precipitation and help alleviate some future problems with climate change detection. *There is no need for new programs to achieve this purpose; a simple sharing of available information may produce acceptable results.*

2. Technological changes in the way precipitation is measured or estimated may challenge our future ablity to use both the old and new information about precipitation to assess and detect climate change. Considerable efforts must be focussed in addressing this dilemma. These homogeneity problems are crucial for the future use of modern observational systems in any climate related study, including the climate change detection problem. This is (from our point of view) a major potential impediment to obtaining high quality data from precipitation observing systems and providing archives to the science community for the assessment of global changes in precipitation. *If homogeneity is not addressed before new measurement systems become operational, the two-hundred year period of instrumental precipitation observations will be devalued for future climate change analyses.*

3. No simple solutions for the problem of documenting precipitation changes over the ocean exist. Modern satellite observations (Arkin and Ardanuy, 1989; Spencer, 1993) give a short period of global data coverage. Ship reports together with island stations are a scarce and inadequate source of information but are available for the past 100 years. Since oceans cover nearly three-quarters of the earth's surface, it is imperative that accurate estimates of precipitation time series for the oceans be made available. The experience accumulated in analysis of proxy climate information (cf., paleodata, historic records) provide us a hope that a joint analysis of satellite data combined with surface observations (e.g., the Comprehensive Ocean Atmospheric Data Set) may help clarify some of the major features of climatic changes in oceanic precipitation during the past century although it is doubtful.

4. We want to stress that some studies (whose results are important to the precipitation change problem) are currently well under way and still remain to be completed. Among these programs are the International Solid Precipitation Measurement Intercomparison Project being conducted under the auspices of the WMO (WMO, 1991; Goodison *et al.*, 1992), the creation of an updated version of the Global Historical Climatology Network which is supported by several agencies and programs in the United States (Vose *et al.*, 1992), and the Global Precipitation Climatology Project sponsored by the WMO (Rudolf *et al.*, 1994).

References

Adler, R. F. and Mack, R. A.: 1984, 'Thunderstorm Cloud Height-Rainfall Rate Relations for Use with Satellite Rainfall Estimation Techniques', *J. Clim. Appl. Meteorol.* **23**, 280–296.

Alexandersson, H.: 1994, *Some Comments on Recent Climate Fluctuations – from a Swedish Horizon.* Climate Variations in Europe, Proceedings of the European Workshop on Climate Variations held in Kirkkonummi, Finland, 15–18 May 1994, Publ. of the Academy of Finland, pp. 293–305.

Allerup, P. and Madsen, H.: 1980, 'Accuracy of Point Precipitation Measurements', *Nord. Hydrol.* **11**, 57–70.

Arkin, P. A. and Ardanuy, P. E.: 1989, 'Estimating Climate-Scale Precipitation from Space: A Review', *J. Clim.* **2**, 1229–1238.

Biswas, A. K.: 1970, *History of Hydrology*, North-Holland Pub. Co., Amsterdam; American Elsevier Pub. Co., New York, 336 pp.

Bogdanova, E. G.: 1966, 'Investigation of Precipitation Measurement Losses due to the Wind', *Transact. Main Geophys. Observatory*, Leningrad, **195**, 40–62, (in Russian).

Bodanova, E. G.: 1986, 'A New Map of Atmospheric Precipitation over the World Ocean', *Izv. AS USSR, Seria Geograph.* **1**, 37–49, (in Russian).

Borzenkova, I. I.: 1994, *Climate Change in the Cenozoic*, Presented at the U.S. Russian Workshop on Paleocalibration of Climate Sensitivity, Washington, DC, Aug. 15–17, 1994, 338 pp.

Bradley, R. S., Diaz, H. F., Eischeid, J. K., Jones, P. D., Kelly, P. M., and Goodess, C. M.: 1987, 'Precipitation Fluctuations over Northern Hemisphere Land Areas since the Mid-19th Century', *Science* **237**, 171–275.

Bradley, R. S. and Groisman, P. Ya.: 1989, 'Continental Scale Precipitation Variations in the 20th Century', in Sevruk, B. (ed.), *Precipitation Measurement*, WMO/IAHS/ETH Workshop on Precipitation Measurement, St. Moritz, 3–7 December 1989, pp. 499–503.

Budyko, M. I. and Drozdov, O. A.: 1976, 'On the Causes of Changes in Hydrological Cycle', *Water Resources* **6**, 35–44, (in Russian).

Budyko, M. I. and Izrael, Yu. A. (eds.): 1987, *Anthropogenic Climate Changes*, Gidrometeoizdat, Leningrad, 406 pp., (in Russian; in English: 1991, Univ. of Arizona Press, Tucson, 485 pp.).

Czelnai, R., Desi, F., and Rakoczi, F.: 1963, 'On Determining the Rational Density of Precipitation Measuring Networks', *Idojaras* **67**, 257–267.

Dahlström, B.: 1986, *The Improvement of Point Precipitation Data on an Operational Basis*, Nordic Hydrological Programme. NHP-Report No. 17, 86 pp.

Dahlström, B.: 1994, *Short Term Fluctuations of Temperature and Precipitation in Western Europe.* Climate Variations in Europe, Proceedings of the European Workshop on Climate Variations held in Kirkkonummi, Finland, 15–18 May 1994, Publ. of the Academy of Finland, pp. 30–38.

Diaz, H. F., Bradley, R. S., and Eischeid, J. K.: 1989, 'Precipitation Fluctuation over Global Land Areas since the Late 1800s', *J. Geophys. Res.* **94**, 1195–1240.

Dorman, C. E.: 1982, 'Comments on "Comparison of Ocean and Island Rainfall in the Tropical Pacific" ', *J. Appl. Meteorol.* **21**, 109–113.

Dorman, C. E. and Bourke, R. H.: 1978, 'A Temperature Correction for Tucker's Ocean Rainfall Estimates', *Quart. J. Roy. Meteorol. Soc.* **104**, 765–773.

Dorman, C. E. and Bourke, R. H.: 1979, 'Precipitation over the Pacific Ocean, 30° S to 60° N, *Mon. Wea. Rev.* **107**, 896–910.

Dorman, C. E. and Bourke, R. H.: 1981, 'Precipitation over the Atlantic Ocean, 30° S to 70° N, *Mon. Wea. Rev.* **109**, 554–563.

Eischeid, J. K., Diaz, H. F., Bradley, R. S., and Jones, P. D.: 1991: *A Comprehensive Precipitation Data Set for Global Land Areas*, U.S. Department of Energy Monograph TR051, DOE/ER-69017T-H1, Washington, DC, 82 pp.

Elliott, W. P., Egami, R., and Rossknecht, G.: 1971, 'Rainfall at Sea', *Nature* **229**, 108–109.

Elliott, W. P. and Reed, R. K.: 1973, 'Oceanic Rainfall off the Pacific Northwest Coast', *J. Geophys. Res.* **78**, 941–948.

Folland, C. K.: 1988, 'Numerical Models of the Raingauge Exposure Problem, Field Experiments and an Improved Collector Design', *Quart. J. Roy. Meteorol. Soc.* **114**, 1485–1516.

Folland, C. K. and Wales-Smith, B. G.: 1977, 'Richard Towneley and 300 Years of Regular Rainfall Measurement', *Weather* **32**, 438–445.

Frich, P., Brodsgaard, B., and Cappelen, J.: 1991, *North Atlantic Climatological Data Set (NACD), Present Status and Future Plans*, DMI Technical Report 91–8.

Frich, P.: 1994, *Precipitation Trends in the North Atlantic European Region. Climate Variations in Europe*, Proceedings of the European Workshop on Climate Variations held in Kirkkonummi, Finland, 15–18 May 1994, Publ. of the Academy of Finland, pp. 196–200.

Gandin, L. S.: 1993, 'Optimal Averaging of Meteorological Fields', U.S. Dept. of Commerce, NOAA National Weather Service, National Meteorological Center, Office Note 397, 56 pp.

Gandin, L. S., Celnai, R., and Zakhariev, V. E. (eds.): 1976, *Statistical Structure of Meteorological Fields*, Az Orszagos Meteorologiai Szolgalat, Budapest, 364 pp., (in Russian and Hungarian, resumes in German).

Georgievsky, V. Yu, Zhuravin, S. A., and Ezhov, A. V.: 1995, *Assessment of Trends in Hydrometeorological Situation on the Great Russian Plain under the Effect of Climate Variations*, AGU Proc., Fifteen Annual Hydrology Days, April 3–7, 1995, Fort Collins, Colorado, U.S.A., pp. 47–58.

Golitsyn, G. S., Dzuba, A. V., Osipov, A. G., and Panin, G. N.: 1990, 'Regional Climate Changes and Their Impacts on the Caspian Sea Level Rise', *Doklady USSR Acad. Sci.* **313** (**5**), 1224–1227, (in Russian).

Golubev, V. S.: 1965, 'Some Results of the Studies at the Precipitation Experimental Site at Valdai, Russia', *Transact. State Hydrolog. Instit.* **123**, 81–95.

Golubev, V. S.: 1969, 'Research on Precipitation Measurement Accuracy', *Transact. State Hydrolog. Instit.* **176**, 149–164, (in Russian).

Golubev, V. S.: 1975, 'Results of the Intercomparison of Precipitation Gages', *Transact. State Hydrolog. Instit.* **224**, 38–46.

Golubev, V. S.: 1993, *Experience of Correction of Precipitation Point Measurements and Analysis of Correction Procedures*, Proc. 8th Symposium on Meteorological Observations and Instrumentation, American Meteorological Society, Boston, pp. 325–328.

Golubev, V. S., Koknaeva, V. V., Simonenko, A. Yu.: 1995, 'Results of Atmospheric Precipitation Measurements by National Standard Gauges of Canada, USA, and Russia', *Meteorolog. Gydrolog.* **2**, 102–110, (in Russian).

Goodison, B. E.: 1981, 'Compatibility of Canadian Snowfall and Snow Cover Data', *Water Resourc. Res.* **17**, 893–900.

Goodison, B. E., Golubev, V. S., Gunter, T., and Sevruk, B.: 1992, *Preliminary Results of the WMO Solid Precipitation Measurement Intercomparison*, Proc. of WMO Technical Conference on Instruments and Methods of Observation, 10–15 May 1992, Vienna, Austria, pp. 81–85.

Groisman, P. Ya.: 1991a, 'Data on Present-Day Precipitation Changes in the Extratropical Part of the Northern Hemisphere', p. 297–310 in Schlesinger, M. E. (ed.), *Greenhouse-Gas-Induced Climatic Change: A Critical Appraisal of Simulations and Observations*, Elsevier, Amsterdam, 615 pp.

Groisman, P. Ya.: 1991b, *Unbiased Estimates of Precipitation Change in the Northern Hemisphere Extratropics*, AMS Proc., Fifth Conference on Climate Variations, Denver, CO, pp. 42–45.

Groisman, P. Ya.: 1992, *Studying the North American Precipitation Changes During the Last 100 Years*, Proc. of the 5th International Meeting on Statistical Climatology, Toronto, 22–26 June 1992, pp. 75–79.

Groisman, P. Ya.: 1993, *Towards Unbiased Estimates of North American Precipitation*, AMS Proc., Eighth Symposium on Meteorological Observations and Instrumentation, pp. J43–J47.

Groisman, P. Ya. and Easterling, D. R.: 1994, 'Variability and Trends of Precipitation and Snowfall over the United States and Canada', *J. Clim.* **7**, 184–205.

Groisman, P. Ya., Koknaeva, V. V., Belokrylova, T. A., and Karl, T. R.: 1991a, 'Overcoming Biases of Precipitation Measurement: A History of the USSR Experience', *Bull. Amer. Meteorol. Soc.* **72**, 1725–1733.

Groisman, P. Ya., Mason, E. M., and DelGreco, S.: 1991b, *Metadata for Construction of Homogeneous Long-Term Precipitation Time Series*, AMS Proc., Seventh Conference on Applied Climatology, Sept. 10–13, 1991, Salt Lake City, Utah, pp. 119–122.

Groisman, P. Ya. and Legates, D. R.: 1994, 'The Accuracy of United States Precipitation Data', *Bull. Amer. Meteorol. Soc.* **75**, 215–227.

Groisman, P. Ya., Quayle, R. G., and Easterling, D. R.: 1994, *Reducing Biases in Estimates of Precipitation over the United States*, AMS Proc., Sixth Conference on Climate Variations, pp. 165–169.

Groisman, P. Ya., Easterling, D. R., Quayle, R. G., Golubev, V. S., Krenke, A. N., and Mikhailov, A. Yu.: 1995, 'Reducing Biases in Estimates of Precipitation Over the United States: Phase Three Adjustments', *J. Geophys. Res.* (in press).

Hanssen-Bauer, I. and Førland, E. J.: 1994, 'Homogenizing Long Norwegian Precipitation Series', *J. Clim.* **7**, 1001–1013.

Heino, R.: 1994, *Climate in Finland During the Period of Meteorological Observations*, Finnish Meteorological Institute Contributions, No. **12**, 209 pp.

Hendrick, R. L. and Comer, G. H.: 1970, 'Space Variations of Precipitation and the Implications for Raingage Network Design', *J. Hydrol.* **10**, 151–163.

Huff, F. A.: 1979, *Spatial and Temporal Correlation of Precipitation in Illinois*, Circular 141, Illinois State Water Survey, Urbana, IL, 14 pp.

Huff, F. A. and Shipp, W. L.: 1969, 'Spatial Correlations of Storms, Monthly and Seasonal Precipitation', *J. Appl. Meteorol.* **8 (4)**, 542–550.

Hulme, M.: 1992, 'A 1951–80 Global Land Precipitation Climatology for the Evaluation of General Circulation Models', *Clim. Dynam.* **7 (2)**, 57–72.

Hulme, M.: 1994, 'The Cost of Climate Data – a European Experience', *Weather* **49**, 168–175.

Hulme, M.: 1995, 'Estimating Global Changes in Precipitation', *Weather* **50**, 34–42.

Intergovernmental Panel on Climate Change (IPCC): 1990, *Climate Change. The IPCC Scientific Assessment*, Houghton, J. T., Jenkins, G. J., and Ephraums, J. J. (eds.), Cambridge University Press N.Y., 362 pp.

IPCC: 1992, *Climate Change 1992. The Supplementary Report to the IPCC Scientific Assessment*, Houghton, J. T., Callander, B. A., and Varney, S. K. (eds.), Cambridge University Press N.Y., 200 pp.

IPCC: 1995, *Climate Change. The IPCC Second Scientific Assesment*, (Draft).

Jacobs, W. C.: 1968, 'The Seasonal Apportionment of Precipitation over the Ocean', in Court, A. (ed.), *Eclectic Climatology*, Oregon State University Press, Corvallis, Oregon, 63–78.

Jaeger, L.: 1976, 'Monatskarten des Niederschlags für die Ganze Erde', *Ber. Deutsch. Wetterdienstes* **18**, 38 pp.

Janowiak, J. E.: 1992, 'Tropical Rainfall: A Comparison between Satellite-Derived Rainfall Estimates with Model Precipitation Forecasts, Climatologies, and Observations', *Mon. Wea. Rev.* **120**, 448–462.

Janowiak, J. E., Arkin, P. A., Xie, P., Morrissey, M. L., and Legates, D. R.: 1995, 'An Examination of the East Pacific ITCZ Rainfall Distribution', *J. Clim.*, (forthcoming).

Kagan, R. L.: 1972, 'Planning the Spatial Distribution of Hydrometeorological Stations to Meet an Error Criterion', in *Casebook on Hydrological Network Design Practice*, WMO Publ. 324, Geneva, III-1.2-1-III-1.2-8.

Kagan, R. L.: 1979, *Averaging of Meteorological Fields*, Gidrometeoizdat, Leningrad, 213 pp., (in Russian).

Karl, T. R., Groisman, P. Ya., Knight, R. W., and Heim, R. R., Jr.: 1993a, 'Recent Variations of Snow Cover and Snowfall in North America and Their Relation to Precipitation and Temperature Variations', *J. Clim.* **6**, 1327–1344.

Karl, T. R., Quayle, R. G., and Groisman, P. Ya.: 1993b, 'Detecting Climate Variations and Change: New Challenges for Observing and Data Management Systems', *J. Clim.* **6**, 1481–1494.

Karl, T. R., Williams, C. N., Jr., Quinlan, F. T., and Boden, T. A.: 1990, *United States Historical Climatology Network (HCN) Serial Temperature and Precipitation Data*, NDP-019/R1, Carbon Dioxide Information Analysis Center, Oak Ridge National Laboratory, Oak Ridge, Tennessee, 83 pp. (plus appendices).

Kay, P. A. and Kutiel, H.: 1994, 'Some Remarks on Climatic Maps of Precipitation', *Clim. Res.* **4**, 233–241.

Khrgian, A. H.: 1948, *Essays on the Development of the Meteorological Science*, Gidrometeoizdat, Leningrad, 352 pp., (in Russian).

Kurtyka, J. C.: 1953, *Precipitation Measurements Study*, Investigation No. 20, Department of Registration and Education and the State Water Survey Division, Urbana, Illinois, 163 pp.

Larson, L. W.: 1971, *Precipitation and Its Measurement, a State of the Art*, Water Resources Series **24**, Water Resources Research Institute, Wyoming University, Laramie, Wyoming, 74 pp.

Larson, L. W. and Peck, E. L.: 1974, 'Accuracy of Precipitation Measurements for Hydrologic Modeling', *Water Resourc. Res.* **10**, 857–863.

Lavery, B. M., Kariko, A. P., and Nicholls, N.: 1992, 'A High-Quality Historical Rainfall Data Set for Australia', *Aust. Meteorol. Mag.* **40**, 33–39.

Legates, D. R.: 1987, 'A Climatology of Global Precipitation', *Publ. Climatol.* **40** (1), 84 pp.

Legates, D. R.: 1992, *The Need for Removing Biases from Rain and Snowgage Measurements*, Proc. Snow Watch '92, Niagara-on-the-Lake, Ontario, Canadian Climate Centre, World Meteorological Organization, and Institute for Space and Terrestrial Science, pp. 144–151.

Legates, D. R.: 1995a, 'Global and Terrestrial Precipitation: A Comparative Assessment of Existing Climatologies', *Int. J. Climatol.* **15**, 237–258.

Legates, D. R.: 1995b, *Precipitation Measurement Biases and Climate Change Detection*, AMS. Proc., Sixth Symposium on Global Change Studies, Dallas, TX, pp. 168–173.

Legates, D. R. and Willmott, C. J.: 1990, 'Mean Seasonal and Spatial Variability in Gage-Corrected, Global Precipitation', *Int. J. Climatol.* **10**, 111–127.

Lettenmaier, D. P., Wood, E. F., and Wallis, J. R.: 1994, 'Hydro-Climatological Trends in the Continental United States, 1948–88', *J. Clim.* **7**, 586–607.

Manabe, S., Stouffer, R. J., Spellman, M. J., and Brian, K.: 1991, 'Transient Responses of a Coupled Ocean-Atmosphere Model to Gradual Changes of Atmospheric CO_2. Part 1: Annual Mean Response', *J. Clim.* **4**, 785–818.

Manabe, S., Spellman, M. J., and Stouffer, R. J.: 1992, 'Transient Responses of a Coupled Ocean-Atmosphere Model to Gradual Changes of Atmospheric CO_2. Part 2: Seasonal Response', *J. Clim.* **5**, 105–126.

Metcalfe, J. R. and Goodison, B. E.: 1992, *Automation of Winter Precipitation Measurements: The Canadian Experience*, Proceedings of the WMO Technical Conference on Instruments and Methods of Observation, 11–15 May 1992, Vienna, Austria, WMO/TD, No. 462, pp. 81–85.

Mitchell, J. F. B., Manabe, S., Meleshko, V. P., and Tokioka, T.: 1990, 'Equilibrium Climate Change – and Its Implications for the Future', Ch. 5 in *Climate Change, The IPCC Scientific Assessment*, Publ. WMO/UNEP, pp. 131–174.

Murphy, J. M. and Mitchell, J. F. B.: 1995, 'Transient Response of the Hadley Centre Coupled Ocean-Atmosphere Model to Increasing Carbon Dioxide. Part II: Spatial and Temporal Structure of Response', *J. Clim.* **8**, 57–80.

Nicholson, S. E.: 1995, 'Variability of African Rainfall on Interannual and Decadal Time Scales', in *Natural and Climate Variability on Decade-to-Century Time Scales*, National Academy Press, (in press).

Nørdø, J. and Hjortnaes, K.: 1967, 'Statistical Studies of Precipitation on Local, National and Continental Scales', *Geofisiske Publikasjoner* **26** (12), 46 pp.

Parthasarathy, B., Rupa Kumar, K., and Kothawale, D. R.: 1992, 'Indian Summer Monsoon Rainfall Indices: 1871–1990', *Meteorol. Magazin* **121**, 174–186.

Parthsarathy, B., Minot, A. A., and Kothawale, D. R.: 1994, 'All-India Monthly and Seasonal Rainfall Series: 1871–1993', *Theor. Appl. Climatol.* **49**, 217–224.

Peck, E. L. and Schaake, J. C.: 1990, 'Network Design for Water Supply Forecasting in the West', *Water Resourc. Bull.* **26** (1), 87–99.

Quayle, R. G.: 1974, 'A Climatic Comparison of Ocean Weather Stations and Transient Ship Records', *Marine Wea. Log.* **18**, 307–311.

Rao, M. S. V. and Theon, J. S.: 1977, 'New Features of Global Climatology Revealed by Satellite-Derived Oceanic Rainfall Maps', *Bull. Amer. Meteorol. Soc.* **58**, 1285–1288.

Reed, R. K.: 1979, 'On the Relationship between the Amount and Frequency of Precipitation over the Ocean', *J. Appl. Meteorol.* **18**, 692–696.

Reed, R. K.: 1980, 'Comparison of Ocean and Island Rainfall in the Tropical North Pacific', *J. Appl. Meteorol.* **19**, 877–880.

Reed, R. K. and Elliott, W. P.: 1973, 'Precipitation at Ocean Weather Stations in the North Pacific', *J. Geoph. Res.* **78**, 7087–7091.

Reed, R. K. and Elliott, W. P.: 1977, 'A Comparison of Oceanic Precipitation as Measured by Gage and Assessed from Weather Reports', *J. Appl. Meteorol.* **16**, 983–986.

Reed, R. K. and Elliott, W. P.: 1979, 'New Precipitation Maps for the North Atlantic and North Pacific Oceans', *J. Geophys. Res.* **84**, 7839–7846.

Rodda, J. C.: 1971, *The Precipitation Measurement Paradox – the Instrument Accuracy Problem*, World Meteorol. Organ., WMO/IHD Report No. 16, WMO #316, Geneva, 42 pp.

Rudolf, B., Hauschild, H., Reiss, M., Schneider, U., and Henning, D.: 1992, 'Contributions to the Global Precipitation Climatology Centre', *Meteorolog. Zeitschrift*, Neue Folge, Heft **1**, 7–84, (in German, +English summaries).

Rudolf, B., Hauschild, H., Ruth, W., and Schneider, U.: 1994, *Management and Analysis of Precipitation Data on a Routine Basis*, Proc., Intl. Symp. on Prec. and Evap., Slovak Hydrometeorological Institute, Bratislava, pp. 69–76.

Schutz, C. and Gates, W. L.: 1971, *Global Climatic Data for Surface, 800 mb, 400 mb: January, July, April, October*, Rand, Santa Monica, R-915-ARPA, 173 pp.

Sevruk, B.: 1982, 'Methods of Correction for Systematic Error in Point Precipitation Measurement for Operational Use', *Oper. Hydrol. Rep.* **21**, Publ. 589, World Meteorol. Organ., Geneva, Switzerland, 91 pp.

Sevruk, B.: 1988, 'Towards the Universal Precipitation Gauge of the Future', *Vaisala News* **113–114**, 12–14.

Sevruk, B. (ed.): 1992, 'Snow Cover Measurements and Areal Assessment of Precipitation and Soil Moisture', *Oper. Hydrol. Rep.* **35**, Publ. No. 749, World Meteorol. Organ., Geneva, Switzerland, 283 pp.

Sevruk, B. and Hamon, W. R.: 1984, *International Comparison of National Precipitation Gauges with a Reference Pit Gauge*, Instruments and Observing Methods Report No. 17, WMO/TD-No. 38, World Meteorological Organization, Geneva, Switzerland, 86 pp.

Sevruk, B., Kirchhofer, D., Tihlarik, R., and Zahlavova, L.: 1993, 'Precipitation Corrections in Switzerland', in Sevruk, B. and Lapin, M. (eds.), *Precipitation Measurements and Quality Control*, Proc. of Symposium on Precipitation and Evaporation, Vol. **1**, Zurich, pp. 155–156.

Sevruk, B. and Klemm, S.: 1989, *Catalogue of National Standard Precipitation Gauges*, WMO, Report No. **313**, 50 pp.

Shver, Ts. A.: 1976, *Precipitation over the USSR Territory*, Gidrometeoizdat, Leningrad, 302 pp., (in Russian).

Shver, Ts. A.: 1984, *Regularities in Distribution of Precipitation Amount over Continents*, Gidrometeoizdat, Leningrad, 285 pp., (in Russian).

Simpson, J., Adler, R. F., and North, G. R.: 1988, 'A Proposed Tropical Rainfall Measuring Mission (TRMM) Satellite', *Bull. Amer. Meteorol. Soc.* **69**, 279–295.

Spangler, W. M. L. and Jenne, R. L.: 1990, *World Monthly Surface Station Climatology*, National Center for Atmospheric Research, Scientific Computing Division, Boulder, Colorado, U.S.A., 14 pp.

Spencer, R. W.: 1993, 'Global Oceanic Precipitation from the MSU During 1979–91 and Comparisons to Other Climatologies', *J. Clim.* **6**, 1301–1326.

Spreafico, M., Weigartner, R., and Leibundgut, Ch. (eds.): 1992, *Hydrological Atlas of Switzerland*, Landestopographie, Bern.

Tabony, R. C.: 1980, *A Set of Homogeneous European Rainfall Series*, U.K. Meteorological Office, 18th Branch Memorandum 104, Bracknell, U.K., 244 pp.

Tabony, R. C.: 1981, 'A Principal Component and Spectral Analysis of European Rainfall', *J. Climatol.* **1**, 283–294.

Tucker, G. G.: 1961, 'Precipitation over the North Atlantic Ocean', *Quart. J. Roy. Meteorol. Soc.* **87**, 147–158.

U.S. Department of Commerce: 1968, *Climatic Atlas of the United States*, Environmental Science Service Administration, Environment Data Service, 80 pp.

Vinnikov, K. Ya., Groisman, P. Ya., and Lugina, K. M.: 1990, 'Empirical Data on Contemporary Global Climate Changes (Temperature and Precipitation)', *J. Clim.* **3**, 662–677.

Vose, R. S., Schmoyer, R. L., Steurer, P. M., Peterson, T. C., Heim, R. R., Jr., Karl, T. R., and Eischeid, J. K.: 1992, *The Global Historical Climatology Network: Long-Term Monthly Temperature, Precipitation, Sea Level Pressure, and Station Pressure Data*, Carbon Dioxide Information Analysis Center, Oak Ridge National Laboratory, U.S. Dept. of Energy, Environmental Sciences Division Publication No. 3912, 99 pp., (with Appendices).

Wilheit, T. T., Chang, A. T. C., and Chiu, L. S.: 1991, 'Retrieval of Monthly Rainfall Indices from Microwave Radiometric Measurements Using Probability Distribution Functions', *J. Atmos. Oceanic Technol.* **8**, 118–137.

Willmott, C. J. and Legates, D. R.: 1991, 'Rising Estimates of Terrestrial and Global Precipitation', *Clim. Res.* **1**, 179–186.

Willmott, C. J., Robeson, S. M., and Feddema, J. J.: 1994, 'Estimating Continental and Terrestrial Precipitation Averages from Rain-Gauge Networks', *Int. J. Climatol.* **14**, 403–414.

World Meteorological Organization (WMO): 1973, *Annotated Bibliography on Precipitation Measurement Instruments*, WMO-No. 343, Geneva, Switzerland, 278 pp.

WMO: 1979, *Climatic Atlas of North and Central America*, Vol. 1, WMO, UNESCO, Geneva, Switzerland, 39 pp.

WMO: 1991, *Final Report of the Fifth Session of International Organizing Committee for the WMO Solid Precipitation Measurement Intercomparison*, World Meteorol. Organ., Geneva, Switzerland, 19 pp.

World Water Balance and Water Resources of the Earth (WWB): 1974–1978, Gidrometeoizdat, Leningrad, (1974 in Russian; 1978 in English).

(Received 23 January, 1995; in revised form 7 July, 1995)

INDEXES OF LEADING CLIMATE INDICATORS FOR IMPACT ASSESSMENT

W. E. EASTERLING[1] and R. W. KATES[2]

[1] *Department of Agricultural Meteorology, University of Nebraska-Lincoln, Lincoln, NE 68583-0728, U.S.A.*
[2] *Independent Scholar, Trenton, ME 04605, U.S.A.*

Abstract. Could users of climate information for impact assessment be overlooking an important source of information in climate indicators? We argue that indexes of leading climate indicators of impacts may be usable knowledge for consumers and may provide guidance to the global climate observing community concerning the types of data and information that users need. Five classes of indexes are suggested: Climate Extremes Index (CEI) and Greenhouse Climate Response Index (GCRI) – such are already available from scientists at the U.S. National Climatic Data Center – plus proposed indexes of Hazard Warning, Ecosystem Health, and Energy Demand and Renewable Natural Resources. We conclude that the CEI and GCRI possess several necessary attributes to become usable knowledge; the other indexes have the potential to become usable knowledge, but remain to be implemented with climate data and fully evaluated.

1. Introduction

Usable knowledge (Lindblom and Cohen, 1979) of weather and climate evolves as a function of scientific capability, observational tradition, and practical utility. Over time, such knowledge becomes codified and institutionalized, often in the form of indices, popularized in the media and widely understood by users and the general public. In addition to the usual daily descriptors of weather and seasonal descriptors of climate, indices such as, for example, wind chill factors, heating and cooling degree days and growing degree days are widely used. Regional indicators include, for example, frost-free zones, depth of snow pack, Palmer Drought Severity Index, and Crop Moisture Index. Not all attempts to create indicators that are usable knowledge succeed; indexes of the combined effect of temperature and humidity on human comfort were attempted (e.g., Thom, 1959), but were never widely disseminated or used. In the U.S., a specialized cable television station known as *The Weather Channel* broadcasts a number of climate-related indices such as a lawn watering index, an hay fever index and an influenza index, but we do not know how such indices are perceived and used by viewers.

Recently, Karl, Knight, Easterling and Quayle (1995) have developed two climate change indices – a Climate Extremes Index and a Greenhouse Climate Response Index – which, from their early reception (Kerr, 1995) and media attention (radio broadcasts and newspaper articles too numerous to list) hold some promise of becoming widely used. Not meant to be predictions, the two indices instead are integrations of certain climate time series carefully selected to incorporate the latest scientific consensus on expected features of climate change; they are meant to capture climate trends that may signal significant fluctuations (defined as persistent but temporary departures from current climate normals) or changes (defined as evolving shifts in climate normals, with or without associated changes in variability). Trends in such fluctuations and changes may foreshadow or 'lead' certain impacts on the environment and society.

It is important to note that we use the terms 'foreshadow' or 'lead' broadly to imply that the climate indexes developed from historical trends have anticipatory value encouraging actions aimed at averting impacts before they materialize. Such anticipations may be derived from conventional risk and probability analysis of index values or from more intuitive processes prompted by critical index values. An example of the former is the test of significance used by Karl *et al.* (1995) to assess the likelihood of greenhouse warming. An example of the latter is the collective decisions of many Nebraska farmers (reported in spring 1995 newspaper articles) to plant less acres of drought-resistant grain sorghum and more acres of relatively riskier corn based on the observed absence of significant droughts in the region over the past half-decade.

In this paper, we argue for extending the approach used by Karl *et al.* (1995) systematically to create and disseminate leading climate indicators to foreshadow the impacts of climate fluctuation and change. We suggest that indexes of leading climate indicators can serve as bridges between climate observers who must determine the attributes of climate data and information to monitor and archive for the long-term and users whose activities require them to manage climate risks and opportunities. Such bridging is needed if the products of a climate observing system are to become usable knowledge and if, in turn, long-term public support is to be garnered for an expanded global climate observing system.

We begin with a review of the experience with economic indicators and then develop the conceptual basis for indexes of leading climate indicators of impacts, including a set of necessary attributes of all such indexes. Five broad classes of climate-related problems are identified, each of which warrants the development of a set of leading indicators. The classes are: *climate extremes, greenhouse climate response, hazard warning, ecosystem health, and energy demand and renawable natural resources.* We review the results of the effort by Karl and his colleagues (1995) at the U.S. National Climatic Data Center to develop broad climate indicators of extreme events and greenhouse climate response. For each of the remaining classes: hazard warning, ecosystem health, and energy demand and natural resources, we then suggest a small illustrative group of indicators of impacts. Each of the proposed indexes is evaluated with respect to its necessary attributes. These suggested classes of problems and the indicators thereof are intended as first approximations that likely will benefit from further evaluation and, where necessary, respecification.

2. Leading Economic Indicators

When the U.S. government forecasts major upturns or downturns in the national economy, it relies mostly on an index of leading economic indicators maintained by the National Bureau of Economic Research (NBER) as the primary forecasting tool. This index of leading economic indicators has been used to forecast business cycles since the late 1930s based on the pioneering work of Mitchell and Burns (1938).

The complexity of the economy forbids exhaustive tracking and analysis of all facets of enterprise that contribute to overall economic performance. Hence, the reliance is on a subset of selected indicators that captures general economic trends better than a single measure and is more manageable and understandable than a simulation model. Leading economic indicators are empirical time series that usually experience a turning point before the general business cycle but rarely experience one if no business cycle turning point were imminent (i.e., few false signals).

Currently, the NBER Index of Leading Economic Indicators consists of twelve components of different weights: (1) average workweek of production manufacturing workers; (2) average weekly unemployment insurance claims; (3) new orders for consumer goods; (4) vendor performance reflected by companies receiving slower deliveries; (5) net business formation; (6) contracts and orders for plant and equipment; (7) new homes building permits; (8) change in inventories; (9) change in producer prices for selected crude and intermediate materials; (10) stock prices for 500 common stocks; (11) money supply (deflated to a base year); and (12) change in credit outstanding.

The NBER index often is used to indicate possible near-future changes in sensitive measures of macroeconomic performance such as the Federal Reserve Board's index of industrial production or the unemployment rate. Information generated by the government's index of leading economic indicators is useful to virtually all economic sectors and a wide range of decision makers with a stake in the anticipation of near future economic trends, including stock market analysts, financial planners, plant managers, government analysts, politicians, and especially the public. Though often criticized that it is crude and lacking in theoretical basis (e.g., Auerbach, 1982), the index has been accepted by public and private sector planners, in part, because it has been around long enough to demonstrate that it 'leads' business cycles frequently (though not always). It is usable economic knowledge despite its shortcomings.

The notion of indexes of indicators is gaining wide application in disciplines other than economics. Ecologists have long used indicators to anticipate trends in ecosystem health and integrity (e.g., Rapport, 1989; Cairns et al., 1993). The ecologists are also at the forefront in studying processes in ecotonal regions which may serve as indicators of global environmental change (discussed generally by Bella et al., 1994).

3. Toward Leading Climate Indicators

What if day-to-day weather reflected unvarying climate means with only minor and predictable oscillations around those means? Were this the case, climate would then pose no particular obstacle to human activity. Only the well-known spatial limitations climate imposes on what can be produced where would be of interest. Hare (1985) noted that when climate performs reliably it can be ignored – human activities are generally well-adapted to expected weather conditions. A farmer plants a mix of crops that has been proven profitable in the long-run under expected climate conditions that include a few years of bad weather for crops along with the good ones; engineers design a certain amount of excess capacity into reservoirs to account for the natural variability in long-term precipitation and, thus, are able to meet demands for water under most climate conditions.

Yet, when climate strays markedly from long-term averages, its likelihood of posing extraordinary risks and/or benefits to society increases. Riebsame *et al.* (1991) estimated that the 1988 drought in the U.S. caused about $15 billion worth of crop losses. The floods of 1993 in the U.S. midwest covered approximately 41,000 km^2 of farmland and caused about $12 billion in damage to property. Thus, climate fluctuation is a major risk factor in many human activities, and improved information about the variation of climate across spatial and temporal scales is needed better to manage climate risk.

However, mechanistic prediction of the impact of climate fluctuation and change on complex human systems is not possible. The understanding of the chains of causality beginning with climate fluctuation and change, and ending in human impact and response is weak. There is no generalizable methodology or theoretical model that enables full accounting of the environmental, social, cultural and economic costs/benefits of all forms of climate fluctuation and change. Though considerable progress has been made in climate impact assessment (Kates *et al.*, 1985) and more recently in the linkage of climate with biophysical and economic systems (for example, the use of climate-driven mechanistic crop growth models, hydrologic models, and forest growth models which are then linked to economic models, e.g., Bowes and Crosson, 1993; Easterling *et al.*, 1993; Rosenzweig and Parry, 1994), such modeling schemes are highly stylized and constrained by simplifying assumptions that weaken their predictive capabilities. Difficult to quantify social and cultural impacts are often ignored in such schemes. Part of the problem of establishing strong causal linkage between dynamic climate and specific types of human activities has been the poor match between climate data and measured attributes of social systems. Often, the poor match is the result of major differences in temporal and spatial scale resolution between the climate data and the affected aspect of the environment or society (Clark, 1989).

Climate observing systems are in a period of self-examination and re-evaluation of the kinds of data and information that are needed in order to better understand climate system dynamics and to deliver better information with which to manage climate risks. This paper is a testament to such, being stimulated by discussions at a 1995 workshop on long-term climate data needs in Asheville, NC sponsored by the World Meteorological Organization's Global Climate Observing System effort. Users of climate data and information for impact assessments need to be engaged in the planning of the attributes of future climate observing systems.

Part of the difficulty in communicating the needs of users to the broader climate observing community is the wide range of specific data and information needs among users. For climate observations to be usable knowledge, farmers need information (growing degree-days, frost-free periods, effective precipitation) that is tailored to the localized assessment of climate influences on biological productivity, hydrologists need climate information (evapotranspiration, temperature, precipitation, windspeed, humidity) to compute water balances across different sized stream basins, and load managers for electric utilities need climate information (heating, cooling degree-days) to estimate seasonal cooling and heating requirements across power districts. Even within a sector the types of climate information needed to make a comprehensive assessment can be overwhelming. Added to this list are the general needs and desires of the public to comprehend trends and changes that might affect them.

Such wide-ranging needs must be balanced against limited resources and capabilities for collecting climate data and supplying useful climate information to impact assessors. The ever deepening complexity of weather and climate information needs along with the bewildering array of new climate data and information argues strongly for seeking climate indicators that reflect trends and patterns in selected climate and climate-related measures. Thus, we make our case for the development of indexes of leading climate indicators as a useful step toward improved integration of the climate observing community with the users of climate data and information for impact assessment and response.

4. Necessary Attributes of Leading Climate Indicators

What attributes are necessary for an index of leading climate indicators of impacts? First of all, the set of impacts that are signalled by an index must be well-defined, though not so circumscribed as to preclude those impacts that were unanticipated. Careful definition of impacts is an inductive and deductive process aimed at clearly specifying the linkages between certain index values and certain impacts (Kates in Kates *et al.*, 1985). Guidance from previous climate impacts research is critical, including the long tradition of applied climatology.

Second, the index should 'lead' impacts. Leading indicators are not predictions of future states of climate, but rather providers of strong signals of imminent impacts were observed climate trends to continue. The key is identification of inertial or cumulative properties in the carry-over of climate to impacts on ecosystems and society. The point here is that climate is the cumulative expression of many shorter-term dynamic weather features and, once embarked on a departure from current normals, climate often lags behind the return to normal of the constituent short-term weather features. These inertial properties essentially *predetermine* future impacts in some cases and *condition* future impacts in other cases. For example, accumulated hydrologic drought will predetermine future problems for certain natural resources that are maintained principally by steady replenishment of moisture (e.g., groundwater levels, streamflow, lake levels), and will condition future problems for certain aspects of social welfare (e.g., accessible food supplies for the long-term) where moisture availability is only one of potentially many controlling factors (e.g., economic, political, cultural). Thus, the embodied inertia implied by climate indicators becomes useful in anticipating future problems of low flow or famine without ever directly modeling the affected system. This notion of inertia applies to short-term climate fluctuations and to long-term climate change.

Third, the index should be usable knowledge: widely accepted and used in making decisions, policies, and adapative responses. One criteria of such is that the index should be accepted broadly by the scientific community, resource users and the public. Acceptance by the scientific community helps assure the underlying scientific credibility of the index. Acceptance by resource users and the public helps assure that the index has utility. Another criteria is that the index should engender user responses and, in some cases, have a policy constituency; otherwise, the index is little more than the object of idle curiosity. User responses may range from *ad hoc* household decisions aimed at maximizing or minimizing some climate-affected outcome to public policies aimed at reducing societal risk. Failure to achieve this last criteria has been an abiding reason why long-range climate forecasts have yet to succeed as usable climate knowledge.

It is also important to point out that the utility of climate indices must extend beyond that to resource managers and other professionals to include that to the general public. As with economic indicators, there is, for many people, a broad curiosity and desire to understand general trends, even those that may not immediately affect them. Climate indices should provide a comprehensible way of tracking those trends.

Fourth, indicators comprising an index must easily be measureable and available on appropriate spatial and temporal scales. The period of record used to compute individual indexes should extend as far back as reliable instrumental records permit. The averaging period used for each climate indicator will vary among impact-related problems as will the spatial scale of climate information included in the indicators. Furthermore, the indicators may require transformation from raw climate data to an amalgamation of climate and climate-related measures to be useful. For example, some drought indices require the integration of climatic, hydrological and biological data (e.g., the PDSI).

Ideally, as each index is introduced below, it is evaluated by how well it measures up to each of the necessary attributes from above. However, a full evaluation is not possible since none of the indexes have been extant long enough to have an established track record – some were developed in recent months and others are proposed here for the first time. While we do sift what insights we can to project the likely attributes of each index in our discussion below, time and much experimentation is needed for a conclusive assessment.

5. Classes of Problems Worthy of Indexes of Leading Climate Indicators

Any attempt to classify climate impact problems is bound to generate debate. Such is the complexity of how dynamic climate interacts with society and the environment. For the sake of argument here, however, we propose that five general classes of problems account for the majority of concerns over the impacts of climate fluctuation and change on environmental and human systems: (1) *the occurrence of climate extremes*; (2) *the anticipation and detection of impacts of greenhouse warming*; (3) *the development of hazard warning capabilities*; (4) *the assessment of climate effects on ecosystem health*; and (5) *the involvement of climate in the determination of the demand for energy and the quantity and quality of renewable natural resources.* Each class demands a slightly different set of leading indicators, though there is considerable overlap of necessary climate information among classes. The individual climate indicators are typically expressed in terms of the percentage of total area of prescribed regions (political or physiographic) that experience a pronounced departure from normal. We adopt the departure-from-normal criteria used by Karl *et al.* (1995) in which the tenth and ninetieth percentiles represent the equivalents of 'much below' and 'much above' normal climate respectively.*

Different climate indexes will require different lengths of averaging periods to establish normals against which individual climate observations are compared (i.e., climate indicators of extreme events may require relatively short averaging periods of a few years while indicators of greenhouse warming may require averaging periods that span the available time series). Furthermore, Lamb and Changnon (1981) found that longer averaging periods do not guarantee better capture of the current and near-future state of the climate. Angel *et al.* (1993) examined all possible averaging periods of 30 years in length or shorter, and found that eleven years consistently best predicted temperatures the year ahead (1930–1987) in Illinois. Such findings suggest evaluating a range of averaging periods (longest period possible, optimal predictive period – even year-to-date accumulations) in the construction of the indexes proposed below. In keeping with procedures used by Karl *et al.* (1995), the specific climate indicators are expressed as the percent area of a country or some meaningful region that experiences a significant departure

---

* We recognize the arbitrariness of these percentiles in relation to climate sensitivity thresholds of the various systems considered in this paper, however, fine tuning of such percentiles will have to come in future efforts.

TABLE I

Index of leading climatic indicators of extreme climatic events (after Karl *et al.*, 1995)

Climate Indicators

The annual average of the sum of (relative to year-to-date, optimal predictive averaging period normals and long-term climate normals):

1. – Percent of national area with maximum temperatures much below normal.
   – Percent of national area with maximum temperatures much above normal.
2. – Percent of national area with minimum temperatures much below normal.
   – Percent of national area with minimum temperatures much above normal.
3. – Percent of national area in severe drought based on the PDSI.
   – Percent of national area with severe moisture surplus based on the PDSI.
4. – Twice the value of the percent of national area with a much greater than normal proportion of precipitation derived from extreme 1-day precipitation events (more than 2 inches or 50.8 mm).
5. – Percent of national area with much greater than normal number of days with precipitation.
   – Percent of national area with much less than normal number of days with precipitation.

from normal. The indicators then are summed to arrive at a single index value for a given region and time-step (usually annual).

6. Existing Climate Indexes

The two indexes developed by Karl *et al.* (1995) are included in our set of indexes of leading climate indicators of impacts. We distinguish them from other indexes proposed below by the fact that they have actually been calculated for 80-year periods with climate data from the U.S. Because they have been actually implemented, albeit recently, these two indexes provide the best opportunity for evaluation with respect to the suggested attributes.

6.1. CLIMATE EXTREMES INDEX

Extreme climate events (e.g., droughts, severe storms, heat/cold waves) tend to occur over smaller temporal and spatial scales than other forms of climate hazard. An index of extreme climate events does not foreshadow or lead individual localized events, but rather should indicate changes in the frequency and intensity of such events over large regions or political jurisdictions.

The Climate Extremes Index (CEI) developed by Karl and colleagues (1995) at the U.S. National Climatic Data Center consists of five indicators of extremes of minimum and maximum temperature, drought and moisture surplus, and frequency and intensity of precipitation, all in one index value. In cases where either the equivalent of much above normal (ninetieth percentile) or of much below normal (fifth percentile) but not both components of an indicator are included in an index, the weight of that indicator is doubled if any other indicators in that index do have both components (see Table I). The need for only one component arises when there is a lack of symmetry in the expected impact (e.g., strong impact of much above normal proportion of extreme 1-day precipitation events versus lack of impact – positive or negative – of much below normal proportion of days with extreme 1-day precipitation events). Such prevents the index from being too heavily weighted toward those indicators comprised of both components. The expected value of the index is 20% since 10% of the values of each indicator over the long-term should fall in the two extreme percentiles (with the exception of more than normal extreme 1-day precipitation events which is weighted double because its obverse is not meaningful). An example of the CEI computed for 80 years for the United States is shown in Figure 1. It contains peaks in the 1930s, the 1950s and a recent sustained period of increased extremes beginning in 1976 as the frequency and intensity of El Niño events increased relative to previous decades.

6.2. INDEX OF GREENHOUSE CLIMATE RESPONSE

The anticipation of the impacts of climate change on human and environmental systems is clouded by uncertainty about how the change is likely to occur. Consensus building exercises such as the Intergovernmental Panel on Climate Change (Houghton et al., 1990) have focused on a small set of predicted features of future climate change that might be attributed to greenhouse warming. Examples of such features in temperate latitudes include increased precipitation in winter, summer dryness in continental interiors, and increase in the frequency of extreme events (mostly precipitation events) along with observed regularities that are partly consistent with expectations (e.g., minimum temperature warming bias). Karl et al. (1995) constructed a Greenhouse Climate Response Index (GCRI) that we shall summarize here as an example for the U.S. that could be modified and applied to other geographic locations to form a global index of greenhouse response.

The index developed by Karl et al. (1995) is based on the percent of the U.S. with above normal minimum temperatures, above normal cold season precipitation, below normal summer moisture as measured by the Palmer Drought Severity Index and above normal extreme precipitation events. The specific computations of the Index for the U.S. are shown in Table II. The application of the Index to the recent historic climate record across the U.S. is shown in Figure 2 with an expected value of 10%. In this application, Karl et al. (1995) concluded that the observed increase in the index is consistent with expectations of greenhouse warming but that the increase barely misses being statistically significant (i.e., the possibility that the observed increase is a feature of normal variability within a stable climate is between 5 and 10%).

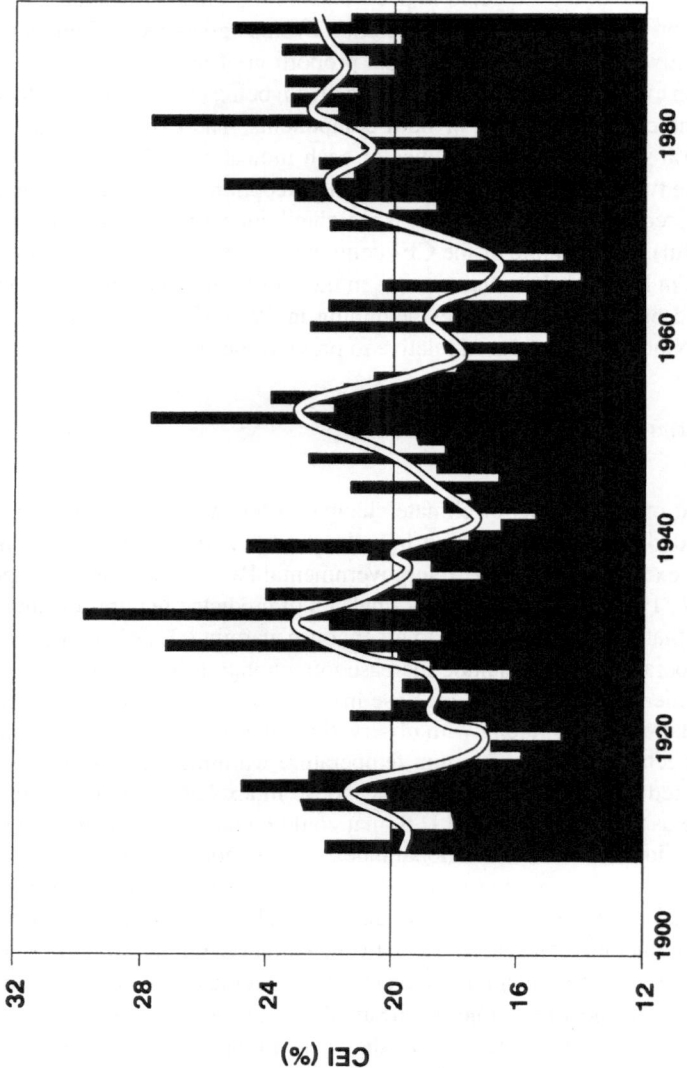

Fig. 1. An annual U.S. climate extremes index (CEI) (redrawn from Karl *et al.*, 1995).

TABLE II

Index of leading climatic indicators of greenhouse warming (after Karl *et al.*, 1995)

Climate Indicators

The annual average of the sum of (relative to long-term climate normals):
1. – Percent of national area with much above normal minimum temperatures.
2. – Percent of national area with much above normal precipitation during the months of October through April (cold season).
3. – Percent of national area in severe drought during the months May through September (the warm season).
4. – Percent of national area with a much greater than normal proportion of precipitation derived from extreme 1-day precipitation events (exce⁻·ding 50.8mm).

6.3. EVALUATION OF ATTRIBUTES OF THE CEI AND GCRI

Since the Climate Extremes and Greenhouse Climate Response Indexes were introduced jointly, many of the observations concerning how well they fulfill the necessary attributes from above apply to both indexes. Thus, we evaluate the CEI and GCRI together here, asking whether these indices possess the suggested attributes of usable knowledge.

Well-Defined Impacts. Primary impacts of climate extremes implicit in the CEI leading indicators are well understood and often documented. Such impacts include loss of life, property damage, and loss of productive capacity. Damage estimates from droughts, floods and severe storms (e.g., tornadoes, hurricanes, hail, blizzards), often with great sectoral detail, are routinely computed by government agencies in many countries. Estimates of impacts on human health (casualties, disease outbreaks, hunger) are also available. Actuarial claims from extreme weather also are tabulated routinely.

Since greenhouse warming has no historical precedent and has yet unequivocally to be detected, the impacts of such can only be approximated through the use of historical analogs and the application of models. However, the literature has converged on major categories of warming impacts primarily on natural ecosystems and sectors of human activity that are climate sensitive such as agriculture and other renewable resources, energy, transportation, and human health.

Anticipatory Value. The climate extremes index can improve knowledge of impacts both temporally and spatially. The measures that constitute the index are usually available more rapidly than measures of actual damage from extreme events. As an aggregate measure not dependent on reports of individual events, the index can readily anticipate answers to such questions as to whether this is a better year or worse year for climate-induced damages and losses. But can knowledge that the Climate Extremes index in any year is high or low actually be used

[503]

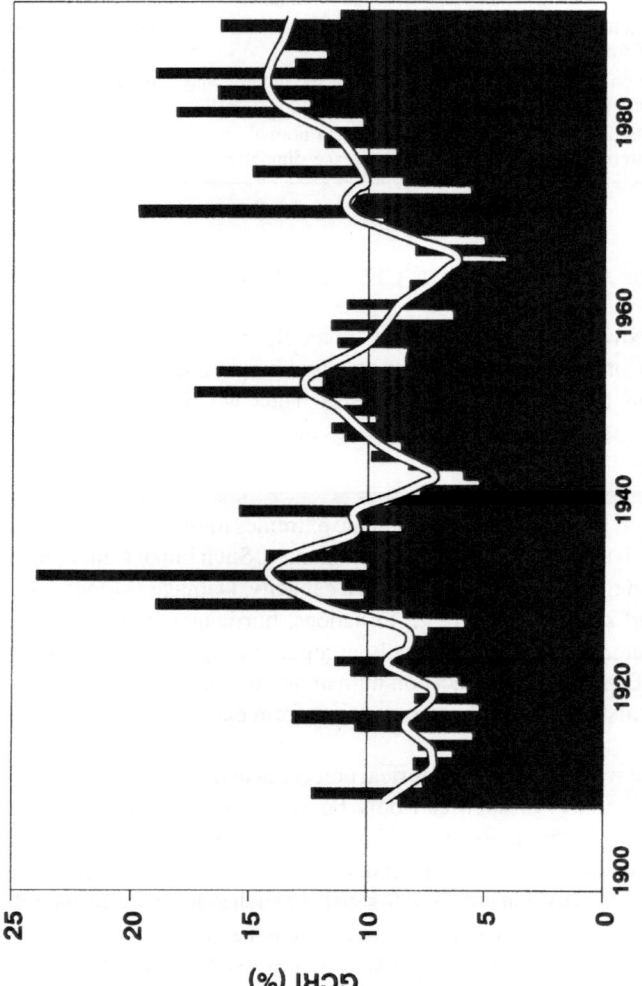

Fig. 2. An annual U.S. greenhouse climate response index (GCRI) (redrawn from Karl *et al.*, 1995).

to anticipate future impacts? Ideally, this could be tested by lagged regressions between composite damage measures and the index. In the absence of such, visual inspection of Figure 1 hints at the persistence of smoothed trends beginning each time the index reaches a new inflection point. The persistence of trends of worsening or improvement appears to range from as little as 3–4 years in several instances to as much as 20–25 years; the most recent 25 years of record has shown a steady increase in the index (i.e., a worsening of expected impacts from extremes).

The question of whether or not the GCRI leads impacts is more problematic than for the CEI. Inter-decadal trends in the GCRI show no clear monotonic signal of warming (Figure 2). Yet, looking over the entire 80-year record there is a trend in the GCRI that is consistent with expectations of greenhouse warming and is very close to statistical significance (Karl *et al.*, 1995). Such realized expectations backed by emerging statistical significance is a potent argument that the GCRI may provide some leading insights into future impacts. The GCRI may particularly be a useful guide in formulating the appropriate spatial and temporal scales at which to search for such impacts.

Usability. Are the CEI and GCRI usable knowledge? The early indications are that both indexes have received the attention of the climate research community documented by the article by Kerr (1995) in *Science* and numerous discussions at scientific conferences including the GCOS meeting referenced above and in a 1995 review of the U.S. Global Change Research Program by the U.S. National Academy of Sciences. While such attention does not imply scientific endorsement of the indexes, it does suggest that they are scientifically intriguing and may fulfill a needed niche in the search for evidence of climate change in the observational record. As noted above, the public has expressed nominal interest in the CEI and GCRI, at least to the extent that the media reflects the public's need to know. It is too early to know whether the resource users have expressed interest in utilizing the two indexes.

There are clear implications of the CEI for user response and policy. In the case of the insurance industry, turning points in the index could be highly influential in the computation of actuarial risk tables. There is a sophisticated natural disaster response capability at all levels of government in most countries which could benefit from the Climate Extremes Index. Congressional interest in the index has widespread implications for the provision of disaster relief.

The implications of the GCRI for user response and policy are as much topics of the research community as matters of practical application. Users in this case may be researchers seeking to understand the best strategies for adapting to the warming. Considerable research is also focusing on appropriate policies to limit greenhouse gas emissions. At the same time, there is serious discourse among governments over the imposition of emissions limitations. It would be surprising were such activities not to take notice of the GCRI.

Measureability. Are the climate indicators that comprise the CEI and GCRI easily measured at appropriate spatial and temporal scales? In the U.S. and most industrialized nations, they are readily available in sufficient spatial and temporal detail for easy computation of the indexes. Periods of record also are sufficiently long to establish statistical significance in evolving trends in the index. As is probably true of most of the indexes discussed in this paper, the likelihood of adequate data to compute them in developing nations is highly variable and in need of the attention of the global observing systems community.

7. Proposed Climate Indexes

We now propose the construction of a number of other indexes of leading climate indicators of impacts. The indexes are not implemented and tested with actual climate data. The climate indicators comprising the indexes are suggested *a priori* from knowledge of specific climate effects on ecosystems and society and are not calculated or tested with actual climate data. Thus they are illustrative of the kinds of climate information that are needed within the broad classes of proposed indices.

7.1. HAZARD WARNING INDEX

The climate extremes index does not warn of likely hazard events rather it warns of aggregate potential for impacts, particularly losses and damages. But climate precursors to certain natural hazard events (e.g., certain types of droughts, floods, severe storms) may permit the development of an Hazard Warning Index. This special subset of extreme events may be presaged by other climate-related anomalies. Such precursors may include, for example, El Niño Southern Oscillation (ENSO) events, drought persistence, falling ground water tables, large snow packs on stream catchments and saturated soils in inhabited low-lying areas.

ENSO events have known teleconnections to surface weather in a variety of locations worldwide, but especially so in the Tropics and less so in certain mid-latitude locations (Barnston *et al.*, 1994b). Sea surface temperature (SST) anomalies in the east-central tropical Pacific Ocean are thought to be the best predictor of ENSO outbreaks (Barnston *et al.*, 1994b). Given the well-established teleconnections of ENSO outbreaks with certain Tropical surface climate anomalies (e.g., drought in northeast Brazil and northern Australia) and less well-established yet identifiable teleconnections with mid-latitudes surface anomalies (exemplified by Barnett *et al.*, 1993 in Figure 3), we argue that ENSO events are a useful leading indicator of near future impacts in those regions.

Drought demonstrates the inertia of accumulated moisture deficiency that carries over to impacts. The effects of drought often persist long after precipitation returns to normal. The Palmer Drought Severity Index (PDSI) is an example of a measure of accumulated hydrologic drought and its highly negative values ('severe' drought or worse) are argued here to be the basis of an indicator of drought persistence.

JANUARY/FEBRUARY PRECIPITATION

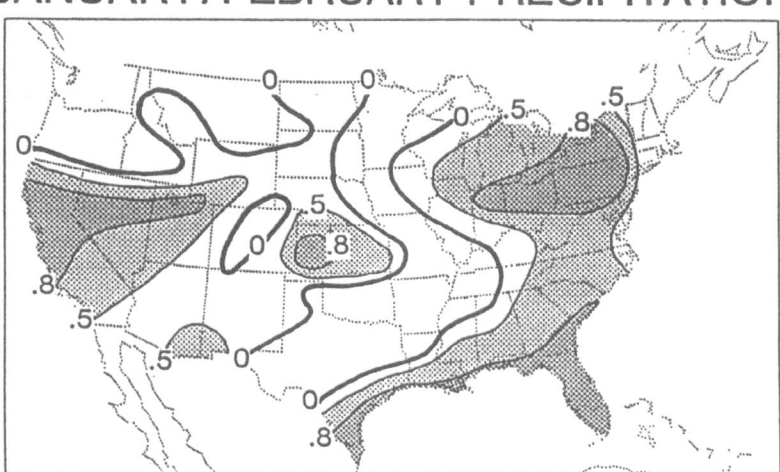

Fig. 3.   Geographical distribution of skill of the Scripps-Max Planck Institute hybrid coupled model forecasts for January-February total U.S. precipitation at two and one-third season lead time for seven strong ENSO events. Skill is expressed as a correlation between model forecasts and observations (after Barnston *et al.*, 1994a).

Falling ground water tables, especially in regions that are highly dependent on such for irrigation and municipal water supply, is a precursor of drought-related problems. As falling ground water tables reach critical levels, relatively small fluctuations in precipitation and/or evapotranspiration can cause droughts where such fluctuations would barely be noticed when ground water tables are adequate and stable.

The inverse of drought persistence is moisture surplus persistence. Moisture surplus persistence plays a key role in determining soil saturation and hence indicates imminent flood risk. Infiltration is severely diminished on saturated soils thus shunting additional precipitation directly to runoff into streams and adjacent low-lying areas. Low-lying areas tend to be favored for agriculture because of their productive soils and are locations of many urban centers (because of their historic strategic location for water-borne commerce). Moisture surplus persistence can be approximated with highly positive values of the PDSI.

Snow pack thickness and extent (i.e., total volume of snow accumulated over the cold season) also can be a strong indicator of drought and flood risk. In regions whose stream drainages receive a high proportion of runoff from snow-melt (mountainous and middle- to high latitude catchments), advance warning of drought and flood potential could be gained by monitoring snow pack thickness and extent. Anomalously small snow packs may indicate imminent drought conditions down-

TABLE III

Index of hazard warning

Climate Indicators

The annual average of the sum of (relative to long-term climate normals):

1. – Twice the percent of East-Central Tropical Pacific (Niño 3) area with much above normal seasonal SST.
2. – Percent of national area in severe drought based on the PDSI.
   – Percent of national area in severe moisture surplus based on PDSI.
3. – Twice the percent of national area with much below normal ground water levels.
4. – Percent of national area with much above normal snow pack.
   – Percent of national area with much below normal snow pack.

stream while anomalously large snow packs may indicate imminent flooding conditions there. From this, we reason that snow pack thickness and extent also should be an hazard warning indicator.

Using these precursors, a hazard warning index is detailed in Table III. This prototype index mainly is focused on the advance warning of the likelihood of drought, flooding and severe storms (as connected to ENSO events). The types of impacts that might be signalled by such an index will vary considerably by location. For example, a high index number driven by a strong ENSO event may signal droughts that strongly condition famine outbreaks in tropical regions while it may signal flood-related problems along the west coast of the U.S.

7.2. INDEX OF CLIMATE INDICATORS OF ECOSYSTEM HEALTH

Interest in the impacts of climate fluctuation and change on the form and function of natural ecosystems has increased recently (e.g., Vitousek, 1994). The knowledge and tools available to sort out such impacts are just beginning to be developed. Holdridge's (1967) simple but well-used scheme of classifying ecological life zones partly on the basis of climate is argued to be useful here in the computation of an index of leading climate indicators of ecosystem health. Accordingly, we define ecosystem health as the relative stability of climate-regulated life zones, recognizing that ecosystems have tended to be adapted to a certain amount of natural climate variability. Yet, the occurrence of protracted climate fluctuations that markedly alter the controlling climate factors of a particular ecosystem, as defined by Holdridge (1967), are likely to result in multiple stresses.

Holdridge's life zones are regions of homogeneous vegetation type forming a global classification system based on climate properties, latitude and altitude (Figure 3). The controlling climate properties of such zones are mean annual *biotemperature*, the ratio of potential evapotranspiration (PET) to total annual

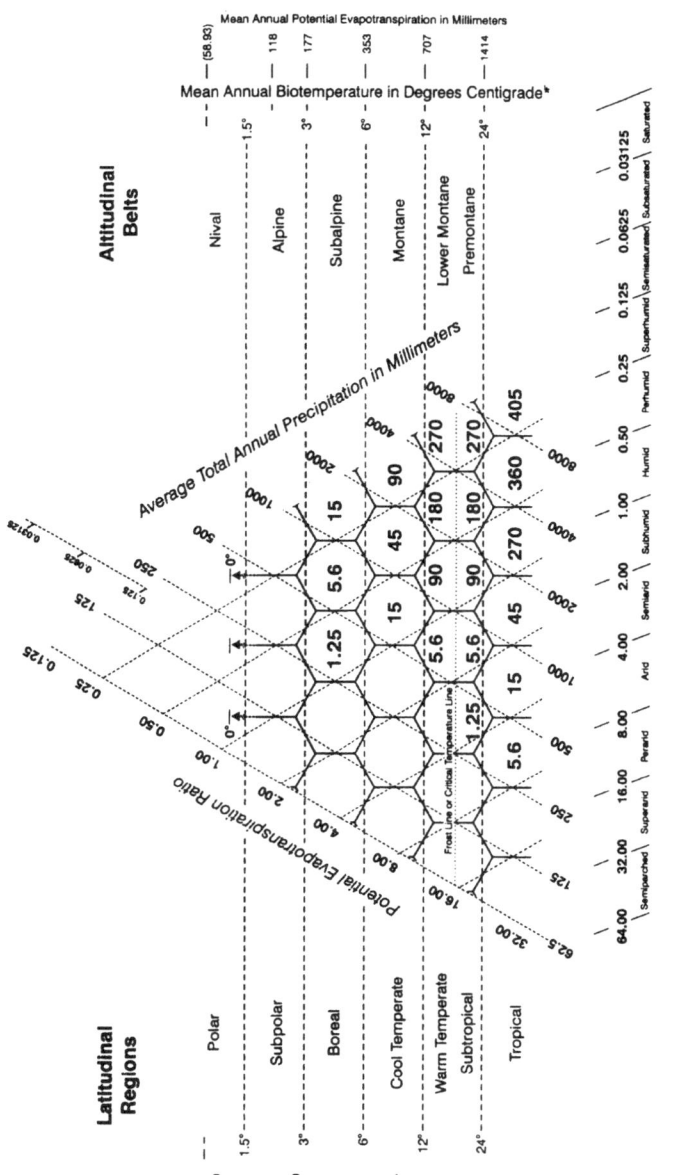

Fig. 4. Holdridge Life Zones (redrawn from Holdridge, 1967)

★ t^{bio} = Mean of unit period temperatures with substitution of 0 and 30 for all temperature values below 0 and above 30 °C, respectively. (The 30 maximum cutoff is tentative pending further investigation.)

TABLE IV

Index of leading climatic indicators of ecosystem health

Climate Indicators

The annual average of the sum of (relative to long-term climate normals):

1. – Percent area of major ecotones with much above normal mean annual biotemperature (defined as the sum over a year of hourly temperatures greater than 0 °C and less than 30 °C divided by the total number of hours in a year, after Holdridge, 1967).

    – Percent area of major ecotones with much below normal biotemperature.

2. – Percent area of major ecotones with much above normal ratio of potential evapotranspiration (calculated by multiplying mean annual biotemperature times the constant 58.93, after Holdridge, 1967) to annual precipitation.

    – Percent area of major ecotones with much below normal ratio of potential evapotranspiration to annual precipitation.

---

precipitation, and total annual precipitation. Mean annual biotemperature is defined as the sum over a year of hourly temperatures greater than 0 °C and less than 30 °C divided by the total number of hours in a year. Holdridge (1967) posited that potential evapotranspiration is highly correlated with mean annual biotemperature and is derived by multiplying biotemperature by the constant 58.93. The shift in vegetative composition across zones is a logarithmic progression of the climate factors, latitude and altitude (Figure 3).

Recognizing the limitations of Holdridge's work pointed out in numerous critical reviews, we maintain that the basic controlling climate properties that Holdridge identified are useful for the examination of general classes of climate stress on vegetative systems. For example, Emanuel *et al.* (1985) recently used the Holdridge life zones to make a first approximation of the effects of climate change on the geographic distribution of certain forest systems (i.e., latitudinal shifts of the boreal forest belt). We argue that prolonged departures of mean annual biotemperature, mean annual precipitation and the ratio of PET to mean annual precipitation from existing long-term normals are likely to excite a cascade of impacts that may stress and possibly alter the form and function of ecosystems. The details of the specific climate indicators comprising the ecosystem health index are shown in Table IV.

The index of leading climate indicators af ecosystem health proposed here may be expected to suggest impending direct impacts of climate stresses (i.e., effects of climate fluctuation on primary productivity) and indirect ones (i.e., effects of climate fluctuation on vulnerability of ecosystems to other stressors like fire or pathogens).

An index of leading climate indicators of ecosystem health need not be constructed for whole ecosystems. Rather, the index may best be focused on climate stresses on spatial margins of ecosystems. Such margins are known as *ecotones* or zones of sharp transition between major ecosystem types (e.g., transition between forest and grassland or forest and tundra). Ecotones are often characterized by steep gradients (climate, soils, terrain). Plant systems and higher trophic levels in ecotones exhibit a high degree of sensitivity to relatively small changes in environmental forcings (Blaikie and Brookfield, 1987; Parry and Carter, 1988). Accordingly, ecotones are gaining popularity in ecological research as bellwethers of climate change (American Association for the Advancement of Science, 1995).

The climate index of ecosystem health impacts proposed here may be useful in the anticipation of changes in other ecological indicators such as the appearance/disappearance of certain bird and plant species. Across large regions and long time periods, the index may signal changes in community structure such as shifts in vegetative composition, and changes in fragile ecosystems such as wetlands and tundra.

7.3. INDICES OF CLIMATE INDICATORS OF ENERGY DEMAND AND RENEWABLE NATURAL SOURCES

Climate is thoroughly involved in the determination of the quantity and quality of many of the world's natural resources. This is particularly true of energy and renewable natural resources such as agriculture, water, forests, fisheries and recreation, and less so for nonrenewable resources such as minerals (Wigley, 1987). At the most basic or first-order level, the interactions of climate with energy and natural resources are played out through the impacts of climate fluctuation and change on energy demand, biological productivity, and water supply. The capture of those basic interactions by indexes of leading climate indicators is required for the anticipation of higher order impacts on energy and renewable natural resources.

Climate indicators are proposed for four sets of resource-related problems: *(1) energy demand for heating and cooling; (2) agricultural productivity; (3) water supply and irrigation demand; and (4) forest productivity.* While several climate-dependent resource sectors are not explicitly addressed by the above indicators, many of the climate features embedded in the above indicators are important to resource user sectors such as fisheries and recreation.

Energy Demand Indicators. Energy consumption for space heating and cooling is highly climate-dependent. There is a well-known relationship between temperature and space heating and cooling requirements. The climate equivalent of space heating and cooling demand is expressed in the form of degree-days. Degree-days are the accumulated difference in absolute values between daily mean temperatures and a base of 18.3 °C. In the case of space heating, heating degree-days are computed as the absolute value of accumulated negative differences between the mean daily temperature and 18.3 °C. For space cooling, cooling degree-days are computed as the accumulated positive difference between the mean daily temperature and 18.3 °C.

[511]

TABLE V

Index of leading climatic indicators of energy and renewable natural resources

Climate Indicators	
	The annual average of:
Heating and Cooling Energy Demand	The sum of (relative to start of heating/cooling season-to-date normals, optimal predictive averaging period normals and long-term climate normals):
	1. – Twice the population weighted percent national area with much greater than normal heating degree-days (absolute value of accumulation of negative difference between mean daily temperature and 18.3 °C).
	2. – Twice the population weighted percent national area with much greater than normal cooling degree-days (accumulation of positive difference between mean daily temperature and 18.3 °C).
Agricultural Production	The sum of (relative to optimal predictive averaging periods normals and long-term climate normals):
	1. – Twice the percent national cropped area with much greater than normal heat stress degree-days (accumulated positive difference between mean daily temperature and 30 °C base).
	2. – Twice the percent national cropped area with much below normal growing degree-days (10 °C base for warm season plants, 5 °C base for cool season plants).
	3. – Twice the percent national cropped area with much shorter than average frost-free period.
	4. – Percent national cropped area with mean growing season crop moisture index (CMI) in severe drought.
	– Percent national cropped area with mean growing season CMI in severe moisture surplus.
Water Resources	The sum of (relative to optimal predictive average in period normals and long-term climate normals):
RUNOFF:	1. – Percent national area with much above normal estimated runoff (calculated as 'effective precipitation': difference between annual precipitation and estimated annual evaporation).
	– Percent national area with much below normal estimated runoff.
IRRIGATION DEMAND:	2. – Twice the percent national irrigated cropped area with Net Irrigation Requirement (difference between crop specific estimated evapotranspiration for major crop types and precipitation) much above normal. (irrigated crop area is all cropped area with mean annual net irrigation requirement > 300 mm based on Peterson and Keller, 1990).
Forest Resources	The sum of (relative to long-term climate normals):
	1. – Percent of national forested area with much above normal five-year running average of mean annual biotemperature (defined above).
	– Percent of national forested area with much below normal five-year running average of mean annual biotemperature.
	2. – Twice the percent of national forested area with much below normal five-year running average of mean potential evapotranspiration to precipitation ratio (defined above).

The specification of heating and cooling degree-day indicators is shown in Table V. It should be noted that the indicators are computed by first weighting the degree-days by population. Such is necessary since degree-day totals in sparsely populated areas will elicit less energy use than in densely populated areas. Examples of impacts potentially led by the index include fuel prices and problems associated with large-scale disruptions of fuel supplies.

Agricultural Productivity Indicators. The starting point in developing leading climate indicators of impacts on agricultural production is the linkage of climate with primary productivity of major crops. Though livestock are directly affected by climate (heat/cold stress primarily), the main effect of climate on them is through the availability of feed and range grass.

The major climate determinants of plant stress that should be reflected by our proposed indicators are: (1) heat stress caused by prolonged periods of temperatures above thresholds where plant growth is known to cease; (2) overly slow or rapid accumulation of warmth needed for plant development during the active growing season; (3) growing season abreviated by early or late freezes; and (4) moisture stress, especially wet conditions during planting and dry conditions during plant reproductive periods. The specific attributes of the climate indicators of these above stresses are shown in Table V. It is recommended that the percent of national area included in the agricultural indicators be restricted to major cropped zones.

A wide array of possible impacts may be indicated by the Agricultural Index varying by location and circumstance. In less developed countries, the index could be linked to the mean nutritional status of the population at risk of hunger or to the ratio of food imports to exports. In industrialized countries, the index may best be linked to average value of agricultural production or to government transfer payments to farmers. Changes in average crop and livestock productivity may be indicated by the Agricultural Index in both developed and less developed countries.

Water Supply and Irrigation Demand Indicators. Frederick (1995) alluded to the great difficulty of measuring the effect of climate fluctuation and change on the adequacy of water supplies across areas the size of the U.S. The large spatial and temporal variability of precipitation detracts from the use of long-term precipitation normals alone in the assessment of the adequacy of regional water supplies. Direct measurement of streamflows may also be misleading because of the large influence of human intervention (e.g., dams, diversions and withdrawals) on major river systems. We argue, however, that widespread and significant departures of runoff – estimated by taking the difference between mean annual precipitation and evaporation (Schaake, 1990) – from long-term normals should provide a first approximation of large-scale adequacies of water supply as affected by climate. The computation of such in an index is shown in Table V.

Climate-induced demand for irrigation water provides another useful indication of impacts on water supply. Irrigation of crops, where it is in widespread practice, is by far the largest consumptive user of water (81% of all consumptive uses in the U.S. according to Waggoner and Schefter, 1990). The demand for irrigation water is regulated mostly by climate conditions. To measure the influence of climate on irrigation demand, we suggest reliance on computed Net Irrigation Requirement (NIR) described by Peterson and Keller (1990). Peterson and Keller (1990) calculated NIR as being the difference between crop-specific evapotranspiration (based on known crop coefficients) and usable precipitation, noting that irrigation is not widespread in regions with a mean annual NIR of less than 300 mm. Thus, our climate indicator of irrigation water demand is computed on the basis of cropped area with 300 mm or greater of mean annual NIR (Table V).

The proposed water resource index may indicate changes in water use to safe yield ratios from major surface water bodies (major resewoirs and rivers). It may also indicate change in groundwater depths in major aquifers with recharge rates fast enough to reflect interannual to decadal climate fluctuation and change.

Forest Resources Indicators. Like agricultural resources, the main impact of climate on forest resources is on net primary productivity, although other factors which regulate forest productivity are also affected by climate (e.g., pests and pathogens, fires). However, unlike most annually-renewing agricultural cropping systems, forests are perennial systems. As such, forest systems tend to integrate climate stresses over longer time periods than most major crops since most crops are planted and harvested within the same growing season. The growth environment (i.e., soil fertility, pest management, water availability) of forests is much less intensively managed than that of crops. In the above regard, forests are more like natural ecosystems than agricultural systems and it is more likely that forest response to climate stress is better captured within the framework of the climate controls of Holdridge's life zones classification than with the agroclimate indicators discussed above.

As with climate indicators of ecosystem health, we propose that the main climate indicators of forest resources include biotemperature, mean annual precipitation and the ratio of PET to annual precipitation (Table V). The major difference between the ecosystem and forest indicators is the use of five-year running means to compute averaging periods for the climate indicators of forest growth versus simple annual means for ecosystem health. This distinction between forests and ecosystems seems reasonable since, as alluded to above, the maturation of trees into harvestable timber occurs over time measureable in decades and climate stresses on trees tend to be cumulative while ecosystems are composed of a mixture of plant and animal species (perennials, annuals, mammals, insects) with life cycles varying among several temporal scales (e.g., days and weeks to decades or longer).

A number of possible impacts may be indicated by the forest resources index, including variations in large-area estimates of maximum sustainable yield, variation in prices for forest products and in proportion of forest products exported versus consumed domestically.

7.4. EVALUATION OF ATTRIBUTES OF THE PROPOSED INDEXES

The speculative nature of the proposed indexes of hazard warning, ecosystem health, and energy demand and renewable resources limits our evaluation of them. Indeed, in the absence of their implementation, evaluation of attributes is somewhat tautological since we suggest them because we think that they possess the necessary attributes. However, we know more about some attributes than others.

Well-Defined Impacts. The primary impacts of climatic hazard events are well-understood. The impacts associated with the ecosystem health index are perhaps the least well-defined of the five indexes discussed in this paper. Attention has only recently focused on the identification of impacts of climate fluctuation and change on ecosystem health, although rapid progress is being made in the linkage of high-resolution satellite imagery with ecosystem models to estimate climate impacts on land cover (reported elsewhere in this volume). The impacts of climate fluctuation on energy demand and systems of renewable resources are relatively well-known. For energy demand, the empirical relationship between temperature and seasonal heating and cooling requirement is not questioned. In the case of renewable resources, many of the biophysical impacts (e.g., runoff, crop productivity) can be simulated mechanistically. The higher-order socioeconomic impacts of climate fluctuation on energy demand and renewable resources are better defined than those of other sectors like industry, commerce and transportation,

Anticipatory Value. The main asset that distinguishes the hazard warning index from the CEI is the linkage to climate precursors. We reason *a priori* that the hazard warning index has strong anticipatory value with respect to impacts. The anticipatory value of the ecosystem health index with respect to impacts is not known. However, again we reason *a priori* that persistent fluctuations in the climate determinants of the Holdridge Classification are bound to presage a variety of ecosystem impacts that we can only guess at here. Evaluation of the anticipatory value of the index of energy demand and renewable natural resources is complicated by the relatively strong influence of many social and economic trends (e.g., technological progress, economic development, population change) which condition climate impacts. However, most of the individual climate indicators that comprise the index currently are used to monitor and project conditions based on known key climate sensitivities within a year or growing season. How well the individual climate indicators may anticipate long-term impacts needs investigation.

Usability. We do not know whether these proposed indexes could become usable knowledge. They were chosen because of the scientific consensus as to their widespread importance and the interest by resource users in these domains. But if implemented, would users be better off with indexes than with convention-al descriptive climatological information (e.g., mean temperatures, precipitation) would need to be demonstrated.

Measureability. Relatively few of the climate indicators necessary for construction of the proposed indexes of hazard warning, ecosystem health and energy demand and renewable resources are routinely observed and/or computed at the necessary locations and scales, even within industrialized countries. Considerable effort will need to be expended by global climate observing systems to obtain data necessary to compute such indicators as biotemperature, crop stress days and runoff for appropriate areas. The lack of such effort will surely impede the development of new indexes proposed here and, hence, may diminish efforts to develop better usable climate knowledge.

8. Conclusions

Climate observing systems are at a cross-roads in planning the architecture and attributes of future climate data and information retrieval systems. If public support is to be obtained and sustained, then the fruits of such systems need to be usable knowledge. The development of indexes of leading climate indicators of impacts on environmental and human systems is suggested here to help guide climate observing systems in the collection of data and information that are important to users; such may provide a new way to anticipate general trends in largescale stresses of climate fluctuation and change, much like the NBER index of leading economic indicators is used to anticipate broad business cycles.

Little doubt exists that the indexes we propose here could be enriched and tailored to specific locations. Furthermore, we make no attempt to examine the indexes in great detail with actual data. Rather, our intention is to generate critical debate on the utility of leading climate indicators of impacts and to foster detailed efforts aimed at testing and operationalizing such.

We conclude that the CEI and GCRI are on the verge of becoming usable climate knowledge and efforts should be made to implement such indexes widely. They should be further evaluated with post hoc analyses to determine how well they may lead impacts. The proposed indexes of hazard warning, ecosystem health and energy demand and renewable natural resources hint at the promise of becoming usable knowledge and already possess some of the required attributes. We look forward to future collaborative efforts to explore their feasibility and utility.

Acknowledgements

We would like to thank our colleagues who participated in the Climate Impacts discussion group at the WMO-GCOS workshop in Asheville, especially Drs. Melvin Briscoe and Peter Robinson, for stimulating this paper. We also wish to thank Tom Karl for his encouragement. Finally, we extend our deep appreciation to Ms. Deb Wood for her graphic arts assistance and Ms. Deanna Batty for her rapid and accurate word processing.

References

American Association for the Advancement of Science: 1995, 'Is a Warmer Climate Wilting the Forests of the North', *Science* **267**, 5704, 1595.

Angel, J., R., Easterling, W. E., and Kirsch, S. W.: 1993, 'Towards Defining Appropriate Averaging Periods for Climate Normals', *Climatol. Bull.* **27**, 2, 29–44.

Auerbach, A. J.: 1982, 'The Index of Leading Indicators: "Measurement without Theory", Thirty-Five Years Later', *Rev. Econ. Statis.*, 589–595.

Barnett, T. P., Latif, M., Graham, N., Flugel, M., Pazan, S., and White, W.: 1993, 'ENSO and ENSO-Related Predictability: Part 1 – Prediction of Equatorial Pacific Sea Surface Temperatures with a Hybrid Coupled Ocean-Atmosphere Model', *J. Clim.* **6**, 1545–1566.

Barnston, A. G., van den Dool, H. M., Rodenhuis, D., Ropelewski, C., Kousky, V., O'Lenic, E., Livezey, R., Ji, M., Leetmaa, A., Zebiak, S., Cane, M., Barnett, T., and Graham, N.: 1994a, 'Long-Lead Seasonal Forecasts – Where Do We Stand?', in *Collected Papers on Seasonal Forecasting*, U.S. Department of Commerce, National Oceanic and Atmospheric Administration, National Weather Service, National Centers for Environmental Prediction, Climate Prediction Center, Washington, DC.

Barnston, A. G., van den Dool, H. M., Zebiak, S. E., Barnett, T. P., Ji, M., Rodenhuis, D. R., Cane, M. A., Leetmaa, A., Graham, N. E., Ropelewski, C. R., Kousky, V. E., O'Lenic, E. A., and Livezey, R. E.: 1994b, 'Long-Lead Seasonal Forecasts – Where Do We Stand?', *Bull. Amer. Meteorol. Soc.* **75**, 11, 2097–2114.

Bella, D. A., Jacobs, R., Li, H.: 1994, 'Ecological Indicators of Global Climate Change: A Research Framework', *Environm. Manage.* **18**, 4, 489–500.

Blaikie, P. and Brookfield, H.: 1987, *Land Degradation and Society*, Methuen, London.

Bowes, M. D. and Crosson, P. R.: 1993, 'Paper 6. Consequences of Climate Change for the Mink Economy: Impacts and Responses', in Rosenberg, N. J. (ed.), *Towards an Integrated Impact Assessment of Climate Change: The MINK Study*, *Clim. Change* **24**, 131–158.

Cairns, J. Jr., McCormick, P. V., and Niederlehner, B. R.: 1993, 'A Proposed Framework for Developing Indicators of Ecosystem Health', *Hydrobiologia* **263**, 1–44.

Clark, W. C.: 1989, 'Scales of Climate Impacts', *Clim. Change* **7**, 5–27.

Easterling, W. E. III, Crosson, P. R., Rosenberg, N. J., McKenney, M. S., Katz, L. A., and Lemon, K. M.: 1993, 'Paper 2. Agricultural Impacts of and Responses to Climate Change in the Missouri-Iowa-Nebraska-Kansas (MINK) Region', *Clim. Change* **24**, 23–61.

Emanuel, W. R., Shugart, H., and Stevenson, M.: 1985, 'Climatic Change and the Broad-Scale Distribution of Terrestrial Ecosystem Complexes', *Clim. Change* **7**, 29–43.

Frederick, K. D.: 1995, 'America's Water Supply: Status and Prospects for the Future', *Consequences* **1**, 1, 14–23.

Hare, F. K.: 1985, 'Climatic Variability and Change', in Kates, R., Ausubel, J., and Berberian, M. (eds.), *Climate Impact Assessment*, SCOPE 27, Wiley and Sons, Chichester, pp. 37–68.

Holdridge, L. R.: 1967, *Life Zone Ecology*, Tropical Science Center, San Jose, Costa Rica.

Houghton, J. T., Jenkins, G. J., and Ephraums, J. J.: 1990, *Climatic Change, the IPCC Scientific Assessment*, World Meteorological Organization (WMO), Cambridge University Press, Cambridge.

Karl, T. R., Knight, R. W., Easterling, D. R., and Quayle, R. G.: 1995, 'Trends in U.S. Climate During the Twentieth Century', *Consequences* **1**, 1, 3–12.

Kates, R.: 1985, 'The Interaction of Climate and Societies', in Kates, R., Ausubel, J., and Berberian, M. (eds.), *Climate Impact Assessment*.

Kates, R., Ausubel, J., and Berberian, M. (eds.): 1985, *Climate Impact Assessment*, SCOPE 27, Chichester, Wiley and Sons.

Kerr, R. A.: 1995, 'U.S. Climate Tilts Toward the Greenhouse', *Science* **268**, 21, 363–364.

Lamb, P. J. and Changnon, S. A. Jr.: 1981, 'On the "Best" Temperature and Precipitation Normals: The Illinois Situation', *J. Appl. Meteorol.* **20**, 1383–1390.

Lindblom, C. E. and Cohen, D. K.: 1979, *Usable Knowledge: Social Science and Social Problem Solving*, Yale Unievrsity Press, New Haven, CT.

Mitchell, W. C. and Burns, A. F.: 1938, *Statistical Indicators of Cyclical Revivals*, NBER, New York.

Parry, M. J. and Carter, T. (eds.): 1988, *The Impact of Climatic Variations on Agriculture*, Kluwer, Dordrecht.

Peterson, D. F. and Keller, A. A.: 1990, 'Irrigation', in Waggoner, P. E. (ed.), *Climate Change and U.S. Water Resources*, Wiley and Sons, New York.

Rapport, D. J.: 1989, 'What Constitutes Ecosystem Health?', *Perspectives in biology and Medicine* **33**, 1, 120–132.

Riebsame, W. E., Changnon, S. A., Jr., and Karl, T. R.: 1991, *Drought and Natural Resources Management in the United States*, Westview Special Studies, Boulder.

Rosenzweig, C. and Parry, M.: 1994, 'Potential Impact of Climate Change on World Food Supply', *Nature* **367**, 13, 133-138.

Schaake, J. C.: 1990, 'From Climate to Flow', in Waggoner, P. E. (ed.), *Climate Change and U.S. Water Resources*, Wiley and Sons, New York.

Thom, E. C.: 1959, 'The Discomfort Index', *Weatherwise* **12**, 2, 57–60.

Vitousek, P. M.: 1994, 'Beyond Global Warming: Ecology and Global Change', *Ecology* **75**, 7, 1861.

Wigley, T. M. L.: 1987, 'The Impact of Climate on Resource Use and Availability', in McLaren, D. J. and Skinner, B. J. (ed.), *Resources and World Development*, Wiley and Sons, Chicester, pp. 79–99.

Waggoner, P. E. and Schefter, J.: 1990, 'Future Water Use in the Present Climate', in Waggoner, P. E. (ed.), *Climate Change and U.S. Water Resources*, Wiley and Sons, New York.

(Received 19 June, 1995; in revised form 8 August, 1995)